Michèle Fieux

The Planetary Ocean

Illustrations by Chantal Andrié and Michèle Fieux
Translation by Ferris Webster

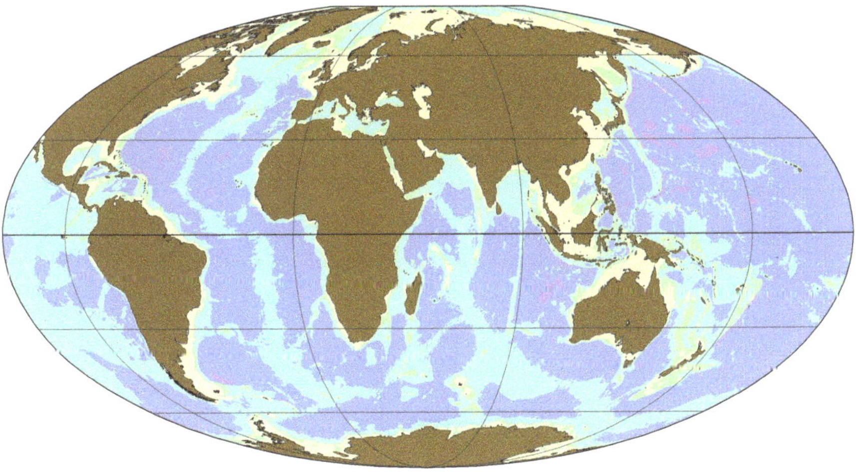

CURRENT NATURAL SCIENCES

EDP Sciences

Cover illustration: The planetary ocean.

Printed in France

ISBN (print): 978-2-7598-2070-2 - ISBN (ebook): 978-2-7598-2150-1

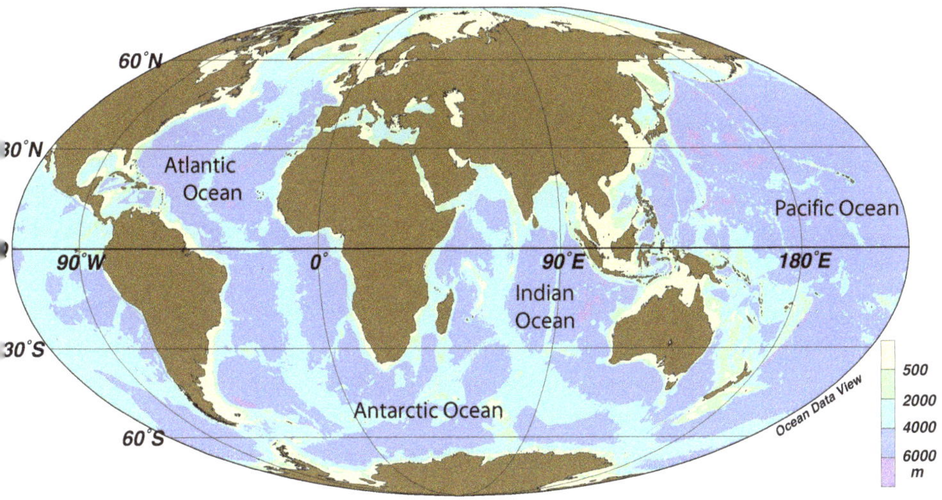

60°N
30°N
Atlantic Ocean
Pacific Ocean
90°W
0°
90°E
180°E
Indian Ocean
30°S
Antarctic Ocean
60°S
Ocean Data View
500
2000
4000
6000
m

Table of Contents

II. The Antarctic (or Austral) Ocean

III. The Atlantic Ocean

IV. The Indian Ocean

V. The Pacific Ocean

VI. Conclusions

© photo Michèle Fieux

"Une fraîcheur, de la mer exhalée,
Me rend mon âme... Ô puissance salée!
Courons à l'onde en rejaillir vivant!"

Paul Valery: *Le cimetière Marin*

"A freshness, exhalation of the sea,
Restores my soul... Salt-breathing potency!
Let's run at the waves and be hurled back to living!"

Paul Valery: *The graveyard by the sea*
Translated by C. Day Lewis

Prolog

The ocean is our habitat of origin.

When I discovered the sea for the first time, at the age of seven, it was a revelation. I was immediately conquered; it overwhelmed me. I felt so inspired as I watched it that I could spend hours admiring it without tiring. I was on the coast of Brittany, near Saint-Malo, where the tides take the water far from land. But the sea always returns, with the sound of the waves, so soft to my ears and that odor of seaweed, which always reminds me of that first meeting. All that has never left me. I never tire of it.

Later, I discovered at the end of my teaching license that a diploma of Advanced Studies in Physical Oceanography (*Diplôme d'Études Approfondies en Océanographie Physique*) had just been created. I did not hesitate a second. I immediately rushed to enroll, trying to get nearer to that vast medium, so enticing and yet so strange, which did not easily reveal its secrets. Later, I admired the sea most often from the deck of a research vessel but the feelings were the same.

I would like to share my passion with you, by leading you to its discovery across the planetary ocean.

I have tried to present the state of the ocean such as we currently know it by attempting to assemble the maximum of information. I am well aware that I have not treated it all.

This work has been a task of long duration. I did it with pleasure, even though I had badly estimated what it represented. I had dreamt of it for a long time, but daily work prevented it.

I would like this text to be accessible to the greatest number of readers. In the first part, I remind the reader of important and simple factors and processes which allow us to understand a lot about how the ocean functions. I have limited equations and have chosen illustrations rather than giving long explanations. That means that the reader should "read" the figures as well as the text. Then I treat each ocean basin separately, beginning with those factors most useful to understanding its circulation and oceanographic features. Finally, I pose a few questions on variability and the effect of climate change.

This text is aimed primarily at people beginning in oceanography so that they may acquire a basic understanding of ocean processes and the oceanographic properties of each of the ocean basins with their similarities and particularities.

I hope that this book will be useful and will also interest those who, without being specialists, would like to discover what is going on beneath their boat.

Preface to the French Edition

Michèle Fieux offers us a work of regional oceanography that is well researched and magnificently illustrated. It begins with a part entitled "Generalities" which describes the water masses and currents of the ocean in relation to the effects of wind and of heat and water exchanges between ocean and atmosphere. These relations are explained progressively beginning with the dynamical equations. This section also provides a review of observation techniques, essentially *in situ*, of ocean properties and currents. The work continues with parts dedicated to the great oceans: the Antarctic, the Atlantic and its marginal seas, the Indian Ocean, and the Pacific. Each current system is described; the causes of its existence are detailed. The conclusion of the book places these regional descriptions in the planetary context of water masses and ends with probable changes to the ocean induced by climate change.

This book updates the basic texts on regional oceanography, those used to educate masters or doctoral students. It relies on the most recent reports of currents and water masses, and draws extensively on the results of the international World Ocean Circulation Experiment (WOCE). During ten years this program mobilized oceanographers from 44 countries, greatly increased the volume of high quality observational data, and brought a number of new techniques to maturity. These data, as well as satellite observations, have notably permitted quantitative estimates of the transport of water, heat, and salt in the ocean, whereas many of these estimates were not available at the beginning of the nineties. Thus the text of this book is rich in precise figures.

This update is notable also for its illustrations. These are numerous, well chosen, and evocative; I even find them elegant. They are systematically referenced: we find especially the results of the past fifteen years. Of course, the great classics are also here. This indispensable framework has served as the foundation for the oceanic system that has been progressively constructed during recent decades.

Michèle Fieux's experience can clearly be seen in this book, especially her contributions to WOCE and the study of the Indian Ocean and its seasonal tropical currents. The section dealing with observational techniques well illustrates this, with an interesting panorama of photographs having a strong human presence, most often personal. We sense here that an oceanographic cruise is highly technical experimental work, shared by all members of the team. Michèle Fieux is recognized for her qualities of cruise leader and her ability during a limited-time cruise to bring out the best from crew and scientists on board to meet all challenges.

Michèle Fieux thus updates the climatic and seasonal description of a complex ocean that students should learn. We are delighted to see this since our discipline needs oceanographers motivated by this human and environmental approach. May this work of Michèle Fieux provide a synthesis useful to both students and experienced oceanographers, and convince students that inquiry, adventure, and scientific discovery await them.

Jean-François Minster

Scientific Director, TOTAL (2006–2016)
Scientific Director-General, CNRS (2006)
President Director-General, IFREMER (2000–2005)
Author of *Les Océans* (1994) and *La Machine Océan* (1997)

Acknowledgments

This work would not exist were it not for the constant, attentive, and constructive help of Chantal Andrié. She wanted to take charge of finding or designing quite a lot of the illustrations and was the first and indispensable proofreader, who accepted to plow through the first version of each part and who always encouraged me.

I would like to dedicate this work to masters and friends who have left us too soon:

Paul Tchernia, my professor who inspired me with his course, gave me a leg up as my thesis director and asked me to continue teaching his course at the École Nationale Supérieure des Techniques Avancées (ENSTA). With Henri Lacombe, he was willing to have confidence in accepting a woman on board an oceanographic ship for the first time.

Paul Tchernia, Henry Stommel, John Swallow, and Fritz Schott gave me much during my life as oceanographer. With them I shared the joy of being at sea, the passion for oceanography, and a deep friendship.

This work let me discover the generosity of a number of colleagues (old and young), all those of LODYC (which became LOCEAN) and of many other laboratories, who were always willing to help me. It brought me happy encounters and much joy each day. I send my very sincere warm thanks:

To the painstaking and enthusiastic proofreaders, so precious to me, who took the time to help improve all or part of the text: Jérôme Sirven, Christophe Herbaut, Rosemary Morrow, Marie-Hélène Radenac, Pascale Delecluse, Young-Hyang Park, Claude Millot, Catherine Rouault, and Claudie Marec.

To all those who, at many levels, helped me with friendly participation in this long-term task by producing maps, or by sending me articles, comments, illustrations, or photos: Clément de Boyer Montégut, Bernard Barnier, Edmée Durand, Isabelle Durand, Gilles Garric, Sébastien Masson, Alice Pietri, Didier Swingedouw, Sabrina Speich, as well as Agus Amatdipoera, Michel Arhan, Christian Begler, Mathieu Belbeoch, François Bellec, Karine Béranger, Gilles Bessero, Bruno Blanke, Jacques Bourgois, Bernard Bourlès, Xavier Capet, Michel Crépon, Nathalie Daniault, Jacques Darchen, Julie Deshayes, Vincent Echevin, Jean-Luc Fuda, Patrick Geistdoerfer, Stuart Godfrey, Joseph Gonella, John Gould, Yves Gouriou, Eric Greiner, Michel Hontarrède, Marie-Noëlle Houssais, Greg Johnson, Gérard Jugie, Annie

Kartavtseff, Johannes Karstensen, Ariane Koch-Larrouy, François Lacan, Jacky Lanoisellé, Benoit Legrésy, Claire Lévy, Gurvan Madec, Christophe Maes, Catherine Maillard, Carlos Meija, Jacques Merle, Robert Molcard, Anne Petrenko, Frédérique Rémy, Gilles Reverdin, Phil Richardson, Fabien Roquet, William Schmitz, Isabelle Taupier-Letage, Pierre Testor, Sergey Varlamov, Bruno Voituriez, Susan Wijffels, Toshio Yamagata, Igor Yashayaev, Jong-Hwan Yoon, and all those who encouraged me.

To Jean-François Minster who, in spite of his heavy responsibilities, agreed to prepare the preface with much enthusiasm and kindness.

To all those courageous unknowns, explorers, sailors, and oceanographers who worked to obtain precious observations, often under difficult conditions, at the price of immense effort. To all the participants in the worldwide WOCE cruises who worked in the shadow, day and night, whose data are now so easy to use.

Many illustrations in this book come from international data centers or from Internet sites providing software for importing data and plots. We thank in particular:

- The National Oceanographic Data Center (NODC) of NOAA, for access to the WOCE database with the collection of the WOD cruises, the global climatology *WOA05*, and Levitus, 1994.
- Reiner Schlitzer of the Alfred Wegener Institute (AWI), Bremerhaven, Germany for the free use of the *Ocean Data View* software for analyzing oceanographic data and for the reproduction of the *WOCE eAtlas* sections online.
- The International Research Institute for Climate and Society (IRI) of NASA for access to its software for plotting hydro data with its easy links to WOCE data, Levitus 1994, and the Silva Atlas 1994.
- The Jet Propulsion Laboratory (JPL) of NASA for allowing use of the POET software "PODAAC Ocean ESIP Tool", for access to temperature and surface satellite data (AVHRR), and for plotting facilities.
- Objectively Analyzed air-sea Fluxes for the global oceans (OAFlux), NOAA/OCO/Woods Hole Oceanographic Institution.
- International Satellite Cloud Climatology Project (ISCCP), NASA Goddard Institute for Space Studies, Columbia University.
- Global Precipitation Climatology Project (GPCP) WCRP, University of Maryland, for data and maps of atmospheric flux and precipitation.

Some illustrations were imported from Internet sites, and we thank in particular:

- Ocean Remote Sensing Group, Johns Hopkins University, Applied Physics Laboratory Space Department for satellite photographs of the Gulf Stream.
- Kuroshio Extension Study System (KESS), University of Rhode Island, for altimetric JASON plots of the Kuroshio extension.

- Marine and Environmental Education and Research (MEER), University of California, for the world map of Herodotus.
- The JCOMM In-situ Observing Platform Support Centre (JCOMMOPS), for ARGO program profiles (http://wo.jcommops.org/cgi-bin/WebObjects/Argo).

We would also like to thank the authors and publishers of works which have been important to us and from which we have used illustrations:

- *Regional Oceanography: An Introduction*, by Matthias Tomczak and J. Stuart Godfrey, Pergamon, London, 1994.
- *Regional Oceanography*, by Paul Tchernia, École Nationale Supérieure des Techniques Avancées (ENSTA), Paris, 1978.
- *Ocean Circulation*, by the Open University Course Team, Pergamon Press, Milton Keynes, 1989.
- *Atlas de Christophe Colomb et des Grandes Découvertes*, by Kenneth Nebenzahl, Bordas, Paris, 1991.
- *Cartes Géographiques Anciennes. Évolution de la représentation cartographique du monde : de l'Antiquité à la fin du XIXe siècle*, by Ivan Kupcik, Gründ, Paris, 1981.

My sincere thanks go also to France Citrini and Sophie Hosotte who took charge of organizing the publication of the english version.

My sincere and warm thanks go to all those who were with me during numerous data-collecting expeditions. We overcame the sea state, technical problems, and difficulties of many kinds. The more the elements were against us, the more the symbiosis between crew and scientists strengthened. That left a deep impression with me. I confess that I am a bit nostalgic for those rare moments.

Michèle Fieux, 2017

Addendum to the English Edition:

I would like to heartily thank my longtime friend Ferris Webster, of the College of Earth, Ocean, and Environment, University of Delaware. When Ferris received a copy of this work in French, he decided with enthusiasm to begin the difficult task of its translation. He had not counted the hours it would require to remain faithful to the French text while seeking and checking each phrase. Translation acts as an excellent sieve that does not support imprecision in the transition from French to English. Thanks to him, a number of ambiguously expressed passages were revealed and we have together tried to improve them. His enormous task allowed me to correct the French text. I am extremely grateful for his admirable work. This book would never have existed in English without him, without his competence, his patience, his enthusiasm, and his generosity.

Thank you so much, Ferris.

Michèle Fieux, 2017

The Author

Michèle Fieux is professor of regional oceanography at the École Nationale Supérieure des Techniques Avancées in Paris and member of the Marine Academy. Her career at CNRS (The French National Center for Scientific Research) in the Laboratoire d'Océanographie Physique of the Muséum National d'Histoire Naturelle, then in the Laboratoire d'Océanographie Dynamique et du Climat at the University Pierre and Marie Curie allowed her to live her passion through research and by leading numerous cruises on oceanographic ships. At the end of the sixties, she began studies of the north-western Mediterranean Sea on board the R/V Jean Charcot. In the seventies, the chance to work with Henry Stommel as guest investigator at the Woods Hole Oceanographic Institution had a great influence on her career. There, she studied the climatic evolution of sea-surface temperature and Indian monsoon variability. Work that followed principally focused on the Indian Ocean, its seasonal variability, and its connection with the Pacific Ocean through the Indonesian Archipelago. She participated in developing and carrying out large international programs of ocean observation, principally in the Indian Ocean, and chaired an international working group (of the Committee on Climatic Changes and the Ocean) on the Indian Ocean. Teaching today allows her to continue transmitting her passion for the sea to young people.

The Translator

Ferris Webster received a PhD in Geophysics at the Massachusetts Institute of Technology, then his career took him to the Woods Hole Oceanographic Institution (WHOI), the US National Oceanic and Atmospheric Administration (NOAA), and the College of Earth, Ocean, and Environment of the University of Delaware, where he is currently Professor Emeritus. His work has centered on the Gulf Stream, oceanographic variability, large-scale ocean-atmospheric interaction, and oceanographic data systems. He has had extensive involvement in international cooperative studies of oceanic and atmospheric phenomena. While at WHOI, he studied French with Michele Fieux as a teacher.

Ferris Webster and Michèle Fieux after the last corrections of the manuscript in Paris

Introduction

Earth is the only planet in the solar system where water is present in its natural state in each of its three forms: solid, liquid, and gaseous. This unique property permitted rudimentary life to appear and then to evolve into its most elaborate forms.

The ocean[1] contains the major part (97%) of the free[2] water at Earth's surface **[Figure 0-1]**. Surface terrestrial waters (lakes, rivers, the water table, permafrost,[3] glaciers) contain only 3% and atmospheric water only 0.001%.

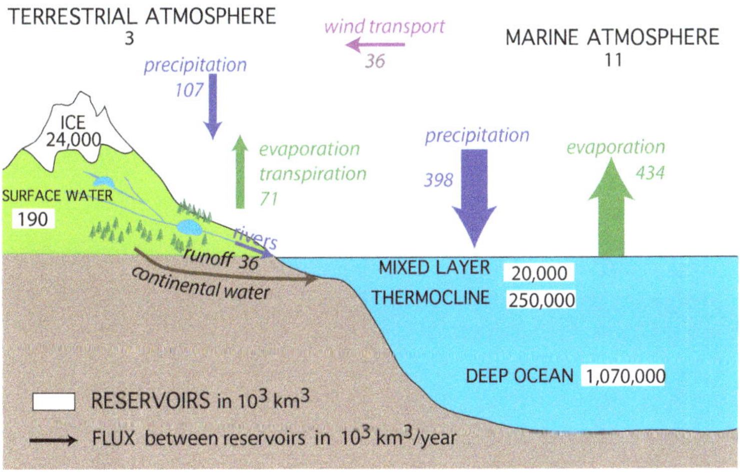

FIG. 0-1 – Mean water distribution among reservoirs on the global surface with the order of magnitude of average annual fluxes among them.

Though it represents only a miniscule part of free water, atmospheric water has a critical role in understanding climate, and its principal reservoir

[1] In Greek legend, Oceanus, the immense river that encircled the terrestrial world, was a Titan, son of Uranus (the sky) and of Gaia (the Earth).

[2] An immense amount of water exists in rocks, liberated in minute quantities during volcanic eruptions.

[3] Permafrost is subsoil permanently frozen, for example in northern Siberia.

is in the ocean. Ocean and atmosphere are great fluid bodies inseparably involved in the perpetual water cycle.

The ocean-atmosphere system represents an immense thermal machine whose principal source of energy is the sun.[4] Solar energy is not distributed uniformly on the globe because of Earth's spherical form, which produces energy-transfers from the equator toward the poles by the two fluids in a coupled and complex manner. The sea-surface represents 71% of the "floor" of the atmosphere and is the site of constant fluxes.

Since the physical characteristics of the two gaseous and liquid environments differ, the time and space scales associated with their evolution also differ. The ocean responds slowly to atmospheric changes and slows the evolution of the coupled ocean-atmosphere system. In that way, the ocean plays an essential role in the climate system. Because of its mass (250 times that of the atmosphere) and its great heat capacity, it represents an enormous reservoir of heat for the combined ocean-atmosphere system.

Beneath its appearance of relative stability, the ocean is in perpetual movement. Its movements range from small-scale (turbulence) through mesoscale (eddies) up to large-scale (the Gulf Stream or deep circulation).

How does an immense mobile reservoir operate on an Earth which itself turns?

Ocean circulation is principally due to exchanges with the atmosphere in the form of heat, water, and momentum.

Differences in atmospheric pressure generate winds, which, while tending to reestablish pressure equilibrium, transfer part of their kinetic energy to the ocean and thereby generate most surface currents.[5] The Coriolis force, due to Earth's rotation, turns the motions generated by these exchanges. Wind influence on circulation produces regions of convergence, where surface water sinks under lighter water and regions of divergence, where subsurface water rises to the surface.

Heat and water fluxes with the atmosphere produce heating, cooling, evaporation, precipitation, freezing, and melting and modify temperature and salinity of surface seawater. These modifications induce density changes. When surface water becomes denser than the water beneath it, it has a tendency to sink and mix, bringing about density variations from one region to another, which in turn generate internal circulation known as thermohaline circulation.[6] In large part due to this process the ocean conserves climatic variations in "memory". A climatic anomaly stored in a region of the ocean may be transmitted in time and space.

---

[4] Earth's interior contains immense quantities of heat, which "escapes" through plate junctions and by active sub-surface volcanoes. This contribution of heat, as well as the heat flux emitted by the sea floor, is crucial for the development of life at great depth but seems to be negligible in the functioning of the ocean.

[5] We are not considering tidal currents.

[6] Because it depends on variations of temperature (thermo) and salinity (haline).

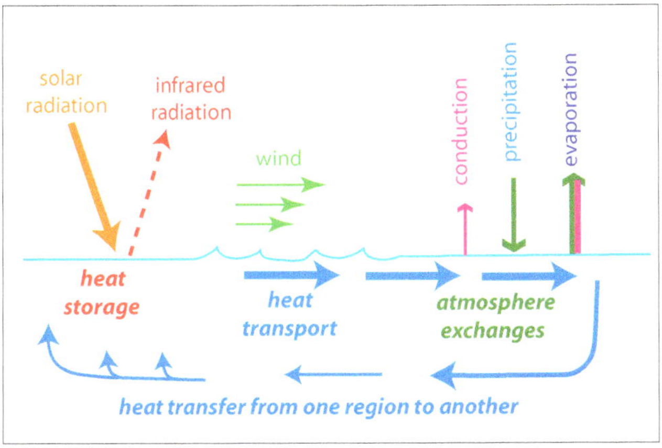

FIG. 0-2 – The ocean's roles in exchanges with the atmosphere.

We describe the roles played by the ocean **[Figure 0-2]** in exchanges with the atmosphere that generate ocean-surface circulation and the formation of the oceanographic conditions responsible for thermohaline circulation. In this text the oceanographic analysis of each ocean basin is followed using an approach that is first global and then regional.

Part I
Generalities

Chapter 1
Ocean characteristics

1.1 Spatial characteristics

Of the 510,100,900 square kilometers of the global surface area, seas and oceans cover 361,059,000 (70.8%). The partition between land and sea is not symmetric in the two hemispheres [**Figure 1-1**]. In the northern hemisphere, the proportion of land is greater, with 39.3% land *vs.* 60.7% sea, while the southern hemisphere has 19.1% land *vs.* 80.9% sea. Consequently, continental effects are more pronounced in the northern hemisphere.

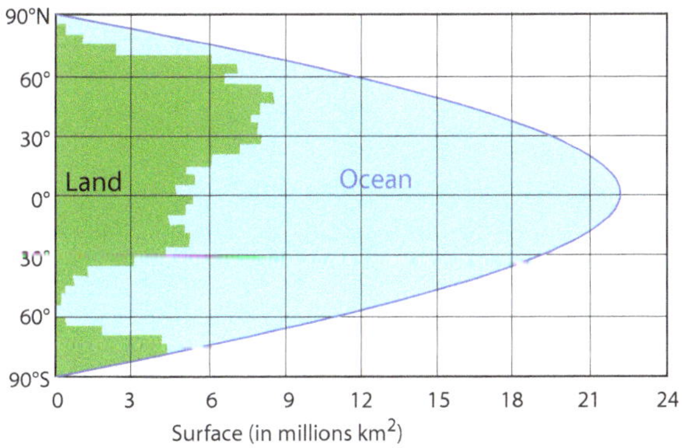

FIG. 1-1 – Areal distribution of land and sea as a function of latitude. (After Tchernia, 1978.)

We can regard Earth from an unusual angle, with a dominantly maritime hemisphere (the pole would be southeast of New Zealand) and a dominantly continental hemisphere (with a pole in western France) [**Figure 1-2**]. In this representation, the ocean would cover 89% of the maritime hemisphere, and 53% of the continental hemisphere.

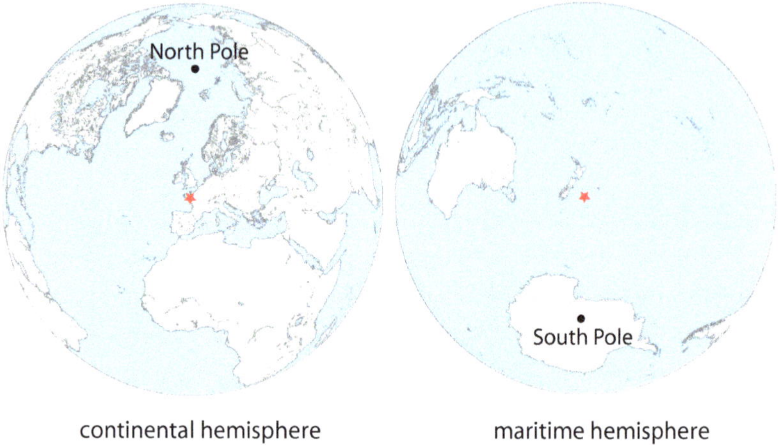

FIG. 1-2 – The continental and maritime hemispheres of Earth. The poles are shown in red. (After Olson, 2008.)

The limits imposed by the continents allow us (for simplicity) to separate the ocean into several basins connected to the Antarctic Ocean: the Atlantic Ocean, linked to the Arctic Ocean by the Norwegian and Greenland seas, the Indian Ocean, closed to the north, and the Pacific Ocean, the largest basin [**Figure 1-3**]. These basins are connected with each other and constitute a single liquid mass. There is continuity from one region to another of the planetary ocean.

The ocean has an average depth of 3,800 m, which is 1/1675th of Earth's radius. The deepest depth, situated in the Mariana trench, reaches 10,994 ± 40 m. (Mt. Everest, the highest point on Earth, has an altitude of 8,846 m and the average elevation of all land above sea level is 840 m.) The ocean thus is a thin skin of saline water on the terrestrial globe. The ocean's vertical dimension is a thousand times less than its horizontal dimension. Extreme exaggeration of the vertical dimension with respect to the horizontal dimension is used to represent the stratification of the ocean.[7]

The transition from the granitic structure of the continents to the basaltic structure of the ocean basins takes place over the continental shelf and the continental slope [**Figure 1-4**].

[7] If we were to keep to the true ratio, a 10-cm-wide representation of a section across the Atlantic Ocean would have a thickness of only 0.05 mm. It is why vertical currents are so small compared to horizontal ones.

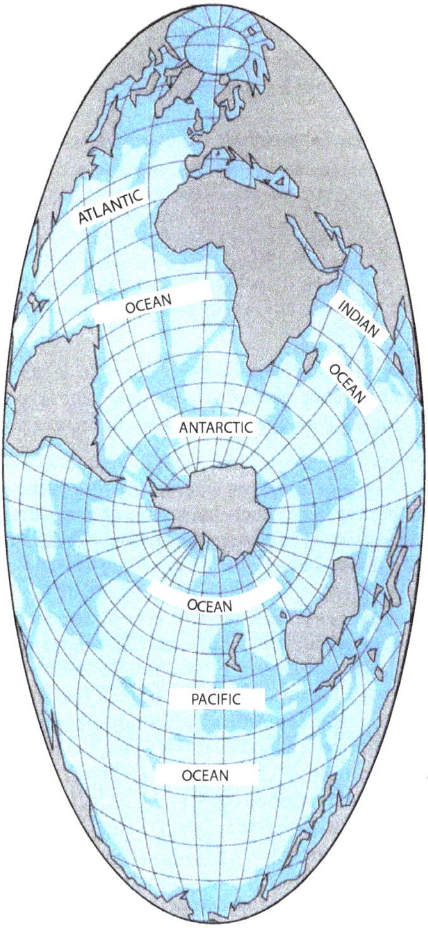

FIG. 1-3 – This view of Earth shows the continuity of the planetary ocean. Depths greater than 4,000 m are in light blue. (After a drawing of Fritz Fuglister, WHOI.)

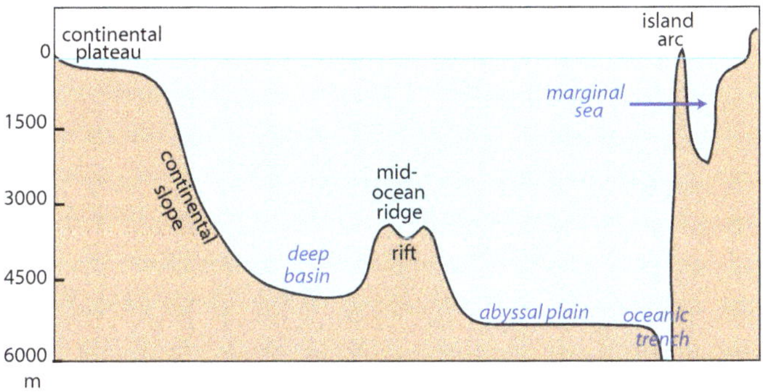

FIG. 1-4 – Types of oceanic depths.

The continental shelf extends the slope of the neighboring continent up to a shelf-break zone, where the depth is about 200 m (between 100 and 500 m). Continental shelves represent 7.6% of the ocean surface. The shelf is often slashed by submarine canyons, which prolong the rivers that eroded the shelf during preceding periods of glaciation. This zone records the great variation in sea level between glacial and inter-glacial periods and receives sediment from continental erosion and weathering. It is subject to tidal currents, and displays strong variations in physical and chemical properties of seawater, imposed by river output. In this zone marine life is the richest and most varied. Off mountainous coastlines the continental shelf is nearly inexistent. It can reach 600 to 800 km offshore of low-lying lands, such as Siberia.

The continental slope links the continental shelf with the oceanic deeps by a slope (15.3%) steeper than that of the continental shelf. The slope represents, with the continental shelf, the continental basement, or continental margin, the transition between the continental crust and the oceanic crust, and has a vertical extent of about 3,500 m. The deep abyssal depths, where the average slope is slight, extend between 3,000 and 6,000 m and represent about 76% of the ocean.

The deep ocean is perturbed by seismically active zones such as oceanic ridges that can rise 1,000 to 2,000 m above the abyssal plains (cut by numerous fracture zones that offset the axis of the ridge) or by isolated volcanoes that do not always reach the ocean surface (guyots). Sometimes a deep and narrow trench occurs between the slope and the abyssal plains, for example off Peru, or off strings of islands isolating a marginal sea as off Japan, which follow the line of subduction of an oceanic plate and, in general, are zones of strong seismic activity.

Oceanic ridges constrain the deep-ocean circulation, but fractures across these ridges sometimes allow the passage of more or less deep water from one basin to another, while the deep trenches[8] do not have an oceanographic influence.[9]

The form and extent of an oceanic basin as well as the topography of the surrounding land, and its position on the Earth's surface have a strong influence on its climate, on oceanic exchanges with the atmosphere, and on the characteristics and circulation of the water masses that form there.

---

[8] 1.2% of the ocean floor has a depth greater than 6,000 m.

[9] Oceanography is the science that deals with physical and chemical properties of marine water. Hydrology is the science that deals with continental water systems. Hydrography is the applied science that deals with the measurement and description of the form of the sea and the coastal zone. Note that these terms do not have the same definition in French.

1.2 Physical characteristics of pure water

The unique properties of a water molecule come from the special position of the two atoms of hydrogen with respect to the atom of oxygen (the angle H-O-H is 104.5°) that provides the molecule with strong polarization and great stability. The polarization is responsible for the existence of links between water molecules called *hydrogen bonding*. Hydrogen bonding confers a remarkable cohesion to the structure of water compared with other liquids. A water molecule tends to surround four other water molecules, forming a tetrahedron. In ice, the bonding is weaker, since the atoms occupy less than half the total volume. During warming, the molecules tighten up again. During the melting of ice, density increases suddenly, and continues to increase from 0 °C up to 4 °C, the temperature of maximum density.[10] Above 4 °C the effect of dilation takes over and density diminishes. Rather surprisingly, in a frozen fresh-water lake, the water is warmer at the bottom (4 °C) than at the surface where the temperature is 0 °C.

Since freezing of water diminishes its density by about 10%, ice floats on water, which has considerable climatic consequences.

Because of hydrogen bonding, water has a specific heat or mass heat (the quantity of heat necessary to raise the temperature[11] of 1 kilogram of water by 1 K, namely 4,186 J kg^{-1} K^{-1} at 20 °C),[12] four times greater than that of air. The latent heat of fusion (the quantity of heat necessary to melt 1 kilogram of ice, namely 330×10^3 J kg^{-1})[13] is elevated, as well as the latent heat of vaporization of water (the amount of heat necessary to transform 1 kilogram of water into water vapor, namely 2.25×10^6 J kg^{-1}).[14]

The strong latent heat of vaporization is the basis for important energy transfers with the atmosphere. Condensation of water vapor in clouds releases heat taken up from the ocean. The surface tension of water (which corresponds to the difficulty of "breaking" the air-water interface) (72.75×10^{-3} N m^{-1} at 20 °C) is the greatest of all liquids. Surface tension plays a role in the formation of waves and droplets and thus also in the transfer of water and salt to the atmosphere. The thermal conductivity of water is weak (0.6 W m^{-2} K^{-1} at 20 °C) as well as its dynamic viscosity (1 Pa s at 20 °C). The dielectric constant of water is high (due to molecular polarization) and most terrestrial chemical elements are soluble in the ocean (including atmospheric gases, oxygen, carbon dioxide, nitrogen...).

---

[10] The density of pure water at atmospheric pressure equals 1,000 kg m^{-3} at 4 °C.

[11] K is the unit of temperature in the Kelvin scale, where zero equals –273.16 °C (the temperature at which there is no more molecular movement). T (in K) = T (in °C) + 273.16.

[12] That is, 1 cal g^{-1} K^{-1}.

[13] That is, 83 cal g^{-1}.

[14] That is, 585 cal g^{-1}.

Their concentration depends on their level of solubility and on temperature. For example, surface water at 2 °C contains more oxygen than it does at 25 °C. Water has low compressibility. However, if water was not compressible at all, sea level would be 30 m higher.

Thanks to its great thermal capacity, water absorbs a large proportion of incoming solar radiation, and transforms it into internal energy in the upper few meters.

The exceptional properties of water give the ocean, with a volume of 1,370 million cubic kilometers, the role of an immense thermal regulator, which "resists" temperature variations. It is an excellent agent for storage and transport of energy [**Figure 0-2**].

The ocean tempers variations of temperature on the planet. The ocean, thanks to its thermal inertia, warms and cools much more slowly than do the continents. Oceanic climates have diurnal and seasonal temperature ranges, whose amplitude is much less than the temperature ranges of continental climates. Thus the distribution of the amplitude of seasonal air-temperature variations clearly shows the position of the continents, at least away from the equatorial zone where the incidence of solar radiation varies little [**Figure 1-5**].

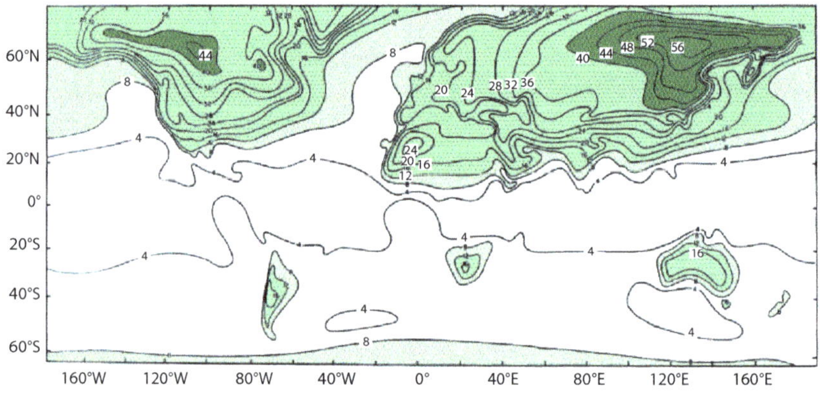

FIG. 1-5 – Amplitude of the seasonal variation of surface air-temperature in °C. (After Monin, 1975.)

1.3 Chemical and physical properties of seawater

Planet Earth was formed about 4.5 billion years ago with a central core composed of heavy elements (iron, nickel) surrounded by a mantle (iron, magnesium, silicon) itself covered by a thin crust (silicates, alumina...). At that time, the temperature was high and intense volcanic activity allowed the escape of gases that formed an atmospheric envelope around the planet composed principally of nitrogen, argon, carbon dioxide, water vapor... The planet was sufficiently voluminous and dense to retain this atmosphere. 400 or 500 million years later water vapor condensed during a strong period of cooling. Liquid water then appeared on the planet and the ocean was formed.[15]

The presence of salts in the ocean goes back to the same period, through the dissolution of atmospheric gases and the scrubbing of rock by primitive acidic terrestrial waters. The principal salts dissolved in seawater are sodium chloride (77.8%), magnesium chloride (10.9%), magnesium sulfate (4.7%), calcium sulfate (3.6%), potassium sulfate (2.5%), calcium carbonate, magnesium bromide, and 0.5% of other salts. The polarization of the water molecule acts to keep these salts in ionic form, of which six (Cl^-, Na^+, K^+, Ca^{2+}, Mg^{2+}, SO_4^{2-}) account for 99% of the dissolved salts.

In spite of the great chemical complexity of seawater, its composition does not depend on the region being considered. The concentration may be variable, but the relative proportions among the different salts remains the same (Dittmar's law[16]). For example, sodium chloride always represents 77.8% of the total mass of salts.

The remarkable equilibrium of oceanic composition is due to regulating processes, the principal of which is oceanic circulation, which assures a perpetual mixing throughout the whole ocean in three dimensions. Other regulating factors are the coefficients of solubility, which precipitate some salts in excess, and biologic activity of living creatures, which fixes some soluble salts and thereby transforms them into insoluble salts, which later are incorporated into sediments.

In 1902, Sorensen, Knudsen, and Forch linked salinity to the amount of ions present in seawater. The mass of all dissolved matter is difficult to obtain, and the residue after heating is highly hygroscopic. Thus, they put in place a technique consisting of desiccation at 480 °C, in a chlorinated medium, up to a constant weight, in which organic materials are decomposed, carbonates are transformed into oxides, and bromide and iodides into chlorides. From this, the definition of salinity equals the ratio between the mass of all solid dissolved material and the total mass of the seawater sample (carbonates converted into oxides, bromides and iodides replaced by chlorides and all organic matter oxidized). This is why the unit *parts per thousand* was used (for example 35 g of salts in 1,000 g of water would be 35‰).

---

[15] Then, later, the appearance of blue algae introduced oxygen into the atmosphere.

[16] Dittmar discovered this law in 1884 during the analysis of 77 samples obtained around the world by the British oceanographic expedition on board the *HMS Challenger*.

> Using Dittmar's law, it is sufficient to measure[17] one of the constituent ions of seawater (in this case Cl^-, representing 55% of mass of the dissolved salts) in order to calculate the total mass of dissolved salts. In practice, the measurement of salinity is delicate. Sorensen *et al.* proposed obtaining it from the empirical relationship: S‰ = 0.030 + 1.805 Cl‰, where Cl‰ is the chlorinity, more easily measured. The chlorinity of a sample is the mass of halogens in grams (except for fluorides which are not precipitated by silver nitrate) contained in 1 kg of seawater, the Br^- and I^- ions being replaced by their equivalences in chlorides.
>
> In 1959, Carritt and Carpenter showed that deviations from Dittmar's law could reach 0.04‰ in salinity. In 1962, the simpler relationship: S‰ = 1.80655 Cl‰ was proposed, with the advantage of associating zero salinity with zero chlorinity. This relationship is rigorously equivalent to the preceding one for a salinity of 35‰ and is very close for usual salinities.[18]

Nowadays the salinity of a seawater sample is obtained by the (much more rapid) measurement of its electric conductivity, which is compared to that of a solution at the same temperature, whose salinity (or chlorinity) is known.[19] Conductivity varies with temperature and it is thus necessary to know the temperature to an accuracy of a hundredth of a degree. Since the measurement is a ratio of salinities, it has been decided internationally to quote salinity as a value with no dimension.[20]

Generally, salinity at a point in the ocean is stable in time, particularly at depth. The strongest variations are observed at the surface, under the effect of exchanges of water, through evaporation, precipitation, and inflows from rivers and glaciers, and heat exchange through the formation or melting of sea ice. These depend on seasonal climate variation and circulation. The process of freezing seawater produces an increase in salinity through the extraction of fresh water. Sea ice, or pack ice, is slightly saline, however, because it traps small brine pockets within a structure of pure ice.

The presence of salts in water has numerous consequences, in particular the lowering of the freezing point of water (a property used for salting roads in winter). For a salinity of 35, at atmospheric pressure the freezing point is −1.9 °C.

---

[17] The Mohr-Knudsen method consists of precipitating halogens by silver nitrate in rigorously equal volumes of "normal" water and of seawater, using a Knudsen pipette, which allows a precision of 5×10^{-4}. From 1902, "normal" water prepared at Copenhagen in sealed ampoules of known chlorinity (about 19.4‰) has allowed measurement of chlorinity as a relative value, with better precision. However, this method depends on the atomic weights of silver and of chlorine whose values have slightly varied due to progress in measurement techniques. Furthermore, the salinity of reference normal water evolved slightly between 1902 and 1937.

[18] The difference between the two formulas only becomes noticeable for weak salinities that depart from Dittmar's law.

[19] Using an algorithm involving temperature.

[20] The unit psu (practical salinity unit) for salinity is sometimes still used.

The density of seawater (ρ in kg m^{-3}) is a function of temperature, salinity, and pressure.[21] Dissolving salts in water decreases the temperature of maximum density.[22] For salinities above 24.7, when temperature decreases, seawater reaches the freezing point before reaching the temperature of maximum density [**Figure 1-6**].

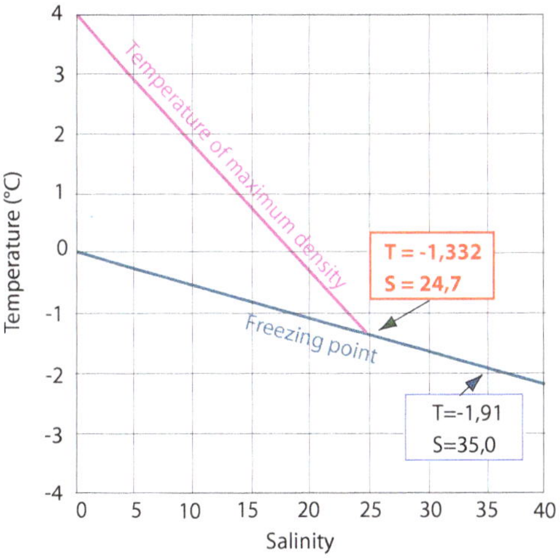

FIG. 1-6 – Variation with salinity of: (violet) the temperature of maximum water density and (blue) the freezing-point temperature of water at atmospheric pressure.

Salinities encountered in the ocean are always above 24.7 outside of estuaries and epicontinental seas, such as the Baltic Sea. Thus, as temperature decreases, the freezing point is always reached before the temperature of maximum density and there is no temperature inversion due to this effect in the ocean. For a given salinity, the coldest water is the densest and the deepest.

It is impossible to measure density directly *in situ* or aboard a ship with sufficient precision. The measurement is indirect, based on temperature, salinity, and pressure.

Density is calculated using the equation of state of seawater, $\rho = \mathrm{f}\,(T, S, p)$ which is not a simple function. It has been empirically determined in the laboratory. The equation has 15 terms and is non-linear.[23] The non-linear effects are especially important for low temperatures near the freezing point.

---

[21] Density increases with salinity and pressure and decreases as temperature increases (*i.e.*, a negative coefficient of thermal expansion).

[22] The freezing point also decreases with pressure (having an impact on Antarctic ice sheets).

[23] UNESCO Technical Paper in Marine Science, No. 38, 1981.

One can use the following approximate equation:

$$\rho - \rho_0 = a(T - T_0) + b(S - S_0) + cp$$

Where
a = coefficient of thermal expansion at T_0 (-0.15 kg m^{-3} K^{-1})
b = coefficient of contraction due to salinity (0.78 kg m^{-3})
c = coefficient of compressibility (4.5×10^{-3} kg m^{-3} dbar^{-1})
(for T_0 = 10 °C, S_0 = 35, and $p = p_0$ (1 atmosphere), ρ_0 = 1,026.97 kg m^{-3})

With this approximate equation the precision is ± 0.5 kg m^{-3}.

The density range in seawater is only a few thousandths, between 1,022 kg m^{-3} and 1,029 kg m^{-3} at atmospheric pressure. That is why oceanographers use the density anomaly, $\sigma = \rho - 1000$ kg m^{-3} while taking the liberty of calling it "density". Brought to atmospheric pressure, that is, at the surface, σ is called σ_0. The range of σ_0 ranges between 22 and 29.2. The determination of σ to a hundredth or even to a thousandth is necessary in order to detect processes such as the thermohaline circulation [**Section I.4.6**] due to differences in density between regions, particularly at depth, away from the influence of wind. Lines of equal density are called isopycnals. Density increases approximately one unit as temperature decreases by 5 °C, as salinity increases by 1, or as pressure increases by 20 bars (that is, about 200 m).[24]

Hydrostatic pressure equals the weight of the column of water:

$$dp = -\rho g dz, \text{ and } p_z - p_0 = \int_0^z (-\rho g dz),$$

where z is taken positive upward.

[24] Specific volume, α, is also used. $\alpha = \frac{1}{\rho}$ (m^3 kg^{-1}). This is often separated into two terms, $\alpha = \alpha_{35,0,p} + \delta$, where $\alpha_{35,0,p}$ is the specific volume of standard water with S = 35, T = 0 °C, at pressure p, and δ is the specific volume anomaly with respect to standard water.

Chapter 2
Heat and water exchanges between ocean and atmosphere

2.1 Global radiation balance

Solar radiation is the primary source of energy for the ocean-atmosphere system. Averaged over a year, the energy balance over the totality of the globe is zero. That is, incident energy that comes from the sun in the form of short-wavelength radiation is either reflected or absorbed and then transformed and returned to space in the form of long-wavelength (infrared) radiation. The transformations that the incident energy undergoes before returning to space occur in many forms: reflection, conduction, evaporation, and infrared radiation [**Figure 1-7**].

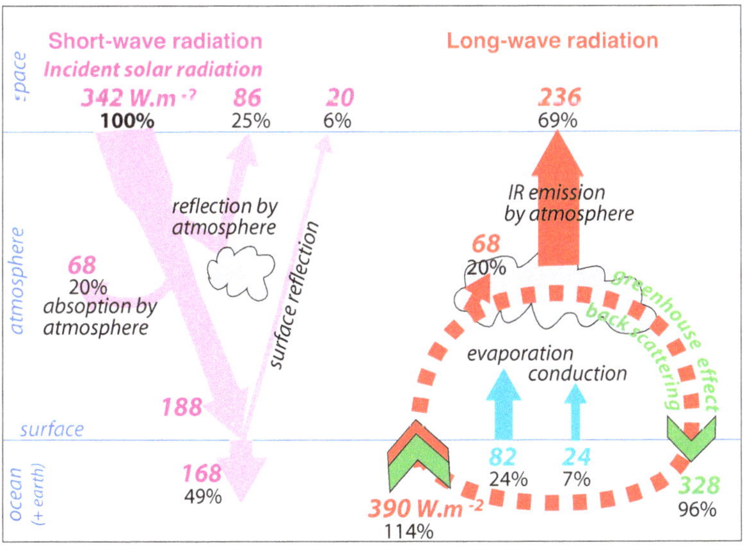

FIG. 1-7 – Average energy balance of the ocean-earth-atmosphere system.

2.1.1 Incident solar radiation and albedo

Sun and Earth emit as blackbody radiators. The energy they emit is a function of their temperature and follows the Stefan-Boltzmann law: $E = \sigma T^4$, where the Stefan-Boltzmann constant, $\sigma = 5.67 \times 10^{-8}$ W m^{-2} K^{-4}, and T is the temperature in degrees Kelvin.

The incident energy which arrives from the sun is that of a black body at 5,780 °C.

The solar constant is the radiation falling on a surface of 1 m^2 perpendicular to the direction of the sun (1,368 W m^{-2}). The total solar radiation (short-wave) arriving at the top of the atmosphere equals $1368 \times \pi R^2$ Wm^{-2} (R is Earth's radius). It is distributed over the terrestrial sphere, with a surface of $4\pi R^2$, and has an average value of 342 W m^{-2} [**Figure 1-7**].

On average, and without a change in wavelength,[25] 86 W m^{-2} (25% of the incident radiation) is directly reflected back to space, by clouds and aerosols, and 20 W m^{-2} (6% of the incident radiation) is reflected directly by the terrestrial surface (earth + ocean) [**Figure 1-7**]. Atmospheric molecules scatter part of the solar radiation in all directions (Rayleigh scattering). Since the amount of scattering is inversely proportional to the fourth power of the wavelength, the shortest wavelengths are scattered the most, thereby giving the sky its blue color.[26]

The albedo of a surface is the percentage of direct reflection of the radiation received by the surface. The albedo of Earth overall is 31% (25% by the atmosphere + 6% by the surface).

Sea-surface albedo varies according to the incident angle of radiation and the sea state. Albedo is higher at sunrise and sunset and in high latitudes. Sea state modifies albedo, particularly when the incident angle of radiation reaches a high value. Solar radiation penetrates more easily into the ocean when the sea is agitated than when it is calm.

In the visible spectrum, water molecules preferentially absorb long wavelengths corresponding to red-orange colors. Short wavelengths in the visible spectrum are less absorbed, which gives the ocean a blue color.[27] The presence of a variety of suspended matter may modify this color. Even in transparent regions, where there is little biological activity, only 1% of the incident energy penetrates to 100 m depth. This depth can decrease to as little as 20 m in zones rich in biological matter or particles. When the sea is ice-covered, the albedo increases greatly and radiation is nearly entirely reflected back to the atmosphere.

---

[25] The range of wavelengths in solar radiation occurs between 0.2 and 4 µm of which 49% are in the visible spectrum, with a maximum near 0.5 µm.

[26] Because of its variable incidence, radiation from atmospheric scattering is reflected proportionally more by the sea-surface than is direct radiation.

[27] Hence the name "Blue Planet" given to Earth by astronauts.

2.1.2 Absorption of incident radiation

68 W m^{-2} or 20% of the incoming solar radiation is absorbed in the atmosphere, by clouds, aerosols, and greenhouse gases [**Figure 1-7**]. For the most part, the atmosphere absorbs solar energy in the ultraviolet in the stratosphere and in the infrared through water vapor.

188 W m^{-2} cross the atmosphere without modification and arrive at the Earth's surface, which absorbs 168 W m^{-2} (nearly half, 49%) of the solar energy arriving at the top of the atmosphere.

Water absorbs luminous energy well. Due to its high specific heat, the first few meters of the ocean can store as much heat as the whole column of air above it.

2.1.3 The greenhouse effect

If we apply the Stefan-Boltzmann law to the long-wave energy sent into space at the top of the atmosphere (236 W m^{-2}) [**Figure 1-7**], we obtain[28] a surface temperature of −18 °C. However, the average temperature of Earth is about +15 °C. Only at very high altitude can we find a temperature of −18 °C. The 33 °C difference comes from the greenhouse effect[29] due to the presence of atmospheric molecules (water vapor, carbon dioxide, methane, chlorofluorocarbons...). A temperature of 15 °C would correspond to an average radiation of about 390 W m^{-2}, which is greater than the solar energy absorbed directly by the Earth's surface (168 W m^{-2}).

Infrared radiation emitted by the surface does not go directly into space. It is absorbed for the most part by the atmosphere and backscattered toward the surface.[30] Back scattering (328 W m^{-2}) connected with the greenhouse effect[31] is on average about the same size as incident solar radiation (342 W m^{-2}).

2.1.4 Long-wave (infrared) radiation

For the ocean, the net loss of heat by infrared radiation (Q_{ir}) is equivalent to the radiation of a black body at the temperature of the ocean surface[32] (390 W m^{-2}) less the backscattering due to the greenhouse effect of the atmosphere (328 W m^{-2}, depending on the amount of water vapor in the air) or, on average, 62 W m^{-2}.

---

28 $E = 236 = (5.67 \times 10^{-8} \times T^4)$ where $T = 255$ K or −18 °C.

29 By analogy with the trapping of solar radiation by the glass of a greenhouse.

30 The temperature difference felt between a clear night and a cloudy night shows the direct effect of the back scattering by clouds.

31 Earth's unique atmosphere represents a truly protective layer in more than one way. The greenhouse effect keeps us in a "warm cocoon" which allows water to exist in its three states and the ozone layer protects us from ultraviolet radiation.

32 Ocean surface temperature is the temperature of the skin of the ocean of a few microns thickness.

We use an empirical formula to estimate infrared losses:

$Q_{ir} = 0.985\,\sigma\,T^4(0.39 - 0.05\,e_a^{1/2})(1 - 0.6\,n_c^2)$

e_a = water vapor pressure in air at 10 m in millibars or hectoPascals

n_c = percentage of the sky covered by clouds (cloud cover)

2.1.5 Evaporation and conduction

In addition to radiative exchanges, other mechanisms contribute to exchanges of heat between the ocean and atmosphere. Part of the solar energy absorbed by the ocean is sent back to the atmosphere: 7% in the form of sensible heat transmitted by direct conduction and propagated by atmospheric convection,[33] and 24% in the form of latent heat linked to the evaporation process [**Figure 1-7**]. Transfers of heat by evaporation are associated with water loss, mitigated by precipitation.

Evaporation transforms liquid water into vapor from the ocean to the atmosphere. It requires a large amount of energy to free the molecules from their hydrogen bonds. This mechanism involves a loss for the ocean, not only of water, but also of heat. Only during the process of condensation in clouds is the heat finally released into the atmosphere, which is why it is called *latent* heat. It represents an important source of heat for the atmosphere. For example, 1,000 mm of precipitation per year would represent a gain of 79 W m^{-2} for the atmosphere.[34]

Estimated average evaporation amounts to a loss of about 1.2 m of water per year per m^2 of ocean. The corresponding latent heat loss is estimated at 82 W m^{-2} or about 24% of the energy received from the sun. This estimate is difficult to make since direct measurements do not exist.

An empirical formula is used to calculate latent heat loss.[35] Latent heat (L), the heat necessary to transform one kg of liquid water into vapor, varies little with salinity. Heat loss by evaporation varies with wind speed (V) and with the gradient of specific humidity[36] of the air above the sea-surface ($e_{surf} - e_a$):

$Q_e = LE = Lc_d\rho_a\,(e_{surf} - e_a)\,V$

---

33 Convection represents heat transfer in a fluid. The transfer is that of matter rather than of radiation.

34 When water returns to the ocean in the form of precipitation and during condensation of water vapor at the surface, there can be a slight addition of heat through latent heat gain, which is negligible compared to the losses by evaporation.

35 This is the Monin-Obukhov relationship, whose coefficients are frequently updated.

36 Specific humidity is the ratio between the mass of water vapor contained in 1 m^3 of humid air and the mass of the same volume of air. Relative humidity is the ratio between the specific humidity at a given temperature and pressure and the specific humidity at saturation at the same temperature and pressure, for the same volume of air.

L = latent heat = 2.5×10^6 J kg^{-1}

E = evaporation rate in kg m^{-2} s^{-1}

V = wind speed in m s^{-1} at 10 m

e_{surf} = saturation specific humidity at the sea-surface temperature, T_{surf} in kg kg^{-1} (in reality, $e_{surf} = 0.98\ e_{sat}$ at T_{surf} to account for the reduction of the vapor pressure of water due to salinity).

e_a = specific humidity of air at 10 m, which depends on air temperature, T_a and its relative humidity in kg kg^{-1}. It increases rapidly with temperature; the warmer the air, the more water vapor it will contain.

$c_d \cong 1.5 \times 10^{-3}$, the coefficient of turbulent exchange. It depends on atmospheric stability and wind speed and is an empirical constant.

ρ_a = air density.

The warmer the air, the stronger the wind, (the more the atmosphere is turbulent) and thus the stronger the evaporation.

Sensible heat is transmitted by direct conduction between molecules of water and air. If the ocean is warmer than the atmosphere, the ocean loses heat by this process. Losses of heat by conduction (Q_c), which depend on wind speed, are also difficult to estimate.

We use an empirical expression comparable to that for latent heat:

$$Q_c = \rho_a\, c_d\, c_p\, (T_{surf} - T_a)\, V$$

T_{surf} and T_a = sea-surface temperature and air temperature in °C.

c_p = specific heat of air at constant pressure in J kg^{-1} K^{-1}.

c_d = coefficient of turbulent exchange, which depends on the degree of stratification (or inversely, on atmospheric turbulence) and on wind speed. It varies between 1.10 and 0.83×10^{-3} (an empirical constant).

V = wind speed in m s^{-1} at 10 m.

The stronger the wind speed, and the greater the difference between the air and water temperatures, the greater is the loss of heat by conduction. Heat conduction losses are of the order of 24 W m^{-2}, which represent about 7% of incident solar radiation.

2.1.6 Radiative balance

The net radiative balance is the difference between the total emitted infrared radiation and the backscattered radiation due to the greenhouse effect (328 W m^{-2}) [**Figure 1-7**]. It amounts to only 62 W m^{-2} or 18% of the incident solar radiation. If this is added to the fluxes of heat by convection due to evaporation (82 W m^{-2}) and conduction (24 W m^{-2}) it amounts to a net of 168 W m^{-2} absorbed at the surface.

In summary, the atmosphere sends 69% (236 W m^{-2}) of the incident energy back to space in the form of infrared radiation and 31% (86 W m^{-2} + 20 W m^{-2}) is reflected back with no change of wavelength.

Solar radiation is not distributed uniformly on the globe. It decreases from the equator to the poles because of the increasing inclination of solar rays with latitude. For a unit surface, Earth receives on average[37] more heat in tropical regions than in polar regions **[Figure 1-8]**. On the other hand, emitted radiation, which depends on temperature, does not vary so much with latitude. This results in an excess of energy between 35°S and 35°N and a deficit at higher latitudes.

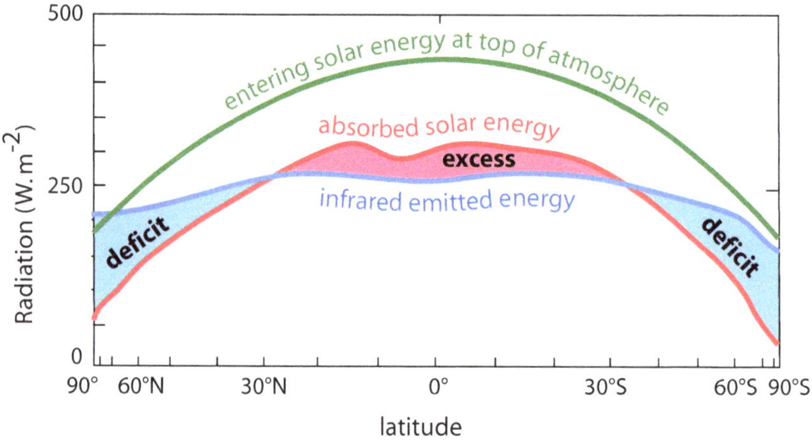

FIG. 1-8 – Average latitudinal distribution of: incident solar energy arriving at the top of the atmosphere (green curve); solar energy absorbed by the terrestrial surface (red curve); and energy emitted in the form of infrared radiation by the surface (blue curve), based on satellite measurements. The scale is proportional to the surface area of the Earth. (After Campbell and Vonder Haar, 1980.)

2.2 Distribution of ocean-atmosphere heat fluxes

2.2.1 Solar radiation

The regional distribution of solar radiation at the ocean surface, in addition to latitude variations, shows differences due to the amount of atmospheric water vapor **[Figure 1-9]**. For example, in the tropical zone subject to strong solar radiation, maxima are found in areas with minimum cloud

[37] This average has strong seasonal variations.

cover (anti-cyclonic centers). Cloud formations in the intertropical convergence zone (ITCZ) and in the area of strong atmospheric convection around Indonesia limit incident radiation.

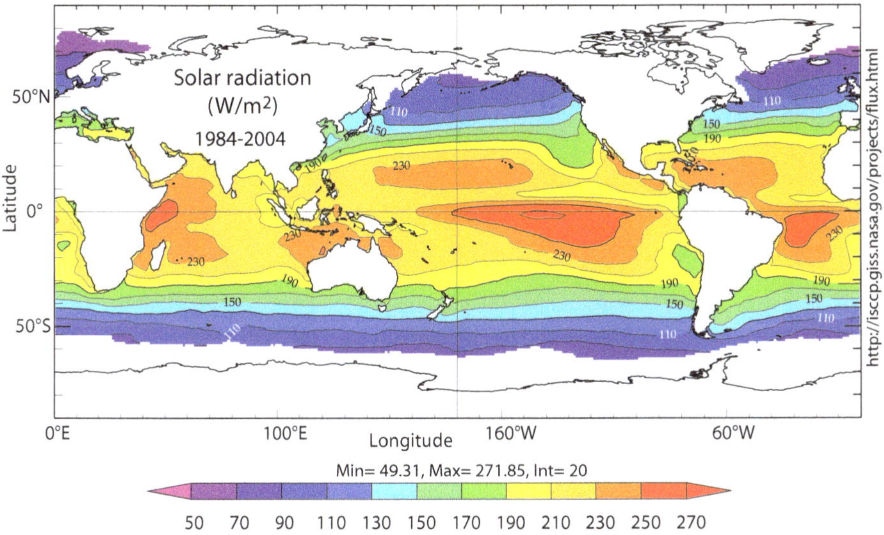

FIG. 1-9 – Solar radiation (W m^{-2}) at the sea-surface. (Courtesy of C. de Boyer Montégut, average data 1984–2004 ISCCP/NASA, Yu *et al.*, 2008.)

2.2.2 Heat loss by infrared radiation

The regional distribution of infrared heat loss [**Figure 1-10**] depends on ocean surface temperature [**Section I.2.4**] and on cloud cover [**Section I.5.1**]. In the Indonesian zone, for example, surface temperature is high but cloud cover is too, resulting in a minimum of infrared losses. Polar regions and upwelling[38] zones, because of their low surface temperature, also show a corresponding minimum in infrared losses compared with warm regions. On the other hand, the anti-cyclonic zones: the Mediterranean, the Red Sea, the Arabian Sea, and the Persian Gulf show maximum losses.

[38] Upwelling is upward movement of water (from below and thus colder) toward the surface.

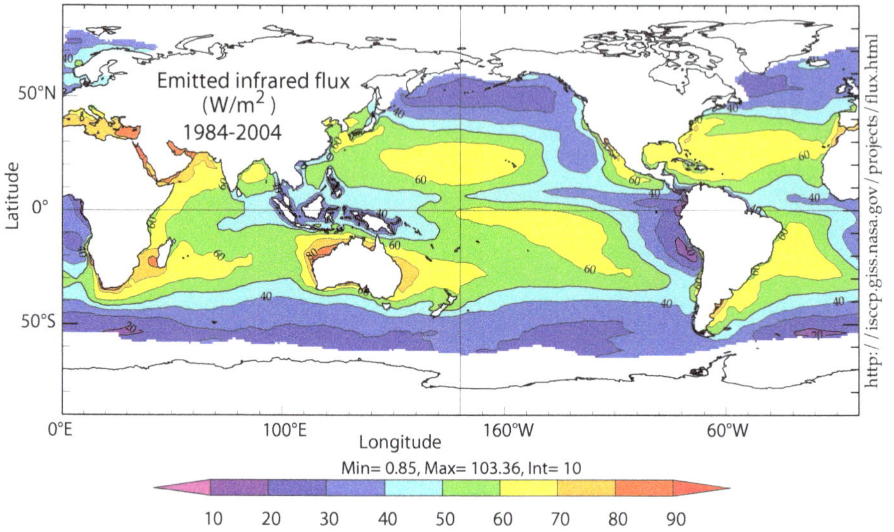

FIG. 1-10 – Infrared energy loss in W m^{-2}. (Courtesy of C. de Boyer Montégut, average data 1984–2004 ISCCP/NASA, Yu *et al.*, 2008.)

2.2.3 Heat loss by evaporation

An annual evaporation maximum exists year-round in subtropical anticyclone regions, where air is warm and dry [**Figure 1-11**]. Strong evaporation occurs in regions of warm poleward-flowing currents, such as the Gulf Stream, the Kuroshio, or east and west of Australia. Winter evaporation is especially strong, when air masses of continental origin are cold and dry. The humidity gradient in the air just above the surface is thereby intensified.

2.2.4 Heat loss by conduction

Heat loss by conduction is intense in ocean currents transporting warm water from tropical regions poleward (such as the Gulf Stream, the Kuroshio, the Agulhas Current, or the East Australian Current) especially in winter [**Figure 1-12**]. On the other hand, in regions where cold water rises to the surface and where the ocean is colder than the atmosphere (coasts of Peru, Somalia...) and in the Antarctic region experiencing subtropical depressions, the ocean tends to gain heat by conduction.

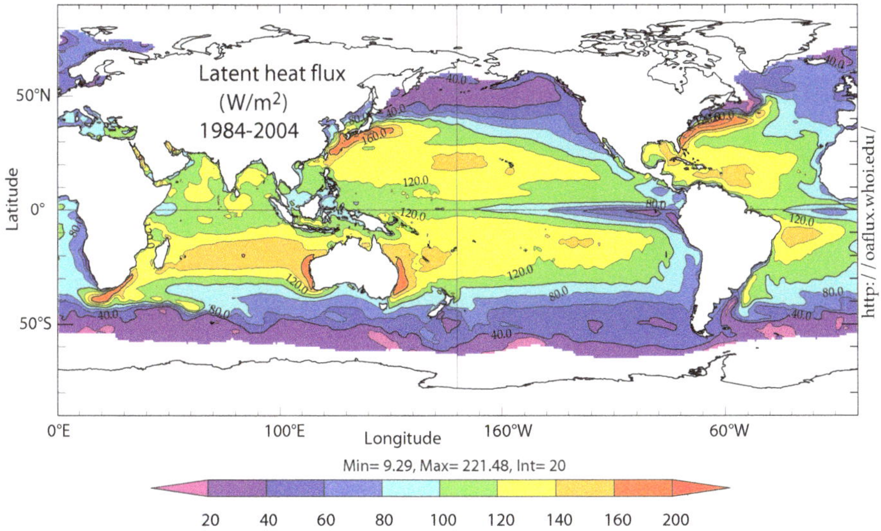

FIG. 1-11 – Latent heat loss by evaporation in W m^{-2}. (Courtesy of C. de Boyer Montégut, average data 1984–2004 OAflux, Yu *et al.*, 2008.)

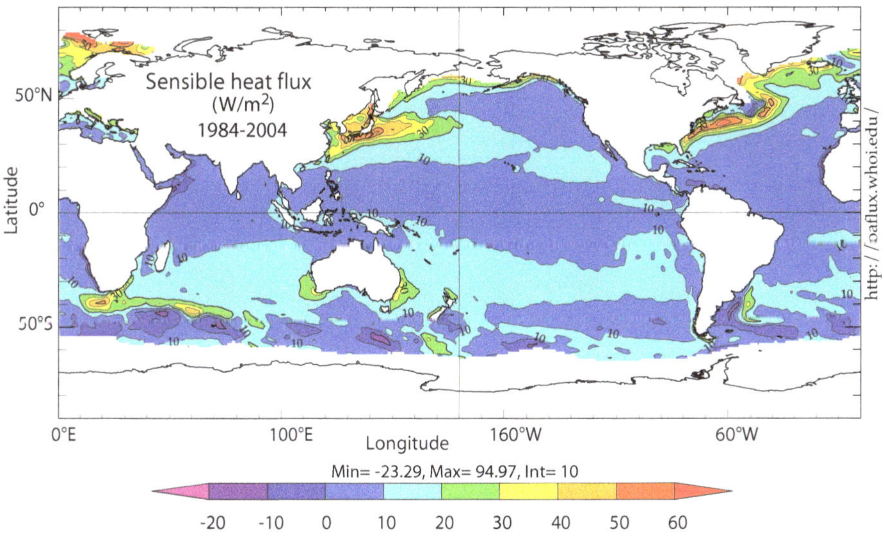

FIG. 1-12 – Heat loss by conduction (sensible heat) in W m^{-2}. Gains are shown in dark blue, violet and rose (they are mostly in the southern ocean). (Courtesy of C. de Boyer Montégut, average data 1984–2004 OAflux, Yu *et al.*, 2008.)

2.2.5 Net heat flux

Given that the ocean has a stable global-average temperature (even though we begin to notice global warming) the gain of solar energy (Q_s) must be balanced by losses to the atmosphere in the form of infrared radiation (Q_{ir}), evaporation (Q_e) and conduction (Q_c) [**Figure 1-7**].

Globally, we may write:

$$Q_s = Q_{ir} + Q_e + Q_c$$

Locally, the net balance[39] of heat flux with the atmosphere may be written:

$$Q = Q_s - Q_{ir} - Q_e - Q_c$$

This flux balance shows strong regional and seasonal variations [**Figure 1-13**].

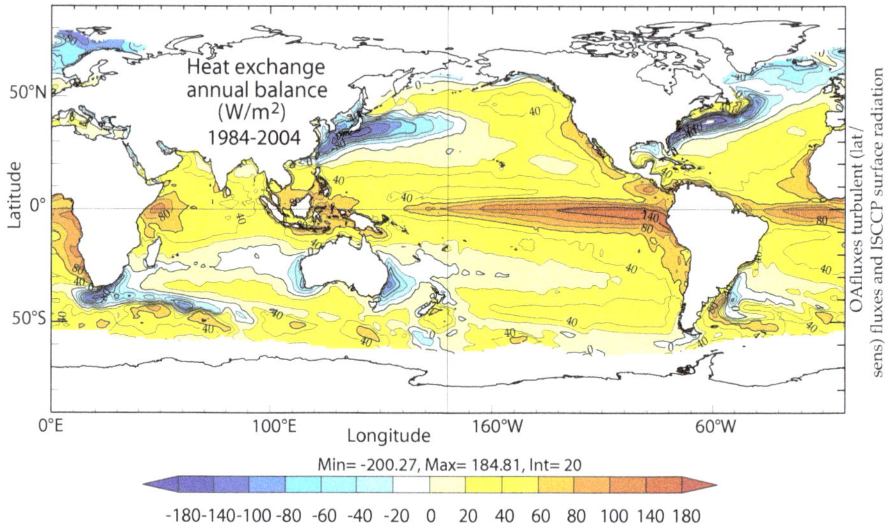

FIG. 1-13 – Annual heat flux balance in W m^{-2} between ocean and atmosphere. (Courtesy of C. de Boyer Montégut, average data 1984–2004 OAflux and ISCCP/NASA, Yu *et al.*, 2008.)

The distribution in **Figure 1-13** gives a good qualitative indication of regional surface energy balance but, because of the difficulty of evaluating all the balance terms, there is great uncertainty in these estimates. Ocean heat losses occur principally in currents transporting warm tropical water poleward, particularly in winter in each hemisphere, when the ocean transmits to the

39 To calculate the balance in a given volume, fluxes by ocean currents must be included (called advection).

atmosphere the heat that it had stored up during the summer [**Figure 1-14**]. The strongest heat gain for the ocean is found in zones of cold water upwelling, such as the eastern equatorial zones of the Pacific and Atlantic, and coastal zones off California, Peru, Mauritania, Namibia, and Somalia.

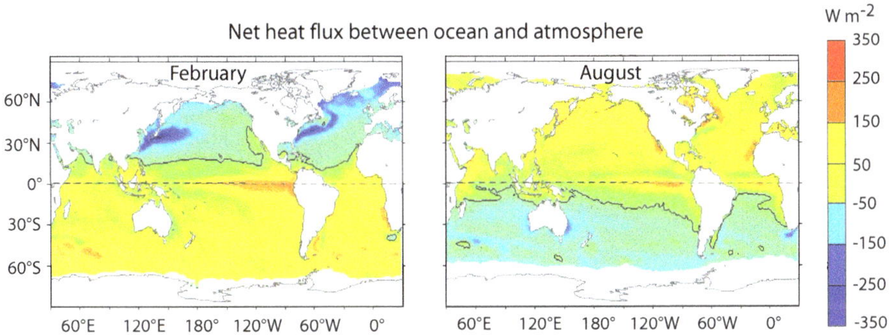

FIG. 1-14 – Seasonal average heat fluxes in W m^{-2} between ocean and atmosphere. Gain for the ocean is positive. (1984–2004 average data, from OAflux and ISCCP, Yu *et al.*, 2008.)

2.3 Atmosphere and ocean heat transport

The disequilibrium in the energy balance [**Figure 1-8**] composed of an energy excess between 35°S and 35°N and a deficit at higher latitudes is compensated by an average meridional circulation in the atmosphere and ocean (fluid envelopes that can move and thus can transport) allowing the transfer of excess heat poleward [**Figure 1-15**].

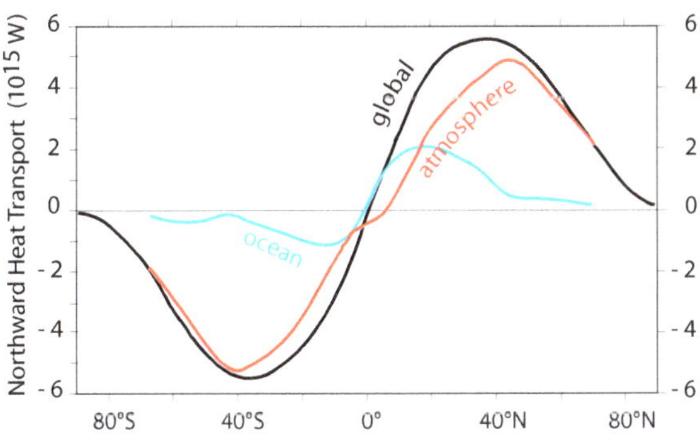

FIG. 1-15 – Total annual mean heat transport in 10^{15} W from the radiation at the top of the atmosphere from ERBE (black) compared with atmospheric heat transport from NCEP reanalyses (red) and derived estimate of the adjusted oceanic heat transport (blue) as a function of latitude (NCEP model, ERBE data from February, 1985 to April, 1989, and ocean data, after Trenberth and Caron, 2001.)

In the ocean-atmosphere engine, the ocean stores heat from the sun, then redistributes it to the atmosphere, which returns a part in the form of kinetic energy (wind) that generates ocean currents.

The ocean, in spite of having slower dynamics than those of the atmosphere, has a much greater thermal capacity, and is thus capable of transporting a large quantity of heat from the equator poleward [**Figure 1-15**]. In the equatorial zone, between around 5°S and 17°N, the ocean dominates the meridional heat transport. At higher latitudes, the atmosphere is the principal agent, with a maximum transport around 45°N and 45°S.

The ocean basins do not all function in the same way. The Pacific transports heat from the equator poleward. The Indian Ocean, closed to the north around 25°N, transports heat accumulated in the north southward. The Atlantic transports heat northward at all latitudes [**Figure 1-16**]. Losses to the atmosphere do not occur in the same way since they depend on the ocean circulation in each basin. However, the ocean returns the heat it had stored in the equatorial regions back to the atmosphere at high latitudes. Thus, thanks to the strong ocean thermal capacity and its circulation, associated with that of the atmosphere, the differences of temperature between equator and poles are limited.

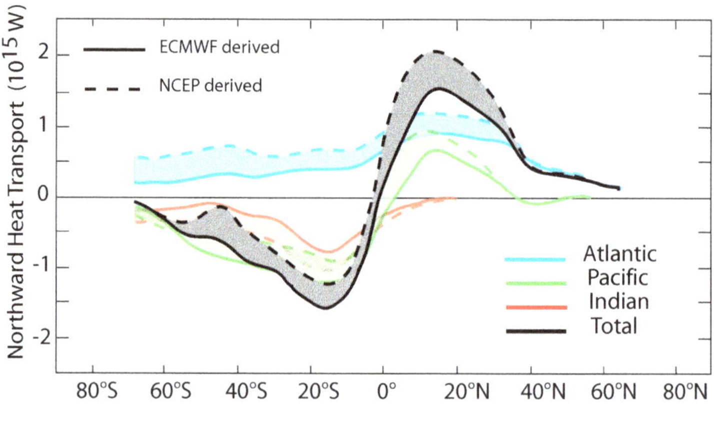

FIG. 1-16 – Annual mean ocean heat transports based on the surfaces fluxes from February, 1985 to April, 1989 for the Atlantic, Indian, Pacific basins and the total, from NCEP and ECMWF atmospheric fields in 10^{15} W. (ECMWF model and NCEP model, ERBE data from February, 1985 to April, 1989, and ocean data, after Trenberth and Caron, 2001.)

2.4 Ocean surface temperature

Due to the increasing inclination of solar radiation with latitude, we might expect a zonal surface temperature distribution. However, neither the absorption of solar radiation [**Figure 1-9**] nor the heat balance [**Figure 1-13**] follows such a distribution because of the influence of atmospheric circulation (cloud distribution) and ocean circulation (modified by the existence of continents).

In the equatorial zones of the Atlantic and Pacific where north and south equatorial currents transport warm water westward then are channeled by the coasts toward higher latitudes, surface isotherms [**Figure 1-17**] open fanlike toward the west. At mid-latitudes, the inverse occurs, and surface isotherms are squeezed in the west and open fanlike toward the east under the effect of circulation. For example, the North Atlantic Drift (the extension of the Gulf Stream) and the North Pacific Drift (the extension of the Kuroshio) transport warm water of tropical origin to the east and northeast.

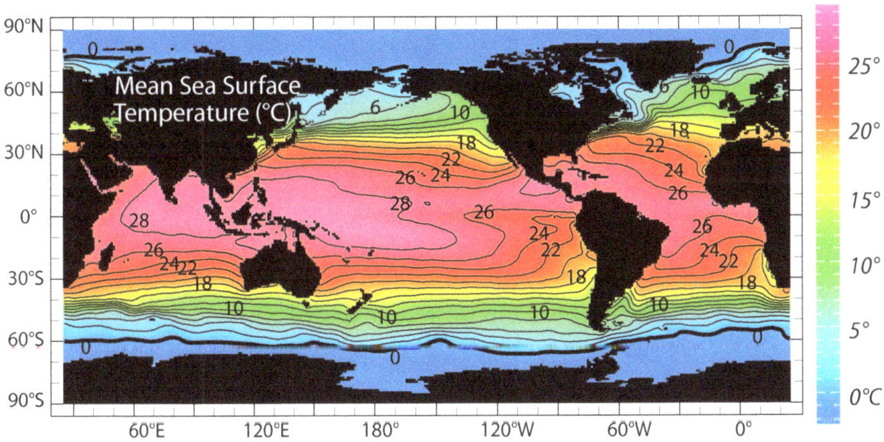

FIG. 1-17 – Average ocean surface temperature. (From IRI/LDEO, WOA05/WOCE NODC data.)

Consequently, temperatures on European coasts are milder than those on eastern American coasts at the same latitude, bathed by the cold Labrador Current coming from the north. Further north, Norwegian coasts are ice-free all year while, at the same latitude, along the eastern coast of Greenland, the current carries pieces of Arctic pack ice. On the contrary, closer to the equator, along the coasts of Mauritania, California, Peru, and Namibia, waters are cold due to wind-driven upwelling of water, as is also the case on the equator in the eastern Atlantic and Pacific. For the same reasons, in the Pacific we note similar disparities between Asiatic and western American coasts.

Regions of strong gradients [**Figure 1-18**] are found around latitudes 45°N and 45°S where warm and cold currents meet. Examples are the Gulf Stream and the Labrador Current in the NW Atlantic, the Kuroshio and Oyashio in the NW Pacific, the Agulhas Current and the Circumpolar Current in the SW Indian Ocean, or the Brazil Current and the Malvinas Current in the SW Atlantic.

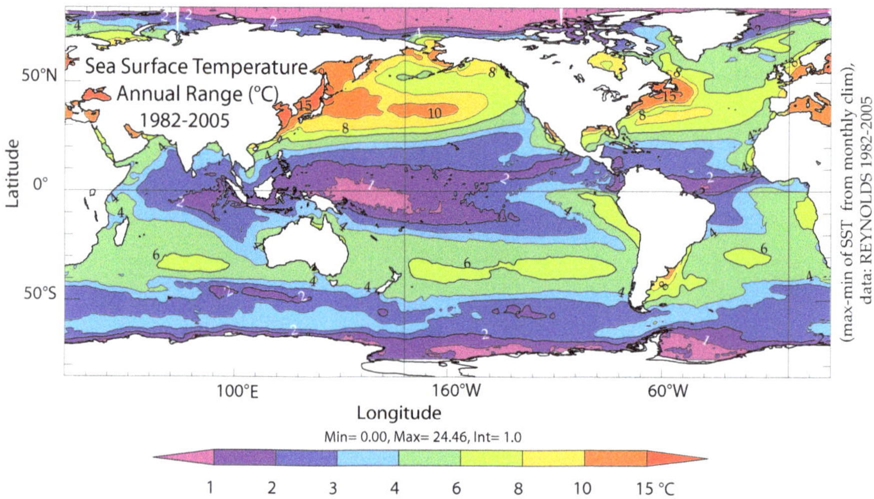

FIG. 1-18 – Annual amplitude of surface temperature. (Courtesy of C. de Boyer Montégut, monthly mean data, Reynolds *et al.*, 2002.)

The region of maximum surface temperature is between the eastern Indian Ocean and the western Pacific Ocean.

The seasonal amplitude of surface temperature [**Figure 1-18**] depends on latitude, on continental influences, and on seasonally varying currents. Maxima are found in the northern hemisphere where air masses with strong continental properties (having strong seasonal variability) flow out over the ocean. The Sea of Japan, the Yellow Sea, the Kuroshio region in the Pacific, and the Gulf Stream region in the Atlantic undergo the strongest seasonal variations. Minima are found in polar and equatorial regions.

2.5 Water fluxes

2.5.1 Evaporation and precipitation

The ocean's average salinity is stable since gains and losses balance in a perpetual cycle amongst the ocean, the continents, and the atmosphere. The ocean loses fresh water by evaporation, and gains by precipitation and inflow (rivers, glaciers). The distribution of zonally averaged precipitation (P) and evaporation (E) as a function of latitude **[Figure 1-19]** shows an evaporation maximum in subtropical zones and three precipitation maxima: in the Intertropical Convergence Zone, north of the equator, and in the zones of the westerlies, centered around 50°N and 50°S, which are zones of convergence between tropical air masses and polar air masses.

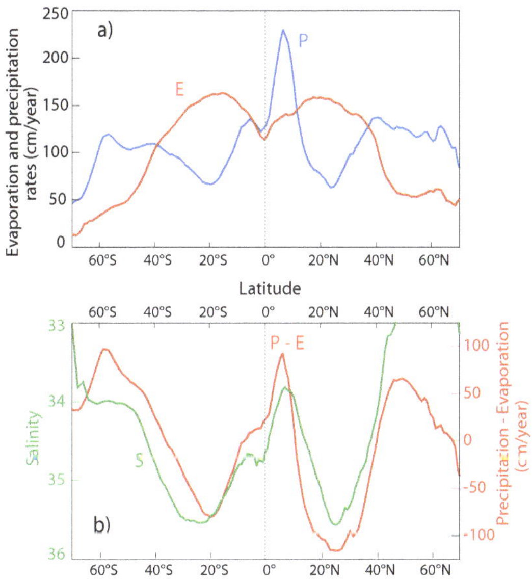

FIG. 1-19 – (**a**) Zonally averaged 1981–2010 precipitation rate (blue) from GPCP (http://www.esrl.noaa.gov/psd/data/gridded/data.gpcp.html), and evaporation rate (red) from OAFLUX (http://oaflux.whoi.edu/evap.html) in cm/year. (**b**) Zonal balance of precipitation-evaporation (P-E, red) in cm/year and variation of the zonally averaged 2010–2015 surface salinity (green) from SMOS (http://www.catds.fr/Presentation). Positive values of P-E represent an oceanic gain of fresh water. (Courtesy of S. Masson, LOCEAN-UPMC-Paris VI University.)

The net zonal balance *P-E* shows a loss of water for the ocean in the subtropical anticyclonic zones, regions of atmospheric subsidence, around 15°–40° N and S and a gain of fresh water in the equatorial zone and around 50°–60° N and S [**Figure 1-19b**]. The pattern correlates with surface salinity variations.

The precipitation pattern [**Figure 1-20a**][40] shows the strongest values in the intertropical convergence zone, with maxima in the western Pacific and the eastern Indian Ocean, where values exceed 3 m of water per year up to 5.5 m per year. Ocean surface temperatures are highest in this region [**Figure 1-17**] and consequently atmospheric convective activity is intense.[41] The latitude and extent of this region vary seasonally [**Figure 1-20b**]. For example, in January the strong precipitation zone in the Pacific extends southeast under the South Pacific Convergence Zone, and in the Indian Ocean it reaches Madagascar.

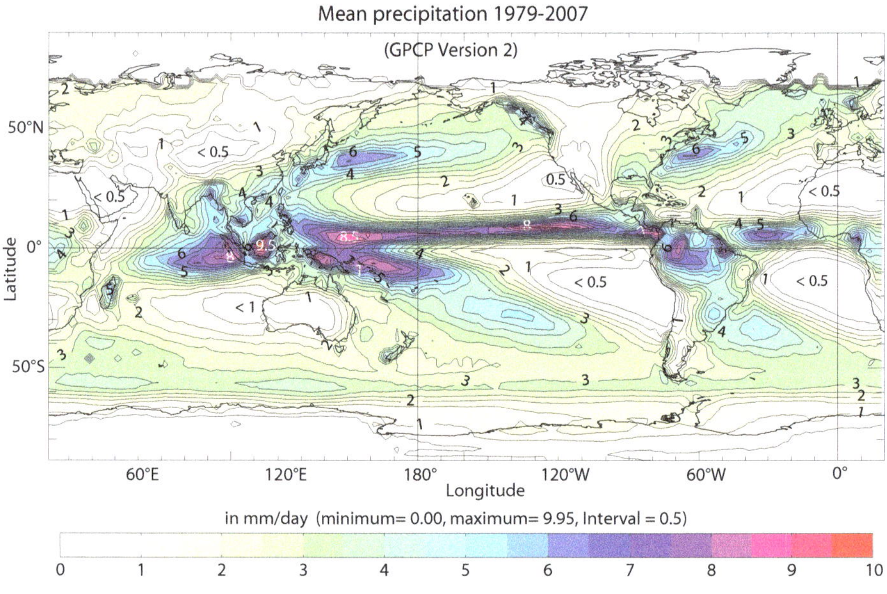

FIG. 1-20A – Precipitation in mm/day (1 mm/day = 36.5 cm/year). (Courtesy of S. Masson, 1979–2007 GPCP/WCRP average data.)

40 Monthly mean precipitation values are obtained with infrared data (the American satellite GOES, the Japanese GMS, and the European Metéosat geostationary satellites, and NOAA polar-orbiting satellites) with microwave data (American SSM/I and DMSP satellites), and with data from more than 6000 meteorological measuring stations.

41 This region is the ocean's most important "furnace".

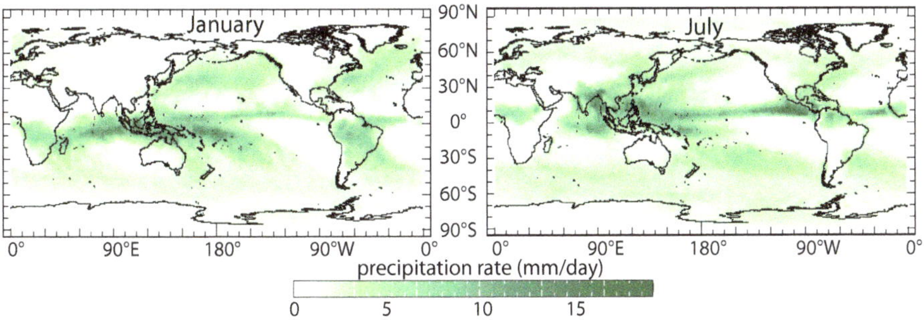

FIG. 1-20B – Seasonal precipitation in mm/day (a) January, (b) July. (From IRI/LDEO 1979–1995 NCEP/CAMS-OPI average data.)

The evaporation pattern shows maxima in the anticyclonic regions, especially in the southern Indian Ocean, and in currents transporting warm water poleward. It shows minima in polar regions and in upwelling zones in the eastern equatorial parts of the Atlantic and Pacific, and along the coasts of Peru, California, Mauritania, and Namibia [**Figure 1-21**].

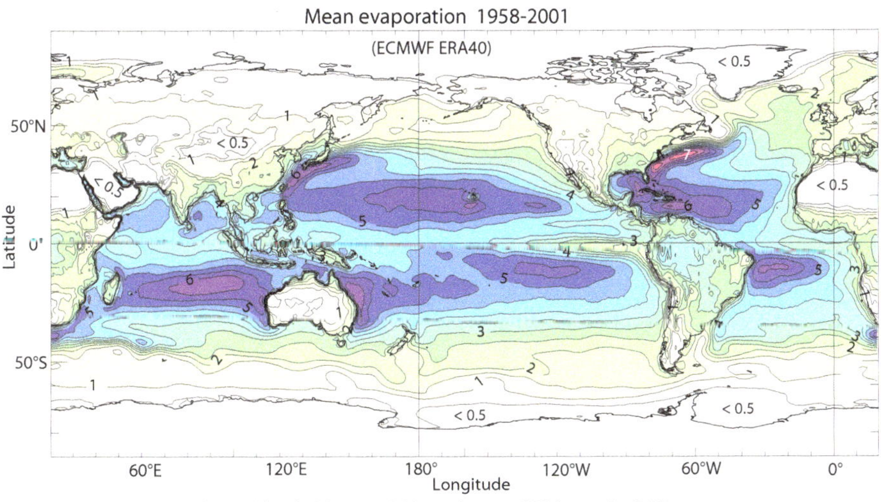

FIG. 1-21 – Average evaporation (mm/day). (Courtesy of S. Masson, average data, 1958–2001 ECMWF, ERA40.)

2.5.2 Water flux balance

The regional distribution of fresh water balance (evaporation-precipitation) [**Figure 1-22**] shows a water deficit in subtropical anticyclonic zones where evaporation is intense [**Figure 1-21**] and a water gain in zones around the equator. The water excess is especially extensive and maximal in the western Pacific and eastern Indian Ocean [**Figure 1-20**]. In the Indian Ocean, a strong asymmetry exists between the NW (Arabian Sea, Red Sea, Persian Gulf) and the NE (Bay of Bengal) due to the influence of the SW monsoon, and to the topography. The Norwegian Sea, the Red Sea, and the Mediterranean Sea have a deficit.

Water losses are stronger in the Atlantic and in the Indian Ocean than in the Pacific, where regions of fresh water accumulation are more extended.

Ocean-mass variations due to gains and losses of fresh water are weak. However, salinity variations that result from them have an important impact on ocean dynamics.

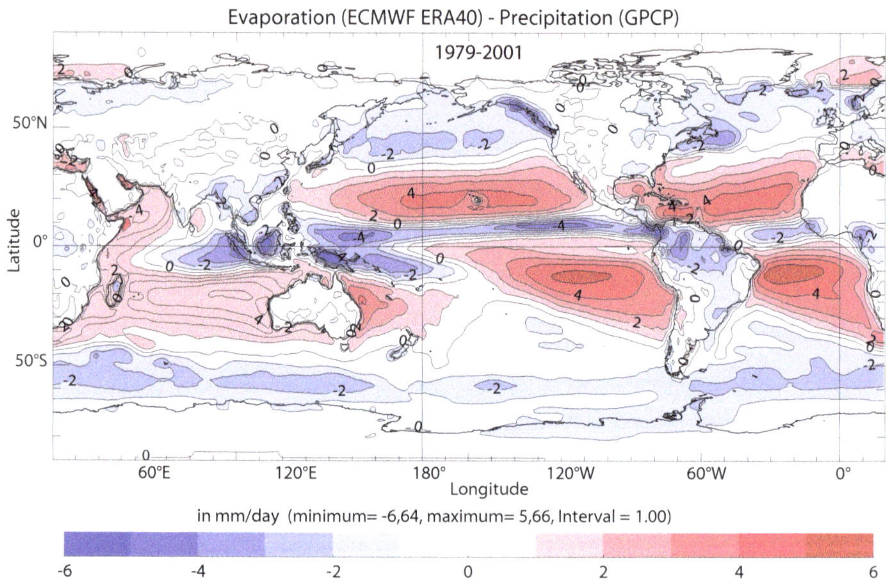

FIG. 1-22 – Evaporation-precipitation balance (mm/day). Ocean water gain is shown in blue and loss is shown in red. (Courtesy of S. Masson, average data 1979–2001 GPCP and ECMWF.)

2.6 Surface salinity

The surface salinity distribution [**Figure 1-23**] is highly correlated with the ocean-atmosphere water-flux balance and represents an aspect of the interaction between the ocean and climate.

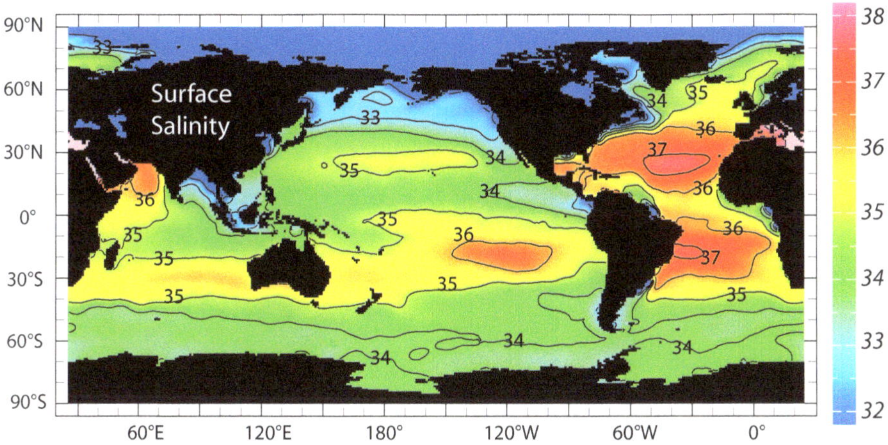

FIG. 1-23 – Average surface salinity. (From IRI/LDEO, WOA05 data.)

Regions of strong precipitation [**Figure 1-20**], such as the western Pacific, the Indonesian region, or the Intertropical Convergence Zone, are characterized by weak surface salinity. Conversely, in regions of strong evaporation such as in the center of subtropical anticyclones or in the Mediterranean Sea, the Red Sea, the Persian Gulf, and the Arabian Sea, salinities are higher. Contributions of fresh water from runoff noticeably reduce surface salinity offshore from rivers with large drainage basins such as the Amazon, the Mississippi, the Congo, the Niger, the Ganges, or the Siberian rivers flowing into the Arctic Ocean. As with temperature, ocean circulation influences the distribution of surface salinity. For example, the entrainment to the NE of higher-salinity water of subtropical origin by the North Atlantic Drift can be found as far as the coast of Norway (isohaline 35). Conversely, the Labrador Current transports low salinity cold water of arctic origin to the south along the North American coast.

2.7 Surface density

Surface temperature and salinity can be used to determine surface density[42] [**Figure 1-24**]. The lowest values of density are found in the equatorial band, where salinities are low and temperatures are high. High-density values are found in the Antarctic region and in the Nordic Seas where temperatures are low[43] as well as in those basins with very high salinity (Mediterranean, Red Sea, or Persian Gulf).

[42] In fact, it is the density anomaly that is used. See **Section I.1.3**.

[43] In spite of low temperature, density is low along the Siberian coast where river contributions are important and along the coast of Greenland where currents transport low-salinity water.

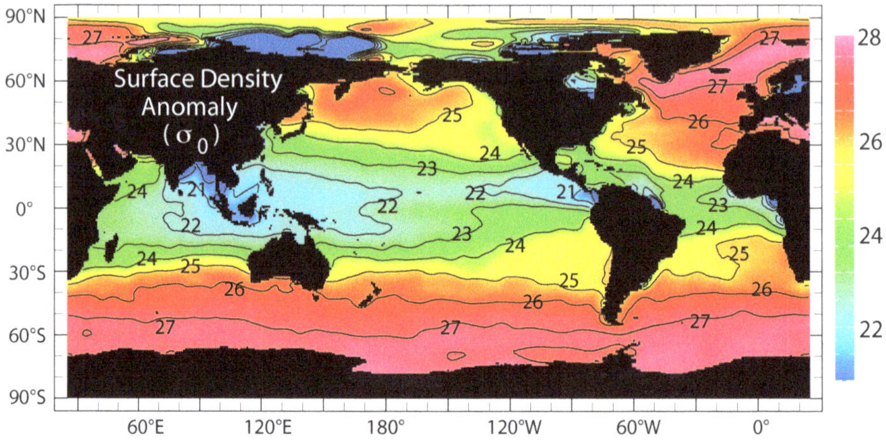

FIG. 1-24 – Average surface-density anomaly (σ_0) (kg m^{-3}). (From IRI/LDEO, Levitus data 1994.)

2.8 Heat and salt transfers to the ocean interior. The thermocline

The ocean absorbs solar radiation in the upper few meters, and only a thin layer of surface water is directly warmed. Heat fluxes, which modify temperature, and water fluxes, which modify salinity are strictly superficial.

How are these changes transmitted deeper?

Ocean temperatures reach their strongest values under dead-calm sea conditions (without wind, which is rare) and in regions of high solar radiation. In the Indian Ocean, between monsoons, temperatures can reach 31 °C to 32 °C at times, and temperatures in the Persian Gulf can even exceed these.

When the wind blows, agitation by wind-generated waves increases turbulence and creates a mixed layer, which homogenizes characteristics. The thickness of the surface mixed layer depends on wind conditions and stratification.

Under the effects of winter cooling, surface density increases and stability decreases. When the density of surface water reaches a level greater than the water just below, it mixes with the water below, bringing about convective movement. The result is a homogenization of water properties, deepening during the winter until it reaches the depth of the thermal gradient maximum, called the main thermocline. Mixing reaches its maximum depth at the end of winter. In those regions where stratification is weak, such as the polar or subpolar regions, winter surface density variations (temperature and salinity variations) bring about convective mixtures that can attain great depths. Thus characteristics acquired at the surface (temperature, salinity, oxygen, and other tracers...) are entrained at depth.

Seasonal warming progressively creates a warm layer separated from the layer not touched by the warming by a zone of strong vertical temperature gradient, called the seasonal thermocline [**Figure 1-25**]. The more the ocean warms, the more the vertical gradient of temperature (and of density) increases, the more the ocean becomes stable, and it becomes difficult for heat to penetrate to depth. The temperature maximum is reached at the end of the summer (August–September for the Northern Hemisphere). Warming stabilizes the ocean, as opposed to the atmosphere, which is warmed from below.

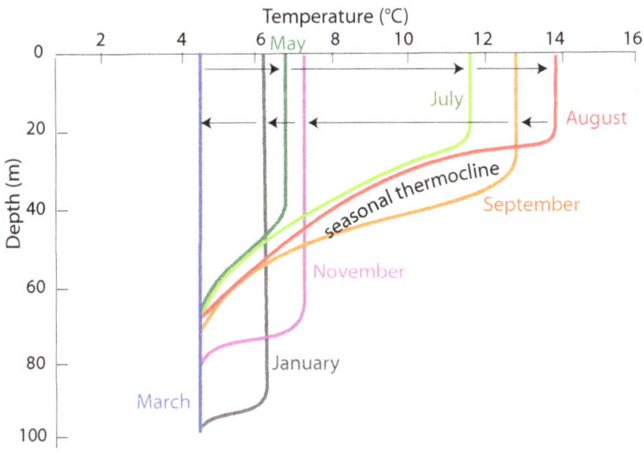

FIG. 1-25 – An example of mid-latitude temperature-profile evolution during the year, leading to the formation of a seasonal thermocline.

In general, temperature decreases from the surface to depth, but in a manner depending on the region [**Figure 1-26a**].

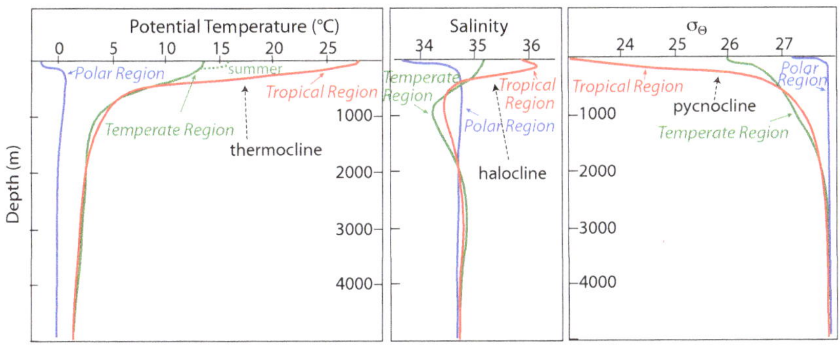

FIG. 1-26A – Regional average profiles of potential temperature (°C), salinity, and potential density (kg m^{-3}). (After Tchernia, 1978.)

In polar and subpolar regions the thermocline is absent since temperature differences between the surface and depth are weak. We often find a surface layer colder than the deep layer. This colder layer is stable, however, since its density is low because of reduced surface salinity (except in winter when salinity can increase through the formation of sea ice).

In equatorial regions, the surface temperature is elevated (25 °C to 30 °C) then decreases sharply with depth. Seasonal temperature variations are slight. There is a rather thin warm surface layer, whose characteristics depend essentially on wind effects and on the currents thus generated. It is separated from deep water by a sharp thermocline whose water comes from subtropical latitudes. Below this, the variation with depth is much weaker.

In mid-latitudes, the main thermocline is located at the limit of the winter mixed layer, around 400 m to 800 m depending on the ocean basin. At these latitudes, seasonal temperature variations are important and entrain the formation, in summer, of a seasonal thermocline shallower than the main thermocline.

The mixed layer depth varies with the season and with the region. The thickest mixed layer, which can reach more than 600 m, is found in the high latitudes and during the winter period, where the sea-surface temperatures are the lowest and the sea-surface densities the highest, resulting in a weak vertical gradient (**Figure 1-26b**). In the equatorial regions where the thermocline is strongest, the mixed layer is never very thick and its variations are mostly due to the wind speed.

Though it does not compare with the main and seasonal thermoclines, a diurnal thermocline sometimes occurs. It is shallow, resulting from heat gained during the day. It disappears overnight. Solar radiation has a diurnal cycle, while infrared radiation losses vary little.

To illustrate the general characteristics of the ocean, we here show Atlantic Ocean average meridional sections (averaged across all longitudes) from Antarctica to the Arctic Ocean [**Figures 1-27, 1-28,** and **1-29**].

The temperature section [**Figure 1-27**] shows the main thermocline between the subtropical fronts around 40°–45° north and south. The thermocline nears the surface in the equatorial zone under the effect of equatorial upwelling, giving it a meridional 'W' shape.

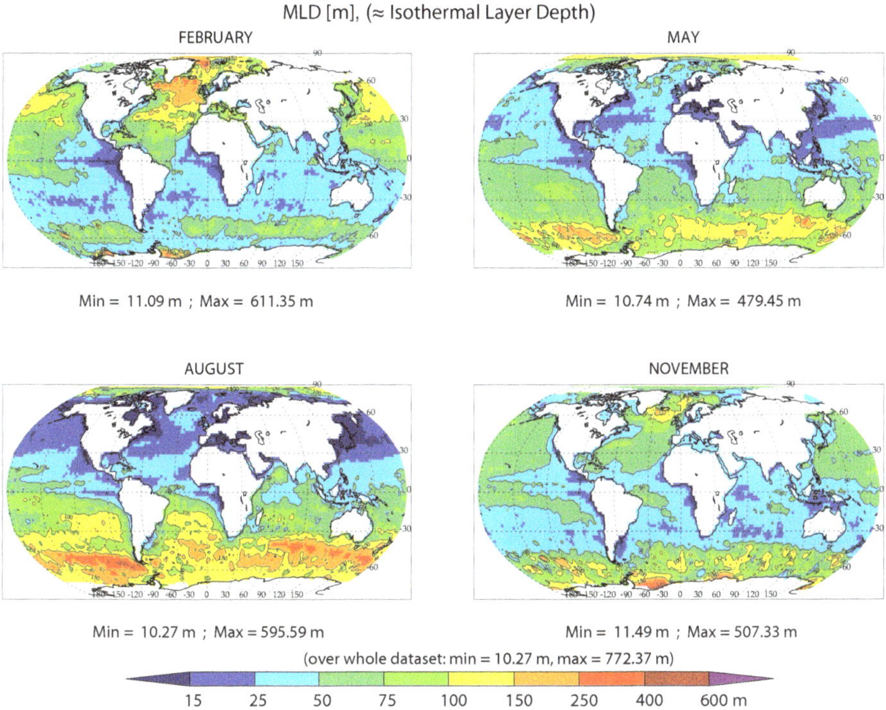

FIG. 1-26B – Mixed Layer Depth in meters (Isothermal Layer Depth) at different seasons calculated from measured temperature profiles with Delta-T criterion = 0.2 °C between the surface (10 m depth) and the bottom of the mixed layer. (Courtesy of C. de Boyer Montegut *et al.*, 2004, IFREMER/LOPS Mixed Layer Depth Climatology, web site: http://www.ifremer.fr/cerweb/deboyer/mld.)

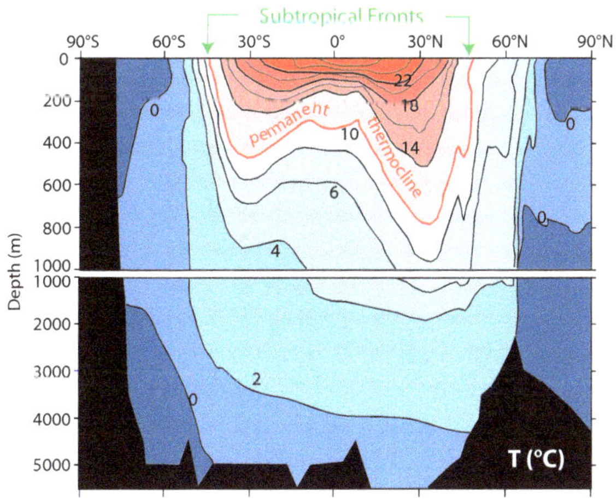

FIG. 1-27 – Average temperature along an Atlantic meridional section. The scale changes at 1,000 m depth. (After Levitus, 1982.)

The salinity section [**Figure 1-28**] shows a structure that is different from temperature in zones around 50°–60° N and S where low salinities extend down to 1,000 m in the south and only 400 m in the north. In the surface layer, low salinities are found in the equatorial zone and in the polar regions where *E-P* is negative. Zones with high salinity are found around 30°S and 30°N in regions of strong evaporation, where *E-P* is positive [**Figure 1-22**]. To the north, the extension down to 1,000 m of a high-salinity layer is due to the influence of Mediterranean Water [**Section III.5.1**]. A large variation of salinity with depth is called a halocline (the vertical salinity gradient is maximum).

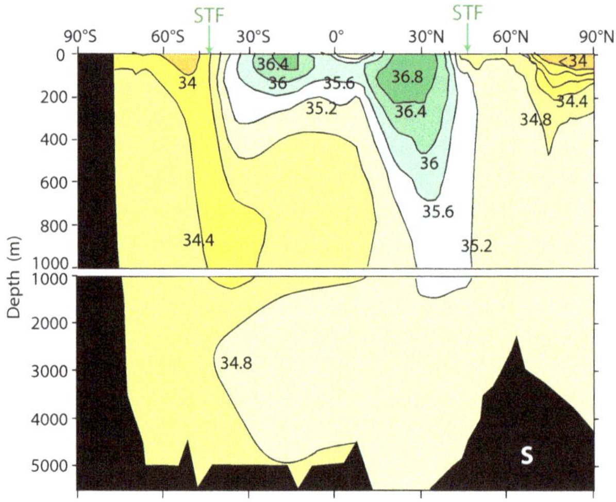

FIG. 1-28 – Average salinity along an Atlantic meridional section. The scale changes at 1,000 m depth. (After Levitus, 1982.)

The density structure [**Figure 1-29**] has the form of bowls piled up in a W-shape, similar to that of temperature. Vertical gradients are weak in polar regions, especially in the Antarctic. These regions favor dense water formation. Around 50°–60°, vertical gradients are still weak, while in subtropical regions the isopycnal 27 drops from the surface down to 600 m, then climbs to around 300 m in the equatorial zone under the effect of upwelling. From about 40°–45° N and S, the vertical density gradient becomes pronounced above 500 m. This is the pycnocline zone [**Figure 1-29**], tied to the thermocline [**Figure 1-27**], which isolates surface layers (the warm sphere) from deeper layers (the cold sphere). The strong vertical density gradient makes vertical transfers difficult.[44] This limit is often chosen in simple circulation models consisting of two layers of different densities.

[44] Called diapycnal transport (across isopycnals).

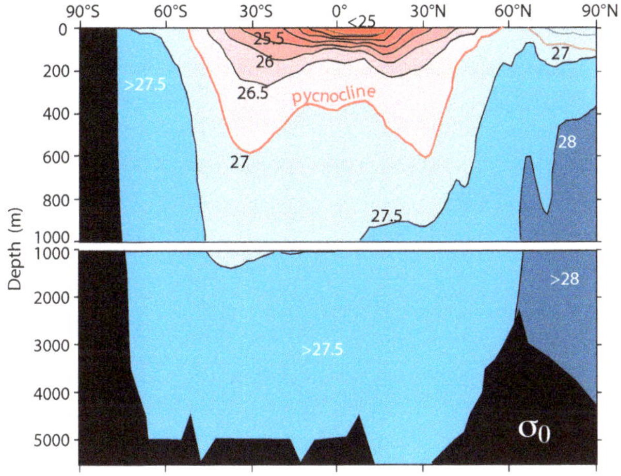

FIG. 1-29 – Average density (σ_0 in kg m^{-3}) along an Atlantic meridional section. The scale changes at 1,000 m depth. (After Levitus, 1982.)

The generally stable structures of temperature, salinity, and density are the result of equilibrium among horizontal and vertical movements (horizontal and vertical advection, Ekman pumping,[45] diffusion...) created by wind action, thermohaline processes,[46] and internal waves (including internal tidal waves).

[45] See Ekman pumping in **Section I.5.3.1.**

[46] Heat and salt fluxes modify the density and thus the stability (see **Section I.4.6**).

Chapter 3
Water masses

3.1 General properties of water masses

The preceding oceanic sections **[Figures 1-27, 1-28,** and **1-29]** show various characteristics which vary depending on level and region inside of which we can define water masses.

Ocean water temperatures range between –1.9 °C and 32 °C.[47] Ocean salinities range from 0 (at river mouths) to 42 (in the northern Red Sea). However, 90% of ocean water is between salinity 34 and 35 and 50% is between 34.6 and 34.7. **Figure 1-30** shows the three-dimensional volume distribution of the temperature-salinity characteristics of deep water masses. This figure covers all world-ocean temperatures less than 4 °C, which is below about 1,500 m depth except in high latitudes.

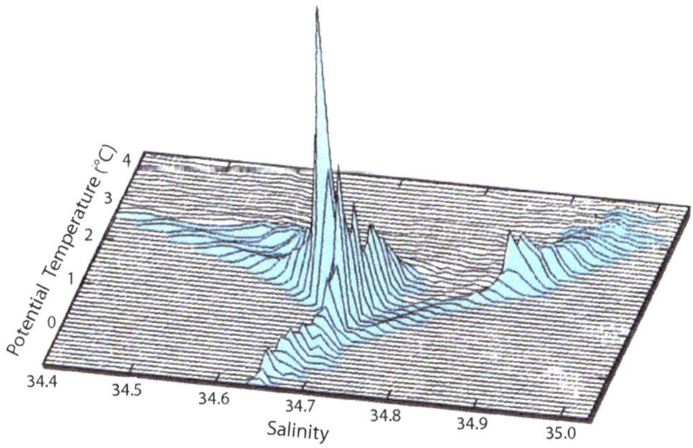

FIG. 1-30 – Three-dimensional histogram (θ, S, and volume) of water with potential temperature less than 4 °C. The highest peak corresponds to a volume of 26 × 10^6 km^3. (After Worthington, 1981.)

[47] Temperatures can be even higher in some coastal regions.

A *water mass* is characterized by a range of temperature and a range of salinity corresponding to a volume of water whose elements have the same history. That is, they were formed in the same manner in the same place.

Throughout this text we capitalize the names of water masses (for example Antarctic Deep Water or North Atlantic Deep Water).

Precipitation diminishes salinity and heating raises temperature, both of which tend to reduce the density of the surface layer, thus stabilizing the water column. On the other hand, evaporation (the loss of water associated with the loss of latent heat) brings about an increase in salinity and a drop in temperature. Heat losses by direct conduction with the colder atmosphere and losses by infrared radiation bring about a decrease of temperature. Freezing of seawater, which separates salts from fresh water, increases salinity of the water just below the ice that is formed. All these processes of heat and fresh water extraction contribute to raising the density of surface water thus diminishing vertical stability, sometimes to the point where stability is destroyed, creating convection (vertical mixing due to density differences). When cooling and extraction of fresh water intensifies during the winter, convective movements (descending and rising) reach layers progressively deeper leading to vertical homogeneity in temperature, salinity, and all other characteristics (oxygenating deep layers, for example). This is the winter mixed layer. The weaker the initial stratification, and the harsher the conditions (cooling and evaporation intensified by violent winds) the more the mixing reaches greater depths. Water masses characterized by temperature and salinity reach their depth of equilibrium according to their density and preferentially extend along corresponding isopycnal surfaces (the path least "costly" in energy).

After leaving the surface, water-mass characteristics can only be modified by mixing with other water masses encountered in their path.

3.2 Mode Waters, Central Waters, and Intermediate Waters

Central waters are thermocline waters [**Figures 1-26** and **1-27**] whose characteristics cover a large range of temperature and salinity. They form in subtropical convergence zones due to wind [**Section I.5.3.1**][48] where waters tend to subduct (sink below). Under the effect of winter cooling (whose intensity varies with latitude) mixed layers are formed through convection, creating homogeneous oceanographic characteristics. They are called Subtropical Mode Waters when they are formed in the subtropical zone. Each type of Mode Water corresponds to a particular density that depends

[48] Convergence zones are regions where wind tends to "drive" surface water masses below lighter water.

on the latitude of its formation. Mode Waters are in turn subducted toward the equator under lighter water layers along isopycnal surfaces corresponding to their density [**Figure 1-31**]. Once subducted, they are characterized by a minimum of potential vorticity [**Section I.5.3.4**]. That is to say they have a minimum vertical density gradient and are often called pycnostads.

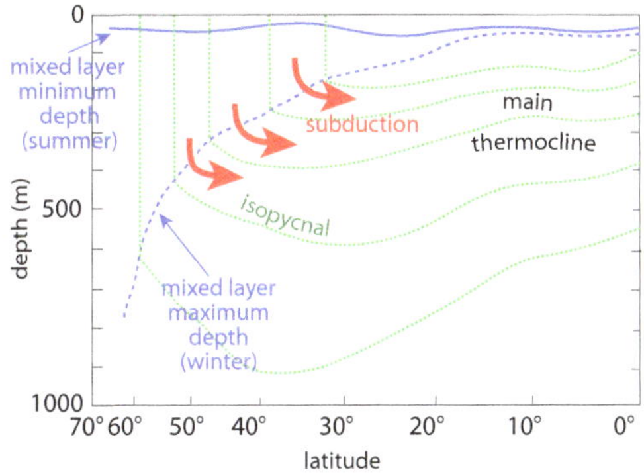

FIG. 1-31 – Schematic of winter convection leading to subduction of Mode Water.

The process is called "ventilation" because it brings oxygen to the thermocline layers. Mixed layers, with characteristics (T, S, σ) depending on their latitude (A, B, C, D) [**Figure 1-32**] follow their isopycnal, and closer to the equator at station E (points A′, B′, C′, D′) they are distributed in the vertical according to their density.

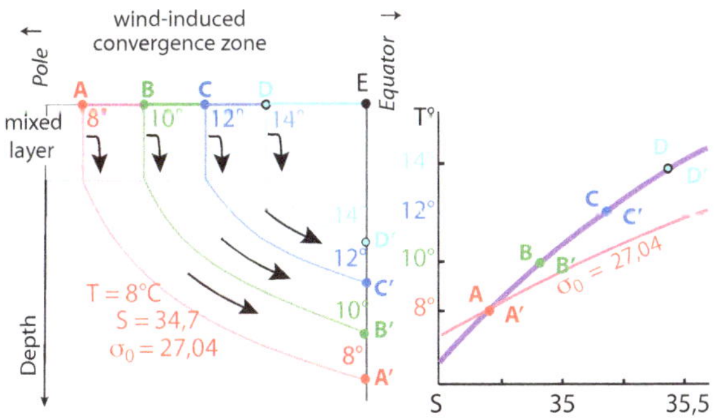

FIG. 1-32 – Schematic of convection and subduction mechanisms as Subtropical Mode Water forms Central Water in the region of subtropical convergence. The surface T-S diagram A, B, C, D is identical to the vertical diagram A′, B′, C′, D′ at station E. (After Tomczak and Godfrey, 1994.)

The temperature and salinity profiles in **Figure 1-33** correspond to Mode Water in formation (Stations 1 and 3) then subducted (Stations 2 and 4).

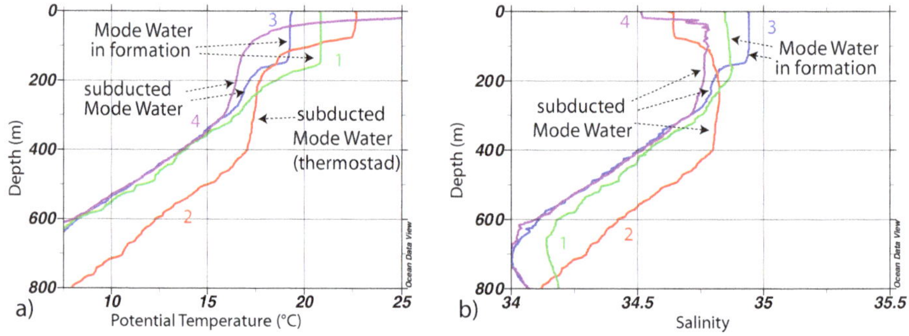

FIG. 1-33 – Profiles of temperature and salinity in a formation region of Subtropical Mode Water (Stations 1 and 3). Stations 2, 3, and 4 profiles show homogeneous Mode Water subducted under lighter water (also depicted as thermostad layer). (With Ocean Data View, ODV Schlitzer, 2008, and WOA05 data.)

Subtropical Mode Waters constitute an important volume of Central Water. They are renewed each winter and those richest in oxygen are the coldest. Spring and summer warming brings about the formation of a seasonal thermocline, isolating the deep part of the winter mixed layers. The characteristics of winter surface water determine the characteristics of subducted water in the thermocline.

Away from their place of formation, Subtropical Mode Waters can be distinguished by a spreading of the isotherms in a vertical temperature section. Maintaining a permanent thermocline between the two zones of subtropical convergence [**Section I.2.8, Figures 1-27, 1-28,** and **1-29**] is the result of equilibrium between the processes of subduction and erosion by mixing and vertical diffusion. **Figure 1-34** shows the distribution of various Mode Waters.

There are also Subantarctic and Subarctic (or Subpolar) Mode Waters, formed in convergence zones (antarctic or arctic), which are sometimes homogeneous over several hundreds of meters. Antarctic Intermediate Water, characterized by a salinity minimum, is the densest form of Subantarctic Mode Water.

The formation of Intermediate Waters in polar convergence zones is composed of a winter-convection phase followed by subduction under the Subantarctic Mode Water (which forms just to the north of the Subantarctic Front and has a slightly lower density than Antarctic Intermediate Water) and the less dense Central Water [**Figure 1-31**]. Convection can reach 600–700 m, particularly in the southeast Pacific, the region of principal formation.

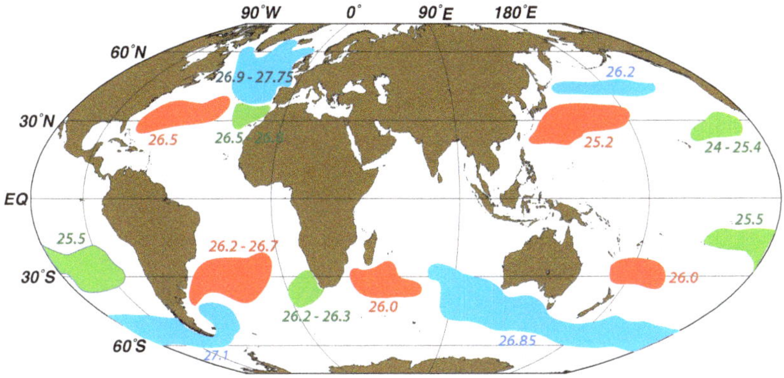

FIG. 1-34 – Mode Waters with their density range: Subtropical Mode Waters associated with western boundary currents (in red). Subtropical Mode Waters associated with eastern boundary currents (in green). Subpolar North Atlantic Water, Subpolar North Pacific Water and Subantarctic Mode Waters (in blue). (After Hanawa and Talley, 2001.)

3.3 Bottom Waters and Deep Waters

Bottom Waters (the densest) form intermittently during the winter season in relatively limited regions (relative to their extent in the global ocean).

The existence of Bottom Water is associated either with the freezing of seawater (formation of pack ice) or with strong evaporation. The effect is to increase salinity and seasonally decrease the temperature. There are two processes of dense water formation: one on the continental shelf, the other in the center of cyclonic gyres in deep basins.

In some shallower continental shelf areas, water and heat losses at the surface take place over a thin layer, producing density modifications that are stronger than they would be in a deeper layer. As a result of seawater freezing in winter, this process creates a dense water mass on the continental shelves around Antarctica. This dense water flows down the continental slope and mixes with Circumpolar Deep Water to form Antarctic Bottom Water. The same process creates dense water around the Arctic Ocean. At higher temperatures, evaporation is the analog of freezing. For example, dense water is formed in the Red Sea (on the continental shelf of the Gulf of Suez) and in the Persian Gulf.

In the Greenland Sea, convection takes place at the center of the cyclonic circuit responsible for the formation of Deep Water that reaches the ocean floor. This is also the case in regions with milder climates such as the Mediterranean basin. Convection is relatively easy to observe in the Mediterranean, since temperatures are less rigorous, there is no ice, and the region is accessible. However, the sea state under the effect of violent winds has for long been an obstacle to observation. High temperatures found at depth (in the Mediterranean and Red Sea) result from regional climate and the existence of a relatively shallow sill that prevents cold ocean water from penetrating into the basin.

Bottom Water formed in seas with sills such as the Norwegian Sea, the Greenland Sea, the Mediterranean, and the Red Sea have, in general, in the formation region, density higher than that of bottom water formed in Antarctica. However, Antarctic Bottom Water predominates on the ocean bottom, since the others lose their high densities through mixing in passing the obligatory sills connecting them to the open ocean.

Under the effect of gravity, the densest bottom waters carpet the ocean bottom, guided by submarine topography. They can generally be found in the open ocean deeper than 4,000 m depth. Deep waters are distinguished from bottom waters by their origin and characteristics. They result most often from a mixture of several water masses formed by convection in different regions, and are found in general between 1,500 m and 3,500 m [**Figure 1-35**].

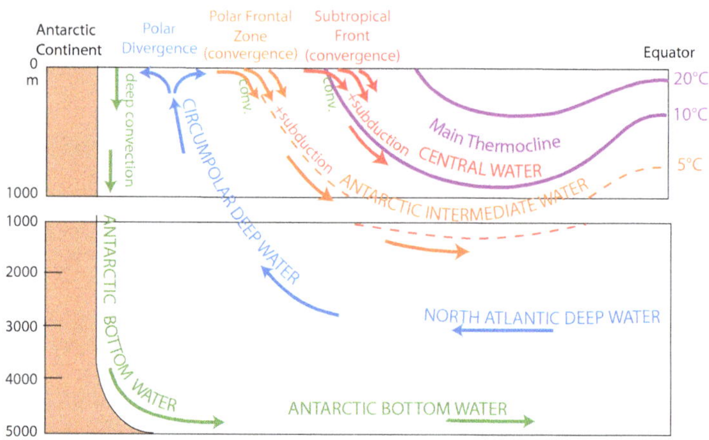

FIG. 1-35 – Schematic of formation processes of Southern Hemisphere water masses.

3.4 Analysis of water characteristics

Water-characteristic analysis is the study of water masses and their evolution in the ocean. The analysis relies on the fact that gains or losses of heat and fresh water, which modify water-mass characteristics, occur at the ocean surface. Once a water mass leaves the surface, it keeps its temperature and salinity, which can only evolve slowly through mixing with other water masses having different characteristics. We call salinity and temperature *conservative properties*. In fact, because of the slight compressibility of seawater, which results in a warming under pressure, *in-situ* temperature is not a conservative property, since it varies under pressure. In order to avoid this "non-conservativeness" a temperature corrected for the pressure effect called *potential temperature* (θ) is used.

3.4.1 Potential temperature and potential density

Potential temperature is the temperature of a seawater sample raised to the surface adiabatically (that is, without heat exchange). Potential temperature is less than in-situ temperature since pressure acts to raise temperature [**Figure 1-36**]. The difference between in-situ temperature and potential temperature is a function of pressure, the absolute temperature of the sample, the acceleration of gravity, seawater density, and the coefficient of thermal expansion. It is inversely proportional to the specific heat of seawater (depending on pressure and temperature). It is not a simple function of the in-situ temperature.[49] For example, the difference between the *in-situ* temperature and the potential temperature is 1.3 °C at 10,000 m in the Mindanao Trench (Philippines). Potential temperature must be used to compare water-mass characteristics at depth.

Knowing potential temperature and salinity at a given pressure, potential density, σ_θ, may be calculated. Potential density brought to surface pressure is called simply σ_0.

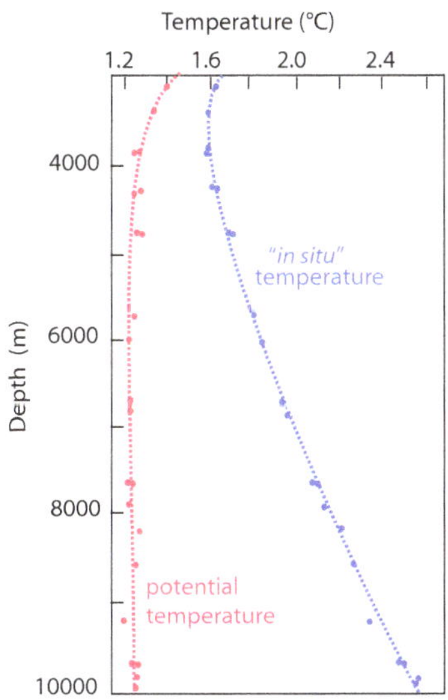

FIG. 1-36 – In-situ temperature and potential temperature between 3,000 m and 10,000 m in the Mindanao Trench. (After Galatea Deep-Sea Expedition, Bruun *et al.*, 1956.)

[49] UNESCO Technical Papers in Marine Science No. 44, 1983.

3.4.2 Potential temperature-salinity diagram

Potential temperature and salinity are conservative properties that represent the water mass "signature", acquired at the surface through exchanges with the atmosphere in the region of formation.

A pair of specific temperature and salinity values characterizes a *water type*. A water type is represented by a point on a θ-S diagram, with temperature as the ordinate and salinity as the abscissa [**Figure 1-37**]. The mixing of two water types of different θ-S characteristics leads to a new water type, whose representative point is located on the straight line connecting the characteristic points of the two original water types. Knowing the θ-S coordinates of the point representing the mixture allows the calculation of the proportion of each water type in the final mixture. Using the equation of state of seawater [**Section I.1.3**] isopycnals may be drawn on θ-S diagrams. These curves of equal potential density provide information on the stratification of water masses [**Figure 1-37a**]. Isopycnals are not straight, especially at low temperature, since the equation of state connecting temperature, salinity, and density is not linear.

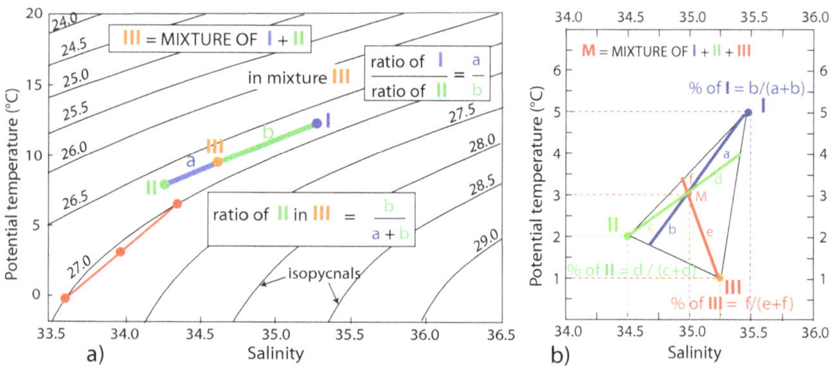

FIG. 1-37 – (**a**) θ-S diagram showing the mixture of two water types. (**b**) θ-S diagram showing the mixture M of three water types I, II, and III. (After *Ocean Circulation*, 1989.)

A point representing the mixture M of three water types lies in the triangle formed by the three water types (I, II, III) as a function of the proportions of its composition of the mixture [**Figure 1-37b**]. For example, the proportion of water mass III in the mixture M equals f/(e + f).

Suppose that three homogeneous water types (A, B, and C) with different θ-S characteristics are superposed as in **Figure 1-38**. The surface water (A) has the same salinity as the deep water (C) but a different temperature. The intermediate water (B) and the deep water have the same temperature but different salinities. The first θ-S diagram corresponds to the situation before mixing where the water types are represented by three points. During the evolution in time and space of these three water types, the salinity minimum is eroded little by little through mixing with the water above and below, both of which are more saline. However, the intermediate water is always recognizable by its salinity minimum, which represents the part of the water type less mixed (the "heart" of the water mass). Changes in temperature between the surface layer and the intermediate layer take place progressively. The mixing that occurs for each water type can be graphically calculated.

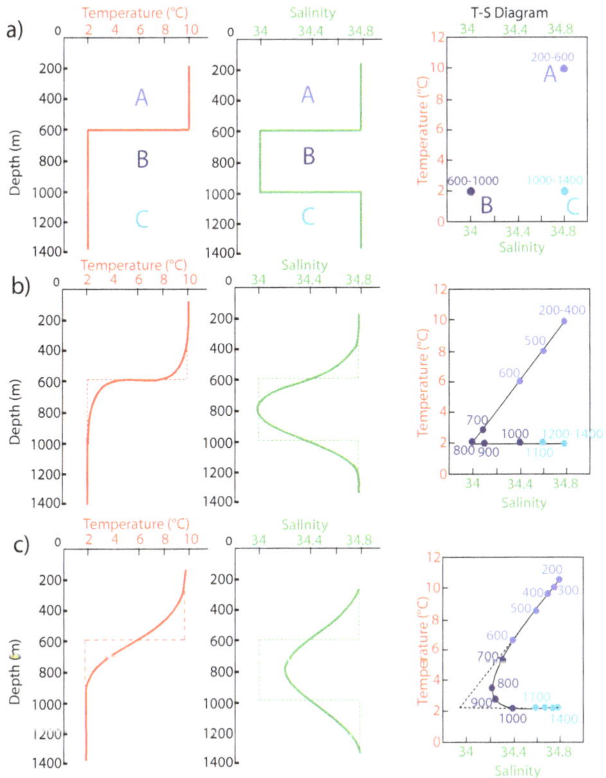

FIG. 1-38 – Evolution (**a**) → (**b**) → (**c**) under the effect of mixing of three water types. (A, B, C) are characterized by temperature (left column), salinity (center column), and T-S diagram (right column, depth in meters). (After *Ocean Circulation*, 1989.)

The characteristics of a hydrographic station may be represented by profiles of potential temperature, salinity, and potential density as a function of depth [**Figure 1-39**] and by the basic tool of oceanographic analysis, namely the two-dimensional θ-S diagram. The θ-S diagram of a station allows us to determine the different water masses present along the vertical of the station according to the form of the diagram (maximum or minimum) and according to temperature, salinity, and density values.

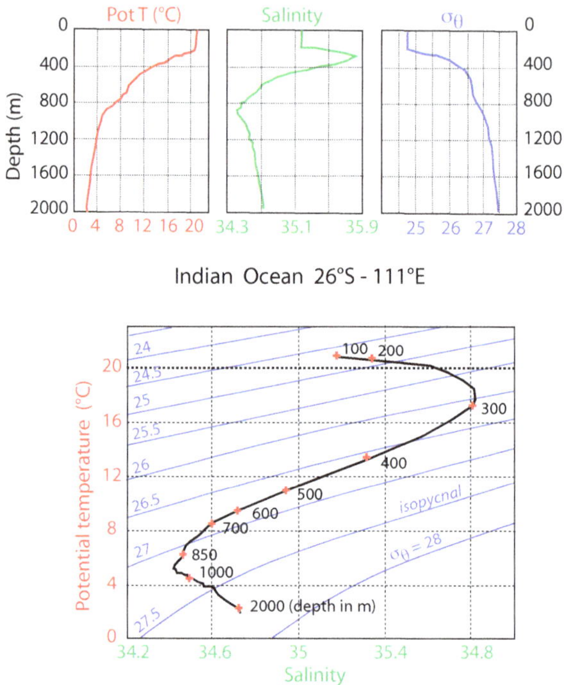

FIG. 1-39 – Profiles of temperature, salinity, and density and the corresponding θ-S diagram for a hydrographic station in the SE Indian Ocean (Red points: depths in meters). (After Fieux, TIP2000 cruise.)

The portion of the θ-S diagram representing the surface layers is more extended than the portion representing the deep layers because of the large variation of temperature and salinity that occurs near the surface due to exchanges with the atmosphere.

On a θ-S diagram taken from a hydrographic station, density at each point increases with depth.

Potential density brought to surface pressure (σ_0) is usually chosen for plotting. Because of the variation of seawater compressibility, sometimes isopycnals are plotted with respect to different pressure levels in deep-water studies (for example, σ_{3000} may be used for a reference at 3,000 db or σ_{4000} for a reference at 4,000 db). **Figure 1-40** shows the advantage of using σ_{3000} instead of σ_0 in order to compare deep water masses located around 3,000 m. It can be seen that Greenland Sea Deep Water is the densest of the region with a density with respect to 3,000 db of 1.042 (σ_{3000} = 42).

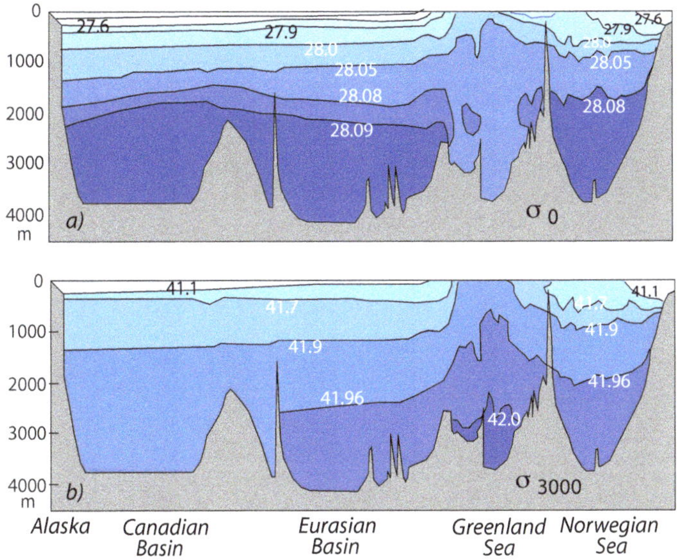

FIG. 1-40 – Potential density anomalies, σ_0 and σ_{3000} along an arctic-basin section. (After Aagaard *et al.*, 1985.)

Since 1997 a new variable, *neutral density* (called gamma γ[50]), has been in use. Neutral density is locally referenced and calculated through an algorithm that includes non-linear terms of the equation of state for seawater and includes those arising from variations of the coefficient of thermal expansion with pressure and the coefficient of contraction due to salinity. Neutral density is a function of both hydrographic and geographical position. Neutral density is thus continuous in the vertical, which is not the case using densities referenced to various levels. Neutral density is quite useful while following water masses that undergo large pressure changes such as bottom waters.

Diagrams of θ-S at different stations of various separations allow following the evolution by mixing of different water masses and the calculation of the percentage of mixing at each station. We observe, for example in **Figure 1-41,** the salinity minimum of Antarctic Intermediate Water, which diminishes with distance from its place of formation in the south until it disappears. Along its mean track, the more a water mass extends from its place of formation, the more it undergoes mixing, and the more its characteristics are blurred. Mixing preferentially takes place between water masses having similar densities, requiring the minimum of energy.

[50] Jackett and MacDougall, 1997

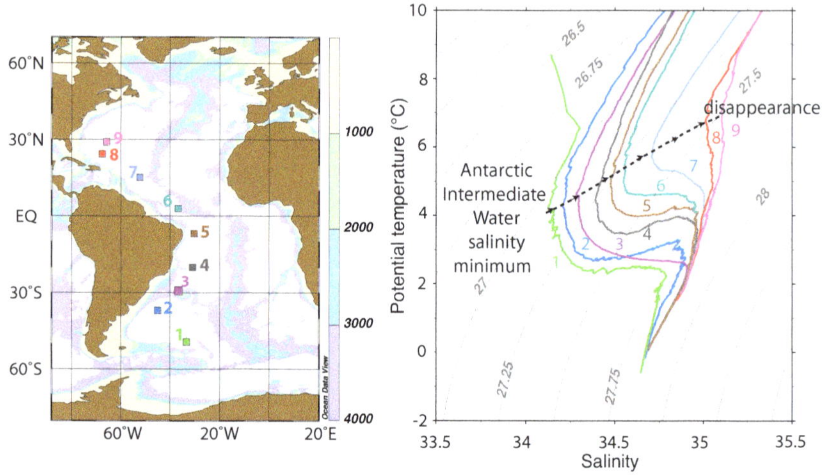

FIG. 1-41 – Evolution of Antarctic Intermediate Water, characterized by a salinity minimum, from its formation region in the Antarctic Ocean to its extinction by mixing around 25°N. (With ODV and WOCE data.)

Because of the non-linearity of the equation of state of seawater, which is particularly strong at low temperatures, it sometimes happens in polar regions that a water mass, resulting from the mixing of two low-temperature water masses, has a density greater than that of the two original water masses and thereby sinks [**Figure 1-37a**]. This process is called *cabbeling*.

3.4.3 Tracers

Potential temperature and salinity are used as *tracers* to track the evolution of water masses from their place of formation (particularly in deep water where direct current measurements are rare). Water-mass characteristics thus provide a means to indirectly arrive at the general character of ocean circulation at all depths, with the aid of simple vertical profiles of temperature and salinity at different points in the ocean. Using this method, the grand outlines of ocean circulation were sketched out early in the last century based on the small amount of oceanographic data collected in the course of early oceanographic expeditions by several countries.

The first great oceanographic expedition was that of the British ship *Challenger* from 1872 to 1876 followed, among others, by that of the French *Le Travailleur* in 1880–1884 and *Le Talisman* in 1883, the Norwegian ship *Fram* in 1883–1896 in the Arctic, the German *Valdivia* in 1898–1899, *Gauss* in 1901, *Planet* in 1906, and *Meteor* from 1925 to 1927, then the British *Discovery I*, *William Scoresby*, and *Discovery II* which made several around-the-world trips, and many others such as those led by Commandant Charcot in the polar regions or by Prince Albert of Monaco at the beginning of the last century.

The concentration of oxygen in a water sample is routinely sampled,[51] most often continuously, with the aid of a sensor on a CTD probe (see **Section I.6.2.2**). In layers near the surface, oxygen rapidly reaches its equilibrium of solubility (solubility decreases as temperature increases). At depth, where biological activity is reduced, oxygen may serve as a tracer. "Old" water, that is water that has left the surface a long time ago is low in oxygen and may be distinguished from water that is young and ventilated. However, variable oxygen levels in the ocean should be used with caution, since they are non-conservative in near-surface layers. For example, independent of their "age",[52] water masses under zones of strong biological productivity are depleted in oxygen by decomposition processes.

Nutrient salts (nitrates, phosphates, silicates) depend strongly on biological processes of photosynthetic production in the surface waters. When the detritic organic matter sinks at depth, the detritic azote and phosporus forms are remineralized as nitrates and phosphates, and the biogenic silica is redissolved as silicates. They may however sometimes be useful in following deep water masses, for example using silicates to track silicate-rich Bottom Waters.

Oceanographers also use conservative chemical compounds as circulation tracers. These are true markers, which precisely follow water-mass trajectories, like ink poured into the ocean.

Atmospheric tracers, dissolved in surface water in more or less large quantities according to their solubility, penetrate into the ocean at the moment of water-mass formation. Tracers in surface waters are generally in solubility equilibrium with the atmosphere.

In this way, the age of deep water has been determined using natural carbon-14 originating in atmospheric CO_2. About a thousand years are needed for North Atlantic Deep Water to reach the North Pacific! Other natural radioactive elements (produced by cosmic rays) such as argon-39 or radon-222 are sometimes used.

Artificial radioactive elements, such as carbon-14 and tritium (^3H) make up transient[53] tracers that have penetrated into the ocean depths following their injection into the atmosphere during the nuclear tests of the 1960's.

CFCs (chlorofluorocarbons or Freons) do not exist in a natural state. Beginning in the 1940's they were created as cooling fluids and thus were introduced into the atmosphere.[54] Though CFCs are not radioactive, they are also transient tracers since they were introduced at a given time and their

---

[51] Oxygen concentration is expressed in μmol kg^{-1}. To a first approximation one may convert ml l^{-1} as μmol kg^{-1} by multiplying by 44.66 and dividing by ρ, ρ being the density of seawater in kg/dm^3 (for example, 1.02788 for $\sigma_0 = 27.88$).

[52] The age of a water mass is the time since it quit the surface.

[53] The concentration of transient tracers evolves with time.

[54] CFCs are partly responsible for the destruction of the ozone layer that protects mankind from ultraviolet radiation.

concentration has evolved with time. Before their appearance, the ocean was free of CFCs. CFCs have weak solubility, enough however to be detected.[55] They can be measured from oceanographic vessels with the same spatial resolution as classic chemical parameters. Their distribution provides a means of following water masses in time and space. Knowing the evolution of their concentration in time since their injection at the ocean surface allows the estimation of average velocities (before the complete contamination of the ocean). Measuring the concentration of different CFCs (CFC-11, CFC-12, CFC-113) provides estimates of mixing during the water-mass trajectory.

Using the isotopic composition of other tracers, such as rare earths, for example neodymium, provides information on the origin of particle sources in the ocean, their evolution (mixing, biologic decomposition, dissolution...) and exchanges between ocean water and its various reservoirs.

[55] Detecting CFCs is a delicate measurement since seawater samples containing small amounts of CFCs are sensitive to local pollution.

Chapter 4
Ocean circulation

4.1 Laws controlling oceanic motion

Ocean movements obey the general laws of mechanics associated with the laws of thermodynamics for a density-stratified saline fluid located on a rotating sphere. They depend on five types of forces:

- The force of gravity.
- Forces due to pressure gradients, produced by density differences, themselves due to heat and water fluxes with the atmosphere, or by the accumulation or depression of water by wind, which augments or diminishes potential energy.
- Forces of internal friction and forces of friction at the interface with the atmosphere and with continental boundaries.
- Astronomical forces that generates tides.
- Coriolis force, due to Earth's rotation about its axis [**Section I.4.2**], which comes into play when there is movement due to the balance of the acting forces.

In the vertical, the force of gravity (equal to the weight of the water column) balances the pressure forces, namely $dp/dz = -\rho g$ (the hydrostatic equation). (p = pressure, ρ = density, which is a function of z, g = acceleration due to gravity, $-z$ = depth on z-axis, z is positive upward.) The ocean is, to a good approximation, in hydrostatic equilibrium.[56] Pressure increases with depth and with water density. In a stable situation, the ocean is stratified in density and lesser densities are found above greater densities.

The equations controlling ocean movements are:

- The Navier-Stokes equations or the "equations of conservation of momentum" linking acceleration to the forces acting on the fluid, as applied to a thin fluid envelope covering a rapidly turning nearly spherical core;

[56] If we neglect vertical velocities which are nearly always weak with respect to horizontal velocities.

- The equations of conservation of mass and salt, which control the evolution of water masses through exchanges and mixing;
- The equation of conservation of heat, which defines the evolution of temperature through heat exchange;
- The equation of state of seawater which links density to thermodynamic parameters (T, S, and p).

We do not propose to study ocean dynamics here or to enter into the details of the mathematical formulations.[57] We give only the results of two particular balances between these forces, allowing us to understand a large part of ocean circulation:

- Ekman balance, in which the frictional force is balanced by Coriolis force, and
- Geostrophic balance, in which Coriolis force balances the horizontal pressure gradient.

4.2 The effect of Earth rotation on motion; the Coriolis force

The Greek philosophers thought the celestial sphere turned around Earth; they understood that the Earth was round. In 1543, Nicolas Copernicus observed that Earth turned around the sun. After the work of Johannes Kepler on planetary movements (Kepler's laws, 1609) and of Galileo Galilei on the principle of inertia (1632), Sir Isaac Newton formulated the principle of universal gravitation in 1686. A reference system fixed with respect to Earth [**Figure 1-42**] is sufficient most of the time to describe and measure the classic movements. Carl Friedrich Gauss and Pierre Simon Laplace were the first to understand the deviation due to the effect of Earth's rotation and calculated it (1776) starting with equations of motion deduced from Newton's laws.

It was only in the 19th century that several scientists conceived experiments to validate the calculation of the deviation of motion linked to Earth's rotation. To do so, they had to isolate, as much as possible, ties to Earth. In 1803 and 1831, experiments with falling balls dropped in sufficiently deep mine shafts allowed measurements of average deviations that were very close to the calculations. These deviations, though weak, showed well that Earth turned during the fall of the balls. The pendulum experiment of Jean Bernard Foucault, installed in 1851 under the dome of the Panthéon in Paris, showed in brilliant fashion the rotation of the Earth on itself while the pendulum oscillated in the same plane with respect to a celestial reference. The experiment allowed the validation of the calculation of the period of rotation of the plane of the pendulum (that is, the time necessary to

[57] See, for example, *Atmosphere-Ocean Dynamics*, Gill, 1982.

return to the beginning plane of oscillation, half the period of its rotation). The speed of rotation of Earth[58] ($\boldsymbol{\Omega}$) is $2\pi/24$ h in radians, or $360°/24$ h or nearly $15°/\text{h}$, or 7.292×10^{-5} rad s^{-1}. The angular velocity of rotation in the oscillation plane of the pendulum is the component of the velocity of Earth's rotation ($\boldsymbol{\Omega}$) perpendicular to the tangent plane at the location, at latitude ϕ, where the pendulum oscillates, namely $\boldsymbol{\Omega}\sin\phi$ **[Figure 1-42]**. At the poles, the pendulum returns to the same plane of oscillation in 12 hours and carries out a complete tour in 24 hours. At the Panthéon (latitude of Paris = 48°50N) the time taken by the pendulum to return to the beginning plane of oscillation is about 16 hours. This result agrees with the expected period of 12 h/$\sin\phi$. At the equator, where $\sin\phi = 0$, the period of pendulum rotation is theoretically infinite.

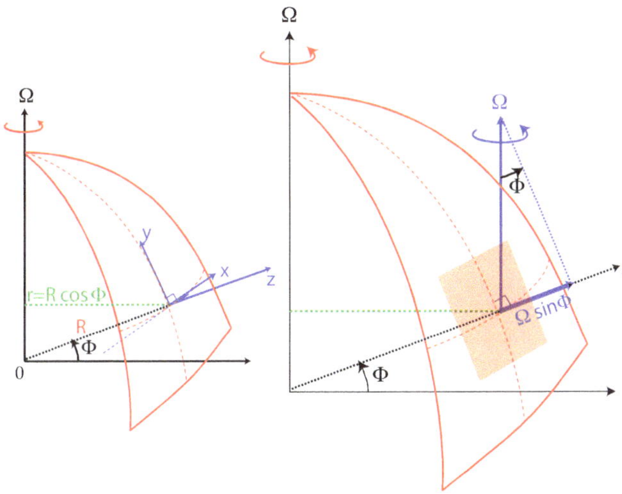

FIG. 1-42 – (**a**) Terrestrial frame of reference at latitude ϕ (x eastward, y northward, and z upward). (**b**) Projection of the rotation vector on the vertical at location $\Omega\sin\phi$.

Earth's rotation has a major effect on large-scale and long-duration motions that are long with respect to Earth's rotation period, such as oceanic and atmospheric currents. When motion occurs on our planet, it is diverted by the Earth's rotation. A point on Earth's surface moves eastward. Its horizontal velocity is proportional to its distance from Earth's axis of rotation:

$$\boldsymbol{V}_e = \boldsymbol{\Omega} \cdot \boldsymbol{r} = \Omega R \cos\phi$$

(R = Earth's radius and ϕ = latitude). This velocity reaches a maximum of 465 m s^{-1} at the equator and is zero at the poles. It is 356 m s^{-1} at 40°N or 40°S.

[58] In reality, we must use the duration of the sidereal day of 23 hours, 56 minutes (and not 24 hours).

Take, for example, the following case [**Figure 1-43**].

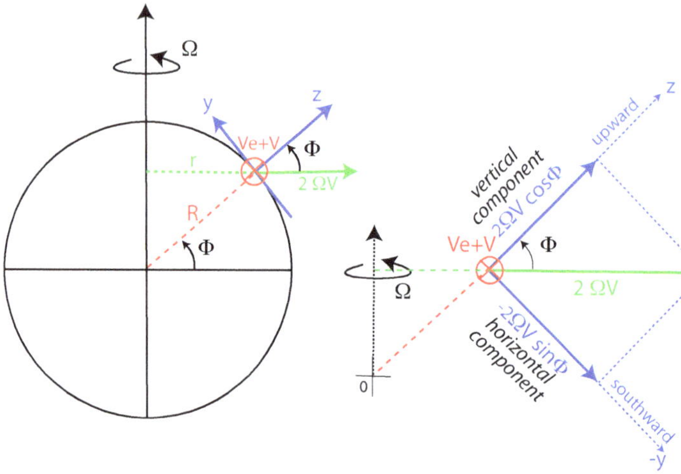

FIG. 1-43 – Centrifugal force in the plane of terrestrial rotation for a velocity $\boldsymbol{V}$ oriented eastward. Vertical and horizontal components of Coriolis force are with respect to the tangent plane.

Consider a point located at latitude ϕ having a velocity $\boldsymbol{V}$ with respect to the Earth (relative velocity) oriented toward the east, and a velocity $\boldsymbol{V}_e$ with respect to a celestial reference (the velocity of Earth's displacement). The centrifugal acceleration, which acts in the meridional plane, perpendicular to the axis of rotation is thus:

$$\frac{(\boldsymbol{V}_e + \boldsymbol{V})^2}{r} = \frac{\boldsymbol{V}_e^2}{r} + \frac{\boldsymbol{V}^2}{r} + \frac{2\boldsymbol{V}_e \times \boldsymbol{V}}{r}$$

The first term on the right represents the centrifugal acceleration on a resting element (with respect to the Earth) and is taken into account in the acceleration of gravity, g.

The second term is very small with respect to the two others, unless $\boldsymbol{V}$ is of the same order of magnitude as $\boldsymbol{V}_e$, which is never the case for ocean velocities.

The third term is the Coriolis acceleration.[59] It may be written as a component of acceleration, which, with $V_e = 2\boldsymbol{\Omega} \cdot r$, equals $2\boldsymbol{\Omega} \cdot V$, perpendicular to the vectors $\boldsymbol{V}$ and $\boldsymbol{\Omega}$.

The horizontal component of the Coriolis acceleration (in the tangent plane at latitude ϕ) equals $-2\Omega V \sin\phi$.

---

[59] Named after the French mathematician Gaspard Gustave Coriolis, who described it in 1835.

The vertical component of the Coriolis acceleration equals $2\Omega V \cos\phi$. It is generally negligible with respect to g (for a current of 1 m s^{-1} at the equator, the ratio is 15×10^{-6}) and with respect to the vertical component of the pressure gradient.

In the general case, if the components of velocity $\boldsymbol{V}$ with respect to a terrestrial frame are u, v, w (u eastward along a parallel, v northward along a meridian, and w upward, perpendicular to the tangent plane) the horizontal components of the Coriolis force are:

$$2\Omega u \sin\phi \quad \text{along the } y\text{-axis.}$$

$$-2\Omega v \sin\phi - 2\Omega w \cos\phi \quad \text{along the } x\text{-axis.}$$

Since the vertical extent of the ocean is small compared to the horizontal extent, the scale of vertical motion is very small compared with horizontal motion. Vertical velocities (w) are thus weak with respect to the horizontal velocities (u, v), and the term $2\Omega w \cos\phi$ is always negligible. If $\boldsymbol{V}$ is eastward (along the x-axis) as in **Figure 1-43**, $v = 0$, and the Coriolis force component along the x-axis is zero.

The term $2\Omega \sin\phi$ is the Coriolis parameter, f, defined as a function of latitude. The horizontal components of the Coriolis force are thus written:

$$-fv \quad \text{along the } x\text{-axis.}$$

$$fu \quad \text{along the } y\text{-axis.}$$

For example, if a water mass moves poleward from the equator, it conserves (through inertia) the velocity of eastward motion (due to Earth's rotation) that it had when leaving the equator. Leaving the equator, it arrives at a latitude where the Earth's eastward velocity is less and thus it appears (in the terrestrial frame) to deviate eastward. That is, the water deviates to the right in the northern hemisphere, and to the left in the southern hemisphere. A point moving toward the equator will be deviated westward. That is, it will always deviate to the right of its motion in the northern hemisphere and to the left in the southern hemisphere.

The deviation increases as the moving point nears the poles, where it arrives at the axis of rotation of the Earth. At the equator, the horizontal component of the Coriolis force is zero. This apparent force is perpendicular to the axis of rotation of the Earth (that is parallel to the equatorial plane passing through the center of the Earth) and acts perpendicularly to the direction of movement.

If a particle with velocity $\boldsymbol{V}$ is subject only to the Coriolis force (perpendicular to $\boldsymbol{V}$) it describes a circle, turning to the right in the northern hemisphere and to the left in the southern hemisphere. This is called the *inertial circle*. Its radius is $V/2\Omega u \sin\phi$. The period of revolution about the inertial circle is $\pi/\Omega \sin\phi$.

At the surface of the Earth, in order for a particle to follow a straight line, there must be some other force, equal to and opposed to the Coriolis force. Any other motion will be the superposition of a linear velocity $\boldsymbol{V}_0$ and an angular velocity, which ends to form an inertial circle (except at the equator, where $f = 0$). The resultant is then a cycloid motion whose radius is inversely proportional to the sine of the latitude [**Figure 1-44**]. In reality, frictional forces come into play to slow and deform the motion.

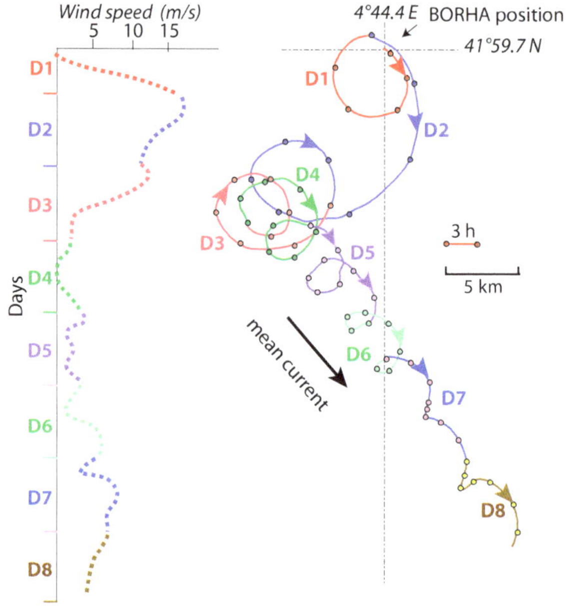

FIG. 1-44 – Cycloidal trajectory of an imaginary water particle following a wind event, from current meter measurements taken from the fixed buoy *BOHRA* in the Mediterranean. (After Gonella, 1969.)

The Coriolis force is not a classic force. It is an inertial force which does not produce any work and which only exists if there is motion with respect to the Earth. It occurs in particular for large-scale motions that are slow with respect to the rotation of the Earth. It is fundamental in understanding ocean circulation.

4.3 Geostrophy: the principal balance of forces in the ocean

In the stationary state,[60] $\partial V/\partial t = 0$ and, neglecting non-linear effects and friction forces, including those at the surface, the horizontal pressure

[60] That is to say, with no acceleration.

gradient is balanced only by Coriolis force. The equations of motion in the horizontal plane then are:

$$\frac{1}{\rho}\frac{\partial p}{\partial x} = fv$$

$$\frac{1}{\rho}\frac{\partial p}{\partial y} = -fu$$

These equations represent geostrophic balance.

Then, using the hydrostatic equation, $dp = \rho g dz$, we may write:

$$V(z) = \frac{g}{\rho f}\int_{z_0}^{z}\frac{\partial \rho}{\partial x}dz + V_0$$

This relation is called the *thermal wind equation.*

x is the horizontal coordinate along the axis passing through the two points between which V is calculated, and V_0 is a constant equal to the horizontal velocity at the reference level z_0.

Geostrophic balance applies to nearly all large-scale ocean circulation in the ocean interior away from the layer influenced by wind, far from coasts where friction is important, and away from the equatorial zone (2°N–2°S) where the Coriolis force is zero. We need only know the vertical distribution of density between two points separated by Δx to obtain the horizontal velocity at each level within an unknown V_0, called the *reference velocity.*

At the surface, to obtain the total current, we must add to the geostrophic current the current due to entrainment by wind [**Section I.5.2**].

A slope at the surface will generate a pressure gradient within the water mass (in the case of a homogeneous medium) directed from high pressure (the highest level of water) to lowest pressure (the lowest level of water) [**Figure 1-45**].

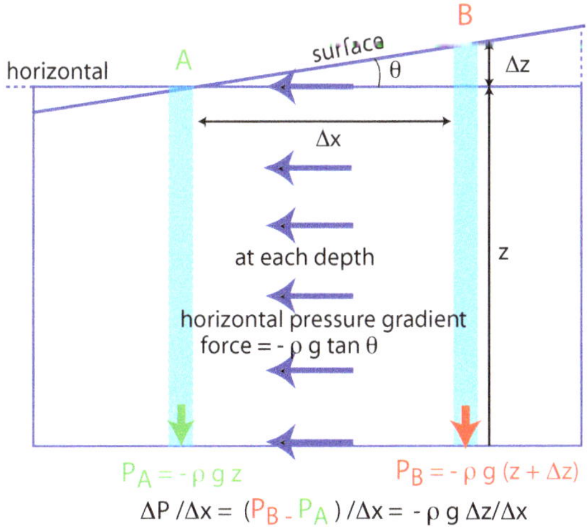

FIG. 1-45 – Pressure difference between points A and B, separated by Δx, produced by a surface slope in a homogeneous ocean of density ρ.

The pressure gradient (perpendicular to the isobars) may be broken down into a vertical component, balanced by gravity (perpendicular to the equipotential surfaces) and a horizontal component [**Figure 1-46**].

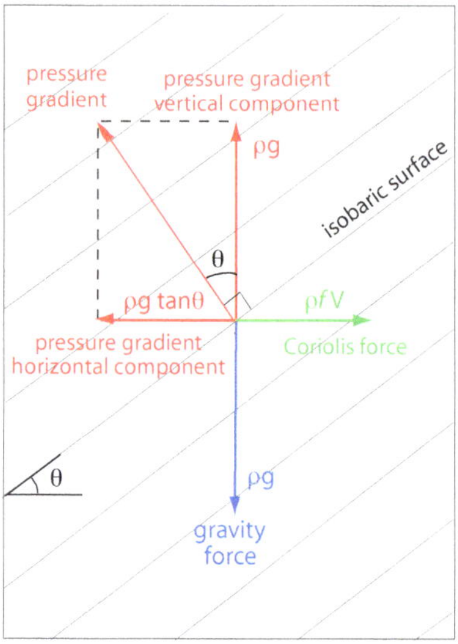

FIG. 1-46 – Components of the pressure-gradient that is perpendicular to the isobars (lines of constant pressure). The vertical component of pressure gradient is balanced by gravity. In geostrophic balance Coriolis force, proportional to the current, balances the horizontal component of pressure gradient. (After *Ocean Circulation*, 1989.)

With constant density, the horizontal pressure gradient, between points A and B, situated at the same depth, and separated by Δx is:

$$\Delta p/\Delta x = -\rho g\,\Delta z/\Delta x, \text{ (z-axis upward)}$$

Now $\Delta p/\Delta x = \tan\theta$, θ being the angle between the ocean surface (or an isobaric surface) and the horizontal surface (which is an equipotential surface). The pressure gradient per unit of mass then is:

$$\frac{1}{\rho}\frac{\Delta p}{\Delta x} = -g\tan\theta$$

The existence of a horizontal pressure gradient generates a motion from high to low pressure [**Figure 1-47**]. When motion appears, the Coriolis force ($f \cdot V$) acts at right angles to the current ($\boldsymbol{V}$) deviating the current to the right in the northern hemisphere or to the left in the southern hemisphere, until it balances the equal and opposite horizontal pressure gradient. For geostrophic balance, we have:

$$\boldsymbol{f} \cdot \boldsymbol{V} = -g\tan\theta$$

This is the geostrophic equation for a unit mass, valid on each isobaric surface inclined at the angle θ. The current is then perpendicular to the Coriolis force and to the pressure gradient. The current becomes zero when the isobars become horizontal. In that case the pressure gradient is zero, that is $\tan\theta = 0$.

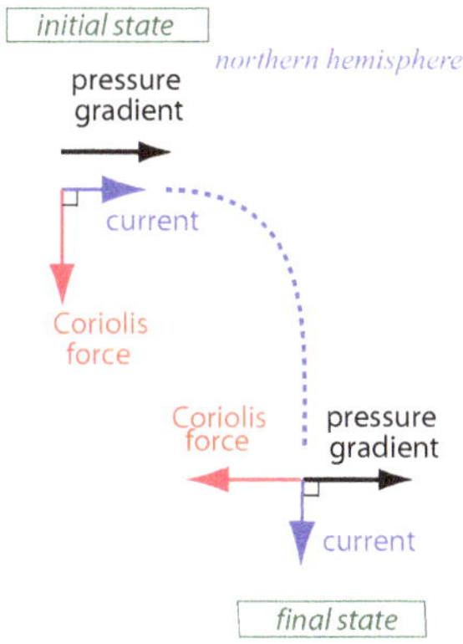

FIG. 1-47 – Current development in geostrophic balance in the presence of a pressure gradient. (After *Ocean Circulation*, 1989.)

In the stationary state the current is proportional to the slope of the isobars ($\tan\theta$), perpendicular to the pressure gradient ($\partial p/\partial x$), and follows the isobars (lines of equal pressure) along a horizontal plane.

If the isobaric surfaces are parallel to the equal-density surfaces (isopycnals) the conditions are said to be *barotropic* [**Figure 1-48**]. Under these conditions the slope of the surface (which is an isobar) determines the current,[61] which is the same throughout the water column.

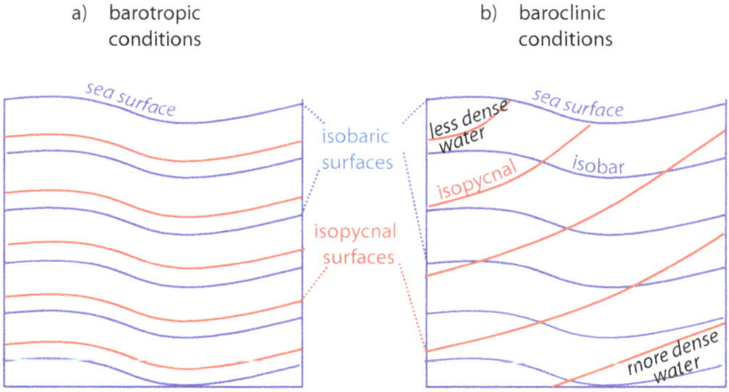

FIG. 1-48 – Isobars and isopycnals in (**a**) barotropic conditions, (**b**) baroclinic conditions. (After *Ocean Circulation*, 1989.)

61 Which are also called "slope currents".

If the density varies horizontally, the pressure is not only a function of the surface slope but also of the density distribution. The isobars and the isopycnals then have different slopes [**Figure 1-48**]. In this case the current varies with depth and the conditions are called *baroclinic*.

The sea-surface is an isobaric surface if we neglect rapid variations[62] of atmospheric pressure (of the order of 2 to 3 hPa, that is, about 0.02 to 0.03 m). In theory, measuring the surface slope should permit calculating surface velocity [**Figure 1-46**] but it is not directly measurable from a ship since the slope is extremely weak and the ship has no means of reference. For example, intense currents like the Gulf Stream have a slope reaching only one meter per hundred kilometers!

4.4 The dynamic method

To calculate currents, we use the dynamic method. This determines currents using only measurements of pressure (p), temperature (T), and salinity (S) and thus density (ρ). The procedure gives an estimate of currents time-averaged over high-frequency variations since it uses the geostrophic equation, which implies a stationary current.

Between two hydrographic stations, the average speed of the current, V, perpendicular to the plane passing through the two stations is given [see **Section I.4.3**] by the geostrophic equation:

$$\boldsymbol{f} \cdot \boldsymbol{V} = -g \tan \theta$$

The current $V = -(g/f) \tan \theta$, with $\tan \theta = \Delta z / L$.

L is the distance between stations A and B [**Figure 1-49**] and Δz is the difference in level of the isobar P at the two stations.

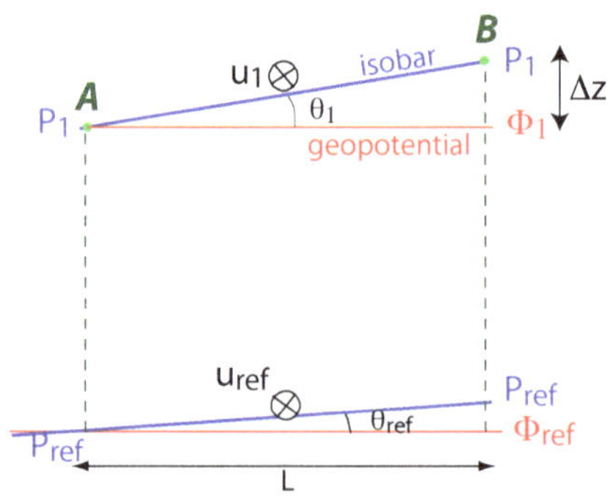

FIG. 1-49 – Relation between geopotential surfaces (Φ_1 and Φ_{ref}, horizontal) and the slopes of isobaric surfaces (P_1 and P_{ref}) corresponding to the velocities u_1 and u_{ref}. (After *Ocean circulation*, 1989.)

---

[62] Which become zero after some time. The surface of the ocean reacts as an inverse barometer.

However, Δz is not directly measurable. Instead, density is precisely measured as a function of pressure. The hydrostatic equation is used to go from dp to dz with aid of density.

Take the isobar $P_{\mathbf{ref}}$ as the reference surface where the velocity is $u_{\mathbf{ref}}$, and the angle with the equipotential surface is $\theta_{\mathbf{ref}}$, then the geostrophic equations at the two levels are:

$$\boldsymbol{f} \cdot \boldsymbol{V}_1 = g \tan \theta_1 \text{ and } \boldsymbol{f} \cdot \boldsymbol{V}_{\mathrm{ref}} = g \tan \theta_{\mathrm{ref}}$$

$$f\,(V_1 - V_{\mathrm{ref}}) = g(\tan \theta_1 - \tan \theta_{\mathrm{ref}})$$

The differences of the levels between the isobars are calculated by integrating the hydrostatic equations:

$$\int_{z_{\mathrm{ref}}}^{z_1} g dz = \int_{p_{\mathrm{ref}}}^{p_1} \frac{-dp}{\rho}$$

Using the specific volume[63] $\alpha = \dfrac{1}{\rho} = \alpha_{35,0,\,p} + \delta$,

where $\boldsymbol{\delta}$ is the specific volume anomaly (with respect to seawater at 0 °C, salinity 35, and pressure p) calculated at each station and at each pressure level.

$$-\left(\frac{1}{\rho}\right) dp = -\left(\alpha_{35,0,p} + \delta\right) dp$$

The distribution of ρ (=1/α) is known along each vertical (from temperature, salinity, and pressure), ρ_{A} and $\boldsymbol{\delta}_{\mathrm{A}}$ in A and ρ_{B} and $\boldsymbol{\delta}_{\mathrm{B}}$ in B.

Then,

$$V_1 - V_{\mathrm{ref}} = \frac{1}{fL}\left\{\int_{p_{\mathrm{ref}}}^{p_1} (-\delta_{\mathrm{B}} dp) - \int_{p_{\mathrm{ref}}}^{p_1} (-\delta_{\mathrm{A}} dp)\right\}$$

We eliminate the constant of integration by calculating the velocity (V_1) on isobar P_1, with respect to the velocity ($V_{\mathbf{ref}}$) on reference isobar $P_{\mathbf{ref}}$, where the velocity is supposed to be zero if $\tan \theta = 0$, or is known.

The geopotential, $\Delta\Phi$, equals the work necessary[64] to raise a unit mass a distance dz:

$$\Delta\Phi = gdz \text{ (in m}^2\text{ s}^{-2})^{65} = -\frac{dp}{\rho} = -\alpha dp = -\left(\alpha_{35,0,p} + \delta\right) dp$$

The geopotential anomaly between the levels $P_{\mathbf{ref}}$ and P_1 at station A equals:

$$\Delta\Phi_{\mathrm{A}} = \int_{p_{\mathrm{ref}}}^{p_1} (-\delta_{\mathrm{A}} dp)$$

---

[63] See **Section I.1.3**.

[64] This is the potential energy per unit mass.

[65] Or joule kg^{-1}.

And $\Delta\Phi_B$ is the geopotential anomaly between levels P_{ref} and P_1 at station B. Then,

$$(V_1 - V_{ref}) = (1/fL)(\Delta\Phi_B - \Delta\Phi_A).$$

Here, L is in m, δ in $m^3\ kg^{-1}$, p in pascals, $\Omega = 7.29 \times 10^{-5}\ s^{-1}$, and $(V_1 - V_{ref})$ is in $m\ s^{-1}$.

Rather than use the geopotential, Φ, oceanographers use the *dynamic height*, $D = \Phi/10$ (in $10\ m^2\ s^{-2}$, or dyn m)[66] so that the numeric value is of the same order as the numeric value of the depth, z, in meters.[67]

Using the dynamic height, calculating the geostrophic velocity, V, with respect to the chosen reference level is:

$$V_1 - V_{ref} = (10/fL)(\Delta D_B - \Delta D_A),$$

where $\Delta D_A = \dfrac{1}{10}\displaystyle\int_{p_{ref}}^{p_1}(-\delta_A dp)$ at station A

and $\Delta D_B = \dfrac{1}{10}\displaystyle\int_{p_{ref}}^{p_1}(-\delta_B dp)$ at station B.2.

Then $$V_1 - V_{ref} = (10/2\Omega \sin\phi L)(\Delta D_B - \Delta D_A)$$

With L in m, P in dbar, and δ in $cm^3\ g^{-1}$, ΔD is in dyn m and $(V_1 - V_{ref})$ is in $m\ s^{-1}$.

At each measured station we calculate the anomaly of dynamic height at each level with respect to a reference level, thus integrating ΔD between the two levels.[68]

Thus, in the dynamic method, the "horizontal" pressure gradient allows the calculation of the "vertical" increments of velocity $(V_1 - V_{ref})$.

The difficulty of the dynamic method is the determination of a reference level where the velocity is zero ($\tan\theta = 0$ and $V_{ref} = 0$) in order to obtain absolute velocities. Currents calculated by this method are always intrinsically uncertain to the extent that the current at the reference level is unknown. In general, the deepest level is chosen, where the currents are supposed to be weak. However, it is apparent from an increasing number of direct measurements that in some regions deep currents are not negligible, putting in question the value of geostrophic estimates. A profile of geostrophic currents may be adjusted if the current is known at some level.

---

66 The unit of dynamic height is the *dynamic meter*: 1 dyn m = $10\ m^2\ s^{-2} = 10\ J\ kg^{-1}$.

67 For example, the geopotential difference between the surface and 100 m is about $980\ m^2\ s^{-2}$, while the difference in dynamic height is about 98 dyn m, a value near 100 m.

68 Usually, the calculation is first made at each level with reference to the surface (of which the slope is unknown) then after having chosen the reference level, the offsets are inverted to the reference level.

Direct current measurements at great depth are few, however. Furthermore, to be of value they should cover a period long enough to average out the variability. The study of the distribution of water-mass characteristics (T, S, and other tracers) of various origins sometimes allows the determination of a depth of current reversal where the current becomes zero, which can serve as a reference level.

4.5 Dynamic topography

Using hydrographic data, maps of dynamic height anomaly are drawn at different levels, chosen with respect to a reference level, for example 0 dbar/2,000 dbar or 100 dbar/1,500 dbar. On these dynamic topography maps, a hump is analogous to atmospheric high pressure and a hollow to low pressure. Lines of current are parallel to lines of equal dynamic height, just as winds are parallel to isobars, and the stronger the slope, the more the isopleths are tight, the stronger the current.

The average dynamic topography of the surface, obtained with the aid of thousands of hydrographic stations [**Figure 1-50**], displays humps and hollows. The circulation is anticyclonic around humps, representing regions of convergence or high pressure. The circulation is cyclonic around hollows, representing regions of divergence or low pressure [**Section I.5.3.1**]. The isopleths tighten in the NW region of ocean basins, indicating the presence of stronger currents. Note the maximum difference in dynamic height, of the order of 2.7 dyn m between the Pacific Ocean and the Greenland Sea.

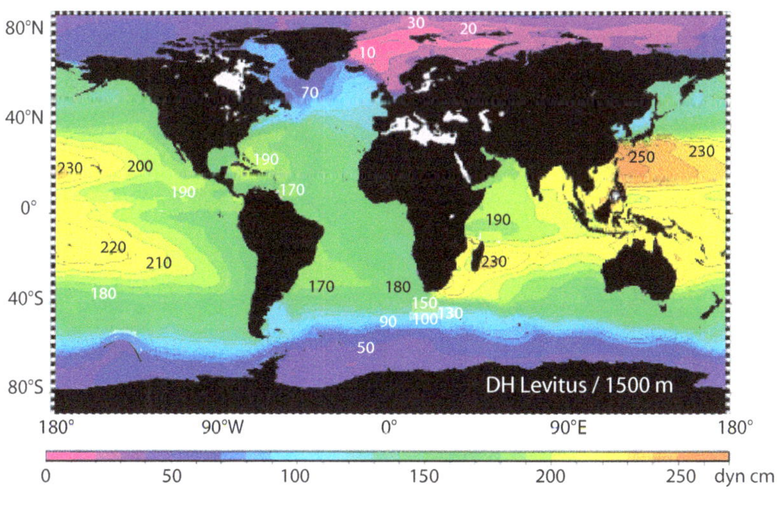

FIG. 1-50 – Average dynamic ocean surface topography with respect to 1,500 m. (After Levitus, 1982.)

4.6 Thermohaline circulation

Due to temperature and salinity variations, the ocean displays regional density variations. They produce horizontal pressure gradients at all depths, which in turn generate large-scale oceanic motion, constituting the *thermohaline circulation*. **Figure 1-51** gives a schematic diagram of the large-scale thermohaline circulation, linked to deep convection processes in the North Atlantic under the effect of cooling and evaporation.

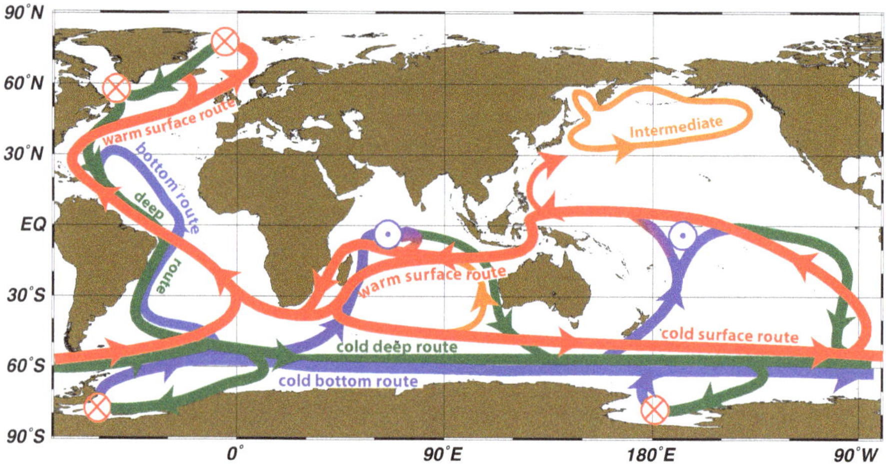

Fig. 1-51 – A schematic thousand-year ocean-water trip. Circled red crosses show dense water formation sites. Circled blue points indicate slow diffusive upwelling. (After Broecker, 1991; Ganachaud & Wunsch, 2000; Gordon, 1986; Maier-Reimer *et al.*, 1996; Rintoul, 1991; Schmitz, 1995.)

After sinking in the North Atlantic (Greenland Sea, Norwegian Sea, Labrador Sea) water flows southward at depth and joins the deep circumpolar circulation transporting dense water formed along the Antarctic continent. Branches from the deep circumpolar circulation feed the Indian and Pacific oceans and rise slowly toward the surface, warming through thermal diffusion. Surface warm water leaves the Pacific Ocean across the Indonesian archipelago and the Indian Ocean, from whence it joins the north Atlantic, either passing south of Africa (the warm route) or through the Drake Passage (the cold route). The water undergoes transformations all along its paths through exchanges with the atmosphere and by mixing and diffusion. The time required for a water particle to make this "long voyage" has been estimated, with the help of carbon-14 measurements, at about a thousand years.

Oceanic general circulation results from a complex combination of thermohaline circulation and wind-driven circulation.

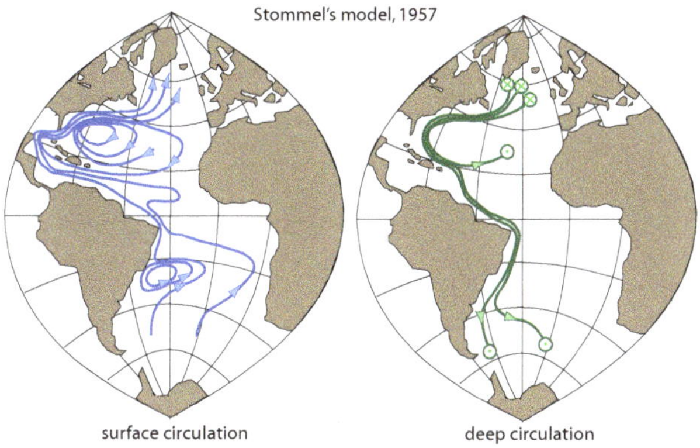

FIG. 1-52 – Diagrams of surface circulation and deep circulation from the two-layer model of Stommel. Convection zones are shown with circled crosses, upwelling zones are shown with circled points. (After Stommel, 1957.)

In a two-layer model of the Atlantic by Henry Stommel [**Figure 1-52**], vertical advection, due to exchanges with the atmosphere that modify density is compensated by a slow upward diffusion of water to the surface.[69] A southward-flowing deep current appears under the surface western-boundary current, a schematic representation of the Gulf Stream.

At about the same time as Stommel's model, John Swallow built the first acoustic floats, which he managed to balance so that they sank to the depth of the "supposed" deep western boundary current (see **Figure 1-104**). And Stommel's theoretical result was brilliantly verified!

[69] Where $w\partial\rho/\partial z = D(\partial^2\rho/\partial z^2)$, D being the vertical coefficient of diffusion (Gill, 1982).

Chapter 5
The role of wind

5.1 Oceanic and atmospheric surface circulation

Schematically, atmospheric circulation is composed of cells having rising and descending branches [**Figure 1-53**].

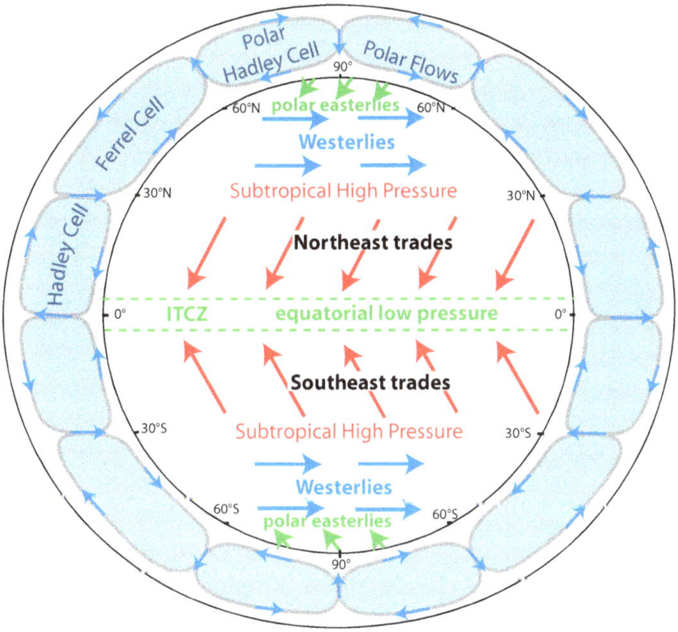

FIG. 1-53 – Relationship between atmospheric cells and surface winds on a theoretical Earth without continents (ITCZ: Intertropical Convergence Zone). (After Tchernia, 1978.)

In the equatorial region, humid air warmed by surface water becomes lighter, rises, and cools, producing condensation of water vapor and strong precipitation. This region, the Intertropical Convergence Zone (ITCZ), lies between the NE trade winds and the SE trade winds of each hemisphere and is a zone of low-pressure. At the poles, cooled air accumulates near the ground and forms a region of high pressure. If the Earth did not turn, air would circulate from the poles toward the equator to re-establish pressure balance at the surface, and from the equator toward the poles at altitude, to form a meridional circulation cell. In fact, the velocity of rotation and the sphericity of Earth bring about three meridional cells (Tropical Hadley, Ferrel, and Polar Hadley) between the equator and the poles [**Figure 1-53**].

For each hemisphere, the tropical Hadley cell (the most intense) is composed of a rising branch at the ITCZ and a descending branch around 30°, a zone of subsidence[70] and subtropical high pressure where dry air produces strong evaporation. Between 30° and the equator, surface winds have a dominant easterly component: the trade winds. The Ferrel cell occurs between the zone of subtropical subsidence (30°) and the zone of subpolar convergence around 60°. In this region, surface winds have a dominant westerly component. The Polar Hadley cell occurs between the pole and about 60°, with an easterly surface-wind component.

In reality, the irregular forms of the continents (which warm and cool more rapidly than the ocean) produce variations from this theoretical scheme. The Intertropical Convergence Zone (also called the "zone of equatorial calms") is characterized by heavy cloud formations, whose position varies with the seasons and with ocean basin [**Figure 1-54**][71]. In the Pacific and Atlantic oceans, the ITCZ is located slightly north of the equator. In January–February, during the NE monsoon it is located clearly south of the equator in the Indian Ocean and in the western Pacific.

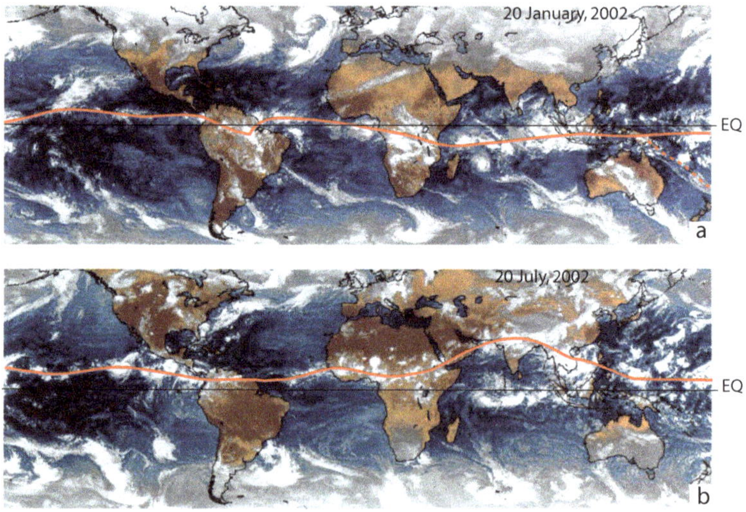

Fig. 1-54 – Photomosaic from five satellites: (a) 20 January, 2002; (b) 20 July, 2001. In the equatorial zone, clouds indicate the location of the ITCZ (marked in red on the figure) which widens over Indonesia and the western Pacific. In January, a cyclonic disturbance is visible east of Madagascar. Anticyclonic zones are cloud-free. (Courtesy of M. Hontarrède, Météo-France images.)

[70] Descending motion.

[71] **Figure 1-54** was obtained from Metéosat 7, Metéosat 5, GOES East, GOES West, and GMS.

The belt of subtropical high pressure is located on average around 30°. It is characterized by cloudless regions, such as the blue ocean regions seen in **Figure 1-54**, and undergoes seasonal variations [**Figure 1-55**]. The trade winds blow from the NE and SE, according to the hemisphere, between the belt of high pressure (broken up in summer by continental low pressure) and the low pressure of the intertropical convergence zone. Westerly winds (perturbed in the northern hemisphere by the mass of Eurasia) blow between subtropical high pressures and subpolar low pressures. Easterly polar winds blow between polar high pressures and subpolar low pressures. Because of continental influences, the strongest seasonal variability occurs in the northern hemisphere and extends beyond the equator in the Indian Ocean and the western Pacific.

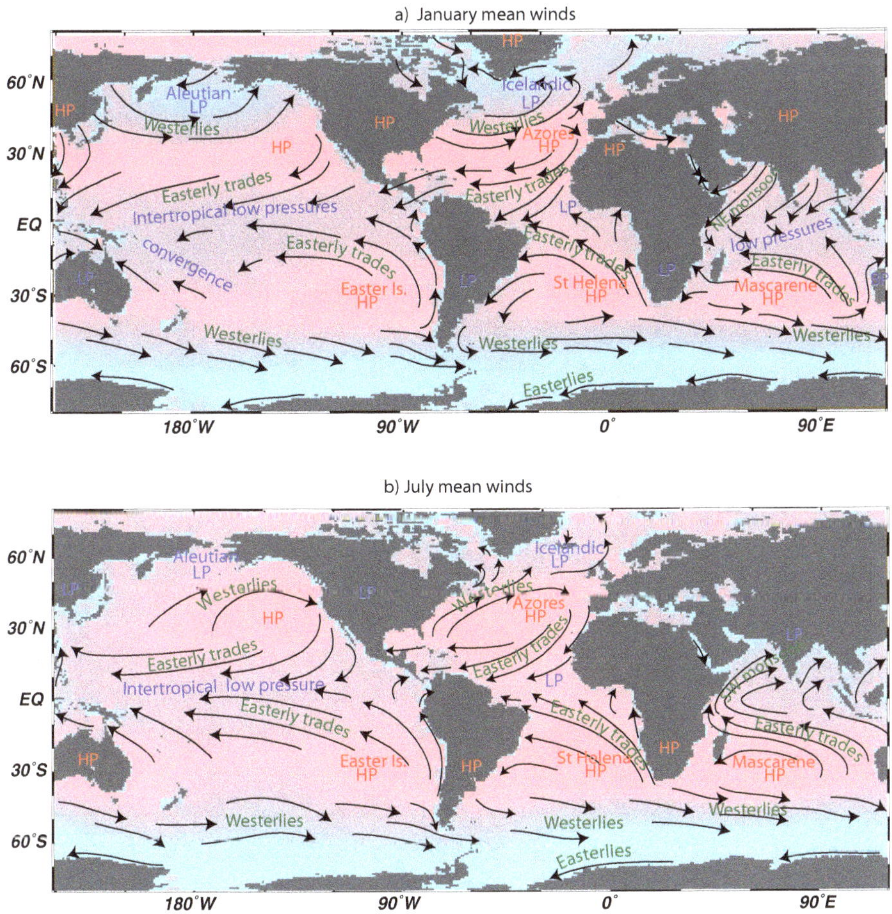

FIG. 1-55 – Surface winds and principal pressure centers (a) January, (b) July. Regions of low pressure and intertropical low pressure are in light blue, high-pressure regions are in pink. (Pressures are from IRI/LDEO; ECMWF/GGA data.)

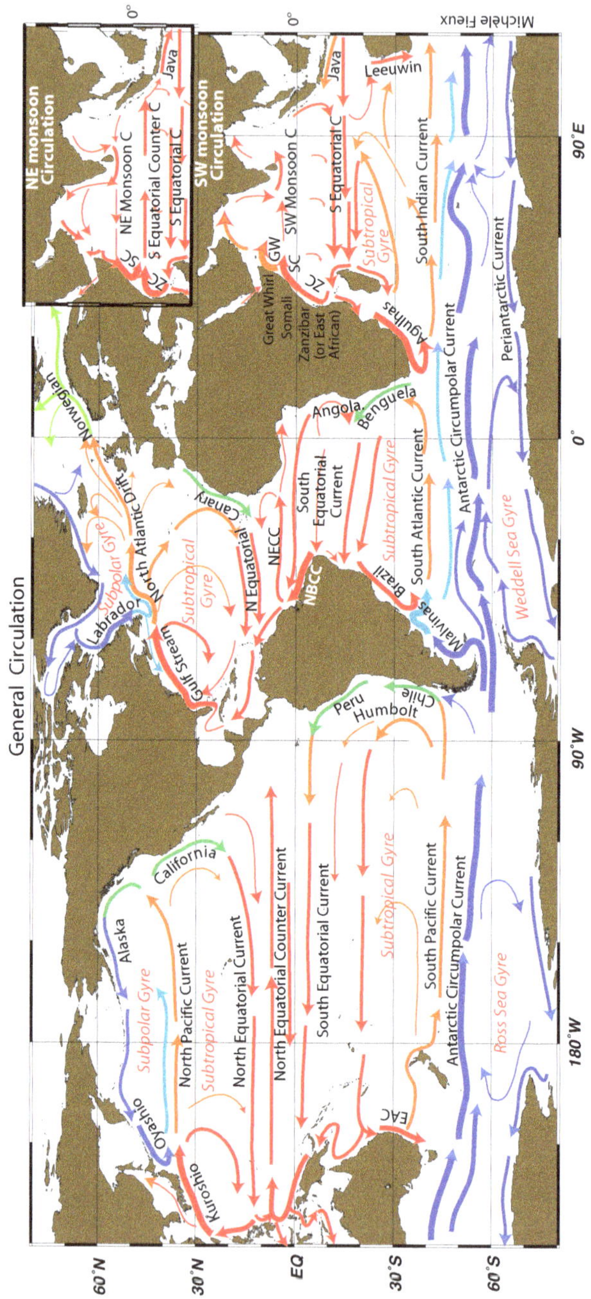

FIG. 1-56 – General surface circulation of the planetary ocean [East Australian Current (EAC), North Brazil Coastal Current (NBCC), North Equatorial Countercurrent (NECC), South Equatorial Current (SEC), Great Whirl (GW), Somali Current (SC), Zanzibar Current or East African Current (ZC), Northeast Monsoon Current (NEMC), Southwest Monsoon Current (SWMC)].

The general surface ocean circulation [**Figure 1-56**] is, in many respects, a "mirror" of the atmospheric circulation [**Figure 1-55**] by which it is principally driven. There are, however, some deformations, due to the form of the ocean basins and to the Coriolis force. Thanks to the compilation and exploitation of observations scrupulously noted by ships' officers in their logbooks, we have been able to determine average ocean currents from ships.[72] By comparing the estimated position, determined from ship's speed, direction, and time of passage, with the observed position using the sun, the stars, coastal landmarks, by radio navigation or by GPS,[73] one can estimate the drift[74] of the ship, due principally to currents.

5.2 Local wind effects: Ekman transport

The atmosphere transmits part of its momentum to ocean layers through wind friction[75] acting on the marine surface, and by pressure forces acting on waves when they form. Wind is the dominant driver of ocean surface circulation. It is what puts the ocean in motion, aside from tidal currents, which are especially pronounced near coasts. The transfer of energy is even more efficient if the sea-surface is rough. The more the wind blows, the more waves build up, the more the transfer of momentum increases.

At the end of the 19th century, Fridtjof Nansen, during his long voluntary voyage on the *Fram*, trapped by arctic ice, noticed that ice floes did not drift down the direction of the wind but moved to the right. He had understood that it was an effect of the Coriolis force, and that the underlying layers deviated even more to the right. He sent his observations to a young Swede, Vagn Walfrid Ekman, who studied the problem theoretically for the case of a homogeneous ocean by supposing that the only imposed constraint was surface wind friction (assumed constant) with a frictional coupling between water layers. The friction imposed by the wind on the surface layer by turbulent viscosity[76] is denoted as wind stress, τ.

The stress, τ, is proportional to the square of the wind speed. It acts in the direction of the wind and equals, for a unit surface:

$$\tau = c_d \rho_a U^2, \text{ in kg m}^{-1}\text{ s}^{-2} \text{ or N m}^{-2} \text{ or Pa.}$$

c_d = coefficient of friction (or drag coefficient), dimensionless, of the order of 1.4×10^{-3} (which varies with wind velocity).

---

[72] Ship positions are no longer secret, as they were at the time of the first discoveries.
[73] Global Positioning System is a system of position determination from satellites.
[74] Sometimes the wind effect on the ship (the "windage") is important and the drift must be corrected for it.
[75] Friction comes into play when there is relative motion between two solid or fluid media.
[76] Molecular viscosity is negligible compared with turbulent viscosity.

ρ_a = air density = 1.2 kg m^{-3}
U = wind velocity at 10 m above the surface in m s^{-1} (the wind profile near the surface being logarithmic,[77] it is necessary to indicate the height at which the wind is measured).

Wind stress is transmitted to layers beneath the surface by Reynolds-stress turbulence (τ_x, τ_y). Assuming that friction results only from the shear of vertical velocity, Reynolds stress is equal to:

$$\tau_x = \rho_0 A_z \frac{\partial u_e}{\partial z}$$

$$\tau_y = \rho_0 A_z \frac{\partial v_e}{\partial z}$$

A_z is the coefficient of turbulent viscosity, ρ_0 is water density, and u_e and v_e are horizontal velocity components of the wind-generated current, called *Ekman velocity.*

The force of horizontal friction, which influences the ocean interior by turbulent viscosity is written:

$$F_x = \frac{1}{\rho_0}\frac{\partial \tau_x}{\partial z} = A_z \frac{\partial^2 u_e}{\partial z^2}$$

$$F_y = \frac{1}{\rho_0}\frac{\partial \tau_y}{\partial z} = A_z \frac{\partial^2 v_e}{\partial z^2}$$

By looking only at the balance between the frictional force of the wind, with components F_x and F_y, and the Coriolis force, which immediately turns the current generated by the wind, with components $-fv_e$ and $+fu_e$, the equations may be written in terms of the two horizontal components to obtain Ekman's equations:

$$A_z \frac{\partial^2 u_e}{\partial z^2} = -fv_e$$

$$A_z \frac{\partial^2 v_e}{\partial z^2} = +fu_e$$

By taking A_z to be constant, Ekman showed that a surface current forms at 45° to the wind, to the right in the northern hemisphere and to the left in the southern hemisphere.

[77] Friction with the ocean surface implies that air speed is zero at the surface.

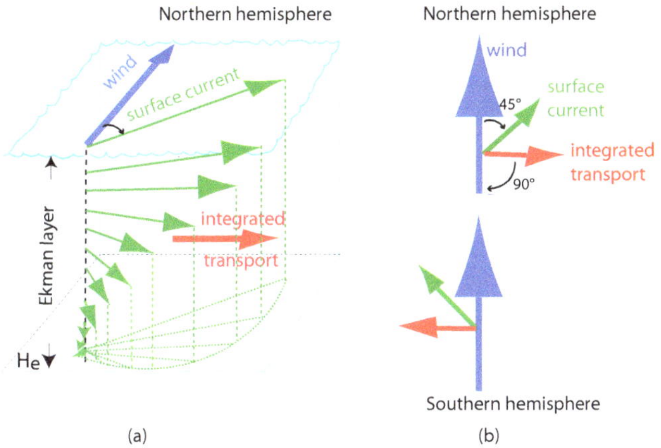

FIG. 1-57 – (a) Ekman spiral, showing current direction and intensity with depth and integrated transport over the wind-influenced layer. (b) Direction of Ekman transport and surface current with respect to the wind in each hemisphere. (After Ekman, 1905.)

The surface layer thus entrains the underlying layer by turbulent viscosity, which is in turn deflected to the right of the surface layer, and so on. There is thus penetration and rotation of the momentum transmitted by the wind. The current decreases exponentially with depth (z is positive upward) as it turns to the right in the northern hemisphere and to the left in the southern hemisphere [**Figure 1-57a**] in what is called the *Ekman spiral.*

$$u_e = \pm V_e e^{(\pi/H_e)z} \cos((\pi/4) + (\pi/H_e)z)$$

(+ for the northern hemisphere, – for the southern hemisphere)

$$v_c = V_e e^{(\pi/H_e)z} \sin((\pi/4) + (\pi/H_e)z)$$

Where V_e is the surface current, $\dfrac{\pi\tau\sqrt{2}}{H_e \rho f}$

At the surface, $u = \pm V_e \cos 45°$ and $v = V_e \sin 45°$

At depth H_e, where the current has turned 180° with respect to the surface current (a characteristic damping depth) the current becomes weak ($e^{-\pi} = 1/23$ of the surface current). H_e, called the *Ekman depth*, is taken as the depth to which the wind influence is felt,

$$H_e = \pi\sqrt{\frac{2A_z}{f}}.$$

The layer between the surface and the Ekman depth is called the *Ekman layer*.[78] Though it is assumed constant in these calculations, in reality the coefficient of turbulent viscosity, A_z, is not constant and varies with wind speed and stratification.[79]

The spiral described by the Ekman current is difficult to observe in the ocean because the assumptions are rarely satisfied, and because strong turbulence in the surface layer tends to destroy the theoretical spiral by homogenizing motions. In temperate latitudes the Ekman layer is often confused with the mixed layer where temperature is uniform.

The Ekman transport (T_e) characterizes the overall wind effect on the Ekman layer. This is the transport due to the wind, integrated from the surface down to the Ekman depth. It is proportional to the wind stress and inversely proportional to the Coriolis parameter. Its components are:

$$T_{ex} = -\tau_y/f, \text{ with } T_{ex} = \int_0^{H_e} \rho u_e \, dz$$

$$T_{ey} = \tau_x/f, \text{ with } T_{ey} = \int_0^{H_e} \rho v_e \, dz$$

Ekman transport occurs at right angles to the wind, to the right in the northern hemisphere and to the left in the southern hemisphere [**Figure 1-57b**].

Since the Coriolis parameter $(f = 2\Omega \sin\phi)$ decreases as latitude (ϕ) decreases, a weak wind in equatorial regions may generate important transports. However, this formula is not applicable between about 2°N and 2°S where f becomes small and disappears at the equator. The thickness of the Ekman layer varies between about ten and a hundred meters according to latitude, the amount of turbulence, and density stratification. Since the transport generated is proportional to the square of the wind speed, storms can have a preponderant effect on surface circulation.

Ekman's theory in part provides an explanation for wind-generated local surface circulation.

The distribution of surface wind stress [**Figure 1-58**] indicates regions of strong winds (in red), since stress is proportional to the square of the wind speed.

---

[78] From observations, Ekman estimated that the velocity of a surface current V, generated by a wind U, is equal to $V = 0.0127U/\sqrt{\sin\phi}$, which in mid-latitudes is about 3% of the generating wind speed.

[79] Values of A_z range between about 0.01 kg m^{-1} s^{-1} and 10 kg m^{-1} s^{-1}.

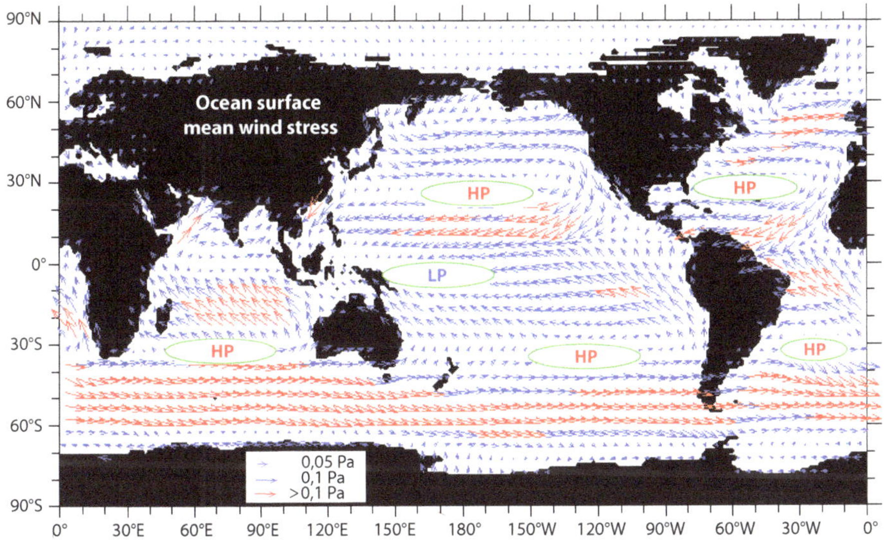

FIG. 1-58 – Average annual surface wind stress (in Pa). In the northern Indian Ocean the SW monsoon dominates the annual average. (After a document of Météo-France.)

5.2.1 Wind acting near a coastline: coastal upwelling

In the northern hemisphere, winds parallel to a coast on the left of the wind [**Figure 1-59a**] move water in the surface layer 90° to the right of the wind, that is, seaward [**Figure 1-59b**]. Due to the coastal boundary, the seaward movement will entrain colder, deeper water up to the surface near the coast, called *coastal upwelling*.

The lowering of sea level at the coast and the upwelling of isopycnals produces, at depth, a pressure gradient at right angles to the coast with high pressure offshore and low pressure at the coast (where the water depth is less) [**Figure 1-59d**]. The isobars (or constant pressure lines) thus slope. As the thickness of the surface layer (of density ρ_1) diminishes, the thickness of the deep layer (of density ρ_2) increases to re-establish the deep pressure balance. If the surface of the ocean slopes, the thermocline (the boundary between the two layers) will slope in the inverse sense to compensate for the pressure differences [**Figure 1-60**]. If the difference in density between the two layers is weak ($\rho_2 - \rho_1 \approx 0.001$ to 0.002) the slope of the thermocline (h_2) is amplified with respect to the slope of the surface (Δh). The slope of the thermocline then is in the ratio of the density differences, $h_2/\Delta h = \rho_1/(\rho_2 - \rho_1)$, and is of the order of 500 to 1,000.

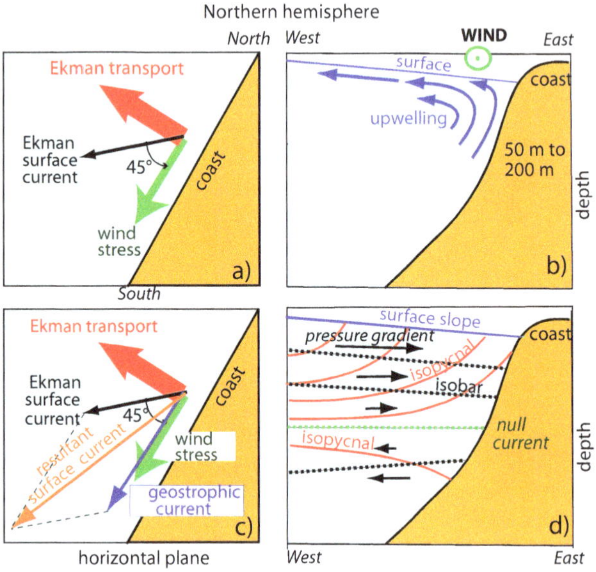

FIG. 1-59 – Wind effects along a coast. (After *Ocean Circulation*, 1989.)

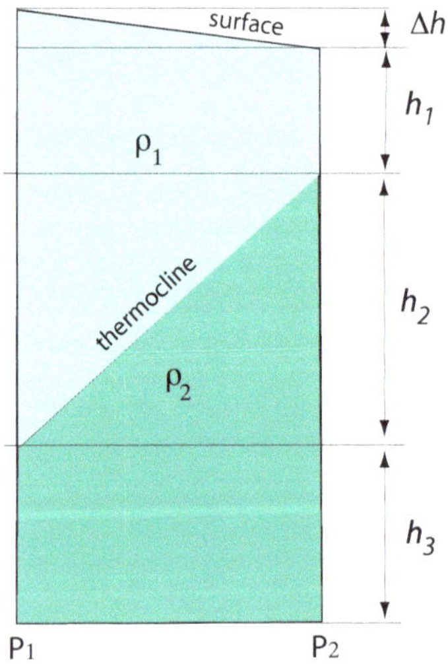

FIG. 1-60 – Compensation at depth of the surface slope in a two-layer ocean with densities ρ_1 and ρ_2. At depth we may write that $p_1 = p_2$. That is, $(\rho_1(h_1 + h_2 + \Delta h) + \rho_2 h_3)g = (\rho_1 h_1 + \rho_2(h_2 + h_3))g$, or $\Delta h \rho_1 = h_2(\rho_2 - \rho_1)$, where $h_2/\Delta h = \rho_1/\rho_2 - \rho_1$. For example, if $\rho_1 = 1.024$, and $\rho_2 = 1.026$, then $h_2 = 500\Delta h$.

While the surface slope is impossible to measure from a vessel, the thermocline slope is easily measurable. The horizontal pressure gradient in the surface layer generates a geostrophic current perpendicular to the pressure gradient, and thus parallel to the coast [**Figure 1-49, Section I.4.3**]. It is generally more intense than the current produced directly by the wind [**Figure 1-59c**] leading to a composite situation. Near the coast, the current is parallel to the coast with a small perpendicular component producing upwelling. Offshore, the transport in the surface layer is perpendicular to the coast. This is observed along the coasts of SW Africa (Namibia), NW Africa (from Morocco to Senegal), Somalia during the SW monsoon, California, and Peru.

Regions of upwelling are productive because they bring up nutrients from richer subsurface layers. They appear as cold anomalies in surface-temperature patterns [**Figure 1-17**].

5.2.2 Wind at the equator: equatorial upwelling, Equatorial Undercurrent

The equator represents a dynamic boundary in ocean circulation because the Coriolis force changes sign there.

The effect of wind explains the surprising existence of a surface-temperature minimum at the equator in the eastern parts of the Atlantic and Pacific. Under the influence of easterly trade winds, Ekman transport entrains surface waters to the right in the northern hemisphere and to the left in the southern hemisphere. This produces an equatorial surface-current divergence, compensated by colder water upwelling to the surface from below and a rise of the thermocline [**Figure 1-61**].

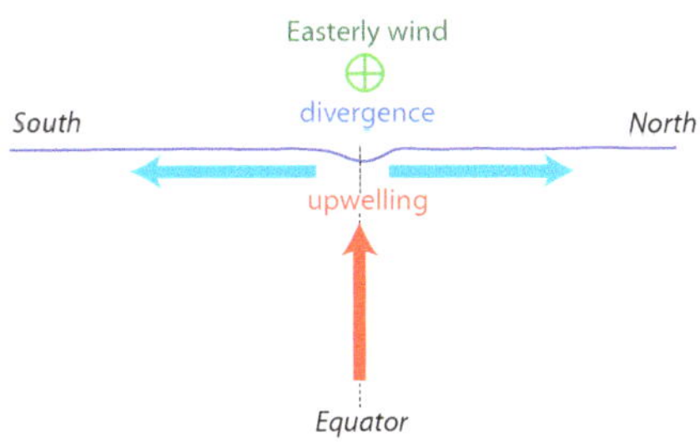

FIG. 1-61 – Diagram of equatorial upwelling generated by easterly winds (SE trades of the Pacific and Atlantic oceans).

At the equator, easterly trades drag and pile up warm surface water toward the west in the Atlantic and Pacific Oceans, thereby establishing a surface slope. In compensation, the surface slope produces an inverse thermocline slope, deeper in the west than in the east [**Figure 1-62**]. Cool water is thus found near the surface in the east, where it can reach the ocean surface.

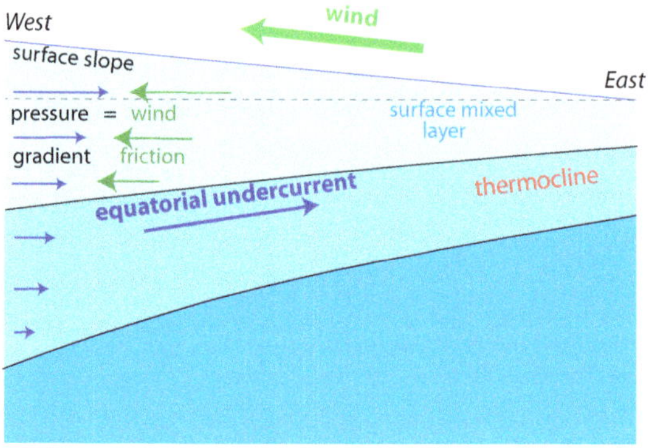

FIG. 1-62 – Structure of an equatorial section in the Atlantic and Pacific. (After *Ocean Circulation*, 1989.)

The pressure gradient generated by the surface slope is balanced by wind stress in the wind-influenced layer [**Figure 1-62**]. Beneath, the pressure gradient, which weakens with depth, but not as fast as the wind effect, is not balanced. It is one of the causes of the existence of an equatorial undercurrent, centered on the equator, travelling eastward, located in the center of the thermocline. Since the Coriolis force goes to zero at the equator, the current flows in the direction of the slope. Other theories explain it in part using the conservation of potential vorticity of a current of constant thickness flowing towards the equator.[80] In the simplified theory of the equatorial undercurrent in a stationary regime (*i.e.*, without acceleration) the pressure gradient is balanced by lateral friction. The Coriolis force begins to act at about half a degree from the equator. That produces, in the presence of a zonal pressure gradient, an equatorward transport which feeds the equatorial undercurrent.

The equatorial Indian Ocean is under the influence of the alternating monsoon regime, during which the wind has a strong meridional component at the equator. Weak winds from the west at the equator exist only in the short periods between monsoons. The equatorial ocean responds rapidly to these westerly winds which generate a current toward the east, or equatorial jet.[81]

[80] See **Section I.5.3.4** and the theoretical work of Pedlosky, 1987 and of Hua *et al.*, 1997.

[81] Also called the Wyrtki jet, from the name of the oceanographer who discovered it in ship-drift logs.

For winds and currents going eastward at the equator, there is equatorial convergence with a tendency to downwelling[82] **[Figure 1-63]**.

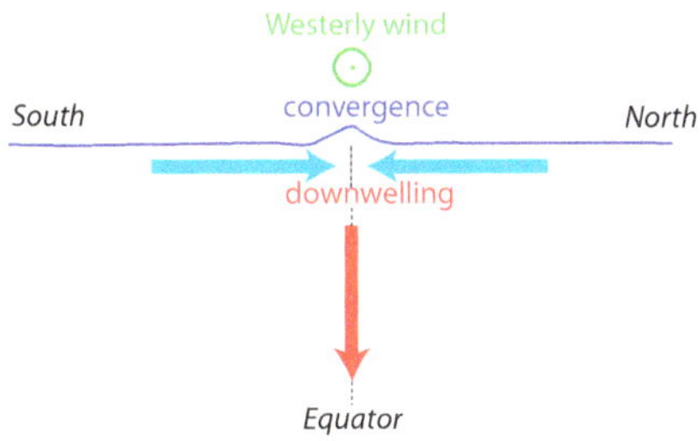

FIG. 1-63 – Schematic of equatorial downwelling generated by westerly winds, an example of the Indian Ocean during the intra-monsoon period.

5.3 Large-scale wind effects

5.3.1 Ekman pumping

Spatial wind variations[83] produce divergences or convergences by means of variations in Ekman transport, which they generate at the surface. These in turn produce vertical motions at the base of the Ekman layer in the ocean interior.

In the case where frictional forces are balanced only by Coriolis force, the equations of motion in terms of the horizontal components are

$$\frac{1}{\rho}\frac{\partial \tau_x}{\partial z} = -fv_e$$

$$\frac{1}{\rho}\frac{\partial \tau_y}{\partial z} = fu_e$$

Where τ_x and τ_y are the surface wind stress components.

By differentiating with respect to x and y, and by assuming that ρ and f are constants:

$$\frac{1}{\rho}\frac{\partial}{\partial z}\left(\frac{\partial \tau_y}{\partial x} - \frac{\partial \tau_x}{\partial y}\right) = f\left(\frac{\partial u}{\partial x} + \frac{\partial v}{\partial y}\right)$$

Using the equation of continuity in an incompressible ocean,

$\frac{\partial u}{\partial x} + \frac{\partial v}{\partial y} + \frac{\partial w}{\partial z} = 0$, we obtain:

$$\frac{1}{\rho}\frac{\partial}{\partial z}\left(\frac{\partial \tau_y}{\partial x} - \frac{\partial \tau_x}{\partial y}\right) = f\left(-\frac{\partial w}{\partial z}\right)$$

82 Downwelling is sinking of water, the opposite of upwelling.

83 Spatial wind variations define the curl of the wind stress (see the next footnote).

By integrating between the surface, where $w = 0$, neglecting the free-surface variations, and the Ekman depth, where the stress, τ, is negligible, the vertical velocity (w_e) at the base of the Ekman layer is:

$$w_e = -\frac{1}{\rho f}\left(\frac{\partial \tau_y}{\partial x} - \frac{\partial \tau_x}{\partial y}\right)$$

So that[84] $w_e = \dfrac{1}{\rho f} \operatorname{curl} \tau$.

The Ekman transport of the surface layer is proportional to the strength of the wind and is directed to the right of the wind in the northern hemisphere and to the left in the southern hemisphere. Suppose that a wind maintains a steady direction but diminishes (or increases) in intensity laterally [**Figure 1-64**]. The Ekman transport associated with the wind will then vary, while keeping the same direction. The variation of the Ekman transport, a function of wind intensity, creates a convergence or a divergence leading to downwelling or upwelling.

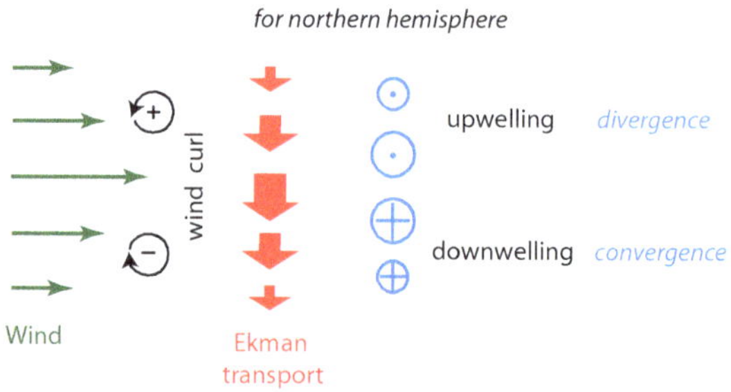

FIG. 1-64 – Divergence or convergence associated with Ekman transport caused by a wind of constant direction and variable velocity, corresponding to a positive or negative wind-stress curl.

On the large scale, with a cyclonic wind pattern,[85] equivalent to positive wind-stress curl in the northern hemisphere [**Figures 1-65a** and **1-66a**], the Ekman transport, directed to the right of the wind in the northern hemisphere and to the left in the southern hemisphere, drives the surface water toward the exterior of the cyclonic circuit (or gyre), creating divergence in the center of the gyre. This in turn creates upward vertical motion, or upwelling ($w_e > 0$) [**Figure 1-65b**]. Because of the divergence, a depression forms at the ocean surface, which is a zone of low pressure since the

[84] The definition of the curl of the wind stress, τ, is $\tau = (\partial t_y/\partial x) - (\partial t_x/\partial y)$, which is the vertical component of the wind-stress curl, the only non-zero component for a horizontal wind.

[85] A cyclonic wind turns counter-clockwise in the northern hemisphere and clockwise in the southern hemisphere.

height of the water is less. This creates a surface slope going from the outer edge of the gyre to the center. This slope generates a pressure gradient between the outer and inner portions of the gyre, bringing about a geostrophic current perpendicular to the pressure gradient and hence parallel to the wind [**Figure 1-65c**]. At depth, in reaction to the surface depression and the upwelling, the thermocline forms a hump to balance the deep pressure [**Figures 1-60** and **1-65b**]. The depth of the thermocline is an oceanic response to wind forcing.

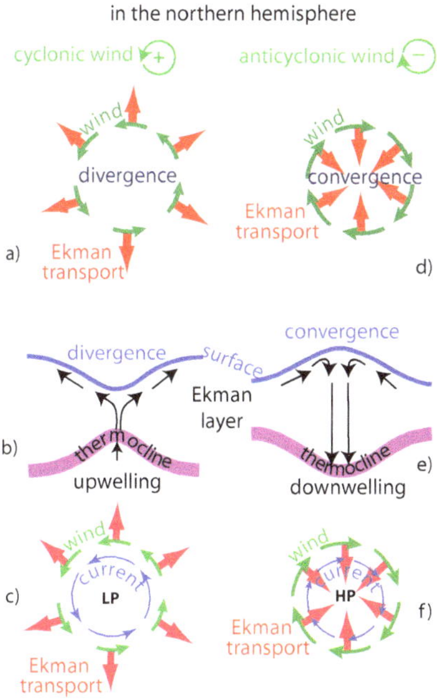

FIG. 1-65 – Ekman transports associated with: (a) cyclonic winds and divergence or (d) anticyclonic winds and convergence; (b) upwelling or (e) downwelling and surface and associated thermocline slopes; (c) and (f) geostrophic currents generated by the resultant pressure gradients. (After *Ocean Circulation*, 1989.)

Under anticyclonic wind circulation[86] (equivalent to negative wind-stress curl in the northern hemisphere) [**Figures 1-65d** and **1-66a**], Ekman transport accumulates surface water in the interior of the anticyclonic gyre, bringing convergence to the gyre center, producing downward vertical motion or downwelling ($w_e < 0$). Because of the convergence, the height of water is higher in the center (it is a hump seen by a satellite altimeter) and is a high-pressure zone [**Figure 1-65e**]. This creates a surface slope away from

[86] An anticyclonic wind turns clockwise in the northern hemisphere and anticlockwise in the southern hemisphere.

the center to the outside of the gyre. This slope generates a pressure gradient between center and the outer edge of the gyre producing a geostrophic current perpendicular to the pressure gradient and thus parallel to the wind [**Figure 1-65f**]. At depth, in reaction to the hump at the surface and to the downwelling, the thermocline forms a depression to balance the deep pressure [**Figures 1-60** and **1-65e**].

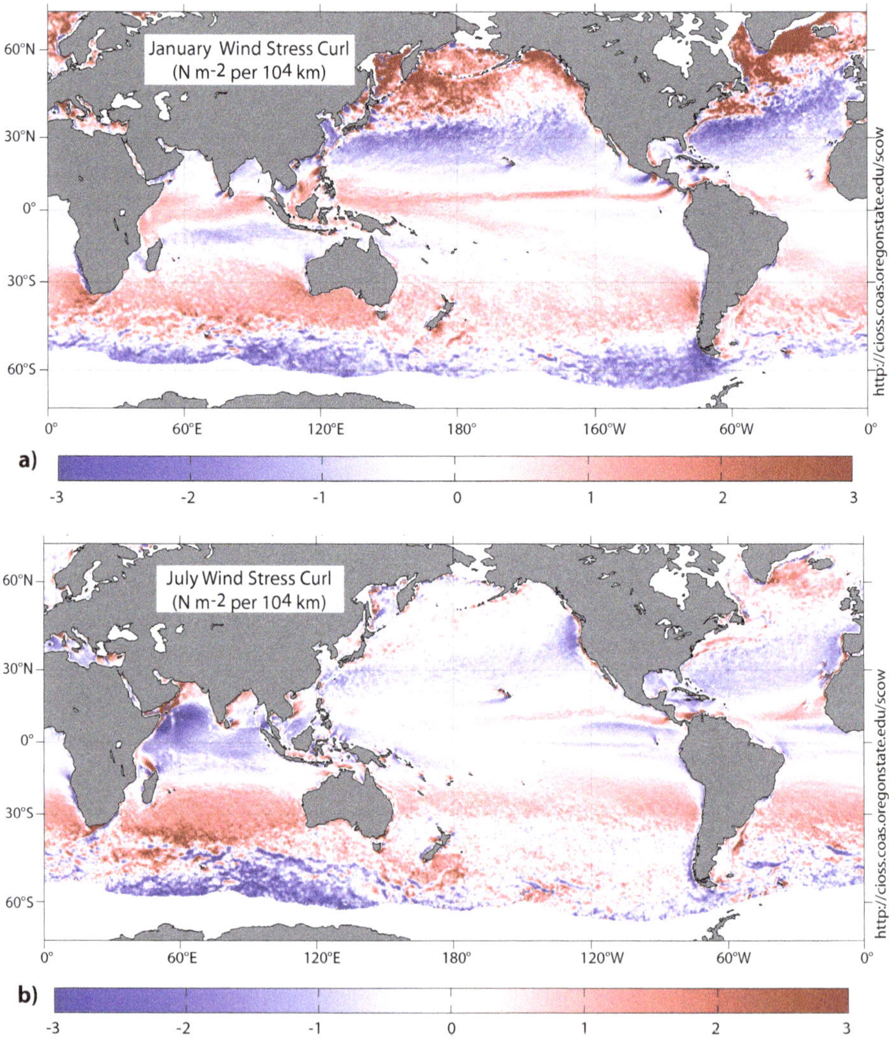

FIG. 1-66A – Wind-stress curl (September 1999–August 2007) from wind measurements by the NASA QuikSCAT scatterometer (Scatterometer Climatology of Ocean Winds (SCOW) Atlas), for **a)** January, **b)** July. Upward Ekman pumping is positive (red) in the Northern Hemisphere and negative (blue) in the Southern Hemisphere. (After Risien and Chelton 2008.) (Available online at: cioss.coas.oregonstate.edu/scow/figures)

In a non-rotating medium, water tends to flow from high to low pressure in order to re-establish balance. On Earth, the Coriolis force comes into play, turning the current to the right in the northern hemisphere and to the left in the southern hemisphere until the pressure gradient and the Coriolis force balance [**Figure 1-47**]. The current then becomes stationary and perpendicular to the pressure gradient, that is, it follows the oceanic isobars[87] [**Figure 1-65**]. Furthermore, there is neither loss nor gain in potential energy. With weak frictional losses, the amount of potential energy accumulated in the center of the large gyres is such that even if the wind stops for months, the current goes on.

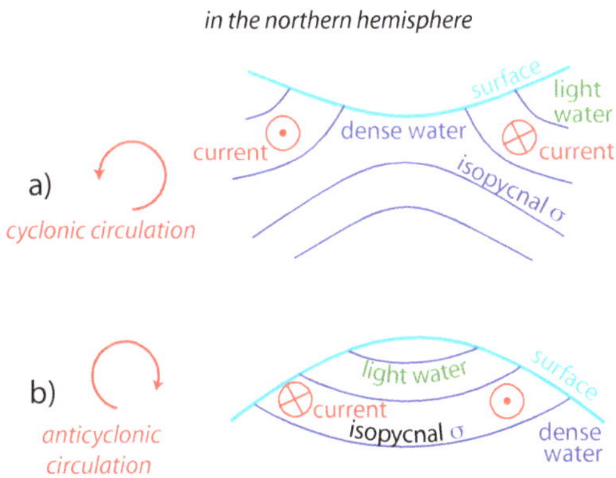

FIG. 1-66B – Density distribution and surface slope in the northern hemisphere: (a) in cyclonic circulation, (b) in anticyclonic circulation. For the southern hemisphere the direction of the current is inversed. Surface slope is greatly exaggerated in this figure.

The pattern of surface density depends on the circulation [**Figure 1-66b**]. In a cyclonic gyre, subsurface water of higher density rises to the center of the gyre, with lighter water sent to the periphery. In an anticyclonic gyre, lighter water is concentrated at the center of the gyre, and higher-density water is at the periphery. The current velocity is a function of the slope of the isopycnals.

The Coriolis force tends to impede the re-establishment of pressure balances since the current turns around a center of pressure (as in the atmosphere). It allows imbalances to continue!

[87] The current flows in the same direction as the wind, which generated the current.

5.3.2 Sverdrup balance

The effect of wind is not confined to the Ekman layer (directly influenced by the wind) but is also reflected in deeper circulation by means of pressure gradients which have effects at depth.

In 1947, Harald Ulrik Sverdrup studied the extension of Ekman's theory to the basin scale. He considered a homogeneous ocean of density ρ_0, separated from the atmosphere by an Ekman layer and took the case of stationarity ($\partial v/\partial t = 0$), frictionless, and without a bottom boundary layer. Starting with simplified equations of motion (a balance between pressure forces and Coriolis force):

$$\frac{1}{\rho}\frac{\partial p}{\partial x} = fv$$

$$\frac{1}{\rho}\frac{\partial p}{\partial y} = -fu$$

By differentiating the first equation by y and the second by x and then subtracting, the term $\partial f/\partial y$ (the variation of the Coriolis parameter with latitude) called β appears.[88] Using the continuity equation ($\partial u/\partial x + \partial v/\partial y + \partial w/\partial z = 0$) and integrating the equation from the ocean bottom (where $w = 0$) to the base of the Ekman layer (where Ekman pumping, w_e, occurs), we obtain:

$$fw_e = \beta \int_{bottom}^{H_e} v dz$$

Since $fw_e = \text{curl}(\tau/\rho_0)$

$$\beta \int_{bottom}^{H_e} v dz = \text{curl}\left(\frac{\tau}{\rho_0}\right)$$

The meridional transport $T_M = \int_{bottom}^{H_e} v dz$ is proportional to the curl of the wind-stress divided by β, the latitudinal variation of the Coriolis parameter:

$T_M = \text{curl}\left(\dfrac{\tau}{\rho_0 \beta}\right)$ is the *Sverdrup relationship.*[89]

For example, anticyclonic winds, whose curl is negative, generate an integrated meridional transport southward in the northern hemisphere, in

[88] A plane where the Coriolis parameter varies linearly with latitude is called a β-plane.

[89] Ocean current transport is expressed in 10^6 m^3 s^{-1} and the name Sverdrup, abbreviated as Sv has been given to this unit of transport.

the negative direction [**Figure 1-67**]. Inversely, cyclonic winds, whose curl is positive, generate an integrated meridional transport northward in the northern hemisphere, in the positive direction. For a given wind stress, the transport increases as latitude decreases. In regions where wind stress curl is zero, there is no meridional transport. The transport there is purely zonal, eastward or westward. These regions are natural boundaries between large wind-driven cyclonic and anticyclonic circulations. For example, the North Atlantic Drift is located between the Azores subtropical anticyclonic circulation and the Icelandic subpolar cyclonic system.

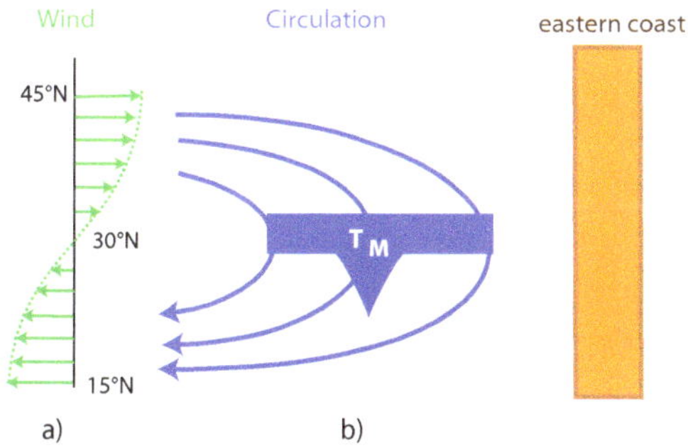

FIG. 1-67 – (a) Diagram of the wind circulation in the Azores anticyclonic system, and (b) associated current paths and meridional transport, T_M. (After Sverdrup, 1942.)

Noting that velocity depends only on x and y and that the ocean is non-divergent, we may write that the integrated horizontal velocity depends on a current function, Φ, such that:

$$\int_{bottom}^{H_e} u dz = -\frac{\partial \Phi}{\partial y}, \text{ and}$$

$$\int_{bottom}^{H_e} v dz = \frac{\partial \Phi}{\partial x}$$

We then get:

$$\beta \int_{bottom}^{H_e} v dz = \beta \frac{\partial \Phi}{\partial x} = \text{curl}\left(\frac{\tau}{\rho_0}\right)$$

The interior meridional ocean transport may be calculated by integrating the curl of the wind stress from the eastern boundary layer (x_{east}):

$$\Phi(x) - \Phi(x_{\text{east}}) = -\int_{x_{\text{east}}}^{x} \text{curl}\left(\frac{\tau}{\rho_0 \beta}\right) dx$$

The Sverdrup relationship applies away from coasts and boundary layers and in regions where ocean stratification is well approximated by a two-layer system consisting of a homogeneous interior ocean, with no bottom boundary layer, and an Ekman layer, at the base of which Ekman pumping, w_e, occurs. The relationship is of great oceanographic value since it permits obtaining vertically integrated transport using only the curl of the wind stress. This view of the wind-driven circulation is effective in explaining circulation in oceanic basins.

Ekman thought that a geostrophic non-divergent flux in the interior should bring about a divergent frictional layer at the ocean floor in order to compensate for convergence in the Ekman layer. With Sverdrup's hypothesis, convergence in the Ekman layer is compensated by divergence of geostrophic flow in the ocean interior.

5.3.3 Western intensification of ocean currents

The large-scale ocean circulation has a lot of resemblance to the atmospheric circulation [**Figure 1-55**]. However, a closer examination of surface currents shows an asymmetry between eastern boundary currents and western boundary currents, which are clearly more intense. However, the wind systems that drive them are relatively symmetric, for example, over the North Atlantic [**Figure 1-68**].

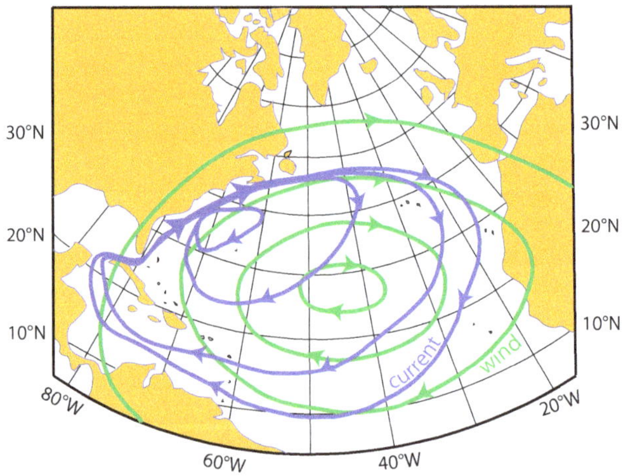

FIG. 1-68 – Diagram of the wind circulation (in green) and currents (in blue) in the North Atlantic, showing the westward offset of the currents (After Stommel, 1972.)

In 1948, Henry Stommel added a bottom linear frictional term to the Sverdrup relationship, which allowed the diffusion of energy at the boundaries of a theoretical basin, which was rectangular, closed, and located in a single hemisphere. As a driving force he used easterly wind stress in the southern part and westerly wind stress in the northern part of the basin.

The result of his calculations showed that the intensification of currents, observed on the western boundary of the basin, is due to the variation of the Coriolis force with latitude [**Figure 1-69**] [**Section I.4.2**].

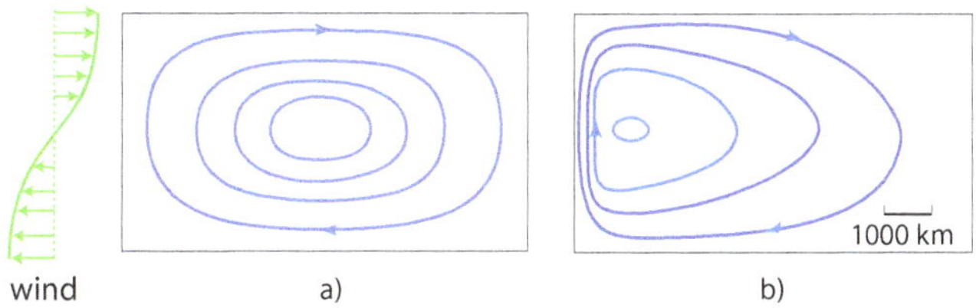

FIG. 1-69 – Stommel's (1948) diagram. It shows the zonal wind component and current paths in a rectangular basin turning (a) with constant f and (b) with f varying linearly with latitude.

Western-boundary intensification notably explains the existence of the Gulf Stream and the Brazil Current in the Atlantic, the Kuroshio and East Australian Current in the Pacific, and the Somali and Agulhas Currents in the Indian Ocean [**Figure 1-56**].

In 1951, Walter Munk combined the results of Ekman, Sverdrup, and Stommel with a more realistic wind distribution and a boundary layer of lateral friction to obtain a more general solution of wind-driven ocean circulation [**Figure 1-70**]. To summarize, in the stable solution, the geostrophic current, defined by a balance between Coriolis force and horizontal pressure gradients, is retarded by the force of lateral friction and reinforced by energy transmitted by surface winds.

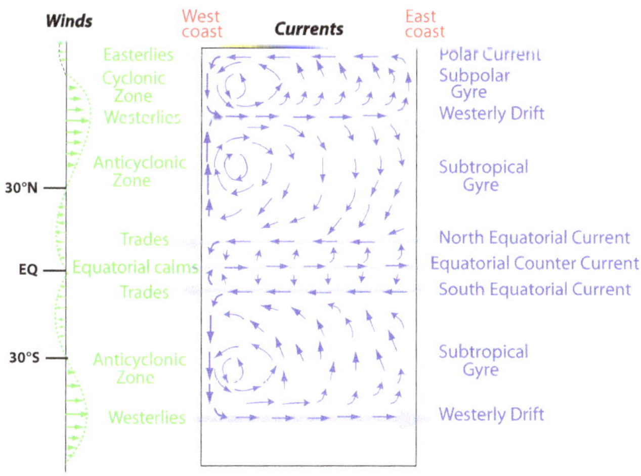

FIG. 1-70 – Munk's (1950) generalized diagram showing the zonal wind component and associated oceanic circulation.

The resulting circulation pattern qualitatively approximates the observed general circulation [**Figure 1-56**]. The equatorial current system serves as a sort of backbone for the ocean circulation accompanied, on both sides, by large subtropical anticyclonic gyres north and south and, at higher latitudes, by subpolar cyclonic gyres. The westward intensification can be seen for each ocean. But Munk's theory does not explain all the characteristics common to western boundary currents. They are narrow, rapid, follow the western continental slope of the ocean basin and extend in depth well beyond the thermocline. When leaving the continental slope eastward, they become unstable and produce many meanders and eddies. Their characteristics result from basin topography and regional wind patterns.

5.3.4 Conservation of potential vorticity

Vorticity is a term in mathematical language dealing with rotation. It allows us to treat ocean circulation differently than with the classical equations of motion. The conservation of *potential vorticity* of a fluid parcel, that is, the conservation of its angular momentum in the absence of friction provides a simple explanation of the westward intensification of ocean currents.

Several types of vorticity can be defined:

Planetary vorticity, f, (the Coriolis parameter) due to Earth's rotation about its axis:

$$f = 2\Omega \sin \phi$$

Relative vorticity, ζ, due to rotational motion in a terrestrial frame, is equal to the curl of the velocity in the horizontal plane:

$$\zeta = \text{curl}\, V = \frac{\partial v}{\partial x} - \frac{\partial u}{\partial y}$$

Absolute vorticity, defined with respect to a celestial frame, is the sum of the potential vorticity and the relative vorticity:

$$= \zeta + f$$

Potential vorticity is the absolute vorticity divided by the thickness, H, of the fluid layer involved:

$$= \frac{(\zeta + f)}{H}$$

Let's take the example of the interior of the subtropical anticyclonic gyre in the North Atlantic.

1. Between the westerly winds in mid-latitudes and the easterly trades in tropical latitudes the winds on average have a negative curl (they turn clockwise) and impart a negative relative vorticity to the surface waters in the area.

2. Furthermore, if a layer of water, whose thickness H is assumed constant, is displaced from the polar region (where planetary vorticity, f, is maximal) toward the equator (where $f = 0$) it will acquire a positive relative vorticity because of its equatorward motion, in order to conserve its absolute vorticity (and inversely).

In the example of **Figure 1-71** a subtropical anticyclonic gyre has constant H, a homogenous medium, and a plane bottom. Boundary friction, proportional to current velocity, generates an increase in positive relative vorticity, both on the eastern boundary with the current flowing equatorward, and on the western boundary, with the current flowing to higher latitudes.[90]

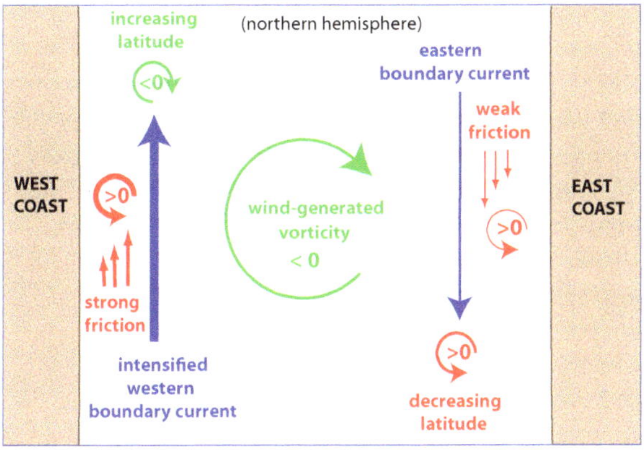

FIG. 1-71 – Vorticity balance on the eastern and western boundaries of an anticyclonic gyre. Negative vorticity is imparted to the ocean by the wind. In the west, the current must be intensified so that the positive vorticity acquired through friction can balance that transmitted by the wind and that acquired through latitude change. (After *Ocean Circulation*, 1989.)

In the eastern interior part of the basin the negative relative vorticity imparted by the wind is easily balanced by the positive planetary vorticity acquired by the equatorward motion, following Sverdrup's relationship. On the other hand, in the west, the negative relative vorticity imparted by the wind adds to the negative planetary vorticity due to poleward motion. The sum of these two negative velocity terms can only be balanced by a positive relative vorticity imparted by friction. Thus currents in the west must be more intense than those in the east (or in the center) of the basin in order to acquire enough positive relative vorticity through friction to balance the two sources of negative vorticity.

90 Mass conservation imposes a return current of equivalent size.

By the simple conservation of potential vorticity, we expect an intensification of the current at the western boundary of basins, as is observed in the Gulf Stream, the Kuroshio, etc. The center of the anticyclonic oceanic gyre is deformed toward the west with respect to the anticyclonic atmospheric circulation (the mirror is a bit twisted!). Energy acquired from the wind is dissipated largely in the form of friction on the western boundary of ocean basins.

Away from the basin boundaries, currents and current shear are generally weak, and the relative vorticity, ζ, is small with respect to the planetary vorticity, f. The potential vorticity is thus nearly f/H and currents follow contours of f/H.

In regions of water-mass formation where density is constant over a depth H_1, relatively greater than in neighboring regions (where H_2 is less) the water mass is characterized by a minimum of potential vorticity ($f/H_1 < f/H_2$). This characteristic of homogeneous waters (called Mode Waters) is a true tracer that allows following a water mass far from its place of formation.

Conservation of potential vorticity, $(\zeta + f)/H$, of a water mass is maintained when its thickness H decreases either by a latitude change, which diminishes f (a displacement equatorward), or by a decrease in its relative vorticity ζ (angular velocity decreases). For example, a current is deflected toward the equator when the ocean floor rises and toward the pole when the ocean floor deepens. An example can be seen in the flow of the Antarctic Circumpolar Current over the Scotia Arc [**Section II.5.2.5**].

Conversely, a meridional movement (latitude change) of a water mass brings about a change in its thickness, H, or of its relative vorticity, ζ.

In this section, we have described the characteristics of the ocean medium and the principal processes involved in exchanges with the atmosphere. Before describing the role of each ocean as an element of the planetary ocean, we next present some ocean observational techniques.

Chapter 6
Observational techniques

Even if, as in every science, research into physical laws and modeling are necessary in order to advance understanding of processes, oceanography is, above all, an experimental science. Each advance in techniques allows us to arrive at new understanding.

The ocean is an immense medium, difficult to access and opaque. Only acoustic waves propagate easily. Measuring instruments must be specially designed to resist physical and chemical attacks resulting from long months of immersion. Without being exhaustive, examples of classical in-situ physical measurements for temperature, salinity, pressure, and current are presented below, though some have become historic. They illustrate the acquisition of the fundamental measurements that are the basis of this text.

6.1 Temperature and salinity measurements

6.1.1 Reversing thermometers

Until the 19th century, the ocean depths remained an immense mystery. Keeping a deep temperature measurement made with a simple mercury thermometer was impossible. A French scientist, George Aimé, in 1845 described the principle of reversing thermometers. The mercury column is composed of a glass reservoir followed by a column of lesser diameter [**Figures 1-72** and **1-73**]. The mercury, shrinking or expanding according to the water temperature, fills the column more or less high. When the thermometer is reversed at the measurement depth, a volume of mercury proportional to the ambient temperature separates off at a narrow point on a curve at the base of the column near the reservoir.[91] The column is graduated to allow

[91] According to this principle, Alphonse Milne-Edwards, professor at the Muséum National d'Histoire Naturelle created a thermometer constructed by Dumaige. In 1875, Negretti, and Zambra produced thermometers of the same type used at the end of the British *Challenger* Expedition in the Atlantic (1873–1876). The German Carl Richter, with the help of the Dane Fridtjof Nansen, perfected them.

reading the temperature (uncorrected for warming due to the temperature at the time of reading).

There are two types of reversing thermometers: pressure-protected[92] and unprotected [**Figure 1-72**].[93] The difference in temperature between the two thermometers is a function of pressure and indicates the depth[94] at which the thermometers are reversed. An auxiliary thermometer, less precise, placed in the interior of the tube of each thermometer gives the temperature at the moment of reading (after stabilization) and allows correcting the reversing (called the principal) thermometer according to the ambient temperature. In general, two thermometers, one protected and one unprotected are placed at each measurement depth [**Figure 1-73**].

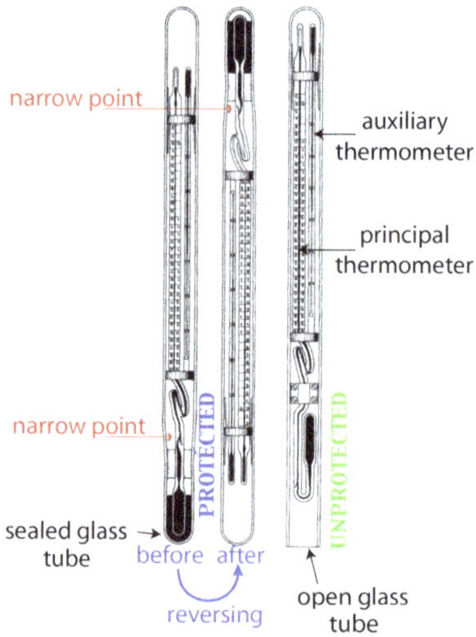

FIG. 1-72 – Richter reversing thermometers. The two on the left are pressure-protected, with a watertight exterior glass tube and are shown in before-and-after reversal positions. On the right, is an unprotected thermometer, with an open end of the exterior tube. Each has an auxiliary thermometer. (After Lacombe, 1965.)

[92] Using this type of thermometer, specially constructed for the range of temperatures in the Mediterranean in winter, a good experimentalist, equipped with a magnifier, was able to read temperatures to a hundredth of a degree!

[93] Reversing thermometers were still used in the 1980's.

[94] Depth is determined within about 10–20 m. This measurement was less precise than that now given by pressure sensors, but was better than using the length of cable paid out.

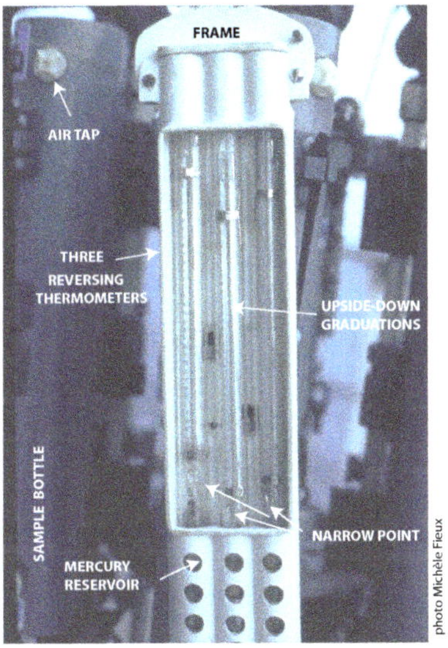

FIG. 1-73 – Three reversing thermometers in their frame in position at the beginning of a station. Two are protected and one is unprotected.

The advent of digital reversing thermometers in the 1980's has contributed to the disappearance of these mercury thermometers.

Technological advances have provided simpler and automated measurement methods. Temperature sensors use thermistors, whose resistance varies with temperature, and are used in many on-board (such as hull-mounted thermal probes...) or over-the-side systems [**Section I.6.2.2**].

6.1.2 Bathythermograph

At Carl Gustav Rossby's suggestion, Athelstan Spilhaus designed the bathythermograph at MIT in 1938 to obtain temperature profiles from a ship underway [**Figure 1-74**]. As the manufacture by the Boston Company was slow, Columbus Iselin assigned Maurice Ewing and Allyn Vine to redesign a newer bathythermograph, and about 200 were made in WHOI shops. This device has been most helpful for undersea transmission methods, particularly for submarine warfare. This 20-kg brass instrument provided a temperature profile[95] up to 275 m depth from a ship underway. [**Figure 1-75**]. With surface calibration by bucket thermometer, precision of the order of 0.1 °C was enough to find the thermocline depth. Because of weight it required a winch to be deployed.

[95] The temperature profile was scratched by a stylus on a glass plate, smoked beforehand with a candle. Once constructed, the instrument cost nothing, except for candles and matches!

FIG. 1-74 – Bathythermograph aboard the research vessel *Jean Charcot* in 1968. (Photo Catherine Maillard.)

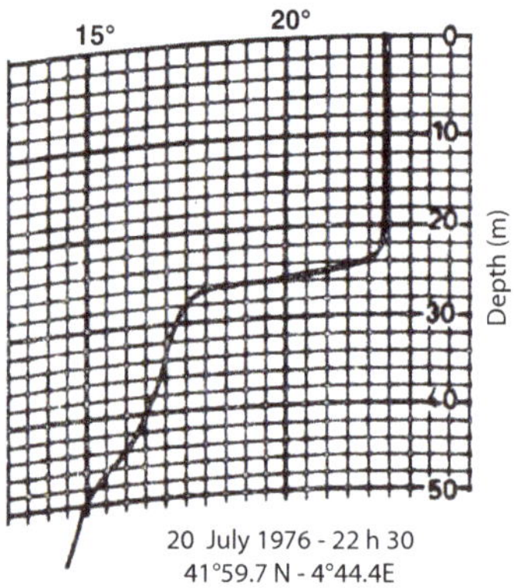

FIG. 1-75 – Example of a bathythermograph trace taken in July, 1976 from the *Bouée Laboratoire*, *BOHRA*, anchored in the Gulf of Lions. The 22-m deep mixed layer is at a temperature of 22.5 °C.

6.1.3 Expendable bathythermograph, or XBT

The bathythermograph was replaced during the 1970's by the XBT (e**X**pendable **B**athy**T**hermograph), a non-recoverable [**Figures 1-76** and **1-77**] thermal probe usable from an underway ship. A thermistor is placed in a heavy metallic head, connected by a fine copper wire wound on a spool attached to the head on one end and on the other end on a spool attached to the base of the XBT that stays in the launching cylinder fixed on board (it can be portable). When the pin is removed, the head descends vertically. The wire in the water unwinds from the head and the part of the wire in the air unwinds at the speed of the ship after launching. The wire is connected to a data acquisition system, which directly tracks temperature with time, converted to depth, and records the launch position. At the end of the drop, the wire breaks and the sensor is lost. Several types of XBT exist, with depth ranges varying between 200 m and 1,800 m. Some acquisition systems send data directly by satellite to international data collection centers that update meteorological prediction models in real time.

FIG. 1-76 – XBT launching from the research vessel *Marion Dufresne.* (Photo Michèle Fieux.)

XCTD probes with temperature and conductivity sensors appeared in the 1980's. They turned out to be more expensive, and lacked precision. They are thus less used than XBTs.

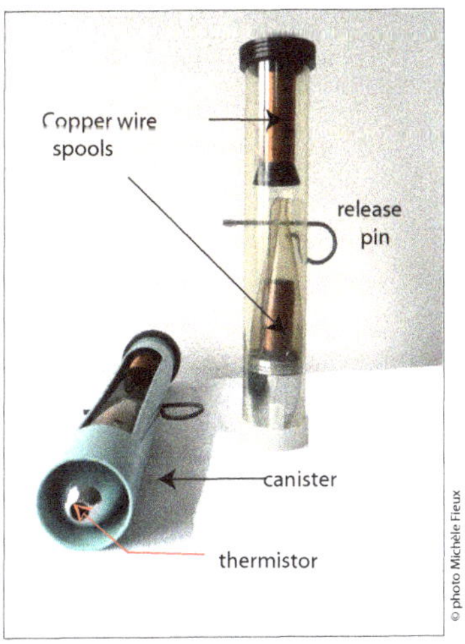

FIG. 1-77 – Expendable temperature-probe Sippican XBT. (Photo Michèle Fieux.)

6.1.4 Salinometer

Mohr and Knudsen developed the technique of seawater salinity measurement at the beginning of the last century on the basis of chlorinity titration [**Section I.1.3**].

This technique was abandoned in favor of measuring electrical conductivity, which varies with the amount of dissolved salts, temperature, and pressure. The salinometer[96] is a shipboard instrument comprised of a cell with platinum electrodes (or with magnetic induction) placed in a temperature-stabilized water bath. It allows measuring the ratio between the conductivity of a seawater sample placed in the cell, and the conductivity of standard seawater[97] whose salinity is known. Tables and algorithms give salinity as a function of the measured ratio, and the temperature at which the measurement was made. On board, it allows verification of values given by conductivity sensors installed on CTD probes [**Section I.6.2.2**].

6.1.5 Thermosalinograph

Most ships are equipped with temperature sensors at the engine-cooling-water input. Some ships have a thermosalinograph system, which continuously measures surface-water temperature and salinity. Samples taken regularly from the circuit that feeds the thermosalinograph, allow after-the-fact calibration of the salinity measurements.

6.2 The hydrographic station and its measurements

During oceanographic expeditions the ship must be stopped to launch instruments to measure seawater characteristics on a vertical profile. We say that the ship is *on station* and speak of the *hydro station* to refer to the temperature and salinity values collected vertically. In bad weather, when the ship experiences strong winds, or when the surface and deep current are very different, it is sometimes difficult to "hold the station".[98]

---

[96] The first salinometers were designed in the 1930's on the same principle.

[97] Standard seawater distributed in sealed ampoules was prepared in Copenhagen. It is now distributed by OSIL at Havant in Great Britain.

[98] Station holding keeps the cable vertical so that it does not touch the hull, in spite of the ship's drift.

6.2.1 Knudsen, Nansen, and Niskin sample bottles

In order to collect seawater samples, Knudsen developed a reversing water-sample bottle, equipped with closing valves at each end holding a frame for thermometers. Usually a dozen of such sample bottles, of 2 to 4-liter volume, are attached to a ballasted cable in order to have a sampling distribution along the vertical. The collection of samples represents a *cast*. Once the bottles have reached the desired depths,[99] the oceanographer waits about ten minutes to allow the thermometers to equilibrate, then drops a 'messenger' down the cable to trip the reversal of the first bottle, which in turn then releases another messenger which goes on to release the next bottle down, and so on until the bottom bottle is reached [**Figure 1-78**]. At the moment of sampling, the thermometers are reversed and the end valves of the bottle close, capturing a water sample at the depth. After waiting the time necessary for the messenger to reach the bottom bottle, the cast is brought back up, removing each bottle in turn. To obtain more than 12 sampling levels, a second cast is made, and sometimes a third, to get 36 sampling levels.

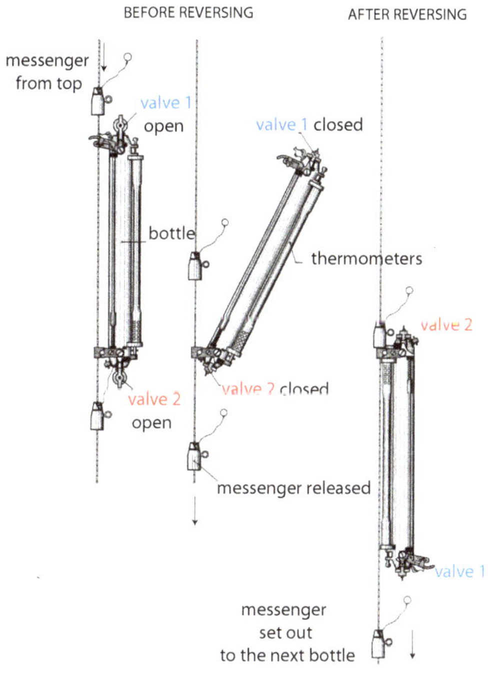

FIG. 1-78 – Diagram of a Knudsen bottle and messenger during reversal. (After Lacombe, 1965.)

[99] The depth is estimated roughly by the length and angle of the cable.

Nansen modified the bottle so that only the thermometer frame is reversed at the moment when the valves closed at the two ends [**Figure 1-79**]. The messenger system remained the same. Later, when CTD probes appeared [**Section I.6.2.2**] the bottles were adapted to a rosette system, but functioned on the same principle (with large valves above and below and a thermometer frame which reversed at the moment of closing). These Niskin bottles could have different volumes according to the need for water.

FIG. 1-79 – Attaching a Nansen bottle to the cable. The closure valves are black and the thermometer frame faces the observer. The photo was taken during a station with a strong Mistral, in winter aboard the research vessel *Jean Charcot*. (Photo Michèle Fieux.)

6.2.2 CTD probe and rosette

During the 1960's, temperature, conductivity, and pressure[100] sensors were developed and installed on a CTD (conductivity, temperature, depth) probe. The CTD probe consists of acquisition electronics placed in

[100] The pressure sensor uses the pressure effect on mechanical deformation, converted to a measurable amount of electricity. This type of sensor has greatly improved on the classical methods of determining measurement depth. Measuring depth using

a pressure-protected case connected to a variety of sensors: temperature, conductivity, pressure, oxygen concentration, (and more frequently now) fluorometer, turbidity meter, and transmissometer). In the early 1970's continuous profiles were possible for the first time, from the surface down to a few meters above the sea floor. The CTD probe is attached to the end of an electric cable [**Figures 1-80, 1-81,** and **1-83**] which can transmit power down and receive sensor signals back up to obtain profiles in real time.

FIG. 1-80 – Preparation of a Seabird CTD probe and sampling bottles in a protective framework before launching. On the right is the pinger, which will be attached below the probe to safely approach the bottom. On the left are cases containing various types of sampling bottles. (Photo Michèle Fieux.)

To obtain water samples at different depths, sample bottles[101] are placed around the CTD probe on a rosette composed of 6, 12, 24, or even 36[102] bottles of various volumes: 4, 8, or 12 liters.[103] Until the beginning of the

the length of cable paid out is fraught with error since the cable is never vertical. Pressure measurement using the temperature difference between protected and unprotected thermometers is not precise. Going from pressure to depth is done using the relation $dp = \rho g dz$, where ρ is the average seawater density along the vertical.

[101] Niskin bottles have caps above and below and are able to accept a thermometer frame.

[102] With a 12-bottle rosette, to obtain more than 12 sampling levels, several casts are made at the same station.

[103] For some measurements that require large-volume filtering (for example to determine the concentration of radioactive elements) special bottles are placed alone on a cable.

21st century, frames for holding reversing thermometers [**Figures 1-80** to **1-84**] were attached to calibrate the CTD temperature sensor. Nowadays temperature sensors have a precision better than a thousandth of a °C, and mercury thermometers are no longer used. Temperature and conductivity[104] sensors are paired to control their eventual drift. Otherwise, sensor calibration is done in a metrology laboratory before and after a cruise. The rosette bottle closing is activated during the instrument ascent,[105] at depths chosen with the aid of profiles obtained during the descent. The collection is often equipped either with a *pinger*[106] or with an *altimeter*[107] to get close to the bottom without touching it [**Figures 1-80** and **1-84**].

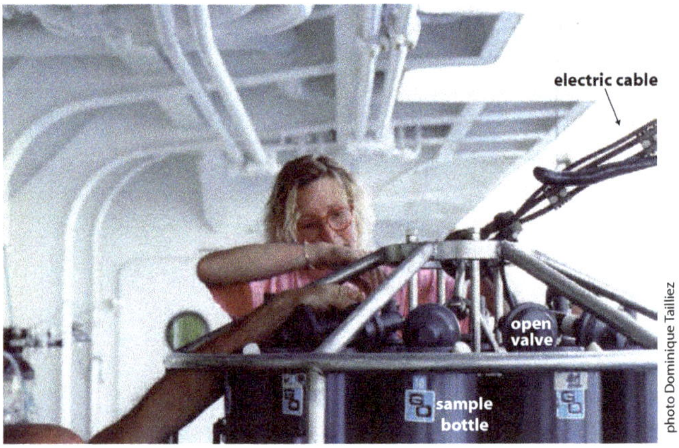

FIG. 1-81 – Setting up a rosette system before making a station. The caps on the sampling bottles are open. (Photo Dominique Tailliez.)

[104] Glass-cell sensors use platinum electrodes, though some still use less-precise induction-cell sensors.

[105] Bottles are not closed during the descent to avoid bottle implosion under the effect of pressure.

[106] The pinger sends out an acoustic impulse. The time difference between the acoustic signal directly received from the pinger and the signal reflected off the bottom gives the distance between the CTD probe and the bottom. This instrument allows measurements very close to the bottom, even over a rough bottom.

[107] The altimeter is a miniature sonar with a limited range (for example 50 m). At 50 m from the bottom, it detects the bottom and its signal (acquired by an auxiliary sensor of the CTD probe) is converted to distance in meters from the bottom by the CTD probe software. It is generally linked to an alarm.

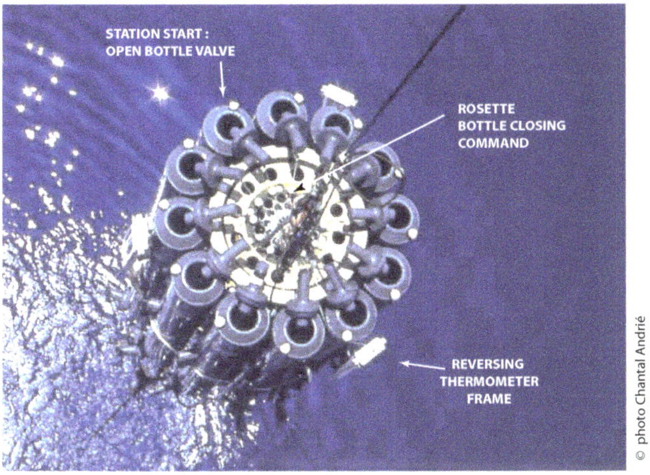

FIG. 1-82 – Launch of a Neil Brown CTD probe with a 12-bottle rosette above. (JADE 1989 cruise, on the research vessel *Marion Dufresne.*) (Photo Chantal Andrié.)

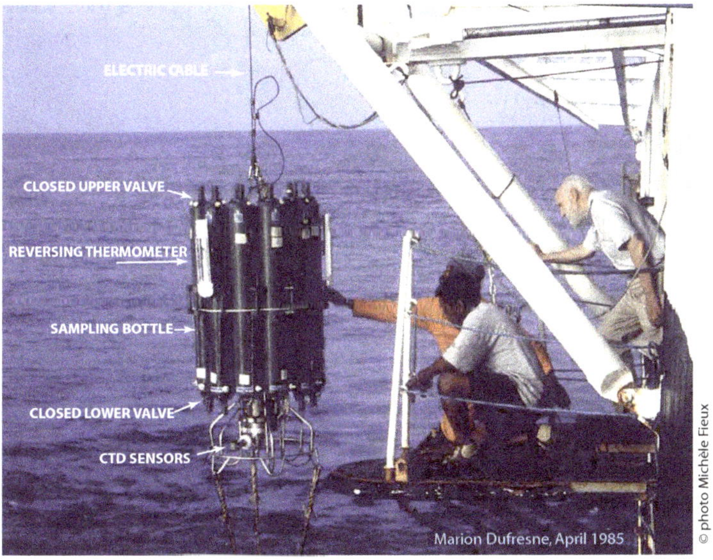

FIG. 1-83 – Recovery of a Neil Brown CTD probe equipped with an 8-sampling-bottle rosette under the careful eyes of John Swallow. The thermometer frames are white. (SINODE cruise 1985 in the Indian Ocean on the research vessel *Marion Dufresne.*) (Photo Michèle Fieux.)

FIG. 1-84 – A 32-bottle rosette with thermometer frames (white) and digital thermometers. The CTD probe (left) and the altimeter (right) are attached to the lower part of the apparatus. (CITHER-I 1993 cruise, on the research vessel *Atalante*.) (Photo Chantal Andrié.)

Seawater samples [**Figures 1-85** and **1-86**] allow calibration of salinity and oxygen concentration[108] measurements given by the CTD probe. They also provide samples for measuring the concentration of a number of other components, either on board ship: CFC's, carbon dioxide, carbonates, nitrates, phosphates, silicates, biological parameters... or in on-shore laboratories: carbon-14, tritium, rare earths...

An oceanographic ship is often a real laboratory where analyses are just as precise as those made on shore. Large international experiments such as *GEOSECS* (**GE**ochemical **O**cean **SEC**tion **S**tudy, 1973–1976), *TOGA* (Tropical Ocean Global Atmosphere, 1980's), *WOCE* (World Ocean Circulation Experiment, 1990's), and a part of *CLIVAR* (**CLI**mate **VAR**iability) were based primarily on measurements made from oceanographic research ships with the help of a CTD probe connected to a sampling rosette.

[108] The Winkler chemical method (1888) is still used (with some improvements) and remains the most precise method to calibrate electrochemical sensors.

FIG. 1-85 – Drawing water into a syringe to measure CFC concentration. (JADE 1989 cruise, on the research vessel *Marion Dufresne*.) (Photo Michèle Fieux.)

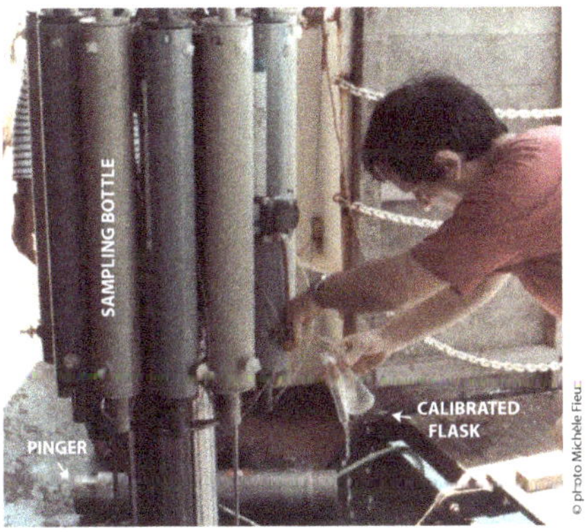

FIG. 1-86 – Drawing water into a flask specially calibrated to measure dissolved oxygen. (SINODE 1985 cruise, on the research vessel *Marion Dufresne*.) (Photo Michèle Fieux.)

6.3 Direct current measurements

6.3.1 Current meters

At the beginning of the 20th century, Ekman invented an ingenious mechanical current meter usable for a few hours of observation, which was lowered overboard at the end of a cable. It was equipped with a vane, which oriented it into the axis of the current and a propeller whose rotation depended on current speed. At each turn of the propeller a little metal ball fell into a round box compartmentalized into 8 sectors. A compass oriented the box. Upon recovery, the balls were counted to obtain an integration of the average current speed and direction.

There followed a number of other current meters designed to meet the need to tolerate immersion in seawater during many months[109] and to record the data. They had also to survive attacks by sharks, attracted by cable vibrations. The classic type is a rotor current meter equipped with a vane and compass. It measures current at a single fixed point in the ocean, an *Eulerian* measurement [**Figure 1-87**]. Some take a series of measurements at programmed time intervals. Others average the current direction and speed[110] over a specific time determined in advance. The measurements were first recorded on magnetic tapes for later data processing, but solid-state devices have superseded tapes.

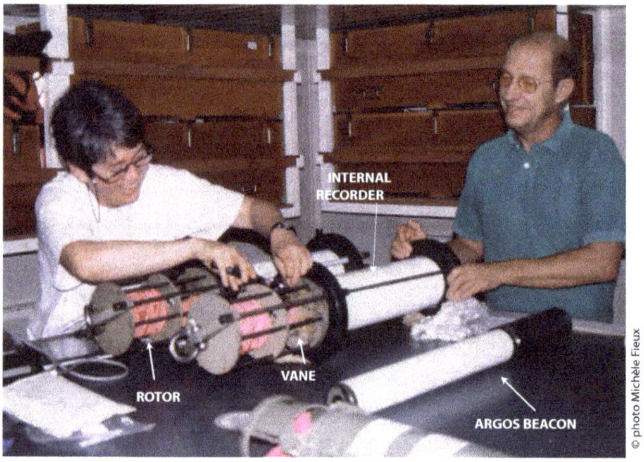

FIG. 1-87 – Preparation of a rotor-and-vane Mors current meter. (JADE 1995 cruise on the research vessel *Baruna Jaya IV.*)

The sampling rate is set by the duration of measurements, power limits (lithium batteries in general are used), and instrument memory capacity. Under some conditions, long-term measurements of one or two years are possible. Instruments of this type have problems in near-surface layers and in warm water. The proliferation of marine organisms can clog up a rotor within a few months. Acoustic and Doppler-effect current meters [**Figure 1-88**] are also more and more used (discussed below) to avoid this problem. Installed on a mooring line, they measure current at a specific location.

[109] To protect the current meter case from electrolytic attack in seawater, zinc or zinc-magnesium anodes are attached depending on the material to be protected.
[110] VACM, Vector Averaging Current Meter.

FIG. 1-88 – Doppler-effect current meter. (Photo Igor Bashmachnikov.)

6.3.2 Acoustic Doppler current profilers

The Acoustic Doppler Current Profiler (ADCP) marks a great step forward [**Figures 1-89** to **1-91**]. These instruments contain transmitters and receivers with frequencies (150 kHz to 75 kHz) chosen depending on the range of the instrument (300 to 500 m). An emitted signal is reflected back by particles carried along by the velocity of the current. The Doppler effect, a function of current velocity, modifies the return frequency received by the instrument.

FIG. 1-89 – Attaching the yoke to connect an ADCP Doppler profiler, showing four transponders and the flotation sphere (orange). The block on the right is the mooring anchor. (JADE 1995 cruise on the research vessel *Baruna Jaya*.) (Photo Claudie Marec.)

FIG. 1-90 – ADCP (placed at the head of the mooring) shown on deck in its flotation (orange) equipped with an Argos beacon along with the mooring anchor, before launching into the sea with the aid of the stern A-frame. (JADE 92 cruise aboard the research vessel *Marion Dufresne*.) (photo Robert Molcard.)

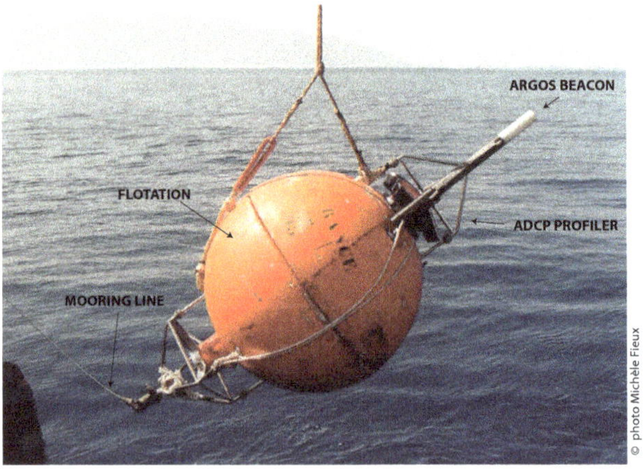

FIG. 1-91 – An ADCP profiler launch. (JADE cruise aboard the research vessel *Baruna Jaya I*.) (Photo Michèle Fieux.)

Placed on a mooring, at a depth that depends on the range of the instrument and the layer to be sampled, transmitting upward (or downward), the ADCP allows measurement of a current profile in the sample layer and effectively replaces several separate current meters. The water column must include particles in order for the return signal to be strong enough. In general, when the range of the ADCP increases, its vertical resolution decreases.[111]

A more compact type of ADCP (L-ADCP or Lowered ADCP) can support pressure up to 6,000 m. It is used on station, installed in the body of the rosette CTD-probe ensemble. A current profile is obtained simultaneously with the CDT-probe [**Figure 1-92**].

FIG. 1-92 – L-ADCP (yellow) inside the frame of a rosette-CTD probe. (EQUALANT 1999 cruise aboard the research vessel *Thalassa*.) (Photo Chantal Andrié.)

Another type of ADCP is frequently installed on the hull of research vessels in order to obtain a current profile along the ship's path.[112] They use frequencies from 150 kHz to 38 kHz, with a consequent range of 300 m to about 1,000 m.

[111] For example, for a frequency of 75 kHz, the range is about 500 m and the vertical resolution is about 16 m. For a frequency of 150 kHz, the range is about 300 m and the resolution is about 8 m.

[112] The data are only usable from 20 to 30 m depth and the quality of the measurements is limited by the turbulence of the medium.

6.3.3 Moored current meters

A mooring with current meters is composed of a cable on which are interspersed current meters, measuring at a point or profiling [**Figures 1-93** to **1-95**]. Flotation elements [**Figure 1-96**] bring instruments up to the surface at the moment of recovery, and in case the cable should be accidentally cut. Generally, a Doppler profiler (ADCP) is at the head of the mooring, with a beacon permitting it to be localized by satellite if, by bad luck, the mooring is found drifting.[113] The cable is fixed to the bottom by an anchor below an acoustic release [**Figure 1-97**]. The acoustic signal allows searchers to find the mooring by interrogating it from a ship. At the end of the experiment, the release mechanism may be fired by a signal from the ship, thus detaching the cable from the anchor to recover the collection of instruments at the surface.[114]

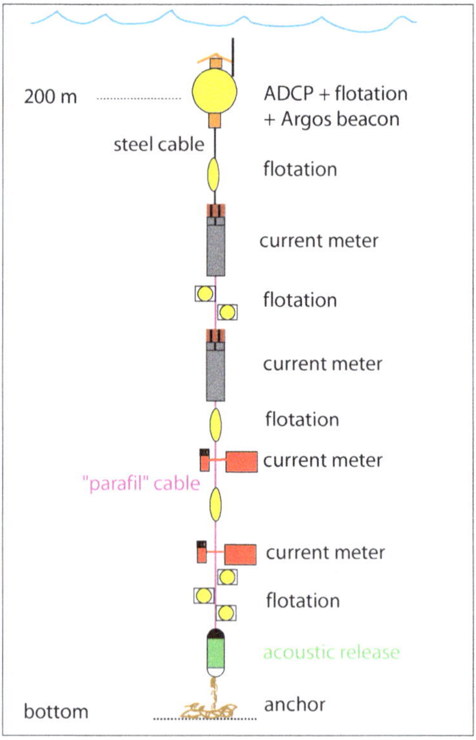

FIG. 1-93 – Schematic of a current meter mooring.

[113] Fishing trawlers sometimes cut the mooring line.

[114] It sometimes happens that fishing long-lines get wrapped around the release, preventing the instruments from returning to the surface at recovery, in which case dragging is necessary.

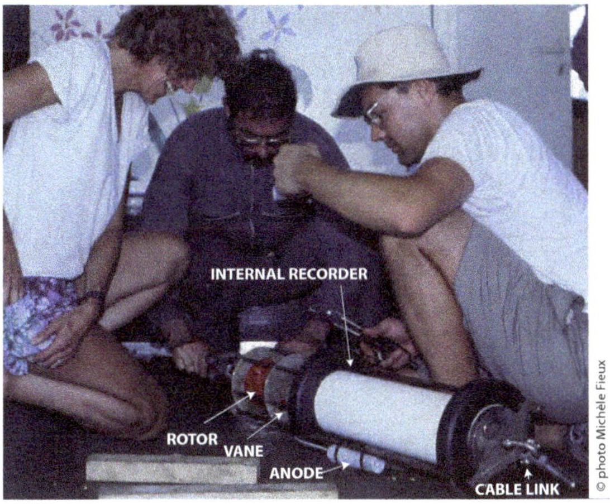

FIG. 1-94 – Installing a Mors current meter with rotor and small vane in the mooring line. (JADE 89 cruise aboard the research vessel *Marion Dufresne.*) (Photo Michèle Fieux.)

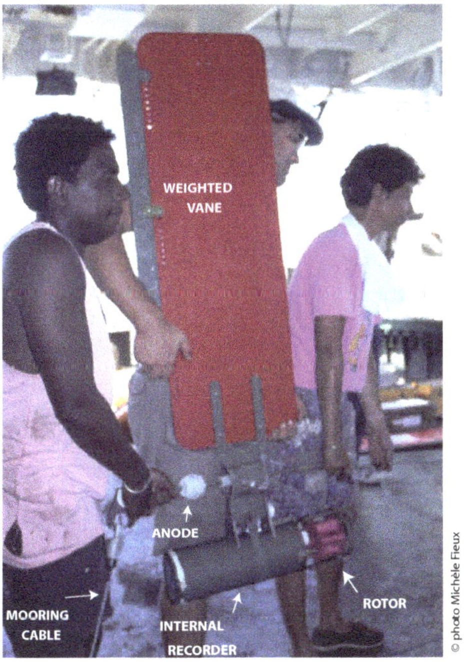

FIG. 1-95 – Launching an Aanderaa current meter, showing the red rotor and large red vane, which orients the instrument into the current. (JADE 89 cruise on the research vessel *Marion Dufresne.*) (Photo Michèle Fieux.)

Launching a mooring is done by first putting the top unit and its flotation into the water, then all the other instruments with their flotation [**Figure 1-96**] separated by the chosen cable lengths, then the release [**Figure 1-97**], finishing with the anchor. When installing a mooring, in 4,000 to 5,000 m depth, the entire mooring with all the instruments associated with it is pulled along at the surface behind the ship (4 to 5 km). The ship at this point has little maneuverability. The current and the time needed for the launching must be well estimated in order to drop the anchor at the planned location. In order not to damage fragile instruments parts during the anchor drop, a parachute is often attached to slow the descent [**Figure 1-98**]. Either a steel cable covered with plastic to protect it from corrosion and/or *parafil*, a synthetic fiber weighing nearly zero in water but more fragile, is used according to the characteristics and location of the mooring.[115]

FIG. 1-96 – Installing flotation in the mooring line. (JADE 95 cruise on the research vessel *Baruna Jaya IV*.) (Photo Michèle Fieux.)

[115] Attracted no doubt by vibrations, sharks' teeth have often been found on steel cables, and rotors have been gashed!

FIG. 1-97 – Attaching the acoustic release in the mooring line. (JADE cruise on the research vessel *Baruna Jaya I.*) (Photo Michèle Fieux.)

FIG. 1-98 – The parachute on the anchor slows its fall during launch. (SINODE 85 cruise on the research vessel *Marion Dufresne.*) (Photo Michèle Fieux.)

Before the arrival of satellite location by GPS (Global Positioning System) in the 1990's, position was taken with respect to the stars at the moment of launching the anchor, in order to be able to return as nearly as possible to the mooring point at the end of the experiment.[116] The GPS has revolutionized the positioning problem.

[116] Moorings are most often subsurface to avoid possible piracy.

An acoustic command is sent to the release at recovery to open its hook and thus to release the anchor so that the instruments rise to the surface with help of the flotation. They are sometimes tangled [**Figure 1-99**] and covered with algae and shellfish, especially those that were at near-surface depths.

FIG. 1-99 – Return to the surface of a mooring with instruments and their flotation. (JADE 90 cruise on the research vessel *Baruna Jaya*.) (Photo Michèle Fieux.)

6.3.4 Surface-moored buoy

Using large surface buoys fixed to deep moorings [**Figure 1-100**], a variety of measurements are continuously carried out at the surface (air temperature, wind velocity and direction, air humidity, precipitation, and solar radiation) and in the water, (temperature, salinity, and pressure) between 0 and 500 m. Data are transmitted in real time *via* satellite by the *Argos* system.

FIG. 1-100 – Launching an Atlas (Autonomous Temperature Line Acquisition System) buoy from the research vessel *Thalassa* during the EQUALANT 99 cruise. (Photo Chantal Andrié.)

6.4 Drifting buoys, floats, profilers, gliders

6.4.1 Drifting buoys

A variety of floats, termed *lagrangian* tracers,[117] have been used to follow surface circulation. Some are rudimentary, such as floating cards [**Figure 1-101**], which were launched in numerous regions by various navies. They were to be returned indicating where the card was found. Others were tracers of opportunity as, for example, the recognizable remains of ships crushed by ice in the Arctic and recovered west of Greenland or tennis shoes lost from a container ship in mid-Pacific! [**Section V.4.3.2**]

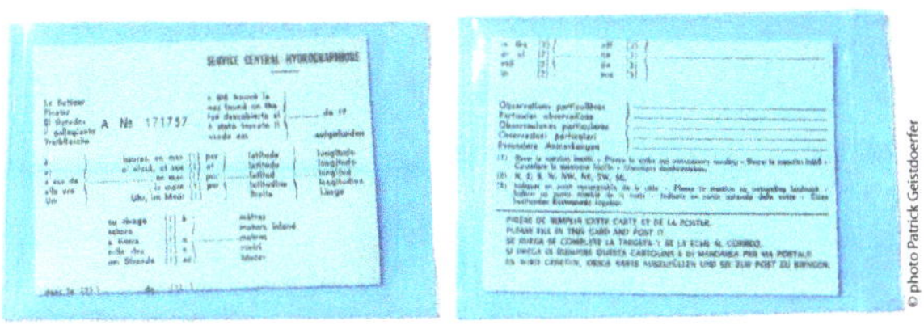

FIG. 1-101 – Floatable card in sealed plastic. The finder is asked to return it, giving the position where it was found. (Photo Patrick Geistdoerfer.)

[117] Lagrangian measurements follow the water mass. Eulerian measurements are made at a fixed point.

With the advent of satellites, trajectories of drifting buoys [**Figures 1-102** and **1-103**] fitted with beacons could be followed to estimate currents. These buoys are often equipped with water-temperature, air-temperature, and atmospheric-pressure sensors. They may be linked to a floating anchor generally composed of a large ballasted canvas (called a drogue) tying them more solidly to the surface layer and diminishing the effect of surface wind on the buoy as much as possible.

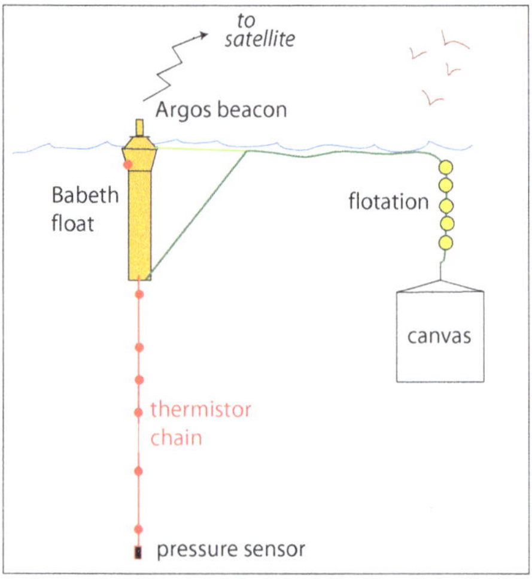

FIG. 1-102 – Diagram of a drifting buoy (Babeth) equipped with a chain of thermistors. A drogue is connected to the buoy by a floating cable in order to limit the tension on the attachment point during heavy weather.

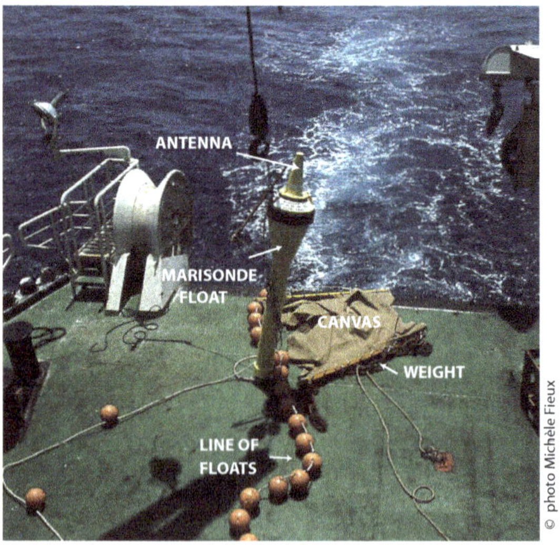

FIG. 1-103 – Drifting buoy (Marisonde) and its drogue. (SINODE cruise in the Indian Ocean on the research vessel *Marion Dufresne*.) (Photo Michèle Fieux.)

6.4.2 Swallow floats and SOFAR floats

In the 1960's, John Swallow invented a drifting float that could be calibrated to stabilize at a determined depth and be followed acoustically.

FIG. 1-104 – First trials by John Swallow, preparing his subsurface float on the *RRS Discovery II* in 1955, under the watchful eye of the ship's cat. (© National Oceanography Center, Southampton, U.K.)

Following this principle, SOFAR (**SO**und **F**ixing **A**nd **R**anging) floats have been used since the mid-seventies to measure deep currents. These floats send an acoustic signal (260 Hz) to listening stations moored on the sea floor. The time taken by the signal to arrive at three (or more) listening stations allows the float to be located and its trajectory to be tracked from data recorded by these stations, recovered at the end of the experiment.[118] The floats can be followed at distances up to 2,500 km.

Since these floats are costly, the technique has been reversed. The stations installed on the sea floor are now acoustic sources and the floats, called RAFOS (SOFAR backwards) are only much less expensive listening devices. For example, the floats *Marvor* and *Provor* of IFREMER, used a lot in the Atlantic, send their data by satellite when they regularly arrive at the surface.

[118] Generally the floats are not recovered, since the ship-time needed would cost more than the price of a float.

6.4.3 Profiling floats

Since the end of the 20th century, programmed profiling floats descend to a chosen drifting depth where they wait until they begin their measurements. At regular intervals they descend to another chosen depth then return to the surface from where they send their data by satellite.

The international *Argo*[119] program, begun in 2000, uses autonomous floats programmed to descend to a drifting depth of 1,000 m and to make profiles from 2,000 m to the surface every 10 days [**Figures 1-105** and **1-106**].

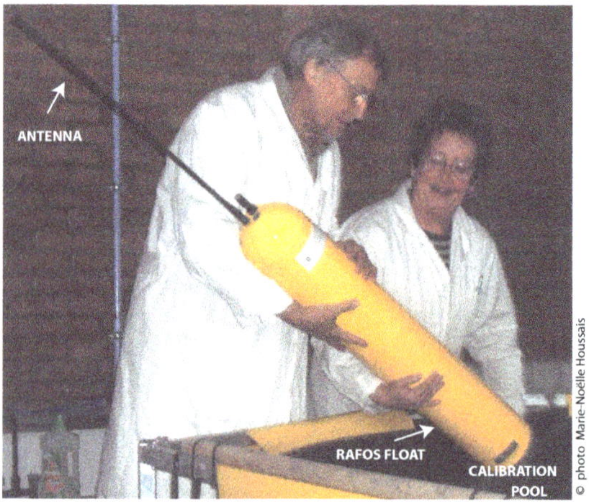

FIG. 1-105 – Laboratory calibration of a Profiling float (RAFOS) used in the Argo program. (Photo Marie-Noëlle Houssais.)

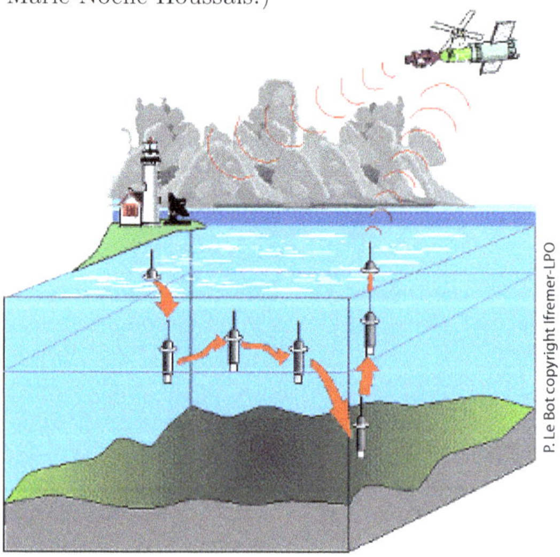

FIG. 1-106 – Operation of a profiling Argo float. (Courtesy of P. Le Bot, © Ifremer-LPO.)

[119] Array of Real-time Geostrophic Oceanography.

During their upward return, they measure temperature, salinity, and pressure for which they send profiles by satellite to a data treatment center upon arrival at the surface. They stay surfaced for several hours then re-descend to their waiting depth. The purpose of the program is to sample the entirety of the global ocean, between 0 m and 2,000 m. In 2007, seven years after the start of the program by IOC and WMO, the aim of 3,000 *Argo* floats was reached. At the end of 2015, the entire ocean was sprinkled with more than 3,900 *Argo* floats [**Figure 1-107**]. These measurements will considerably extend the set of deep observations,[120] impossible to obtain by satellite only.

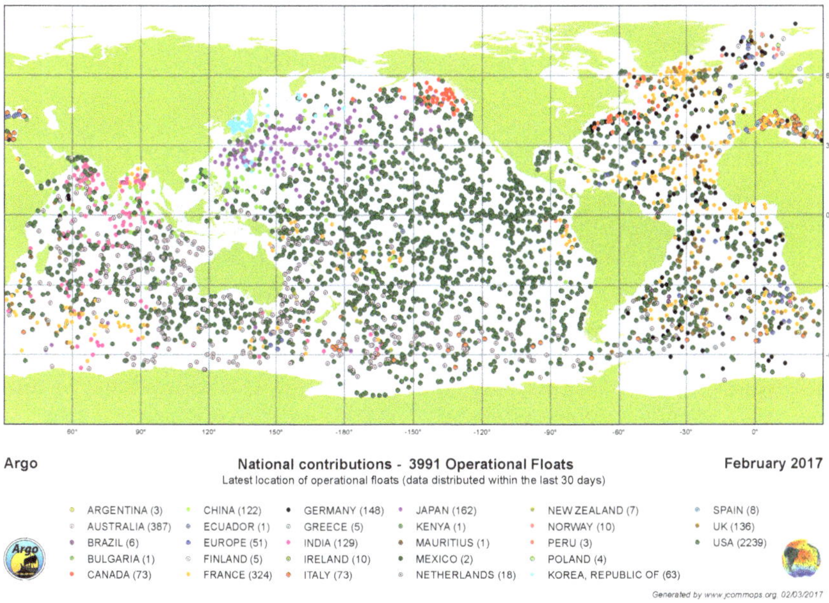

FIG. 1-107 – Distribution of floats in the international Argo program in February 2017. (From Argo Information Center, available online at: www.jcommops.org.)

6.4.4 Gliders

The latest arrival among the autonomous floats is the glider. In 1989, Henry Stommel imagined a revolutionary technique of underwater gliders that he called *The Slocum Mission*.[121] These little robotic autonomous underwater vehicles have been developed and in use since around 2000 [**Figures 1-108** and **1-109**]. They are undersea gliders that use wings and Archimedes' principle to achieve buoyancy-based propulsion. Pumps control

[120] Long-term calibration of the instruments will be needed.

[121] The glider is named after Joshua Slocum, who made the first solo circumnavigation of the globe by sailboat.

flotation and produce slow rise and descent in the water column (up to 1,500 m depth) as well as horizontal motion, requiring little energy.

They can be autonomous for up to several months. Equipped with a GPS that gives their position when they are on the surface, they transmit their extensive data (temperature, salinity, oxygen, chlorophyll, etc.) by satellite. They are telecommanded and while gliding can explore different layers in a region. They differ from *Argo* floats, which make only vertical profiles and are passive floats at their waiting depth.

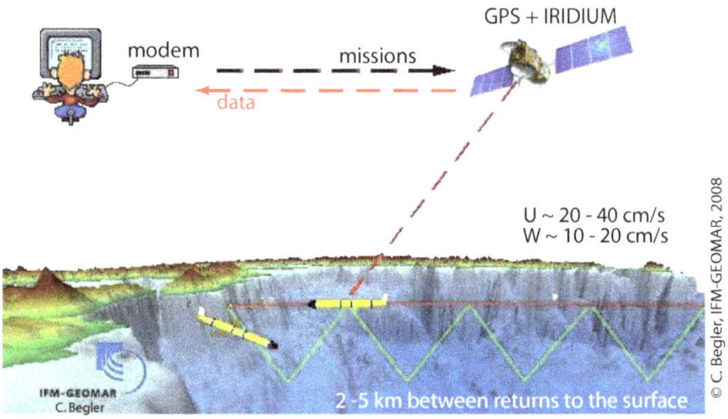

FIG. 1-108 – Schematic of the glider principle. (After C. Begler, IFM-GEOMAR, 2008.)

FIG. 1-109 – A glider on the surface. (Courtesy of Pierre Testor, LOCEAN/ENSTA.)

6.4.5 Animal-borne instruments: How elephant seals can help exploring the ocean

In the early 2000's, with the miniaturizing of satellite beacon and sensors, researchers had the idea of gluing loggers on marine-mammal heads (such as elephant seals) to get information on their way of life as well as on their watery environment [**Figure 1-110a**]. They managed to equip them with conductivity-temperature-depth sensors. The animals thus collect temperature and salinity profiles along their dives which are transmitted through satellite-relay data loggers when they pop up at the surface.

FIG. 1-110A – Female southern elephant seal (mirounga leonina) instrumented in the Kerguelen Islands. (Photo credit: Guillaume Girard.)

This kind of study was started in the Antarctic Ocean from a number of subantarctic islands, such as the Kerguelen Islands (49°20S–70°20E) and along the Antarctic coast, but also in northern regions [**Figure 1-110b**]. Thanks to these animals, information is gathered over large remote areas where it would be extremely difficult to obtain otherwise due to sea ice cover and heavy weather!

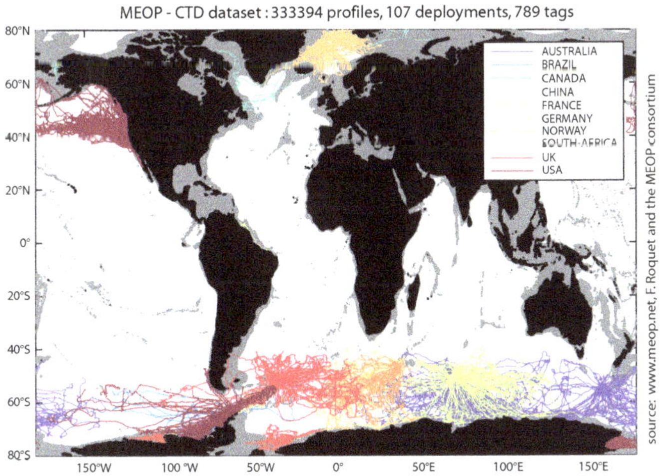

FIG. 1-110B – World map showing the distribution of CTD profiles (*i.e.* vertical profiles of temperature and salinity) currently available in the MEOP-CTD database. (Source: www.meop.net, F. Roquet and the MEOP consortium.)

6.5 Satellite measurements

The advent of satellites was a true revolution in Earth-observation methods. For the first time a view of the entirety of our planet was possible. Oceanography has especially benefitted.

The first satellite sensors for oceanographic purposes were infrared sensors giving the surface temperature, knowing that emitted energy is a function of the fourth power of temperature [**Section I.2.1.4**]. Their precision, though poor at first, was quickly improved. The first time that satellites observed a strong increase in surface temperature in the equatorial Pacific, ships were quickly sent to verify its accuracy. This was the first El Niño event observed by satellite. It demonstrated the large size of the phenomenon that had been thought of as localized off the Peruvian coast!

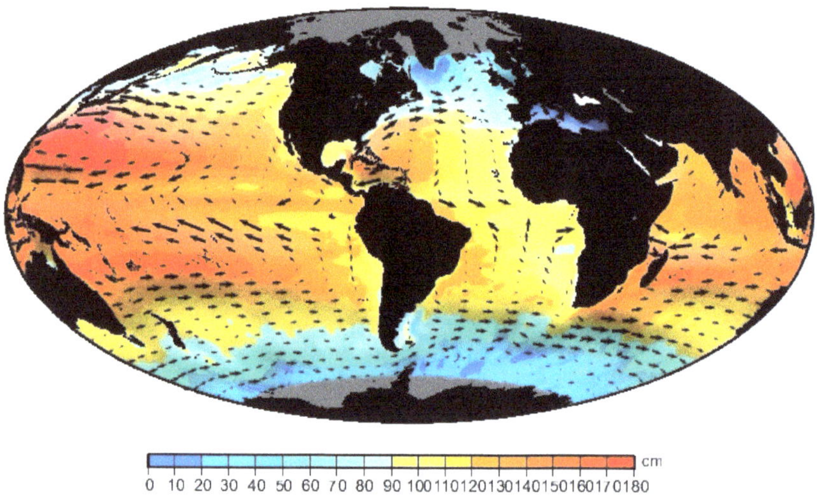

FIG. 1-111 – Average ocean-surface dynamic topography, *i.e.*, the oceanic relief corresponding to the average oceanic circulation. The arrows are proportional to current speed. (After CNES/CLS, 2008.)

Subsequently, a satellite altimeter was installed, which emitted a radar signal and analyzed it after reflection from the surface. The first altimeter, *Seasat*, operated for three months in 1978 and demonstrated useful applications. Then *Geosat* was put in orbit in 1985. The altimeter *Topex-Poséidon* (a cooperative US-French program) was launched in 1992 to "observe and understand ocean circulation". The height of the sea is equal to the difference between the satellite-surface distance (deduced from the round-trip time of the signal) and the position of the satellite with respect to a reference ellipsoid, defined with the help of a high-precision positioning system. The measurement of the anomalies with respect to the *geoid*[122] provides the sea-surface slope, which is directly proportional to ocean currents [**Section I.4.3**]. The surface slope due to currents is very slight, of the order of 10 cm over 10 km in a current like the Gulf Stream. Thanks to the precision of a few centimeters obtained by the *Topex-Poséidon* altimeter, ocean currents can be deduced [**Figure 1-111**]. *Topex-Poséidon* was followed by *Jason I* launched in 2001, and *Jason II* launched in 2008, assuring a continuity of measurement, indispensable for the study of climatic variations.

The analysis of the radar echo also allows the extraction of wave height and wind speed.

Nowadays, there are all kinds of satellite sensors, giving indirect measurements of a variety of parameters (sea color, water vapor, salinity...). These are calculated from signals using algorithms that contain a number of corrections and whose use is always delicate.

Satellite measurements have the great advantage of giving a global synoptic image, but give only surface information, since unlike the atmosphere water only allows penetration of acoustic signals. However, satellites can collect measurements at depth carried out by other instruments (such as *Argo* floats, gliders, animal-borne instruments...), and transmitted from the surface.

This section on observational techniques is far from exhaustive. It gives only a brief overview of past and present techniques. In spite of the immense progress gained from the use of satellites and autonomous devices, specialized ships are still needed.

The research vessel *Le Suroit* in Brest, 1983. (Photo IFREMER, all rights reserved.)

[122] The geoid is a geopotential surface where a plumb bob is everywhere perpendicular.

PART II
The Antarctic (or Austral) Ocean

The Dumont d'Urville scientific base in Adélie Land (Photo Yves Gladu – IPEV.)

Chapter 1
Introduction

The Greek philosophers knew the Earth was a sphere and thought that to "balance" the existing large continental mass north of the equator, there should be an analogous mass in the south. They named the region *Antarktos* (Antarctica) as opposed to *Arktos*[123] (Arctic).

For a long time there appeared a vast continent called *unknown land according to Ptolemy* on maps inspired by the geography of Claude Ptolemy.[124] At the end of the 15th century the mythical continent, Antarctic, was still represented attached to Africa. In 1487, Bartolomeu Dias involuntarily sailed past the Cape of Good Hope during a storm, and in 1520 Fernando De Magellan found a passage to the Pacific south of South America. Cartographers were thus forced to reduce the extent of this continent as shown in the planisphere of 1641 [**Figure 2-1**].

FIG. 2-1 – Planisphere from Novus Atlas, Nova territorii francofurtensis tabula, of Joan and Cornelis Blaeu, tome 1, Amsterdam, 1641 (From *Cartes géographiques anciennes*, Kupcik, 1981).

[123] Arktos is Greek for bear, from the Bear constellation, Ursa Minor, where the pole star indicates the north.

[124] Ptolemy was a Greek astronomer of the second century AD. His *Geography* was taken as authoritative during the Middle Ages and the Renaissance.

In 1616, when the Dutch sailors Willem Van Schouten and Jacob Lemaire went around Cape Horn, without seeing any land to the south, the limits of *Terra Australis Incognita* as written on maps, were pushed farther southward. In 1652 the Dutchman Abel Tasman was charged with exploring the *Great Southern Land* (now Australia) of which only the north and west coasts were known.[125] He discovered Tasmania without seeing Australia and touched northern New Zealand thinking that it was part of the austral continent. The map of 1657 [**Figure 2-2**] shows *Terra Australis Incognita* with some scattered tracks probably corresponding to the limit of pack ice seen in the mist.

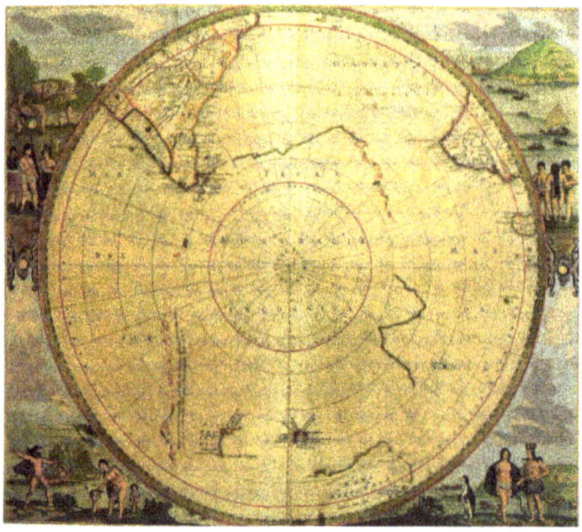

FIG. 2-2 – Map of the Antarctic region by H. Hondius, 1657 (After *Le grand défi des Pôles*, Imbert et Lorius, 2006).

In 1700, the English astronomer Edmund Halley led the first scientific expedition to measure the magnetic declination and to explore the coasts of the *Terra Incognita*. He was blocked by ice at 52°S in the southern Atlantic. During his first voyage (1768–1771) the Englishman James Cook on board the *Endeavour* left Tahiti and sailed to 40°S without finding land. Upon turning westward he discovered the southern island of New Zealand and showed that it was composed of two islands, without connection with the Austral continent. His second voyage (1772–1775) on the *Resolution* and the *Adventure* had the resolution of the *Great South Land* as its principal objective. They crossed the zone of strong westerlies, passed the polar circle and were dazzled by the beauty of the white and blue landscape before their eyes, in the form of pack ice and of "mountains of ice" (icebergs),[126] but they

[125] Longitude was difficult to determine since chronometers allowing 'time-keeping' on board didn't yet exist. The Englishman John Harrison built the first functioning chronometer in 1770.

[126] Iceberg is a term borrowed from Norwegian.

saw no land on the horizon before being blocked by pack ice [**Figure 2-3**]. The two expedition astronomers, William Wales and William Bayly, discovered warmer water under the Antarctic surface water. In 1819–1821, the Russian Fabian Gottlieb von Bellingshausen circled Antarctica without being able to land. He reached 69° 25′ latitude, less than 30 nautical miles[127] from the coast. In 1821, the English seal hunter George Powell also noted that warmer water was found beneath the surface (he found similar conditions in the Arctic). In 1823, another seal hunter, James Weddell, succeeded in reaching 74° 14′ S, in the sea which now has his name, continuing to measure temperatures until all his thermometers were broken!

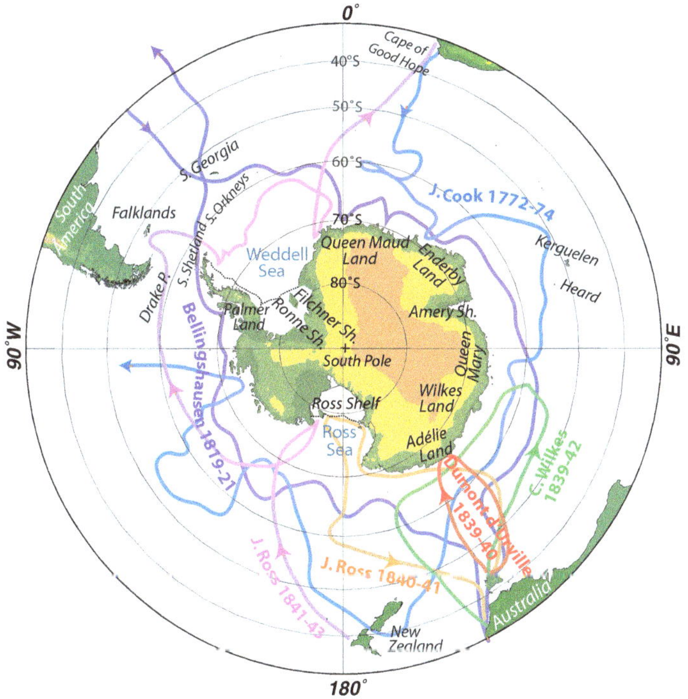

FIG. 2-3 – Tracks of the first Antarctic exploration expeditions in the 18th and 19th centuries. The average extent of sea ice is shown in white. For the topography, see **Figure 2-4**. (After Imbert & Lorius, 2006.)

France, Britain, and the United States simultaneously became interested in this unknown continent and sent out exploring expeditions including scientific programs. The French expeditions aboard the *Astrolabe* and the

[127] A nautical mile equals one minute of latitude or 1,852 m. One degree of latitude equals 60 minutes of latitude, or 60 nautical miles, about 110 km. A knot is one nautical mile per hour or 0.51 m/s or 1.85 km/h.

Zélée, led by Jules Dumont d'Urville, began in 1837. In 1840 he succeeded in reaching Adélie[128] Land, south of Tasmania [**Figure 2-3**] and was the first to set foot on the Antarctic continent. From 1838 to 1842 American expeditions led by Charles Wilkes explored the region that carries his name, west of Adélie land. From 1839 to 1843 British expeditions led by James Clark Ross[129] explored the region east of Adélie Land and the sea afterwards called Ross.

Following the great British *Challenger* expedition from 1872 to 1876, which marked the beginning of deep ocean measurements,[130] interest in Antarctica was revived at the end of the 19th and beginning of the 20th century. Belgium sent the *Belgica* in 1898, Sweden the *Antarctica* (crushed by the ice in 1903 in the Bellingshausen Sea), Scotland the *Scotia* in 1902–1904, Germany the *Gauss* in 1901–1903. The Frenchman Jean Charcot, aboard the *Français* in 1903–1905 and the *Pourquoi Pas* in 1908–1910, explored the western peninsula of Antarctica, wintering there twice. In spite of still-rudimentary technical means, the observations showed a layer of warm water inserted between cold deep water and surface water, close to the Antarctic continent.

Oceanographic knowledge of the Antarctic Ocean is relatively recent because of the immense difficulty of working in violent winds and in the presence of sea ice. In 1911–1912, during the German expedition of Wilhelm Filchner, aboard the *Deutschland*, Wilhelm Brenneke, using reversing thermometers and Knudsens's method of salinity measurement [**Section 1.1.3**], offered the first picture of the oceanographic structure of this ocean. In 1921, aboard the *Gauss*, Wilhelm Meinardus first described the downwelling of Antarctic Intermediate Water. The British, charged with providing physical and biological observations to allow the establishment of international regulations for hunting large cetaceans, crisscrossed the Austral Ocean from 1925 to 1939 on the *Discovery*, the *William Scoresby*, and then the *Discovery II*. In the course of five expeditions over 18 months, 2,500 observation stations were taken. Sir George Deacon, who participated in several of these cruises, made an exhaustive study of them. The Germans Alfred Merz, Georg Wüst, and Gerhard Schott made numerous observations in the Atlantic sector, during the great *Meteor* expedition of 1925–1927. The Norwegians provided their contribution in the Atlantic sector from 1927 to 1930, on the *Norvegia* and in the Pacific sector from 1947 to 1948 on the *Brategg*. In the Indian sector, the Australians carried out oceanographic measurements from 1929 to 1931 on the *Discovery*, and the French, from 1948 to 1950 on the *Commandant Charcot*.

---

[128] The name of his wife.

[129] He was the nephew of Sir John Ross who explored northern parts of the American continent.

[130] In the 19th century, the installation of submarine telephone cables between Europe and America provided the need for deep ocean studies.

During and since the International Geophysical Year in 1957–1958, many oceanographic observations were made by ships providing resupply and relief to scientific bases on the Antarctic continent and on the islands of the Austral Ocean. In 1962, the National Science Foundation and the Lamont Doherty Geological Observatory aboard the *Eltanin* under the direction of Arnold Gordon carried out a systematic study of this ocean. With the arrival of large international programs beginning in 1980 (such as WOCE: World Ocean Circulation Experiment, CLIVAR: CLImate VARiability and Predictability, GODAE: Global Ocean Data Assimilation Experiment, ARGO:[131] Array of Real-time Geostrophic Oceanography) the accumulated oceanographic observations provide a better description of this distant and difficult-to-reach ocean.

The supply ship *Astrolabe*, at Dumont d'Urville, a scientific base in Adélie Land, 2009. (Photo Claudie Marec.)

[131] ARGO is an international measuring program begun in 2000 that includes at least 3,000 autonomous profiling floats (see **Section I.6.4.3**).

Chapter 2
Geographic characteristics

At first glance, looking at a terrestrial globe, an Austral (Antarctic) Ocean is not obvious. One tends to simply extend each ocean basin (Atlantic, Indian, and Pacific) to the Antarctic continent. Furthermore, the International Hydrographic Bureau cuts the world ocean into three sectors, corresponding to the three oceans (Atlantic from 67°16W to 20°E, Indian from 20°E to 146°55E, and Pacific from 146°55E to 67°16W). In reality, the Antarctic Ocean can be distinguished from the other oceans by the relative uniformity of its climate, circulation, and water properties. Its northern limit is not geographic but oceanographic. It's called the Subtropical Front,[132] is located around 40°S [**Figure 2-4**], and is characterized by a tightening of the surface isotherms [**Section II.5.2.1**].

Thus delimited, the Antarctic Ocean forms an annular ocean encircling the Antarctic continent, contiguous with the three other oceans. This oceanic ring has an average width of 2,900 to 3,300 km, including two indentations toward the South Pole: the Weddell Sea and the Ross Sea, and two notable constrictions. One, south of Tasmania (43°S), and New Zealand (47°S),[133] reduces the width to about 2,600 km, not counting the extension of the Campbell Plateau southward. The other, much more marked, south of South America extends southward to 56°S (Cape Horn) and faces the Antarctic Peninsula which extends northward to 63°S, reducing the width of the Drake Passage to 780 km between 500 m isobaths. The southern tip of Africa (Cape Agulhas) is 4,000 km from the Antarctic continent and is located about 10° north of the Subtropical Front.

The area of the Antarctic Ocean thus defined is 77 × 10^6 km^2, or 22% of the total surface of the planetary ocean.

The shoreline is not always distinct. The Antarctic continent, or Antarctica, is topped with a thick cap of ice more than 2,000 m thick, which flows slowly toward the sea [**Section II.4.1.1**] and often obscures the terrestrial shoreline. Seismic and ultrasonic soundings during the International Geophysical Year in 1957–1958 delineated two completely different regions [**Figure 1-5**].

[132] The Subtropical Front used to be called the Subtropical Convergence until the 1980's.

[133] The cape southwest of Stewart Island.

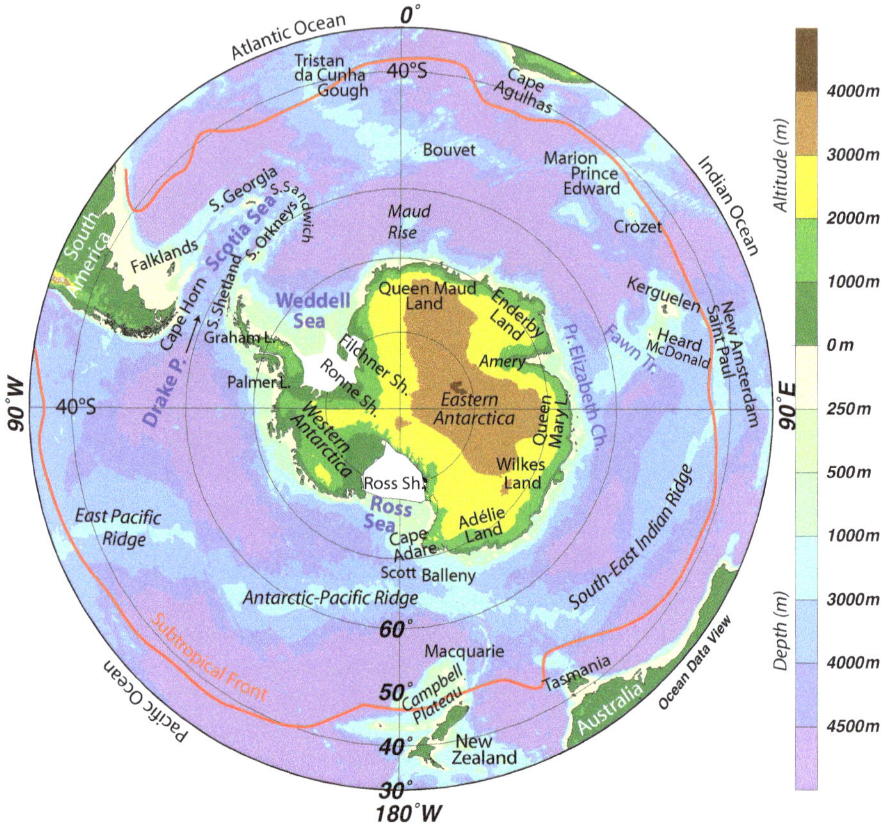

FIG. 2-4 – Map of the Antarctic Ocean, the topography of the Antarctic continent, and the average position of the Subtropical Front. (With ODV: Ocean Data View.)

Eastern Antarctica is composed of large regions where the continental platform is raised, including Queen Maud Land, Enderby Land, the East Antarctica Ranges, and the Transantarctic Mountains along the Ross Sea, from the western limit of the Ross Sea at Cape Adare to southeast of the Weddell Sea.[134]

Western Antarctica is composed of an archipelago of islands, including the Vinson Massif[135], the Antarctic Peninsula,[136] and Marie Byrd Land.

The ensemble of Antarctica is covered with a thick glacier resting on the continental platform, sometimes located more than 1,000 m below sea level [**Figure 2-5**]. Because of the weight of the icecap, the slope break between

[134] The mountain chain includes Mount Erebus (3,740 m), Mount Terror, the Queen Alexandra Chain, and the Queen Maud Chain, with Mount Kirkpatrick (4,530 m).
[135] The highest point in Antarctica is Mount Vinson (4,892 m).
[136] The Antarctic Peninsula is composed of Graham Land, Palmer Land, Adelaide Island, Alexander Island, and Charcot Island, among others.

the continental plateau and the continental slope is located between 400 and 800 m rather than 100 to 200 m as elsewhere. South of the Australian-Antarctic sector, between Cape Adare and the longitude of the Kerguelen Plateau, there is a long depression between the slope break and the continent, parallel to the coast, reaching 1,400 m depth. This depression includes basins separated offshore by sills, which permit the basins to retain below the sill level during the summer dense water formed in winter.

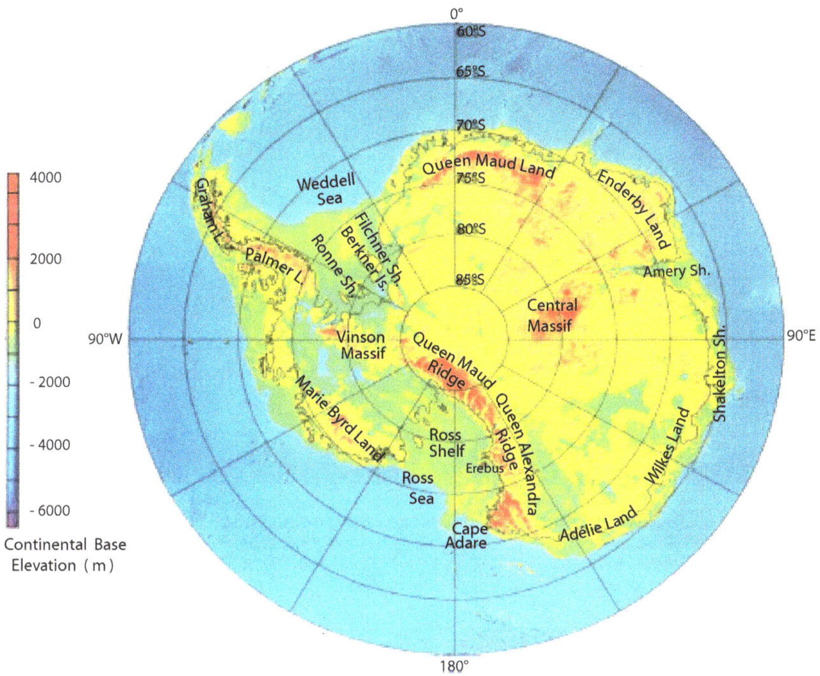

FIG. 2-5 – Under-ice topography of the Antarctic. (After the BEDMAP consortium, Lythe *et al.*, 2001. Reproduced by permission of the American Geophysical Union, © 2001, American Geophysical Union.)

Around the continent, the continental plateau is well developed only in the Weddell, Ross, and Bellingshausen seas [**Figures 2-5** and **2-6**]. North of the Antarctic Ocean, the Falkland Plateau extends southeast from South America and the Campbell Plateau[137] extends south of New Zealand [**Figure 2-4**].

The most important ridge is the Scotia Arc, which prolongs the Andean cordillera. It forms a loop eastward and returns westward to join the Antarctic Peninsula. From south to north, the Scotia Arc is composed of the South Shetland Islands, the South Orkney Islands, the South Sandwich

[137] The Campbell, Auckland, Antipodes, Bounty, and Chatham Islands are on the Campbell Plateau.

Islands, east of which is the Meteor[138] Deep (8,264 m), South Georgia, and the Falkland Islands, situated on the Falkland Plateau. It is the most important barrier for Antarctic circulation. Its depth is generally less than 2,000 m with several openings down to 3,000 m, in particular north of the South Sandwich Islands. The Scotia Sea, located in the interior of this arc, is bounded on the west by the Drake Passage (3,400–4,000 m).

The Mid-Atlantic Ridge turns eastward south of the island of Tristan da Cunha (37°S) in the direction of Bouvet Island (55°S) and is prolonged by the Southwest Indian Ridge [**Figure 2-6**] with depths greater than 4,400 m south of Africa around 25°E–30°E.[139]

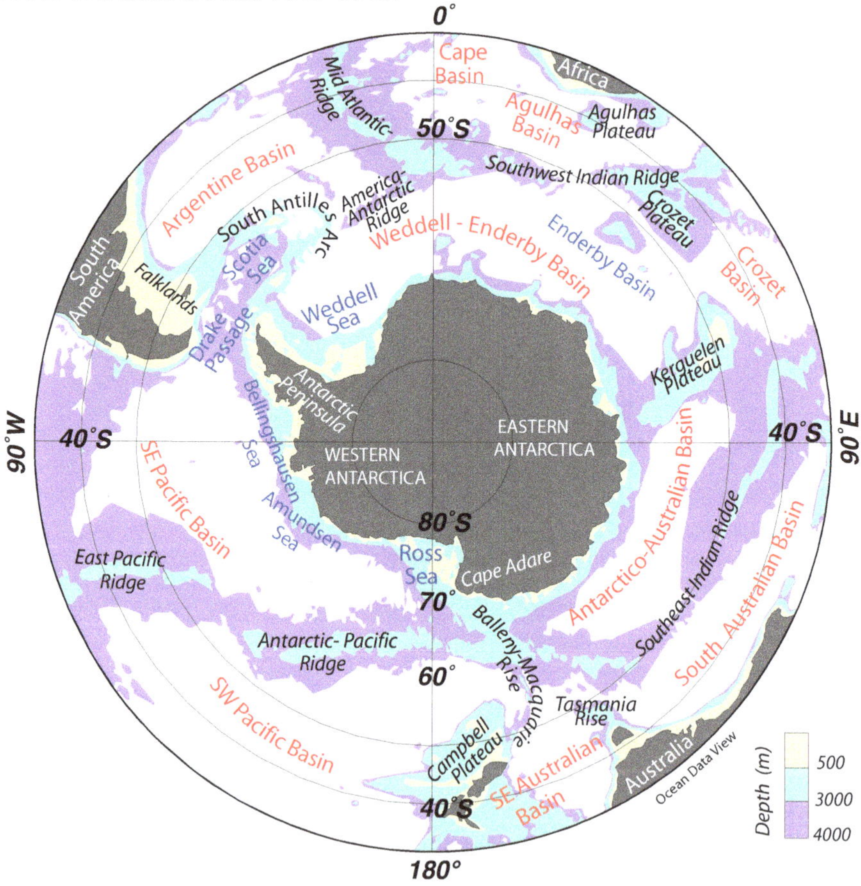

FIG. 2-6 – Simplified bathymetry showing the principal ridges, basins, and seas of the Antarctic Ocean. Depths greater than 4,000 m are in white.

---

[138] The *Meteor* was the German ship that discovered the deep during its 1925–1927 expedition.

[139] The Southwest Indian ridge supports Bouvet, Marion, Prince Edward, and Crozet islands on an eastward extension.

In the Indian sector, the Kerguelen Plateau[140] whose depth varies between 500 m and 2,000 m, is cut around 56°S by the Fawn Trough (reaching 2,650 m depth) and is separated from the Antarctic continent by the Princess Elizabeth Channel (whose sill is about 3,750 m). The Southeast Indian Ridge, which extends from the center of the Indian Ocean as far as the Balleny-Macquarie Rise, includes a region south of Australia where the depths are greater than 3,300 m.

The meridional Balleny-Macquarie Rise between Cape Adare and the Campbell Plateau reaches an average 2,000 m depth, leaving passages of more than 3,000 m north of 59°S. The Antarctic-Pacific Ridge extends to the northeast and is prolonged by the East Pacific Ridge in the direction of Easter Island.

Deep ridges divide the Antarctic oceanic ring into three principal basins [**Figure 2-6**]. The Atlantic-Indian Basin, also called Weddell-Enderby, is located between the Antarctic Peninsula, the southern part of the Scotia Arc, the Deep American-Antarctic Ridge, the Southwest Indian Ridge, and the Kerguelen Plateau. It includes the Weddell Sea on the west, and the Enderby Basin on the east.

The Antarctic-Australian Basin is bounded by the Kerguelen Plateau, the Southeast Indian Ridge and the Balleny-Macquarie Rise.

The Southeast Pacific Basin is located south of the Antarctic-Pacific Ridge, between the Balleny-Macquarie Rise and the Drake Passage. It includes the Ross Sea, the Amundsen Sea, and the Bellingshausen Sea.

On the northern border of the Antarctic Ocean are a number of basins: the Argentine Basin north of the Scotia Arc; the Cape Basin, the Agulhas Basin and the Agulhas Plateau north of the Southwest Indian Ridge; the South Australian Basin north of the Southeast Indian Ridge; the Southeast Australian Basin (or Tasman Sea) north of the Balleny-Macquarie Rise; and the Southwest Pacific Basin north of the Antarctic-Pacific Ridge.

Over this vast marine extent, apart from the Scotia Arc, there are only a few islands, isolated from each other. The climate is humid and cold, the wind is violent, the vegetation is rare, and no tree grows there. But large birds of these windy deserts such as the albatross and the petrel, find refuge here for their young.[141] Since their discovery in the 18th and 19th centuries (by Frenchmen whose names they carry, such as Bouvet, Marion, Crozet, Kerguelen, by Britons, such as Gough or Balleny, by Americans, such as Heard and McDonald, and by Portuguese, such as Tristan da Cunha) the occupation of these islands has always been difficult and intermittent. First there were the hunters, for whales, sea lions, or sea elephants, who settled

---

[140] The Kerguelen Plateau is composed of three plateaus: the northern plateau is the site of the Kerguelen Islands; the central plateau the Heard and MacDonald Islands; the southern plateau is separated from the central plateau by the Fawn Channel.

[141] Penguins, sea lions, and sea elephants, among others, find refuge here as well.

here for a moment to transform their catches into oil. Since 1950, some of these islands house observatories where scientists regularly relieve each other, such as Crozet and Kerguelen. The Austral Ocean is a water desert, particularly between 50°S and 62°S.

Chapter 3
Atmospheric pressure and winds

The high, extremely cold glacial Antarctic plateaus are the seat of high pressure, winter and summer [**Figure 2-7**]. A zone of circumpolar low pressure, located between the high pressure of the Antarctic continent and high subtropical pressure (25°S–35°S), surrounds the Antarctic continent around 65°S.

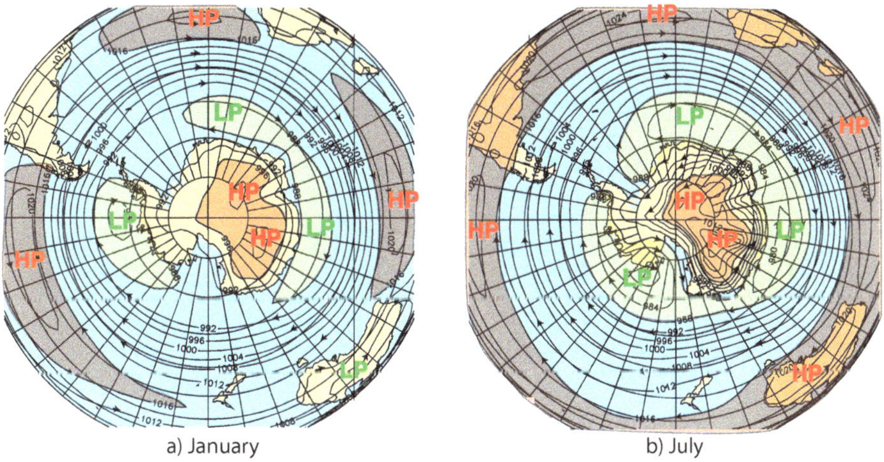

FIG. 2-7 – Average atmospheric pressure in the austral hemisphere, 1979–2001, in (**a**) January and (**b**) July. (Courtesy of D. Swingedouw, based on ERA40.)

Between the subtropical high pressure and the circumpolar low pressure, the Austral ocean is subject to dominant winds from the west that are particularly strong from 45°S to 55°S [**Figure 2-8**] with a maximum in the Indian sector where the gradient between high and low pressure is intensified. Sailors call them the "westerly gales", of force 5 to 6 on the

Beaufort scale[142] (9 to 10 m s^{-1}). They were used by 19th century trade ships[143] led by captains sailing for long sea passages on voyages passing the three capes: Cape of Good Hope, the Southwest Cape of New Zealand, and Cape Horn. To round Cape of Good Hope[144] (34°24S) sailors were on the edge of the roaring forties, which they entered to pass south of Tasmania (43°30S) and south of New Zealand (47°S). They had to face the howling fifties in order to round Cape Horn.

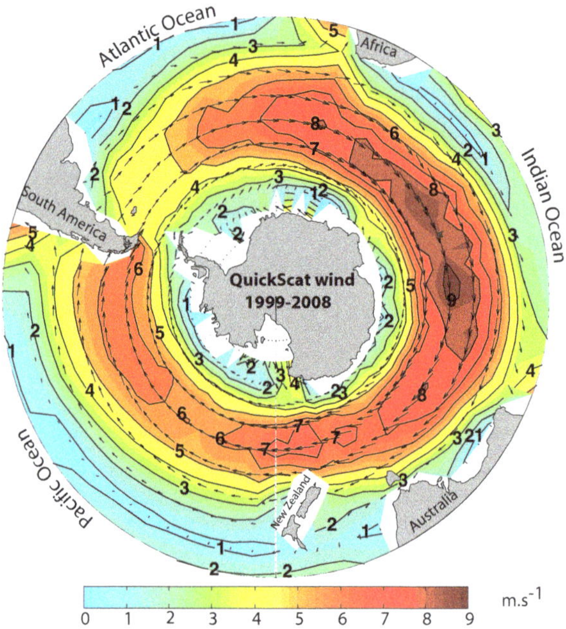

FIG. 2-8 – Speed and direction of average winds over 10 years (1999–2008). Isotachs are in m s^{-1}. There are shadow zones near continents and winds are poorly represented over ice-covered regions. (Courtesy of Y.H. Park, using QuickScat satellite data.)

The zone of the westerlies is dominated by frequent disturbances propagating to the southeast [**Figure 2-9**] in the course of which the winds become storm-force. Since sea-surface wind stress is proportional to the square of

---

[142] Sir Francis Beaufort (1774–1857), British hydrographic admiral, invented the Beaufort scale in 1805 to measure the force of wind using observations of waves and the effect of wind on sails. This scale runs from 0 (calm) to 12 (hurricane). Meteorologists and sailors use the scale to this day.

[143] Clipper ships.

[144] In fact, the southernmost point of Africa is Cape Agulhas (34°52S, 20°E). Cape of Good Hope is slightly to the northwest (34°24S, 18°30E) but forms a more dominant point.

wind speed [**Section I.5.2**], these disturbances have a strong impact on ocean circulation and give it great variability.

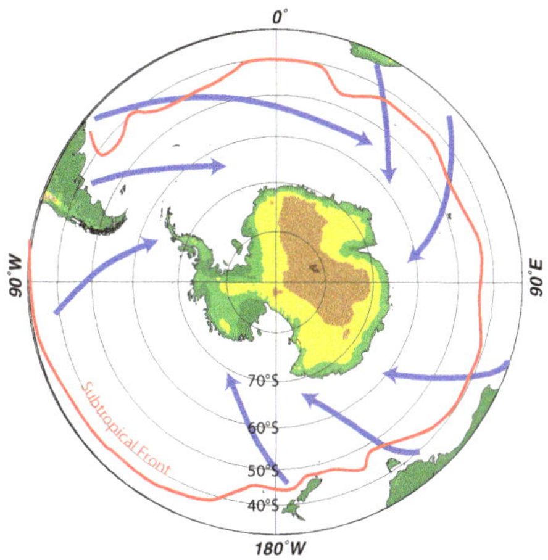

FIG. 2-9 – Paths of disturbances over the Austral Ocean which bring heat and precipitation southward (with ODV, topography with 1,000 m intervals). (After Tchernia, 1978.)

Farther south at around 65°S, between the central polar high pressure and the circumpolar low pressure, an ESE wind regime is established on the continental border [**Figures 2-7** and **2-8**][145] in a restricted zone near the coast, sea ice covered most of the time.

On the ice cap, valleys channel winds. A layer of extremely cold air, thus very dense, flows by gravity along the slopes and accelerates toward the coast. These are the particularly violent 'katabatic' winds whose velocity can reach 180 to 380 km/h at the mouths of the glacial valleys and are found up to 20 to 25 km out from the coast. When they are no longer channeled by topography, they are turned to the left by the Coriolis force (toward the west) and quickly slow down.

Nowhere else does the distribution of land and sea permit a zonal wind circulation so complete, so constant, so intense. The combination of strong winds, blowing non-stop all around the continent,[146] creates the highest waves found in the ocean [**Figure 2-10**]. In the Antarctic region, it happens that some boats founder,[147] producing a rude shock to masts and superstructures.

[145] Isotachs are lines of equal speed.

[146] The fetch (the distance over which the wind blows) is very long.

[147] The boat capsizes forward in the sense of its length under a following wind (often in a situation of flight). The bow descends deeply in the water ahead of the point of equilibrium and the hull turns over completely. This is a frequent occurrence in the Agulhas Current, where some solitary waves are enormous.

The most frequent crossings occur in the narrowest zone (the Drake Passage) between southern South America and the northern point of the Antarctic Peninsula where numerous accessible scientific bases are found.

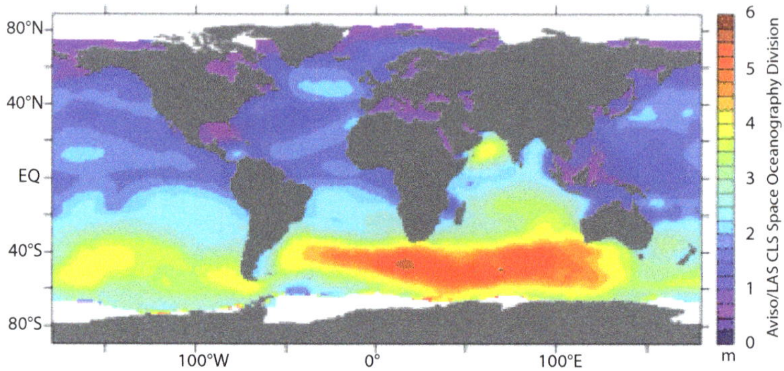

FIG. 2-10 – Average wave height from satellite for July, 2007 over the globe, maximum in red, minimum in violet. (From Aviso/LAS CLS.)

Chapter 4
Climatology

4.1 Glaciers and ice pack

4.1.1 Continental glaciers, ice shelves

The most extreme temperatures have been recorded on the ice cap: –88 °C at the Vostok station (78°30S, 107°E) at 3,488 m altitude in the heart of Eastern Antarctica, the coldest point on the planet. The average annual temperature is –56 °C at Vostok and –51 °C at the South Pole. Antarctic climatic conditions are glacial, uniform, and only diminish at the perimeter. Precipitation, which falls in the form of snow, does not exceed 200 mm per year. Snow piles up and is transformed into a thick layer of ice that covers the continent, known as the ice cap (sometimes called *inlandsis*, from the Danish). Since annual ice accumulation is slight, ice cores in the thick Antarctic glaciers, with thicknesses of 2,000 to 3,000 m can reach down to very old layers. Glaciers are fluids, which flow slowly to the continental periphery where they regularly calve off as icebergs [**Figure 2-11a**].

FIG. 2-11A – Iceberg and pack ice from the research ship *Astrolabe* near the Dumont d'Urville base in Adélie Land. (Photo Claudie Marec.)

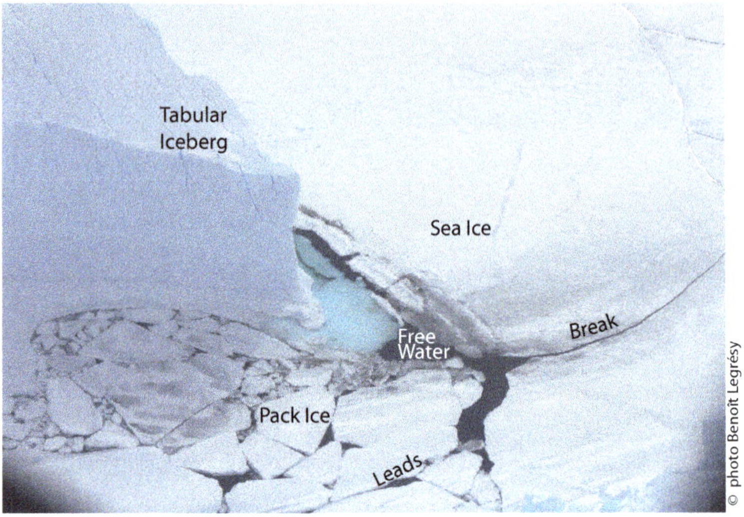

FIG. 2-11B – The two types of Antarctic ice. (Photo Benoît Legrésy.)

Off the extensive continental shelves, glaciers advance without breaking and form immense shelves of ice, floating on the ocean, with thicknesses of 200 to 300 meters. South of the Ross Sea, the Ross Ice Shelf has an area nearly that of France [**Figure 2-4**]. The southern Weddell Sea, covered by the Filchner Ice Shelf east of Berkner Island and the Ronne Ice Shelf to the west, has a comparable area. There are other less-extensive ice shelves, such as the Larsen Ice Shelf east of the Antarctic Peninsula, the Amery Ice Shelf between the eastern limit of Enderby Land and the Antarctic continent, and the Shackleton Ice Shelf off Queen Mary Land. Small glacier tongues extend off shore, such as those west and east of Adélie Land. Some reach islands near the coast, as off western Antarctica in the Amundsen Sea and in the Bellingshausen Sea.

Immense tabular icebergs, sometimes reaching many thousands of square kilometers in area, detach from these ice shelves from time to time [**Figure 2-12a**]. Some icebergs can survive as far north as 35°S in the Atlantic sector. They consist of pure water in the form of dense ice. When they melt, they contribute fresh water to the surface layer, diminishing salinity and augmenting sea level,[148] as well as drawing on the heat necessary for them to melt. They are a potential factor for rising sea level in climate-warming theory.[149]

[148] When the ice pack (which floats and is formed from seawater) melts it does not increase the level of the sea, as opposed to the melting of icebergs, which represent an input of fresh water, similar to that brought by rivers and precipitation.

[149] If all ice shelves were to melt, sea level would rise by around 70 m.

FIG. 2-12A – Antarctic tabular icebergs. (Photo Benoît Legrésy.)

FIG. 2-12B – Drifting pack ice caught in young sea ice. (Photo Claudie Marec.)

4.1.2 The ice pack

On the other hand, the ice pack formed at the sea-surface by seawater freezing, produces climatological and oceanographic impacts more important than do icebergs. The ice pack, also called sea ice, is porous and heterogeneous. Under the effect of surface motion due to wind during its formation it can, on a small scale, hide pockets of brine. On a larger scale, in regions of compression due to winds and currents, it piles up in heaps of ice plates, called hummocks. The thickness of pack ice is quite different from that of icebergs, of the order of 2 or 3 m only, a limit beyond which the seawater is isolated from the atmosphere and freezing ceases. In winter, when the ice pack extends northward to 60°S [**Figure 2-13**], and the ice-covered surface (including the ice cap[150]) increases from 16.2×10^6 km^2 to 32.7×10^6 km^2, it is as if the continental surface had doubled, since exchanges with the ocean become extremely weak.[151]

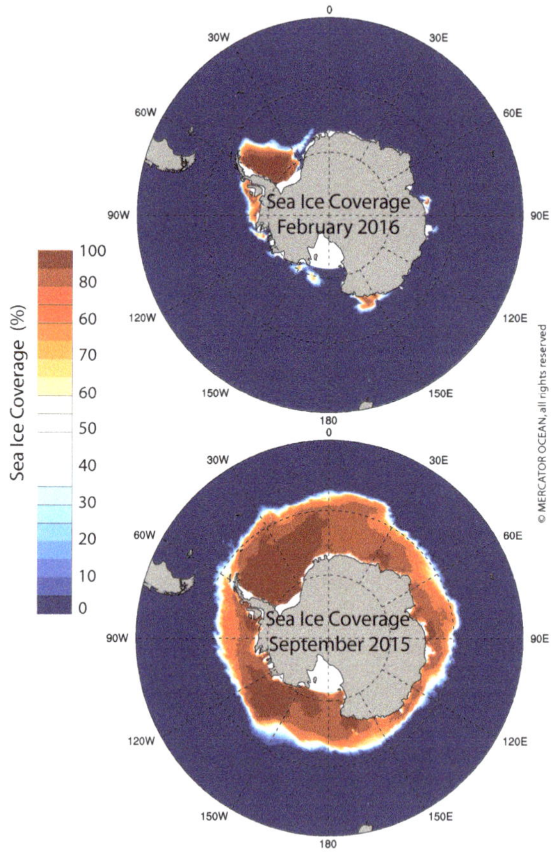

FIG. 2-13 – Sea ice extent a) in summer (2.2×10^6 km^2 in February 2016) and b) in winter (18.7×10^6 km^2 in September 2015). (Image generated using E.U. Copernicus Marine Service Information and interpreted by G. Garric, Mercator Océan. All rights reserved).

[150] The surface of the Antarctic Continent covered with a thick ice cap is 12.5×10^6 km^2 and the combined surface of the Weddell, Ross, and smaller ice shelves is 1.5×10^6 km^2.

[151] Exchanges take place in narrow channels separating sheets of the ice pack.

In spite of the summer melting of more than 16×10^6 km^2, there still remained 2.2×10^6 km^2 (February 2016) of floe ice, pieces of the ice pack [**Figure 2-12b**] entrained westward by the Periantarctic Coastal Current, and piled up east of the Antarctic Peninsula in the Weddell Sea. The ice pack also remains in summer in the Bellingshausen and Amundsen Seas, in the eastern Ross Sea and off Adélie Land. The ice pack prevents sea access to the continent during nine to ten months of the year, except for icebreakers. During the two to three months of summer, even in regions of easier access, the presence of drifting pack ice makes oceanographic measurements delicate.

The ice pack plays an extremely important role in climate by insulating the ocean from the atmosphere. Contrary to seawater,[152] the ice pack reflects a large part of incoming solar radiation and the reflection increases still more when it is snow-covered. Consequently, the variability of pack-ice coverage can bring about strong climatic anomalies.

4.1.3 Polynyas

Terrestrial observing satellites can follow, in a detailed way, the evolution of pack-ice coverage. They reveal that, away from the numerous channels between ice sheets, which represent about 18% of the surface, there sometimes exist clearings in the ice pack called *polynyas*[153] [**Figure 2-14**].

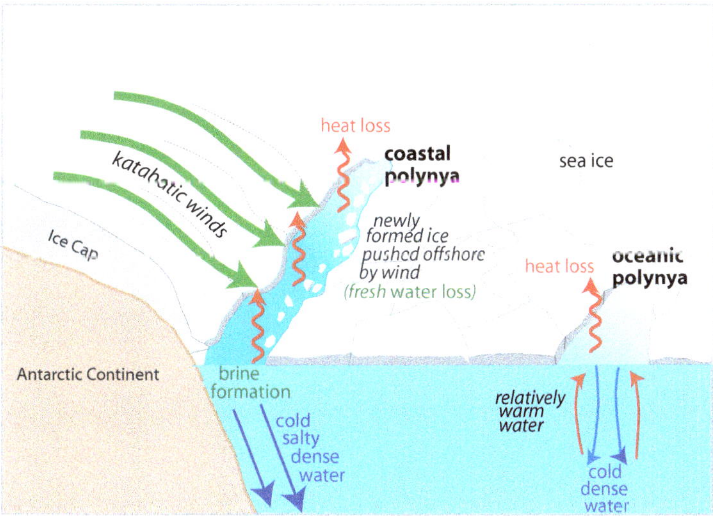

FIG. 2-14 – The mechanism of different types of polynyas.

[152] Seawater absorbs a large percentage of solar radiation. It has a weak albedo.
[153] From the Russian word *polynya*.

There are two types of polynyas: offshore polynyas connected to warmer water below and wind-forced coastal polynyas. Coastal polynyas are found off the outlets of winds coming from the continent. In this type of polynya, when ice begins to form, removing heat and fresh water from the ocean, it is pushed offshore by very cold violent katabatic winds. The open water thus exposed once again loses heat, leading to renewed ice formation, which is again pushed offshore. This process goes on until the wind ceases. Coastal polynyas are truly factories forming ice and dense water.

Offshore, the presence of open water at these latitudes requires a continual inflow of heat, which can only come from upwelling of warm water to the surface. The polynya with the largest opening, found in some years northeast of the Weddell Sea [**Figure 2-15**], appears to correspond to upwelling of Circumpolar Deep Water related to cyclonic circulation [**Figures 2–18** and **1-66**] and possibly undersea topography (the Maud Rise) [**Figure 2-4**].

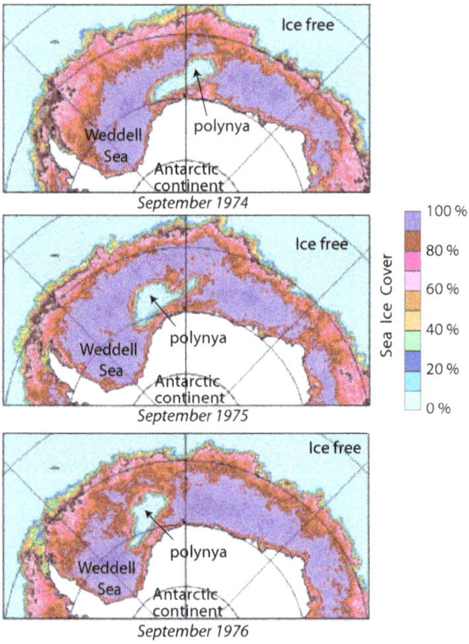

FIG. 2-15 – Sea ice extent in winters 1974, 1975, and 1976, showing the vast polynya (1,000 km × 350 km) northeast of the Weddell Sea. (After Gordon *et al.*, 2007, using SSM/I, NOAA/AVHRR satellite data.)

These ice-free zones have an important oceanographic role since exchanges there between the sea and the very cold atmosphere are particularly intense.

4.2 Precipitation

Measuring precipitation at sea is always difficult. It is even more difficult in the Antarctic Ocean due to the violence of the winds and to infrequent passages in the region. Precipitation exceeds 2,000 mm per year in the south-east Pacific sector, where South America acts as a barrier to humidity-laden westerly winds. Elsewhere, conditions are rather uniform due to the absence of land and thus the few existing measurements can be extrapolated over the entire zone [**Figure 1-20**]. On average, precipitation is of the order of:

- 1,500 to 1,000 mm/year from 40°S to 55°S.
- 1,000 to 500 mm/year from 55°S to 65°S.
- 500 to 200 mm/year from 65°S to 70°S.
- less than 200 mm/year from 70°S to 90°S.

Chapter 5
Surface circulation

5.1 The Antarctic Circumpolar Current

The Antarctic Ocean, with little ship traffic, is the most difficult ocean region to observe in spite of new techniques. Directly measuring currents in the open ocean is not easy. When the current is not known, it is a problem for the sailor. When it is known, it is useful to him. In any case, it needs to be taken into account. To determine it, the sailor compares the position where he estimates he ought to be with the position he has [**Section I.5.1**].

Early explorers who adventured into the Antarctic Ocean noticed that the wind and current came from the west. (Sailors say that a wind coming from the west is a "westerly wind" and that a current that sets their ship to the east is an "eastward current".)

Winds from the west, which blow in the band 40°S–60°S and without obstruction in the latitude range of the Drake passage, and their string of depressions generate a current toward the east, called the Antarctic Circumpolar Current (ACC) or West Wind Drift.

In 1979, during the FGGE (First GARP Global Experiment, of the Global Atmospheric Research Program) a large number of drifting buoys, which follow the current, were deployed in the Antarctic Ocean [**Figure 2-16**]. Tracked by satellite, they revealed many meanders and eddies within the Antarctic Circumpolar Current. The average velocity of the Circumpolar Current is of the order of 0.1 to 0.25 m s^{-1} and can reach 0.5 to 0.75 m s^{-1} in current filaments.

Satellite-mounted altimeters (*Seasat* in 1978, followed by *Geosat*, *Topex-Poseidon*, *Jason 1*, and *Jason 2* in 2008) were able to measure undulations of sea level and their variation, with enough precision to calculate currents.

The dynamic topography shown in **Figure 2-17** was created from an optimal combination of average altimetric data, oceanographic data, and currents obtained from drifting buoys. It provides the most complete representation of the average sea-surface topography.

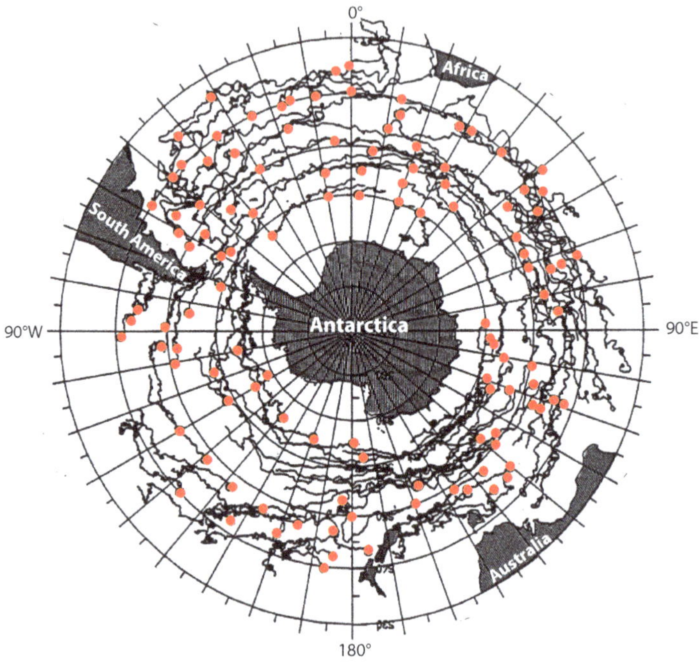

FIG. 2-16 – Trajectories of about 130 satellite-tracked buoys in the Antarctic Ocean during the First GARP Global Experiment in 1979 (launching points in red). (After Daniault, 1984.)

The difference in sea-level height across the Antarctic Circumpolar Current is of the order of 1.5 dyn m with about 0.7 dyn m added across the Agulhas Current and its retroflection. Currents follow contours of equal height. In the southern hemisphere the highest elevation is on the left side of the current [**Section I.4.3**, **Figure 1-49**]. The stronger the surface slope, the tighter are the isopleths and the stronger is the current. This topography shows that the Circumpolar Current [**Figure 2-4**] does not have a uniform structure. Several regions of tighter isopleths where the current is more intense are found: in the Indian Ocean sector from the Agulhas Current retroflection to the Kerguelen Plateau where the circulation is channeled and intensified, in the Drake Passage, in the region of confluence of the Malvinas and Brazil Currents, and in passages across ridges. At ridge crossings, the current first turns left, to the north, when the bottom rises, then right, to the south, when the bottom descends.[154]

[154] The conservation of potential vorticity [**Section I.5.3.4**].

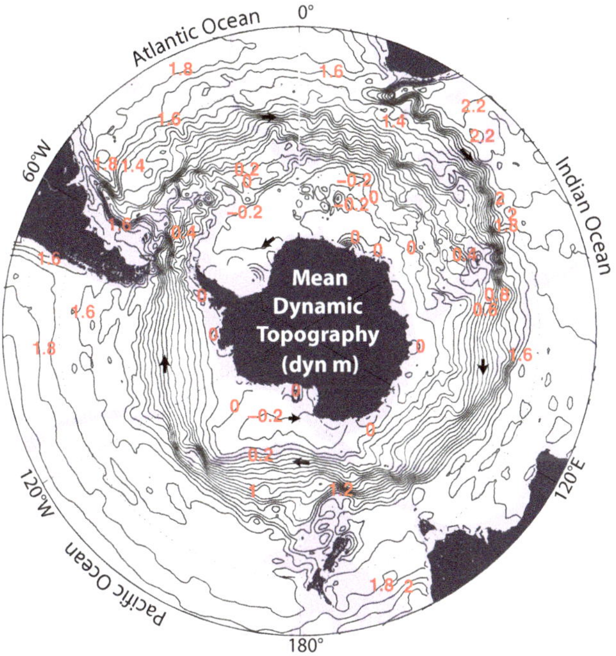

FIG. 2-17 – Average dynamic topography (in dyn m) created by optimal combination of TOPEX/POSEIDON satellite altimetry, hydro station data, and drifting buoy data. Depths less than 3,000 m are in purple. (Courtesy of Y.H. Park, after Rio and Hernandez, 2004.)

A schematic current distribution diagram can be obtained from the average dynamic topography [**Figure 2-18**].

The ACC flows zonally nearly without any continental obstacle all around Antarctica and represents a unique linkage between the three ocean basins. Lying south of the Subtropical Front it is composed of intense current filaments located along three main fronts, strongly steered by bathymetry, in particular by the ridge system, and not always continuous.

In spite of being covered by pack ice during part of the year, three cyclonic subpolar circulations (or gyres[155]) develop south of the ACC: in the Weddell Sea, in the Ross Sea, and a less extended one in the Antarctic-Australian basin, each associated with sites of dense-water formation.

---

[155] The term **gyre** denotes a large-scale (thousands of km) ocean-current circuit.

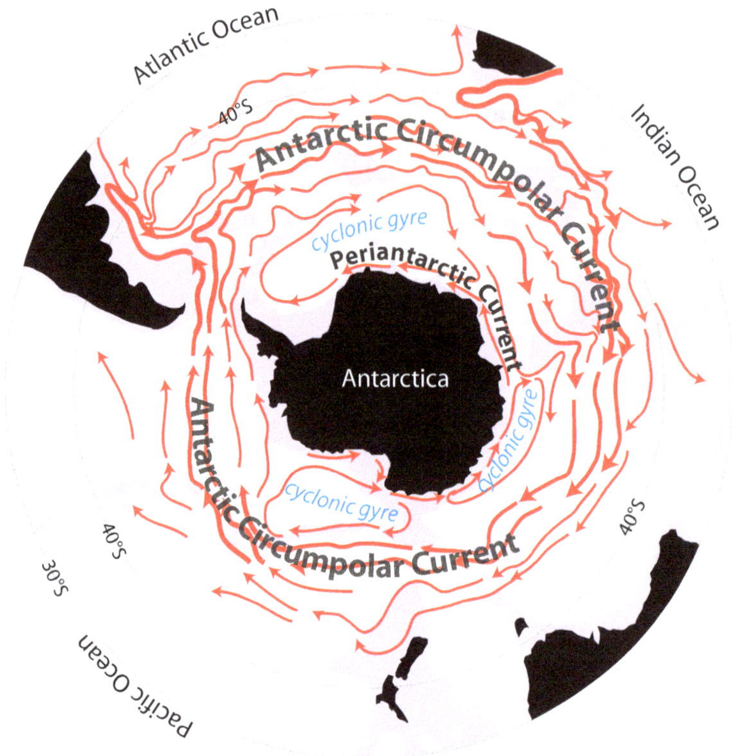

FIG. 2-18 – Schematic surface circulation of the Antarctic Ocean from topographic maps of the sea-surface, showing the Antarctic Circumpolar Current, the Periantarctic Current, and the three cyclonic circuits. The 3,000 m isobath is in purple.

The ACC is not limited to the surface but extends to the deep ocean. North of the westerlies maximum speed, which occurs around 50°S, Ekman pumping is negative, producing convergence, downwelling [**Figure 2-19**], and the formation of a thick surface layer. To the south of the westerlies maximum speed, Ekman pumping is positive, particularly between the zone of the easterlies near the Antarctic continent and the zone of the westerlies, with a divergence around 60°S producing upwelling, and a southward-descending sea-surface slope down to the divergence [**Figure 2-17**]. The resulting pressure gradient, in the direction of the slope [**Sections I.4.3, I.4.4**], produces a geostrophic current perpendicular to the slope, that is, to the east, in the direction of the wind. Because of weak vertical density variations in the Austral Ocean [**Section II.6**] the pressure gradient due to the surface slope extends to great depth. The eastward ACC is associated with a general rise of the isolines toward the south through the entire depth.

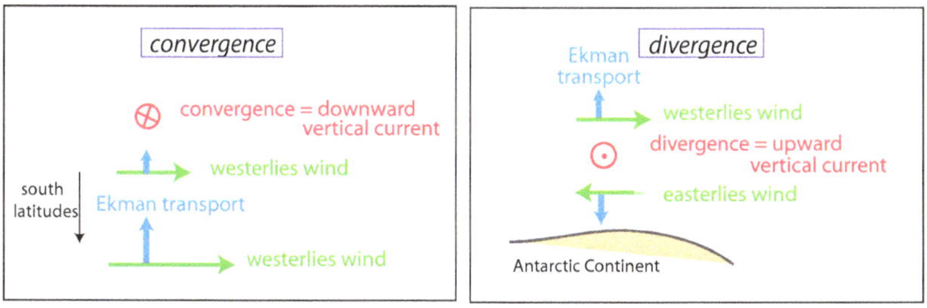

FIG. 2-19 – Convergences (downwelling) and divergences (upwelling) from wind acting on the Austral Ocean.

The Antarctic Circumpolar Current thus extends several thousand meters deep. For this reason, the ACC "feels" the influence of deep topography, especially the current filaments associated with the three oceanic fronts of the ACC. The wind effect is balanced by lateral friction and by bottom roughness. Even though it has no continental obstacle, the ACC is not a smooth and steady current. It is composed of many turbulent current filaments steered by bottom topography and perturbed by eddies.

5.2 The Antarctic Circumpolar Current and oceanic fronts

The Southern Ocean is divided by the Antarctic Circumpolar Current (ACC) into three major regions: the Subantarctic Zone north of the ACC, the ACC transport zone and the zone south of the ACC. The rise of isolines toward the south corresponding to the geostrophic balance of the eastward flowing ACC is not uniform. Increasing observations have shown three fronts where the rise steepens corresponding to the faster current branches[156].

Because of its great horizontal and vertical extent of about 2,800 km width and more than 3,000 m depth, even though velocities are not great, the ACC is the most powerful current on the globe. Its transport has been estimated, using hydrographic measurements calibrated by direct current measurements (Doppler-effect profilers) and compared with satellite sea-level measurements [**Figure 2-17**] at about 130-140 Sv ± 15 Sv. That is of the same order of magnitude as the transport in the Drake Passage. South of Australia, the estimate of ACC transport is about 147 Sv ± 10 Sv. The difference seems to be due to Indonesian "Throughflow"[157] of the order of 15 Sv ± 4 Sv [**Section IV.5.2.3**], which would recirculate around Australia.

[156] The rise of the isolines brings nutrient-rich deep waters closer to the surface southward.

[157] The "Throughflow" is the water transport from the Pacific Ocean to the Indian Ocean Ocean through the Indonesian Archipelago.

A number of fronts are described here: the Subtropical Front (STF), the Subantarctic Front (SAF), the Polar Front (PF), the Southern Antarctic Circumpolar Current Front (SACCF), the Southern Boundary (SB) and the Antarctic Divergence (AD) [**Figure 2-20**]. The SAF, PF and SACCF correspond to the strong branches of the Antarctic Circumpolar Current. Under variable wind action, they show large temporal and spatial variations and the presence of numerous meanders and eddies [**Figure 2-16**].

The fronts constitute boundaries between oceanographic regions, found around the Antarctic Ocean from north to south [**Figure 2-20**].

- The Subantarctic Zone (SAZ), between the Subtropical Front and the Subantarctic Front.
- The Polar Frontal Zone (PFZ), between the Subantarctic Front and the Polar Front.
- The Antarctic Zone (AAZ), between the Polar Front and the Southern Antarctic Circumpolar Current Front.
- The Southern Zone (SZ) between the Southern Antarctic Circumpolar Current Front, and the Southern Boundary.
- The region south of the Southern Boundary forms the Subpolar Zone, which includes the three cyclonic circulations, the Antarctic Divergence and the westward Periantarctic Coastal Current (PACC).
- The Continental zone corresponds to the shelf waters which are separated from the offshore waters by the Antarctic Slope Front (ASF), the northern limit of cold subsurface waters with $T < 0$ °C between 150–500 m.

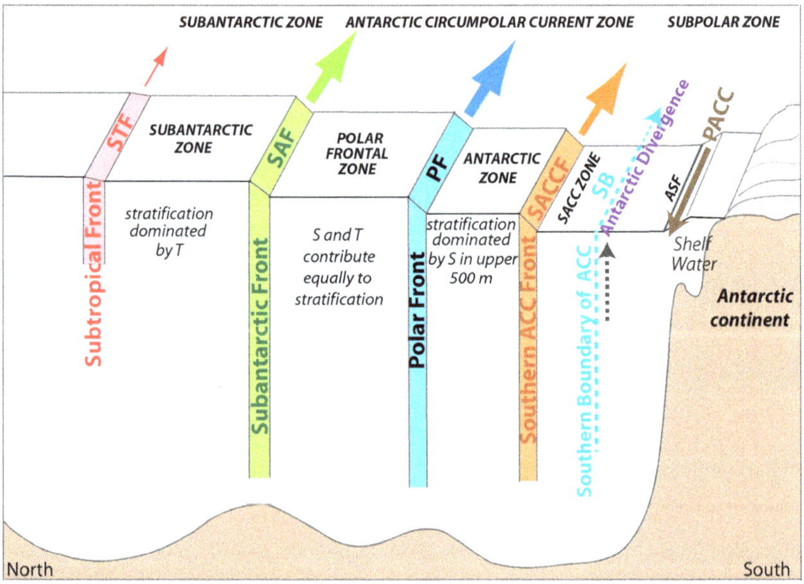

FIG. 2-20 – Block diagram of fronts and oceanic zones of the Antarctic Ocean [Antarctic Circumpolar Current (ACC), Southern Antarctic Circumpolar Current Front (SAACF), Antarctic Slope Front (ASF), Periantarctic Coastal Current (PACC)]. (After references **Figure 2-21**.)

5.2.1 The Subtropical Front and the Subantarctic Zone

The Subtropical Front (STF) defines the northern limit of the Antarctic Ocean. On average, it is located around 40°S [**Figure 2-21**]. It is characterized by the convergence of warm and saline subtropical waters from the

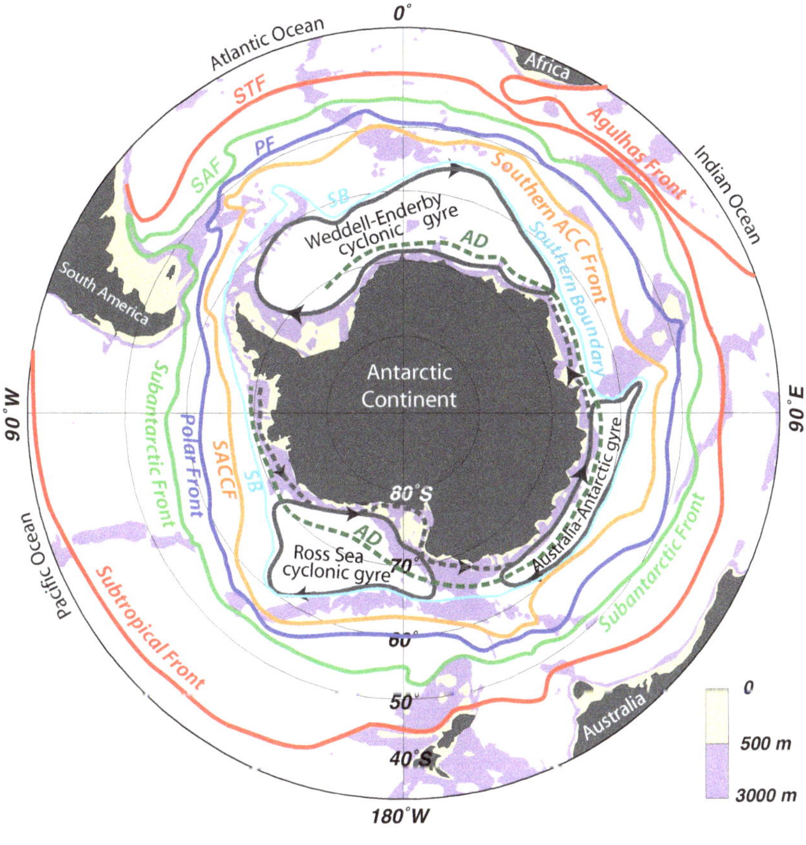

FIG. 2-21 – Average positions of the Subtropical Front (STF), the Subantarctic Front (SAF), the Southern Antarctic Circumpolar Current Front (SACCF), the Antarctic Divergence (AD), as well as the three cyclonic gyres of the Weddell-Enderby Basin, the Antarctic-Australian Basin, and the Ross Sea. (After Deacon, 1937; Gordon *et al.*, 1977, 1978; McCartney, 1977; Nowlin and Clifford, 1982; Whitworth and Nowlin, 1987; Peterson and Whitworth, 1989; Park *et al.*, 1991; Orsi *et al.*,1995; Provost *et al.*, 1995, 1999; Orsi *et al.*, 1999; Vivier and Provost, 1999; Pollard *et al.*, 2002; Park *et al.*, 2008.) Frontal definitions are based on the nomenclature of Orsi *et al.* (1995), and their climatological positions were obtained from selected contours of dynamic height from the MDT_CNES-CLS09 mean dynamic topography climatology (Aviso, CNES, <http://www.aviso.altimetry.fr/>). (Courtesy of F. Roquet, MISU, Stockholm University.)

southern part of the anticyclonic gyres of each basin (Atlantic, Indian, and Pacific) and colder and less saline water from the Subantarctic Zone. It marks the northern extent of subantarctic waters. It is a narrow zonal band where meridional temperature and salinity variations can be rapid, between 9 °C and 14 °C depending on the season [**Figure 2-22**].

The STF is limited to the thermocline layers [**Figure 2-20**]. By convention, the position of the STF is defined as the intersection of the 11 °C isotherm with the 150 m isobath or 10 °C with the 200 m isobath. In the western Atlantic sector, the STF is located at the convergence between the cold Malvinas Current and the warmer Brazil Current, where it reaches its most northern position (up to 35°S–38°S). In the eastern Pacific, the extension southward of the South American continent turns the front northward, where it disappears along the Chilean coast. In this region, the limit of the Austral Ocean has been arbitrarily fixed at 40°S. The STF does not exist in the Drake Passage, which is situated too far south. The STF is most often associated with weak filaments of eastward current since the temperature gradient is partly counteracted by the salinity gradient. Seasonal variations in the westerlies are responsible for variations in the latitude of the STF, which can be as great as 10° between summer and winter especially in the Pacific sector.

The Subantarctic Zone [**Figure 2-20**], between the STF and the SAF, is situated in a region of convergence, which increases vertical mixing and develops thicker mixed layers in winter. Vertical temperature gradients are weaker than they are north of the STF, bringing about a lessening of stratification that is characteristic of the Antarctic Ocean. The Subantarctic Zone delimits the region where isotherms of the permanent thermocline reach the surface. Temperature dominates salinity in the contribution to stratification, which allows a subsurface salinity minimum of Antarctic Intermediate Water to exist. In winter, the reduced temperature stratification results in thick Subantarctic Mode Water pycnostads [**Section I.3.2**].

5.2.2 The Subantarctic Front and the Polar Frontal Zone

Toward the south, a rapid surface-temperature variation [**Figure 2-22a**] determines the zone of the Subantarctic Front [**Figure 2-20**]. It is defined as the zone of maximum meridional temperature gradient between 3 and 8 °C at 300 m depth. The SAF is associated with a strong filament of current (0.5 to 1 m s^{-1}) of the ACC, which develops a steep meridional isoline slope at all depths.

In the Polar Frontal Zone (PFZ), between the SAF and the PF, precipitation and melting pack ice in summer decrease surface salinity. Across the SAF, the salinity minimum of the Antarctic Surface Water is subducted northward from the surface in the Polar Frontal Zone ($S < 34$) to more than 400 m in the Subantarctic Zone with $S < 34.30$ which forms Antarctic

Intermediate Water. It results from northward Ekman transport due to the distribution of the westerlies [**Figures 2-19** and **2-22b**].

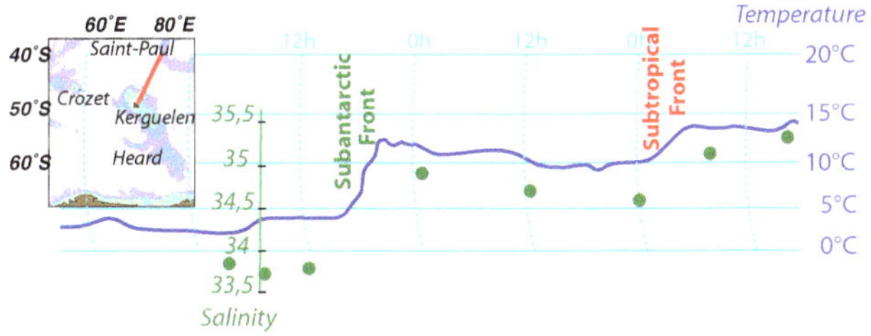

FIG. 2-22 – Continuous sea-surface temperature recording with surface salinity measurements (green points) between Kerguelen and Saint Paul, showing variations across the Subtropical and Subantarctic Fronts. (After Tchernia, 1978.)

In the Polar Frontal Zone, temperature and salinity variations contribute equally to stratification below the mixed-layer. This region, where southern surface waters converge, is weakly stratified, and is subject to strong densification through mixing due to the cabbeling effect (**Section I.3.4.2**). This results in the production of fresh Antarctic Intermediate Water, which is then subducted below sub-tropical gyres in the lower thermocline. The northward advection of Antarctic Intermediate Water can be traced in the Pacific, Indian, and Atlantic basins by its characteristic minimum salinity. The Polar Frontal Zone thus provides a natural gateway for the injection of surface signals into the deep Ocean, such as heat and carbon anomalies (see **6.3.2**).

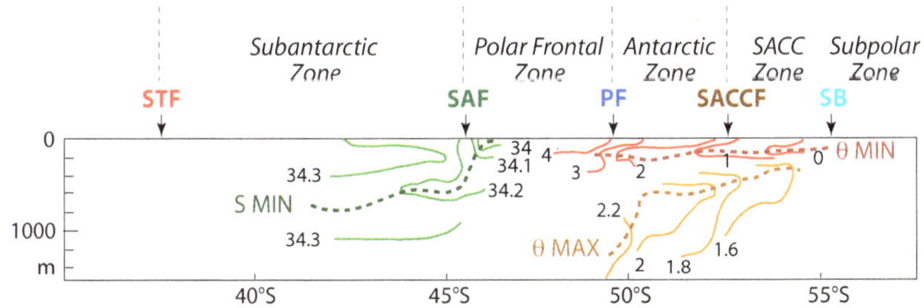

FIG. 2-23 – Green: selected isohalines along the salinity minimum around the Subantarctic Front (SAF) between the Polar Frontal Zone (PFZ) and the Subantarctic Zone (SAZ) corresponding to Antarctic Intermediate Water. Red: selected isotherms (°C) along the subsurface temperature minimum in the Southern and Antarctic Zone (AZ). Yellow: selected isotherms along the deeper temperature maximum corresponding to Upper Circumpolar Deep Water in the region between the Polar Front (PF) and the Southern Boundary (SB). (After Orsi *et al.*, 1995.)

5.2.3 The Polar Front, the Antarctic Zone, and the Southern Antarctic Circumpolar Current Front

The Polar Front (PF) is one of the strong ACC branches. The PF is most often defined by the northern limit of the subsurface temperature minimum less than 2 °C in summer [**Figures 2-21**, **2-22** and **2-23**]. This temperature minimum marks the bottom of the winter mixed layer whose upper part warms in summer. Below this subsurface temperature minimum, temperature increases toward the temperature maximum of the Upper Circumpolar Deep Water (UCPDW), which is a little higher than 2 °C. The Polar Front current branch sometimes splits into two branches and is not always continuous like other Southern Ocean fronts.

In the Antarctic Zone, properly defined between the PF and the SACCF [**Figure 2-20**], as in the Polar Regions where the temperature is less than 2 °C, stratification is dominated by salinity variations because of fresh water input to the surface from melting ice in summer. The downward temperature increase to the Upper Circumpolar Deep Water temperature maximum is compensated by increasing salinity.

The Southern Antarctic Circumpolar Current Front (SACCF) is the third branch of the ACC. The SACCF is found north of the cyclonic gyres of the Weddell Sea, of the Antarctic-Australian Basin, and of the Ross Sea. The SACCF can be defined by a subsurface temperature maximum at 1.8 °C, and is well approximated by the northern limit of a subsurface temperature minimum less than 0 °C in summer. The Southern ACC Front often coincides with winter sea ice extent, as it delimits a zone strongly stratified in salinity favorable to the formation and persistence of sea ice in winter. The strong stratification in salinity south of the SACCF is maintained by upwelling of saline Circumpolar Deep Water bringing the deep maximum in salinity closer to the surface.

5.2.4 The Southern Zone, the Southern Boundary and the Antarctic Divergence

The Southern Zone lies between the SACCF and the Southern Boundary. The Southern Boundary (SB) is defined as the southern limit of the subsurface oxygen minimum of the Upper Circumpolar Deep Water (UCPDW), which upwells in the Antarctic Divergence zone (AD) and mixes with Antarctic Surface Water. The AD zone is located between the westerlies and the easterlies which produce a divergence (*i.e.* upwelling), around 65°S [**Figures 2-19, 2-20** and **2-21**]. The SB is the southern limit of the UCPDW. In some sectors the SB joins the Continental Water Boundary (or Antarctic Slope Front – ASF), in others the SB corresponds to the northern boundaries of the cyclonic gyres.

The region south of the Southern Boundary forms the Subpolar Zone, which includes the three cyclonic circulations and the Antarctic Divergence zone where Upper Circumpolar Deep Water is absent. The AD zone does not appear in the Drake Passage nor in the Scotia Sea. These regions are dominated by the westerlies, with no established easterly polar winds [**Figure 2-8**]. The Antarctic Slope Front (ASF) represents the northern limit of the very cold, dense and weakly stratified Continental Water (or Shelf Water), providing an environment that is nearly barotropic.

5.2.5 The role of topography in the Circumpolar Current

The ACC transport is distributed into branches, which are sometimes fragmented, sometimes confined into few strong branches, depending on the channeling effect of the bathymetry. The current accelerates in the Drake Passage, the narrow region between the Antarctic Peninsula and Cape Horn. Direct current measurements taken in the passage over a number of years show surface velocities up to 1 m s^{-1} and show that 75% of the transport, estimated at 135 Sv $\pm$ 13 Sv (10^6 m^3 s^{-1}) is located in the three frontal zones (PF, SAF, and SACCF).

Coming out of the Drake Passage, a branch of turbulent current composed of waters from the SAF and PF turns sharply northward, flowing around the Falkland Plateau to form the cold Malvinas Current, converging with the warm Brazil current around 35°S [**Figures 2-17** and **2-18**] then returning back southward. The average transport of the Malvinas Current is 41 Sv $\pm$ 12 Sv. Decreasing depth and the Coriolis effect turn the current northward. Passing over the arc of the South Antilles, the current filaments are turned to the left, northward, when depth decreases and to the right, southward, when depth increases.[158] Deviations of this type are observed at each ridge crossing, more or less strong according to the amplitude of depth variations and stratification.[159] South of South Africa, the Agulhas Plateau turns the STF southward. In this region, the powerful Agulhas Current, transporting warm tropical waters, converges with the ACC transporting cold waters. The Agulhas Current [**Section IV.4.2.3**] undergoes a retroflection eastward [**Figures 2-17** and **2-18**] forming the Agulhas Front north of the STF, which represents the southern branch of the anticyclonic subtropical gyre of the South Indian Ocean.

Around 45°E, the Crozet Plateau, an offshoot eastward of the SW Indian Ridge, turns the SAF northward. Further east, the meridional barrier (46°S to 62°S) formed by the Kerguelen Plateau (70°E) turns branches associated

[158] The change of direction is linked to the variation of the thickness of a water mass, applying the principle of conservation of potential vorticity [**Section I.5.3.4**].
[159] The more the vertical density gradient is weak (weak stratification), the more depth variations affect the surface circulation.

with the SAF and the STF northward which, closing to form a large front, concentrate nearly 70% of the transport (100 Sv ± 15 Sv) of the ACC within a zone of 2°–3° latitude thereby reducing the area of the Subantarctic Zone [**Figure 2-21**]. The branch associated with the PF splits into several branches. One branch turns to flow south of the Kerguelen Islands, and then returns northward following the eastern contours of the slope. Another branch passes south of the Heard Plateau. They transport together only about 6 Sv, because of the blocking effect of topography. An important branch associated with the SACCF and part of the PF transports about 40 Sv across a deep passage (the Fawn Trough, 2,650 m) around 56°S to rejoin the Antarctic-Australian Basin then turns northward hugging the eastern continental plateau slope. The current branch associated with the Southern Boundary flows through the Princess Elizabeth Channel,[160] with a transport of 7 Sv, then is deflected northward hugging the Plateau slope, to rejoin the branch coming from the Fawn Trough to form the western boundary current of the cyclonic gyre in the Antarctic-Australian Basin.

South of New Zealand, the branch associated with the STF turns northward and loses itself off Chile. The SAF branch is guided northward along the Campbell Plateau slope. North of the Ross Sea cyclonic gyre, the three branches, associated with the SAF, the PF and the SACCF are deflected in crossing the gap between the Antarctic-Pacific Ridge and the East Pacific Ridge around 120°W, and then flow toward the Drake Passage.

5.3 The Periantarctic Coastal Current

South of 65°S, along the Antarctic continent, in a region nearly continually ice-covered, coastal easterly winds entrain a Periantarctic Coastal Current (PACC) [**Figure 2-18**]. In this region, where access is difficult, rare observations of westward ice drift and ship drift have been episodically observed. Satellite photos permitted tracking some portions of the drift of icebergs[161] along the coast, but the tracks were random. Starting in 1972, satellite-tracked beacons were installed on large icebergs [**Figure 2-24**]. Their trajectories[162] show three regions where parts of the Periantarctic Coastal Current detach northward from the coast to join the Antarctic Circumpolar Current. All three are linked to bottom topography. Since then, numerous iceberg drifts, followed using daily satellite photos, have confirmed these trajectories.

---

[160] With 3,600 m depth, the Princess Elizabeth Channel is between the Kerguelen Plateau and the continent, around 64°S.

[161] Icebergs are entrained by the current due to their deep draft. For example, an iceberg 50 m high has a draft of about 300–350 m.

[162] Average velocities of the icebergs ranged between 8 and 15 km per day (10 to 17 cm/s).

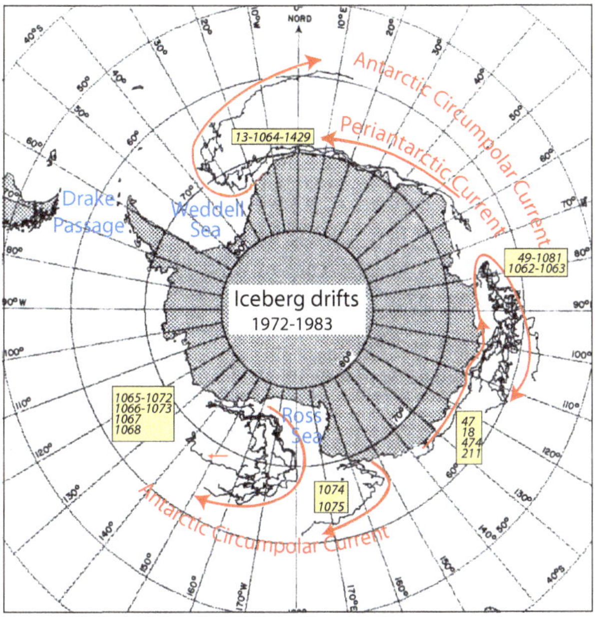

FIG. 2-24 – Nineteen iceberg drifts in the Periantarctic Coastal Current tracked by satellite. Red arrows show average motion; yellow boxes are transponder numbers. This was an historic experiment by Paul Tchernia and the CNES. (After Tchernia and Jeannin, 1983.)

Near the coast, where the vertical structure is most nearly barotropic [**Section I.4.3**], bottom topography has a strong influence on surface currents, which extend down to the bottom. The Periantarctic Coastal Current generally follows the topography of the continental slope. At the eastern entrance of the Ross Sea, it divides into two branches. One branch follows the edge of the continental shelf and the other penetrates far into the Ross Sea, under the ice shelf, and then rejoins the exterior branch near Cape Adare. One part of the Periantarctic Coastal Current is turned northward by the rise of the bottom in the region of the Balleny-Macquarie Ridge south of New Zealand, and then is entrained eastward by the Antarctic Circumpolar Current [**Figure 2-24**].

Similarly, the southern extension of the Kerguelen Plateau turns icebergs northward along its eastern flank then icebergs are entrained eastward by the ACC. The effect of topography, even when deep, [**Figure 2-6**] counteracts the Coriolis effect, which has a tendency to hug the current to the coast.[163]

The other part of the Periantarctic Coastal Current continues westward, through the Princess Elizabeth Channel and along the coast as far as the

[163] Turning currents to the left in the Southern Hemisphere.

Weddell Sea, where it represents the southern part of the cyclonic circuit of the Weddell-Enderby Basin, which turns northward guided by the Antarctic Peninsula, and the South Antilles. The extension of this Pericontinental Current is limited in summer to the drifting pack ice zone, which can extend up to 65°S. Its transport of about 40 to 50 Sv varies according to the sector. The PACC is not found in the Drake Passage as there are no regular easterlies there.

The three branches that detach from the Periantarctic Coastal Current form the western boundary current of the Weddell-Enderby Basin (up 20°E–40°E), the Antarctic-Australian Basin[164], and the Ross Sea (up to 135°W) cyclonic gyres. [**Figures 2-6, 2-17** and **2-18**].

[164] The western boundary current transport of the Antarctic-Australian cyclonic gyre is around 36 Sv.

Chapter 6
Water properties

6.1 Surface temperature

Warm and humid air masses from lower latitudes are entrained by the austral disturbances southeastward over the Antarctic Ocean but barely penetrate the high Antarctic citadel aside from the indentations of the Ross and Weddell Seas. On the other hand, extremely cold winds (katabatic winds) close to the Antarctic continent sweep out over the ocean at whose expense they are warmed.

Surface temperatures show a more or less zonal distribution, modified by ocean circulation [**Figure 2-25**]. Coming out of the Drake Passage, cold water is turned north. North of the Weddell Sea and north of the Ross Sea, very low temperatures extend north and northeast, entrained by the cyclonic circulations of these two regions. Between the region of Kerguelen and the SE Pacific sector, isotherms shifted southward over more than 10° of latitude. Wiggles in the isotherms are due to the presence of many eddies, visible on the trajectories of drifting buoys [**Figure 2-16**]. Aside from the coastal regions of the Antarctic continent, the Weddell Sea, and the Ross Sea, where the temperatures remain below −1 °C, even in summer, seasonal temperature variations increase inversely with latitude and can reach 4 °C to 5 °C near the Subtropical Front.

6.2 Salinity

At the very low temperatures of the Antarctic Ocean, freezing is the principal process affecting salinity. At the moment of pack-ice formation in winter, freezing separates fresh water from seawater, which then becomes more saline, and thus denser. It plunges by gravity, entraining salt to depth.

In winter, precipitation falls in the form of snow onto the pack ice, which then extends up to 55°S–60°S depending on the sector [**Figure 2-13**]. It does not play a role in the water balance until the following summer, when the pack ice melts.

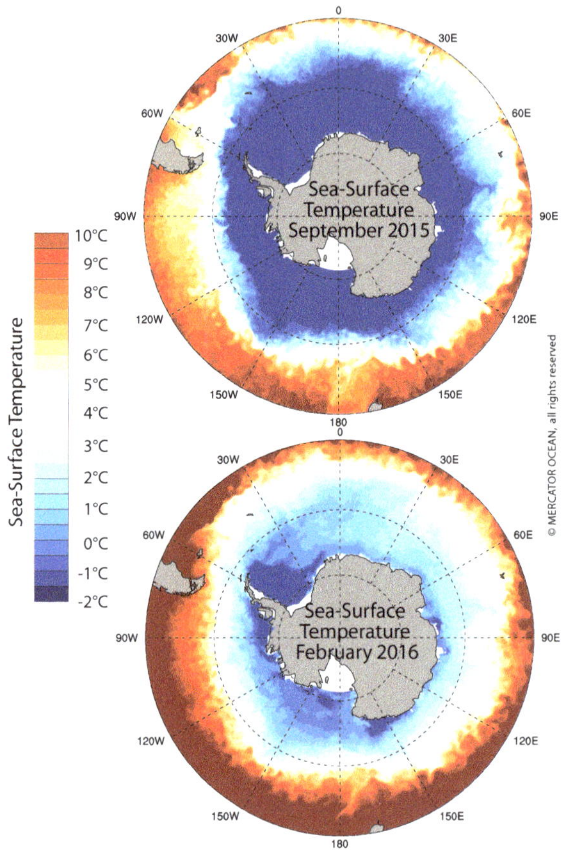

FIG. 2-25 – Average surface temperature of the Antarctic Ocean (**a**) at the end of the austral summer (February, 2016) and (**b**) at the end of the austral winter (September, 2015). The lowest temperatures correspond to sea ice, see **Figure 2-13**. (Image generated using E.U. Copernicus Marine Service Information and interpreted by G. Garric, Mercator Océan. All rights reserved.)

The net result of water fluxes through precipitation, melting, evaporation, and freezing, is manifested in the surface salinity. Salinity at 50 m depth[165] [**Figure 2-26**] shows, as does temperature [**Figure 2-25**], a relatively zonal distribution. It has a minimum ($S < 34$) in the region of the Antarctic Polar Front due to heavier precipitation and to the melting of the pack ice in summer [**Figure 2-27**]. It has a maximum near the coast, due to the influence of more saline Circumpolar Deep Water, which upwells in the Antarctic Divergence region [**Figures 2-21** and **1-35**]. The impact of cyclonic circulation on salinity distribution can be seen in the Weddell-Enderby Basin due to

[165] Obtained with the aid of hydrographic stations during the World Ocean Circulation Experiment (WOCE) in 1990–1998 (at 50 m to avoid surface variability).

Circumpolar Deep Water upwelling in the center of the cyclonic circuit. This higher-salinity water from the Weddell Sea is entrained eastward east of the Scotia Arc. It is also seen off the Ross Sea.

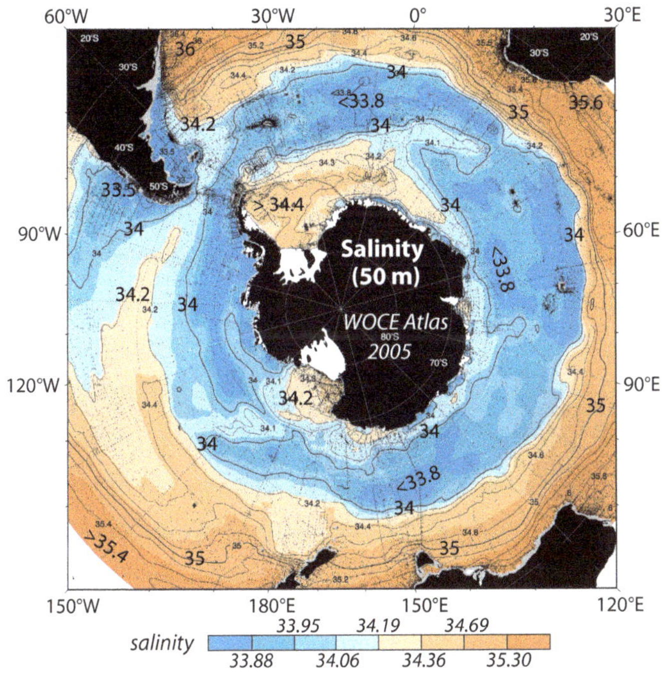

FIG. 2-26 – Salinity at 50 m. (From *WOCE Southern Ocean Hydrographic Atlas*, A.H. Orsi and T. Whitworth, 2005.)

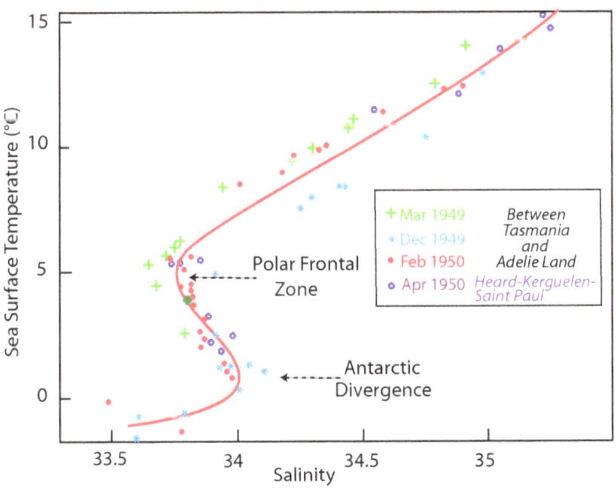

FIG. 2-27 – θ-S diagram from surface measurements taken during four passages between Tasmania and Adélie Land, and one between Heard, Kerguelen and Saint Paul. The Polar Frontal Zone is marked by a salinity minimum and the Antarctic Divergence by a salinity maximum. (After Tchernia, 1978.)

6.3 Antarctic Ocean Water Masses

The block diagram [**Figure 2-28**] was created on the basis of pioneer sections, such as the section taken south of Australia during the 1930's [**Figure 2-29**]. It presents the principal water masses in the Subantarctic and Antarctic regions schematically. It shows a deep water mass advecting toward the Antarctic sector from the north: North Atlantic Deep Water (NADW); the northward expansion of a water mass located on the bottom: Antarctic Bottom Water (AABW); and a water mass in intermediate position: Antarctic Intermediate Water (AAIW) [see also **Figure 1-35**].

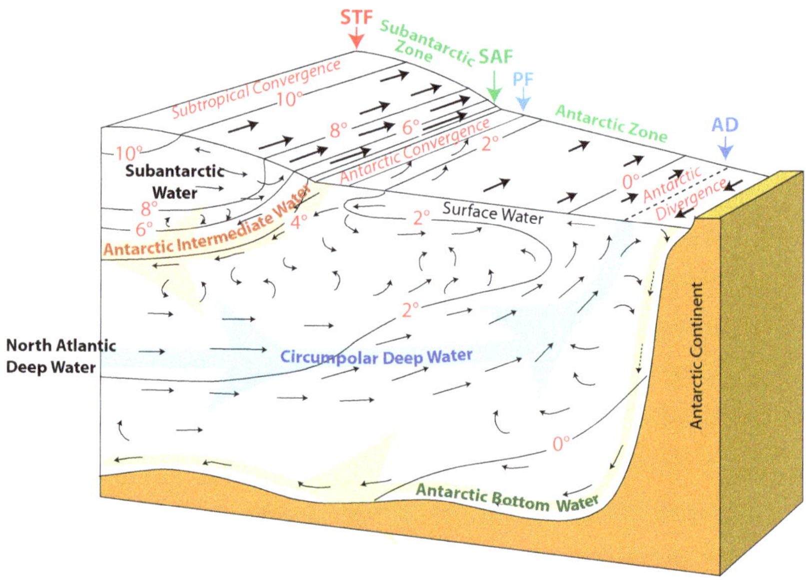

FIG. 2-28 – Schematic block diagram of the Antarctic Ocean showing the vertical motion of the principal water masses. In 1933, after the Discovery II expeditions, Deacon called the region where Antarctic Intermediate Water sinks the Antarctic Convergence, now divided into the SAF and the PF, and he called Subtropical Convergence that which is now called STF. (After Sverdrup, 1942.)

Observations taken with high horizontal resolution and continuous vertical profiles during the international WOCE program show the spatial variability detail necessary for quantifying numerous processes involved in ocean dynamics. Among these processes are the separation of currents branches, eddies, convection, subduction, and mixing. The WOCE observations also define more precisely the oceanographic regions of the Austral Ocean. As an example, **Figure 2-30** shows meridional distributions of temperature, salinity,

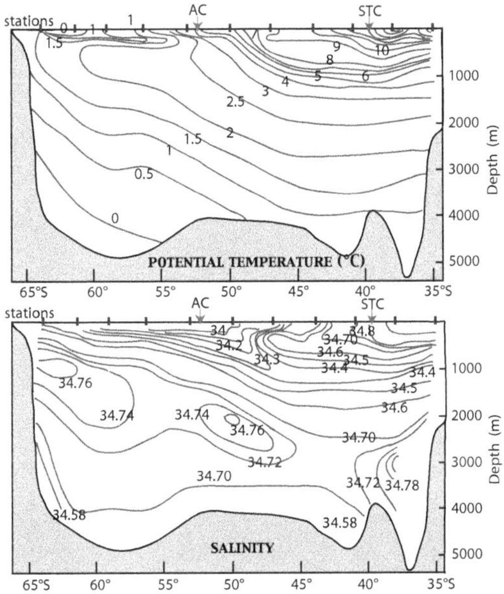

FIG. 2-29 – Temperature and salinity during the 1930's along a section south of Cape Leeuwin, the southwestern point of Australia, at 115°E.
AC = Antarctic Convergence = actual SAF, Polar Frontal Zone and PF. STC = Subtropical Convergence = actual STF. (After Deacon, 1937.)

oxygen, CFC11, and neutral density for three sections located in the SE Atlantic (Section WOCE S2), south of Australia (Section WOCE I9), and in the Drake Passage (Section WOCE S1). The collection of data shows relatively zonally uniform oceanographic conditions, due to the zonal distribution of climatic conditions and circulation. However, in the Ross and Weddell Seas the formation of very dense water and cyclonic circulation, specific to these regions, modify the zonal distribution of properties. It is interesting to compare section WOCE I9[166] **[Figure 2-30]** with that made in the same place sixty years earlier **[Figure 2-29]**. The older section shows the same general structure, but the vertical and meridional gradients are less well defined.

---

166 WOCE I9 was made with a CTD probe giving continuous profiles from surface to bottom, and included much closer station spacing. The WOCE section has 64 stations as opposed to 9 stations and 12 measurement levels for the older section.

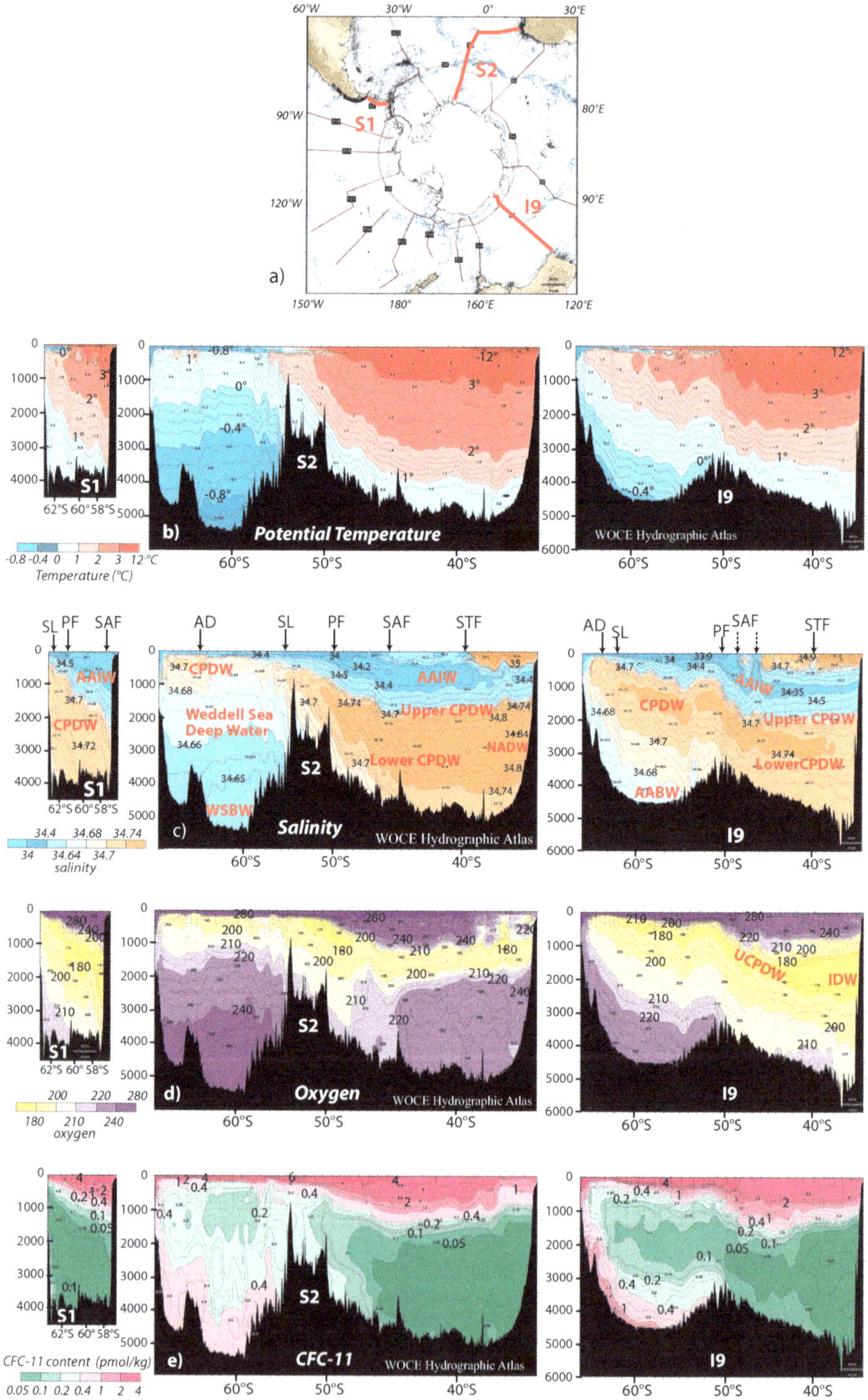
60°W
30°W
0°
30°E
90°W
80°E
120°W
90°E
150°W
180°
160°E
120°E
S1
S2
I9
a)
b)
Potential Temperature
WOCE Hydrographic Atlas
-0.8 -0.4 0 1 2 3 12 °C
Temperature (°C)
62°S 60° 58°S
60°S
50°S
40°S
SL
PF
SAF
AD
STF
CPDW
AAIW
Weddell Sea Deep Water
Upper CPDW
Lower CPDW
NADW
WSBW
AABW
LowerCPDW
c)
Salinity
34.4 34.68 34.74
34 34.64 34.7
salinity
d)
Oxygen
UCPDW
IDW
200 220 280
180 210 240
oxygen
e)
CFC-11
CFC-11 content (pmol/kg)
0.05 0.1 0.2 0.4 1 2 4

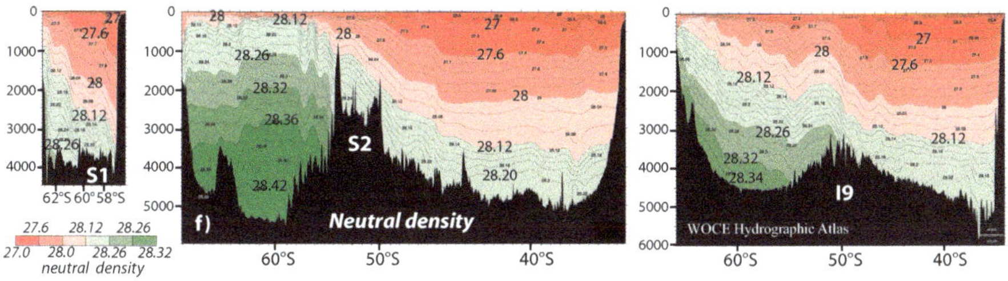

FIG. 2-30 – (**a**) The position of the WOCE meridional sections S1, S2, and I9, and (**b**) potential temperature (in °C), (**c**) salinity, (**d**) dissolved oxygen (in µmol/kg), (**e**) CFC11 (F11) (values in pmol/kg), and (**f**) neutral density [**Section I.3.4.2**]. The position of fronts and water masses are indicated on the salinity distribution. [North Atlantic Deep Water (NADW), Circumpolar Deep Water (CPDW), Antarctic Intermediate Water (AAIW), Antarctic Bottom Water (AABW).] Section S1 in January–March 1990; Section S2 in May–July 1992 north of 48°S and March–May 1998 south of 48°S; Section I9 in December–February 1995. (Sections S1, S2, I9, from *eWOCE Hydrographic Atlas,* Orsi and Whitworth, 2005.)

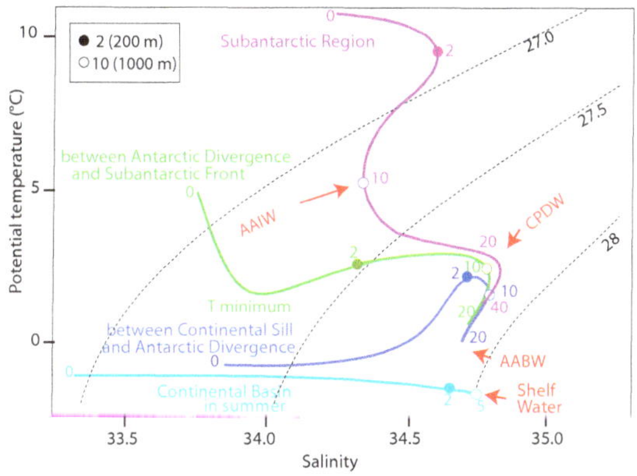

FIG. 2-31 – Evolution of θ-S diagrams in the four principal regions of the Antarctic Ocean. [Antarctic Bottom Water (AABW), Circumpolar Deep Water (CPDW), Antarctic Intermediate Water (AAIW).] Depths are in hundreds of meters. (After Tchernia, 1978.)

The profiles in **Figure 2-31** show the evolution, from south to north, of the water-mass characteristics of the Austral ocean region.

- On the shelves and in the continental basins waters are cold and dense, near the freezing point in winter. In summer, dense waters are trapped in the bottom of the basins below sill level (shelf water) and surface water is desalinated by the effect of pack-ice melting.
- In the Subpolar Zone between the continental shelf (Antarctic Slope Front) and the Antarctic Divergence, an attenuated trace of the maximum temperature of the Circumpolar Deep Water (CPDW) remains.

- In the Antarctic Zone, south of the Polar Front, the profile shows, in turn, a subsurface temperature minimum representing a trace of winter water, then the salinity and temperature maximum of Circumpolar Deep Water and, at the bottom, the temperature minimum of Antarctic Bottom Water (AABW).
- In the Subantarctic Zone, between the Subantarctic Front and the Subtropical Front, above the Antarctic Bottom Water and the Circumpolar Deep Water, Antarctic Intermediate Water (AAIW) appears, characterized by a salinity minimum around 600 m – 1,000 m depth. At the surface, Subantarctic Water is the warmest and often comprises a desalinated surface layer.

The modes of formation and characteristics of each of these water masses are described in the following sections.

6.3.1 Antarctic Circumpolar Deep Water

All the meridional sections in **Figure 2-30** show a salinity maximum ($S > 34.70$) which rises from 3,000 m at the latitude of the Subantarctic Front (SAF) to 150 m at the latitude of the Antarctic Divergence (AD). This salinity maximum follows the passage of North Atlantic Deep Water [**Section II.6.9**], which feeds Antarctic Circumpolar Deep Water (CPDW). North Atlantic Deep Water (NADW) penetrates into the Austral Ocean principally east of the South Antilles, below the sill depth, as a western boundary current [**Section I.5.3.3**] and is then entrained eastward by the powerful Antarctic zonal circulation around the Antarctic continent while slowly transforming through mixing along its path. On its northern side and in its upper part, it receives Indian Deep Water (IDW) and Pacific Deep Water (PDW) whose origin is CPDW transformed in each ocean becoming less dense. On its southern side and below, it mixes with colder less saline Bottom Water formed on the shelves [**Figure 2-28**]. A secondary branch of North Atlantic Deep Water enters the Antarctic Ocean south of Africa. Then turbulent mixing, intensified in the fronts, transforms North Atlantic Deep Water into Circumpolar Deep Water.

The generally rising southward movement of Circumpolar Deep Water is apparent in all water properties. It is connected to the slope of the isopycnals, which expresses the geostrophic equilibrium of the Antarctic Circumpolar Current. This gigantic upwelling corresponds to the distribution of curl (τ/f) [**Figure 2.32**] [**Section I.5.3.1**]. The region where Circumpolar Deep Water upwells to the surface undergoes a positive wind-stress curl, which entrains upwelling.

Furthermore, the formation of very dense water near the continent (through loss of heat and formation of sea ice) which then flows northward on the bottom, implies the need for a balancing input of water to the Austral Ocean. Antarctic Intermediate Water, which forms in the convergence zones by wind-generated subduction, also extends northward, needing a replacement of water.

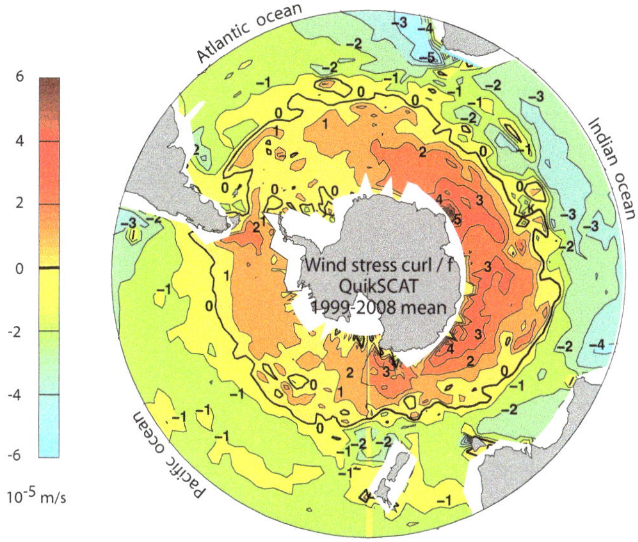

FIG. 2-32 – Mean wind-stress curl divided by the Coriolis parameter (curl τ/f, positive values correspond to divergence). (Courtesy of Y.-H. Park, using QuickScat satellite data, 1999–2008 mean.)

However, between latitudes 56°S and 63°S (the limits of the Drake Passage) the Antarctic Ocean has no continental barrier above the sill depths of the Scotia Arc. Only below sill levels (about 2,500 m–3,000 m) can a zonal pressure gradient be established through pressure acting on the continental edge, thereby providing a meridional water supply. Thus, preferentially, the "deep" water mass coming from the Atlantic basin balances the output of Antarctic Bottom Water and Antarctic Intermediate Water. Above sill levels, turbulent processes play a role in the meridional transport of heat and salt across the Antarctic Circumpolar Current.

Circumpolar Deep Water is characterized by salinity between 34.70 and 34.80 and temperature between 0.5 °C and 2.7 °C [**Figure 2-31**]. Salinities greater than 34.76, in the Atlantic sector indicate the influence of North Atlantic Deep Water. The salinity maximum value diminishes in two ways. First, from west to east, going from the entry region of the source water mass in the South Atlantic over to the eastern Pacific [**Figure 2-33**]. Second, from north to south, by mixing with Antarctic Bottom Water below and with Antarctic Intermediate Water above, both of which are less saline [**Figure 2-31**] and by mixing along the slope with water formed on the shelf.

Circumpolar Deep Water is a thick water mass. Lower Circumpolar Deep Water, the coldest which includes the salinity maximum, lies above Antarctic Bottom Water with which it mixes. The salinity-maximum core of LCPDW occurs above the bottom in the Indian Ocean and on the bottom in the Pacific Ocean, where AABW disappears. Upper Circumpolar Deep

Water, the warmest, poor in oxygen and CFCs, is the "oldest"[167] water of the region [**Figures 2-20d**, **2-29e** and **2-33**] and is influenced Indian Deep Water and Pacific Deep Water.

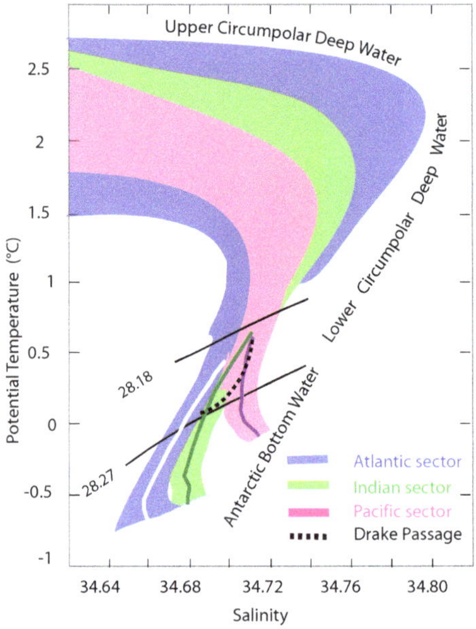

FIG. 2-33 – Average θ-S diagrams (and their associated standard deviations shown by colored regions on either side of the average curves) of deep waters in the Circumpolar Current. The Atlantic sector between 60°W and 80°E is blue, the Indian sector between 80°E and 160°E is green, and the Pacific sector between 160°E and 70°W is pink. The Drake Passage between 70°W and 60°W and 65°S and 55°S is dotted black. Neutral isopycnals 28.18 in the core of LCPDW and 28.27 is the limit between AABW and LCPDW. (After Orsi *et al.*, 1999.)

In the Atlantic sector, the range of variation of the water characteristics of Circumpolar Deep Water is very large [**Figures 2-33** and **2-30,** Section S2]. This is due in the north to the input of North Atlantic Deep Water and, in the south, to the existence of colder and less saline Bottom Water and Deep Water from the Weddell Sea. The amplitude of this variability weakens progressively eastward as far as the eastern Pacific sector where the difference between Lower Circumpolar Deep Water and Antarctic Bottom Water becomes slight. The limit between Antarctic Bottom Water, of a variety of origins, and Lower Circumpolar Deep Water can be fixed at the 28.27 neutral isopycnal. The 28.18 neutral isopycnal is at the core of Lower Circumpolar Deep Water.

[167] Age is the time since a water mass quit the surface at the place where it acquired its characteristics, in particular where it was charged with oxygen and CFCs, through exchanges with the atmosphere.

Below the surface layer, Circumpolar Deep Water, characterized by a salinity maximum[168] and an oxygen minimum, rises to 150–200 m in the region of the Antarctic Divergence. Its influence is visible in the 200 m property distributions [**Figures 2-34a** and **b**].

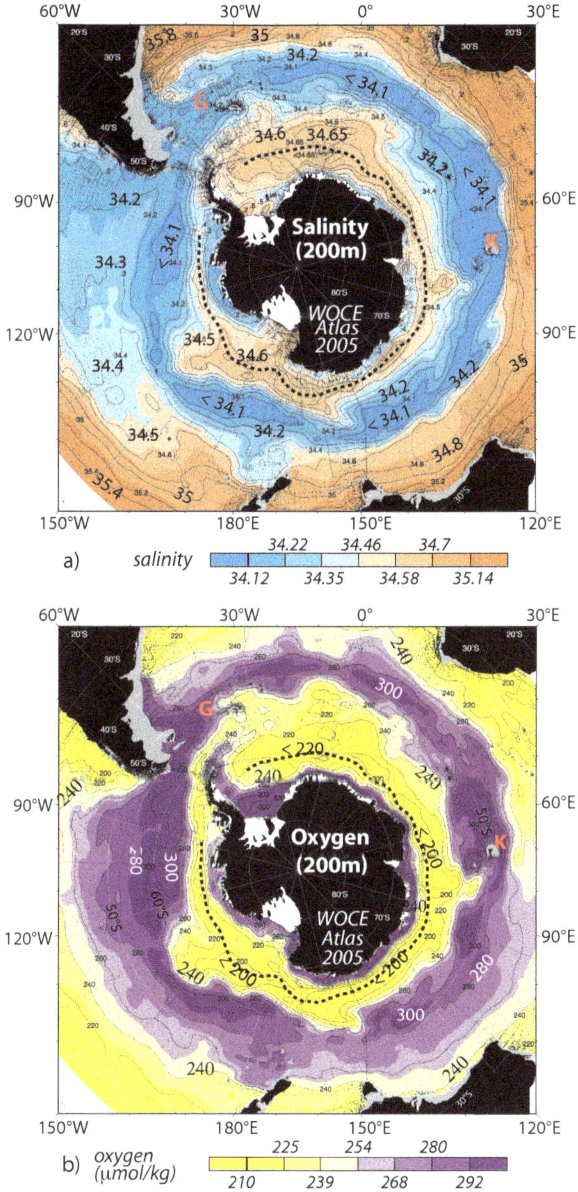

[168] Upper Circumpolar Deep Water is sometimes detectable even at the surface, as in the surface meridional *T-S* diagram [**Figure 2-26**].

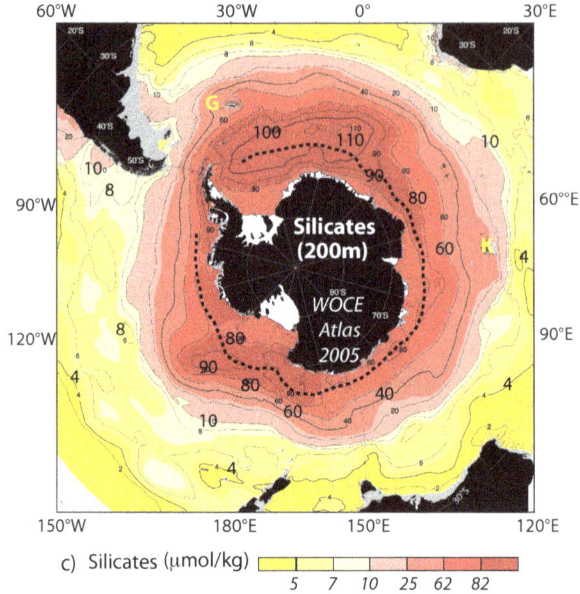

FIG. 2-34 – 200-m distribution of (**a**) salinity, (**b**) dissolved oxygen, and (**c**) silicate concentration. South Georgia (G) and Kerguelen (K) are shown as well as the Antarctic Divergence (heavy dotted line). (From *WOCE Southern Ocean Hydrographic Atlas*, Orsi and Whitworth, 2005.)

The Antarctic Divergence is a region with a strong concentration of nutrients[169] among which are silicates, used as deep-water markers [**Figure 2-34c**]. Further south, Circumpolar Deep Water becomes indistinct through mixing with Pericontinental Water (also called Continental or Shelf Water): very dense, less saline, and weakly stratified, at the origin of bottom waters [**Figure 2-31**].

Circumpolar Deep Water transfers heat at the surface and plays a role in the melting of pack ice in summer, diminishing its own temperature and salinity. Strong winter cooling forms a cold layer from 150 to 200 m thick, a trace of which is found in summer in the form of a subsurface temperature minimum as far as the Polar Front [**Figures 2-30** and **2-31**]. Thus, under the effect of local exchanges with the atmosphere, Circumpolar Deep Water is transformed into Antarctic Surface Water, colder and less saline, which is displaced northward by wind action (Ekman transport).

[169] When the pack ice disappears in summer, millions of diatoms develop in the region of the Antarctic Convergence. These serve to nourish small crustaceans called "krill". Krill stocks provide an immense reserve of nutrition on which the largest animals on the planet feed: large whales.

6.3.2 Antarctic Intermediate Water and Subantarctic Mode Water

In summer, south of the Subantarctic Front in the region of the Polar Front, precipitation and ice-pack melting diminish the salinity of Antarctic Surface Water [**Figure 2-26**]. In winter, the drop in temperature produces a mixed layer of lower salinity through convection. These desalted surface waters from the Polar Frontal Zone are entrained eastward by the ACC and northward by Ekman transport. In winter in the subantarctic region north of the Subantarctic Front [**Figure 2-20**] they contribute to the formation by convection of Subantarctic Mode Water (SAMW).[170] For example, Station 1 of **Figure 2-35** located between the STF and the SAF is homogeneous down to 600 m, indicating recent ventilation. This is an indication of very dense Subantarctic Mode Water, which spreading by subduction under warmer and less deep Subtropical Mode Water [**Figure 2-35**, Stations 2 and 3] mixes but keeps a weak vertical gradient. Temperature profiles have nearly vertical slope. Station 4 shows Mode Water under warmer and less saline surface water.

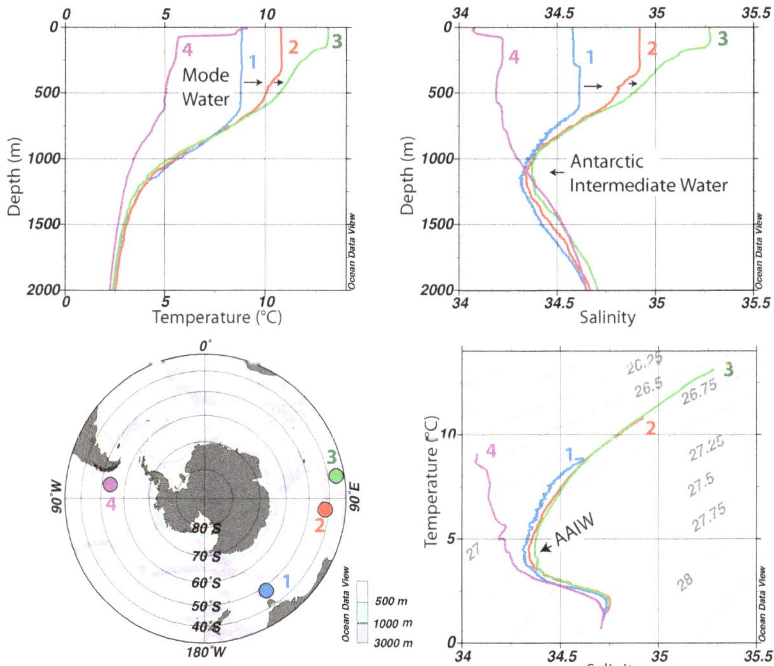

FIG. 2-35 – Temperature and salinity profiles and θ-S diagrams from stations containing Subantarctic Mode Waters. (Station 1 in Sept. 1996, Station 2 in Oct. 1995, Station 3 in Oct. 1995 and Station 4 in Feb. 1993.) (With ODV and data from WOA05.)

[170] Subantarctic Mode Water is characterized by a potential vorticity minimum [**Section I.5.3.4**].

One could say that Antarctic Intermediate Water (AAIW) is the densest and least saline form of Subantarctic Mode Water, characterized by a salinity minimum and a maximum concentration of CFCs and oxygen [**Figure 2-30**]. These characteristics evolve during mixing with overlying water: Central Water [**Section III.6.2**] and underlying water: Upper Circumpolar Deep Water and North Atlantic Deep Water. AAIW density (σ_θ) is in the neighborhood of 27.05–27.1 in the SE Pacific, 27.2 in the SW Atlantic, and 27.25 in the SE Indian Ocean [**Figure 2-36**]. Its salinity minimum occurs between 800 m and 1,200 m depth according to basin.

Antarctic Intermediate Water and Subantarctic Mode Water are entrained by the Circumpolar Current eastward and subducted northward under warmer and more saline lighter water [**Figures 2-28** and **2-30**] by the effect of the wind (Ekman convergence) [**Figure 2-32**] preferentially in the eastern section of the subtropical gyres of each basin. The characteristics of Antarctic Intermediate Water [**Figure 2-36**] and Subantarctic Mode Water [**Figure 2-35**] vary regionally. Going from the Atlantic to the Pacific, Subantarctic Mode Waters become colder, less saline, and denser [**Figure 2-35**] due to the addition of water from the Polar Front region through Ekman transport. The lowest salinity minimums of Antarctic Intermediate Water are found in the SE Pacific sector and SW Atlantic sectors [**Figure 2-36**, Stations 7 and 1].

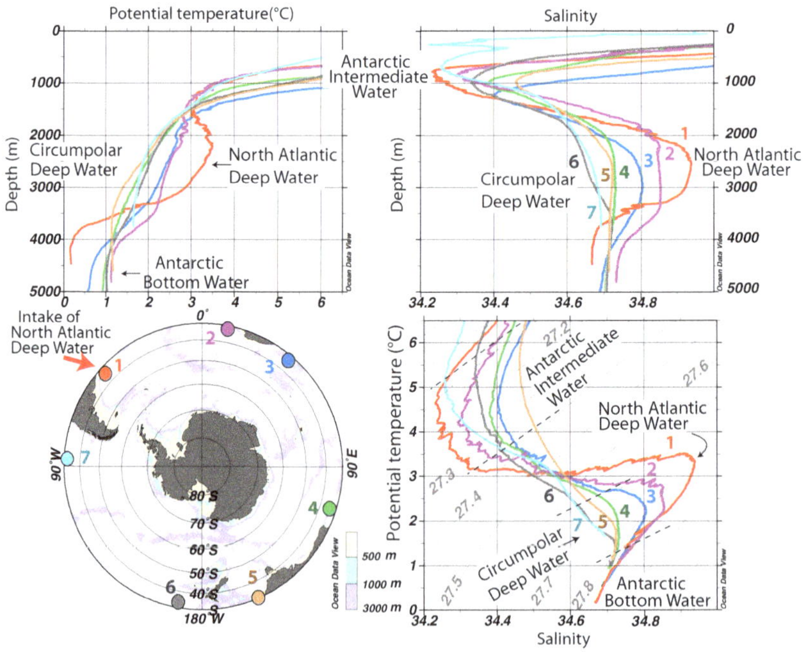

FIG. 2-36 – Evolution of Antarctic Intermediate Water and Circumpolar Deep Water north of the Antarctic Ocean around 34°S. (With ODV and data from WOA05.)

In the different sectors of the Austral Ocean, the variability of observed water properties of Subantarctic Mode Water and Antarctic Intermediate Water is due to several factors:

- Local atmospheric-flux variability (precipitation, evaporation, cooling...).
- Variability of Ekman transport and wind curl associated with northward subduction across the Subantarctic Front.
- Turbulent activity entraining northward diapycnal[171] fluxes.
- Recirculation of water masses in the interior of anticyclonic gyres south of each ocean.

The favored transformation site of Subantarctic Mode Water into Antarctic Intermediate Water seems to extend from the SE Pacific sector from where it is entrained eastward across the Drake Passage and the Scotia Sea [**Figures 2-30** and **2-36**] to the zone of confluence between the Malvinas Current and the Brazil Current around 35°S. Additions of Antarctic Intermediate Water and Subantarctic Mode Water seem to occur from the Indian Ocean to the SE Atlantic by means of mesoscale eddies[172] of the Agulhas Current as well as from the Pacific toward the Indian Ocean south of Tasmania by eddies of the East Australian Current, to which is added a branch of the deep current transporting Antarctic Intermediate Water and Subantarctic Mode Water toward the Indian Ocean, north of the Antarctic Circumpolar Current.[173]

The old vision of regular formation of Antarctic Intermediate Water along the Subantarctic Front all around the Antarctic Ocean is in reality more complex. Rather, it seems that Antarctic Intermediate Water and Subantarctic Mode Water enter into the interior of anticyclonic gyres in the eastern and central sectors of each basin, except in the SW Atlantic. Because of the difficulty of crossing the powerful Circumpolar Current, penetration northward is not continuous.

6.3.3 Bottom Water and Deep Water in the Weddell Sea

The principal process responsible for the formation of very dense water at the origin of the Antarctic Bottom Water is the increase of salinity through freezing of seawater, for which some variations can be distinguished:

- During the Antarctic winter, on the continental shelves, surface-water temperature decreases to the freezing point of seawater of –1.909 °C for a salinity of 35, and –1.853 °C for a salinity of 34 [**Figure 1-6**]. By extracting fresh water, this process increases surface-water salinity and thus its density in a region where stratification is already weak

[171] Diapycnal fluxes are mixtures of water masses of slightly different densities.
[172] The diameter of a mesoscale eddy is of the order of 50 to 100 km.
[173] This branch of current is called the "Tasman Leakage".

[**Figures 2-30** and **2-31**]. The density increase produces convective motion which homogenizes a layer more or less deep during pack-ice formation. When pack ice reaches 2- to 3-m thickness,[174] ice formation stops and dense water formation ceases.

- The formation of dense water by freezing of seawater can combine with additional cooling[175] in basins several hundred meters deep under ice shelves where pressure allows seawater to remain in liquid form at temperatures lower than the surface freezing point [**Figure 2-37**].

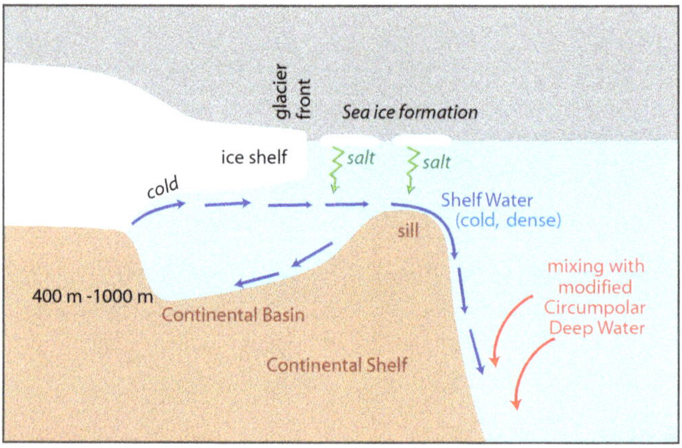

FIG. 2-37 – Sketch showing the influence of ice shelves and continental shelf basins on the formation of very cold Antarctic Bottom Water.

- The extraction of fresh water by freezing can be prolonged in regions of coastal polynyas (revealed by satellites) where ice is continually driven offshore by violent katabatic winds coming from the continent [**Figure 2-14**] thereby amplifying the injection of salt due to freezing. Salt is thus transferred to depth during the period of activity of the coastal polynyas and forms very dense water.[176] The Weddell Sea region is subject to this type of process. Very cold and dense water forms there in winter on the continental shelf where salinity, relatively weak, varies between 34.3 and 34.7 and where the temperature drops down to –2 °C on the bottom of basins located under the Filchner and Ronne Ice Shelves. When the basin fills with very dense and very

[174] Thick ice isolates the ocean from the atmosphere.
[175] Additional cooling increases density.
[176] The loss of heat in coastal polynyas is estimated to reach 200 W m^{-2} enough to form a layer of about 10 cm of ice per day.

cold Shelf Water the water overflows toward the continental slope of the Antarctic Peninsula [**Figure 2-37**] then is turned northward [**Figure 2-38**] by Coriolis force and flows to the floor of the Weddell Sea in a cyclonic movement. During its plunge along the continental slope [**Figure 2-39**, Station 1: plateau water] it mixes with Weddell Sea Deep Water (WSDW) to form Weddell Sea Bottom Water (WSBW) showing a maximum oxygen concentration ($O_2 > 5.9$ ml l^{-1} or 255 µmol kg^{-1}),[177] a salinity minimum ($S = 34.63$ to 34.66), a temperature minimum (–0.9 °C to –0.8 °C), and a potential density maximum, σ_θ of about 27.88 (or a neutral density of about 28.4).

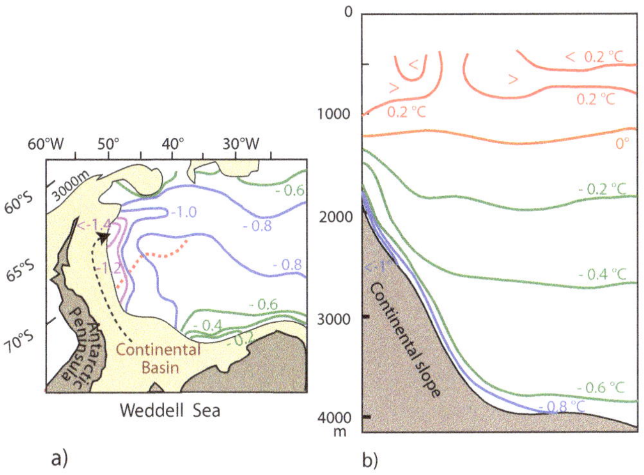

FIG. 2-38 – (**a**) Temperature on the bottom of the Weddell Sea. The Shelf Water trajectory is shown by a lightly black dotted line, the location of Section (b) by a heavy red dotted line, and the 3000m depth by a solid line. (**b**) Vertical section of potential temperature. (After Warren, 1981.)

In the cyclonic gyre of the Weddell sea, Weddell Sea Deep Water (–0.7 °C < T < 0 °C, 220 µmol kg^{-1} < O_2 < 250 µmol kg^{-1}, 34.66 < S < 34.68) is a mixture more than 2,000 m deep of Circumpolar Deep Water and Weddell Sea Bottom Water [**Figure 2-39**].

177 See the units for converting oxygen concentration in **Section I.3.4.3**.

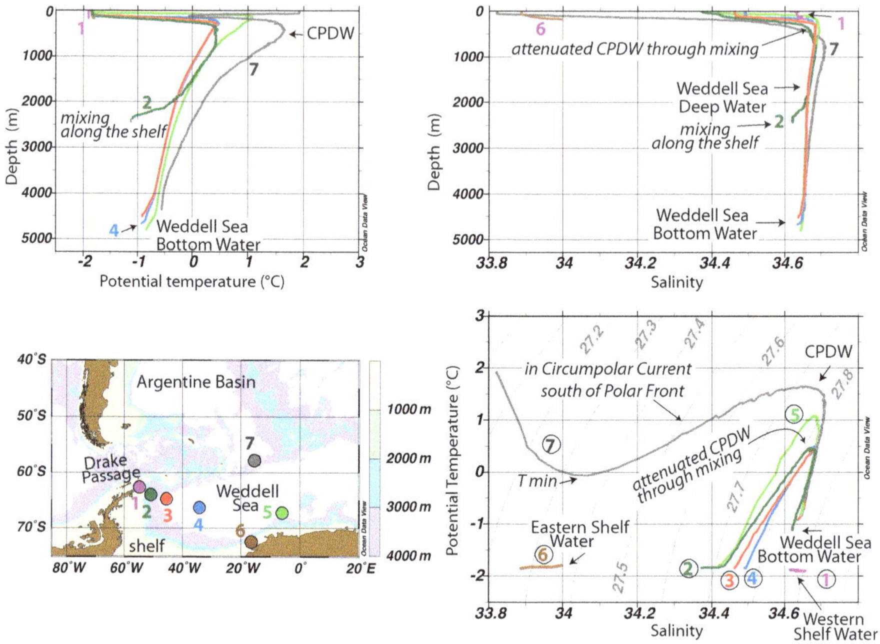

FIG. 2-39 – θ-S diagrams, and temperature and salinity profiles in the Weddell Sea (Stations 1 to 6) and in the Circumpolar Current (Station 7), CPDW: Circumpolar Deep Water. (With ODV and data from WOA05.)

Stratification in the Weddell Sea is weak, especially in the center of the cyclonic circulation where the isopycnals rise[178] toward the surface (27.78 to 27.87 from the surface to the bottom at Station 4) [in blue, **Figure 2-39**]. Much weakened Circumpolar Deep Water, characterized in the region by a temperature and salinity maximum, approaches the surface in the center of the Weddell Sea [salinity profiles in Stations 3, 4, and 5, **Figure 2-39**]. Continental Shelf Water NE of the Antarctic Peninsula (Station 1) is the densest (very cold and relatively saline). Continental Shelf Water east of the Weddell Sea (Station 6) is also very cold, near the freezing point, but is less saline and thus less dense. Station 5 located east of the Weddell Sea has a salinity maximum higher than in the center of the cyclonic gyre (Station 4). Station 7, located in the circumpolar current south of the Polar Front carries a subsurface temperature minimum, a trace of the winter mixed layer, and a salinity and temperature maximum corresponding to Circumpolar Deep Water that is much more marked than in the Weddell Sea.

[178] See **Section I, Figure 1-66**.

Weddell Sea Bottom Water is the coldest, the most oxygenated, and the densest in the region. It is only present in the Weddell Sea [**Figure 2-40**]. Weddell Sea Deep Water is the principal source of Antarctic Bottom Water.

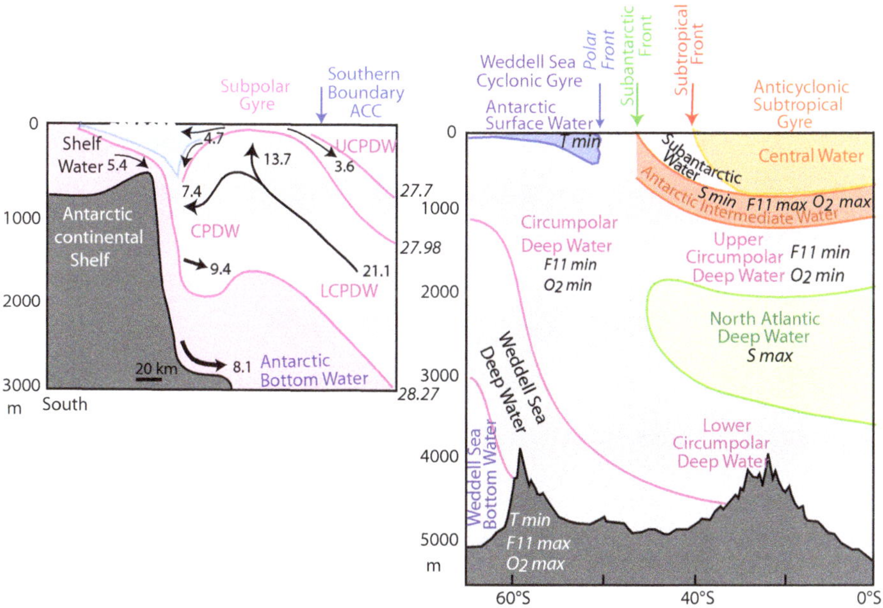

FIG. 2-40 – Antarctic water masses in relation to the oceanic fronts in the South Atlantic sector near the continental shelf, with transport estimates in Sv. (After Orsi *et al.*, 2002, Peterson and Whitworth, 1989.)

6.3.4 Antarctic Bottom Water

Between the Antarctic Divergence and the continent, Ekman transport due to wind has a tendency to displace Antarctic Surface Water toward the continental margin where it is entrained westward by the Periantarctic Coastal Current.

Outside of the Weddell Sea, dense water formation takes place all around the Antarctic continent. Dense water at the origin of Antarctic Bottom Water is formed at the continental margin, from Pericontinental Water whose density stratification is weak. Observations at the moment of formation are rare in regions covered or surrounded by pack ice, where access is difficult. Potential temperature and density distributions on the bottom show [**Figure 2-41**] that the Weddell Sea is the principal source of Antarctic Bottom Water.

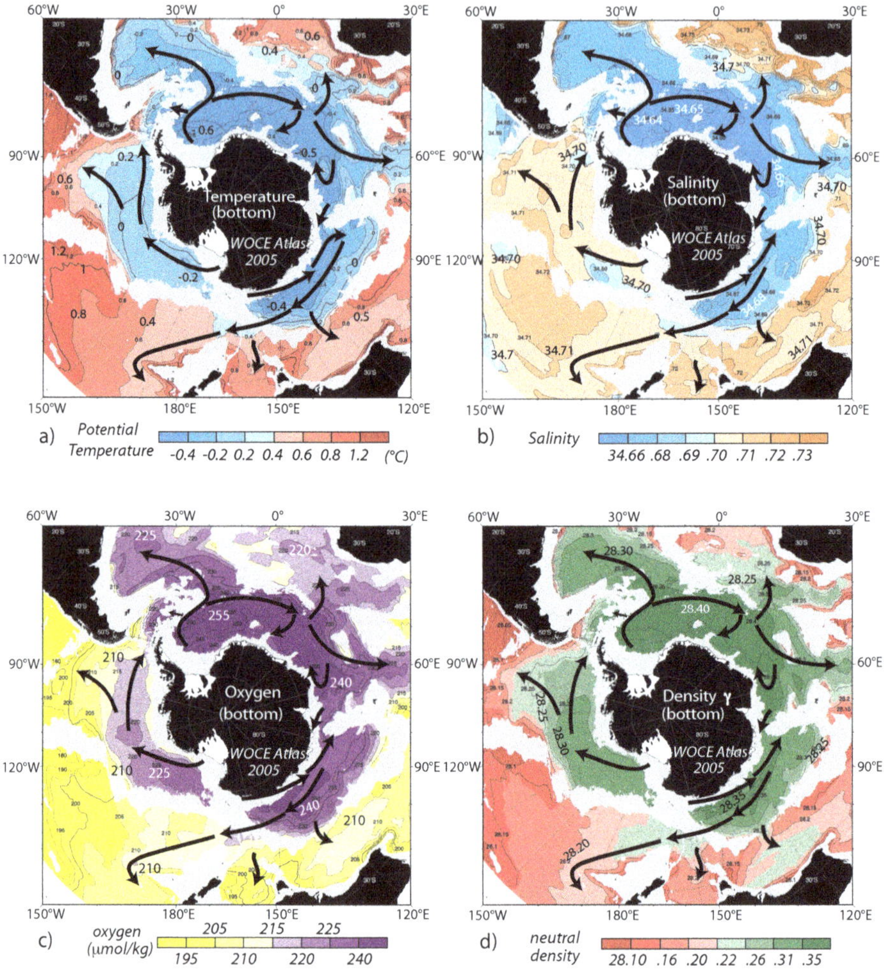

FIG. 2-41 – Antarctic Ocean bottom distribution of (a) potential temperature, (b) salinity, (c) dissolved oxygen, and (d) neutral density from WOCE sections. Arrows indicate the circulation deduced from this distribution. Depths less than 3,000 m are shown in gray. (From *WOCE Southern ocean Hydrographic Atlas*, Orsi and Whitworth, 2005.)

Nevertheless, other regions of formation exist [**Figures 2-42** and **2-43**] around the Antarctic Continent[179], such as off Adélie Land where salinity is slightly higher than in the Weddell Sea (34.7 measured in the continental basin in summer, where winter water is trapped). In the western Ross Sea

[179] The three most important region of dense water production through sea-ice formation are the Weddell Sea (~68%), the Adélie Land coastal region (~24%) and the Ross Sea (~8%).

salinity exceeds 34.8 and temperature is –1.95 °C. In the eastern Ross Sea salinity is lower (34,67). Dense water of the western Ross Sea is diverted westward along the slope, where it receives less saline water formed off Adélie Land. Thus in flowing westward, its salinity decreases, but its oxygen level is increased by the input of Adélie Land water along its path. It covers the bottom of the Antarctic-Australian Basin, mixing with dense water from the shelf of Wilkes Land ($S < 34.68$) and with Weddell Sea Deep Water that has crossed the Princess Elizabeth Channel (–0.4 °C on the 3,650 m deep sill). The thickness of Antarctic Bottom Water of more than 2,000 m in the Weddell Sea (Weddell Sea Deep Water) is reduced to 1,200 m south of the Kerguelen Plateau (76°E, 55°S) and to 800 m on the Balleny-Macquarie Ridge (160°E, 60°S).

θ-S diagrams from the subpolar region south of the southern boundary of the Circumpolar Current show regional variations. Western Ross Sea Water is more saline and that of the Weddell Sea is colder and less saline [**Figure 2-42**].

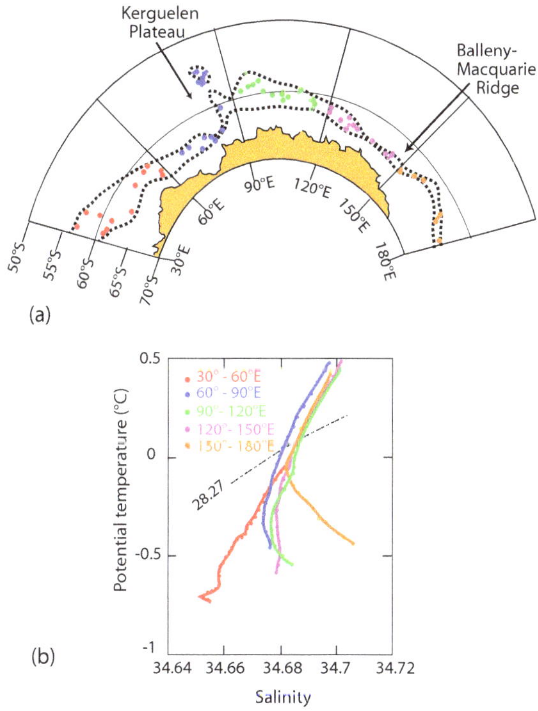

FIG. 2-42 – Average θ-S diagrams around the continent in the region between the southern boundary of the Circumpolar Current and the continental slope for potential temperatures less than 0.5 °C. The colors of the θ-S diagrams correspond to the station groups on the map and the blue-shaded areas around them show the envelopes of the variation in each group of stations. (After Orsi *et al.*, 1999.)

The various dense waters are entrained by the cyclonic circulation in each basin and then extend northward, slowly losing their characteristics through mixing [**Figure 2-43**].

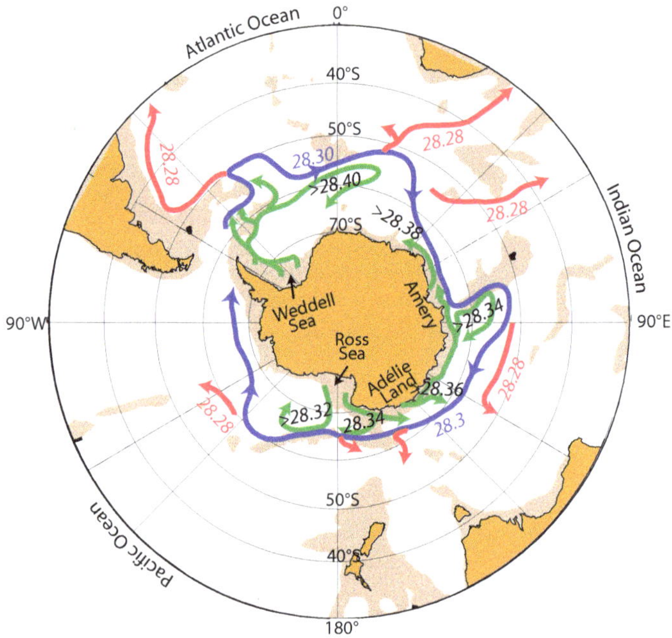

FIG. 2-43 – Diagram showing the trajectories of the various types of dense water formed around the Antarctic Continent from their place of formation, with the evolution of their neutral density (green for the most dense, red for the least dense). The 3,000 m isobath is shown shaded. (After Orsi *et al.*, 1999.)

Part of Weddell Sea Deep Water (WSDW) escapes westward, across the Scotia Ridge, and is responsible for lowering temperature and salinity in the Bransfield Strait,[180] a limited area south of the Scotia Sea [**Figure 2-41**]. Most of WSDW is entrained by the cyclonic circulation of the Weddell-Enderby Basin. Its flow is guided by topography, and turned by Coriolis force [**Figure 2-41**]. It becomes the Antarctic Bottom Water (AABW), which loses its characteristics little by little through mixing or suddenly by topographic blocking. To the east, temperature and salinity of AABW increase by mixing with warmer and more saline Lower Circumpolar Deep Water from the Weddell-Enderby Basin (average temperature –0.3 °C, average salinity 34.66), passing by the Antarctic-Australia Basin (average temperature –0.17 °C, average salinity 34.68), as far as the SE Pacific (average

[180] Bransfield Strait is between the northern point of the Antarctic Peninsula and the South Shetland Islands.

temperature 0.04 °C, average salinity 34.71) while density diminishes. In the Drake Passage, where the narrowing channel makes the current turbulent, Bottom Water salinity goes from 34.71 in the Bellingshausen Sea to 34.66 as a result of intense lateral mixing with leakage from Bransfield Strait [**Figure 2-33**]. Bottom temperature is 0.2 °C while it is –0.8 °C east of the Antarctic Peninsula.

Antarctic Bottom Water extends northward in each basin in the form of a western boundary current [**Figures 2-41** and **2-43**]. It is channeled and, sometimes, blocked by submarine topography. In the Atlantic sector, it skirts the Scotia Arc, east of the South Sandwich Islands,[181] passing under a branch of North Atlantic Deep Water, a western boundary current[182] going in the opposite direction. When encountering a sill, only the less-cold and less-dense water located above the sill depth passes. These passages are marked by a sudden increase of bottom temperature on the other side of the sill, visible in the Indian and Pacific sectors [**Figure 2-41**]. Antarctic Bottom Water penetrates most easily northward in the western basin of the Atlantic Ocean.

---

[181] Passing by the trench of the Sandwich Islands and the Georgia Basin.

[182] The boundary current is intensified by the variation of the Coriolis force with latitude [**Section I.5.3.3**].

Chapter 7
Distinctive features of the Antarctic Ocean

The Antarctic Ocean is an annular zonal ocean that encircles the Antarctic continent, the cold pole of the planet. The zonal circulation of winds and currents is here constant and intense. The extreme climate varies little but is efficient. It entrains a vertical circulation that is especially active under the effect of seawater freezing. This is manifested by weak density stratification with near-barotropic conditions, much different than that of the tropical ocean.

The meridional circulation overlays the dominant zonal circulation of the Antarctic Circumpolar Current, which extends to the bottom. Schematically, the meridional circulation may be summarized as an inflow of warm and saline North Atlantic Deep Water of about 15 Sv, and two less-saline outflows, one on the bottom of very cold Antarctic Bottom Water of about 2.5 to 5 Sv and the other around 800–1,000 m of warmer Antarctic Intermediate Water of about 10 to 12.5 Sv. These extend into the other oceans and influence their oceanographic properties [**Figure 2-44**].

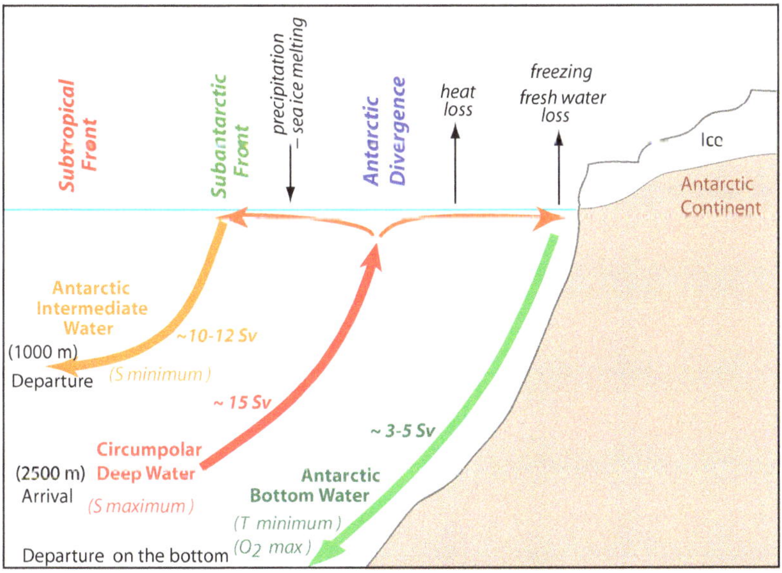

FIG. 2-44 – Water masses with their meridional and vertical movements, and related fronts in the Antarctic Ocean.

In spite of extremely low temperatures, the Antarctic Ocean, by means of Circumpolar Deep Water, provides heat to the atmosphere! Heat losses by the Antarctic Ocean are estimated at 3×10^{14} W. Independent estimates of heat transfer southward by eddies give values between 1 and 5×10^{14} W, an order of magnitude that can balance the calculated losses.

The Austral Ocean is also the regulator of water and heat exchanges between the three other oceans. Seasonal and interannual variations of the pack ice modulated by the Antarctic circulation have an influence on the global climatology. This is a most active ocean, which plays a major role in the balance of the planetary ocean and its variability.

Dinghy in forming sea ice. (Photo Paul Tchernia.)

PART III
The Atlantic Ocean

The research vessel *Atalante*, during the CITHER cruise in 1993. (Photo Chantal Andrié.)

Chapter 1
Introduction

More than 2,000 years ago our ancestors knew about the Atlantic. The works of Herodotus, 450 years before our era, situated the *Atlantis* (later called the *Mare Atlanticum*) west of the Columns of Hercules (Gibraltar) [**Section II.5.1**]. In the *Geography* of the Greek astronomer Claudius Ptolemy [**Figure 3-1**], some Atlantic coasts are indicated. The ancients called it the *Oceanic Sea* to differentiate it from the Mediterranean, their *Interior Sea* (*Mare Nostrum*). The 9th-century Vikings undertook many raids westward from Scandinavia in the search for new lands. They reached Iceland, then Greenland as far as the western shore. From their Greenland base, they crossed the Labrador Sea to reach the coasts of Labrador and even Newfoundland.[183]

FIG. 3-1 – Map of the known world produced in the Middle Ages based on the *Geography* of Claude Ptolemy of the 2nd century AD. (From *Christopher Columbus Atlas*, Nebenzahl, 1991.)

[183] The Vikings were the first Europeans known to have set foot on the American continent. Without doubt cultural inability to adapt to the climate and violent clan disputes put an end to these expeditions and North America was forgotten for some time.

In the middle of the 13th century, with the arrival of the Mongol kingdom in China, land routes to the orient opened up to travelers,[184] merchants, and even Papal envoys. The Silk Road, through the activities of Arab middlemen, brought porcelain, spices, silks, and gunpowder to the West, as well as, among other things, the magnetized needle, paper money, and printing.

When land trade was interrupted by the fall of the Mongol Empire, in 1368, it became necessary to look for maritime routes to Southeast Asia and China. First (1415) the Portuguese, with Henry the Navigator, chose to round the southern tip of Africa and thereby began a long-term maritime adventure. It took 73 years before Bartolomeu Dias rounded the *Cape of Storms* in a raging storm in 1488 (renamed *Cape of Good Hope* by King John II of Portugal) and discovered Brazil while searching for favorable winds. Christopher Columbus in 1492, on behalf of the King of Spain, bet on going west by dropping far enough south to use the trade winds. He had often sailed along the coasts of Europe and Africa and had acquired great experience with the winds. He thus discovered the Antilles and later the coasts of Venezuela, persuading himself that he had found the Indies by going west. Other expeditions soon continued the exploration of the Atlantic coasts, seeking a passage westward. The new information was compiled on ever more numerous charts. For example, in the world map of Jodocus Hondius dating from 1589, the Atlantic Ocean is already well represented [**Figure 3-2**].

Today, the North Atlantic is the most studied ocean. In this region, close to the oldest leading research institutes, the greatest numbers of observations exist. They were first collected in the surface layer for navigation, commerce, and fisheries purposes. Then, with the aid of technology, scientists became interested in studying processes unique to the ocean.

© photo Chantal Andrié

Oceanographic research vessel *Thalassa* during the Equalant 99 expedition in the equatorial Atlantic. (Photo: Chantal Andrié.)

[184] Such as the Venetian Marco Polo, who left at 17 for a long voyage to the East (1271–1295). He brought back much information about the orient, recounted in *Il Milione*, dictated to his cellmate Rustichello while he was a prisoner in the Genoa jails.

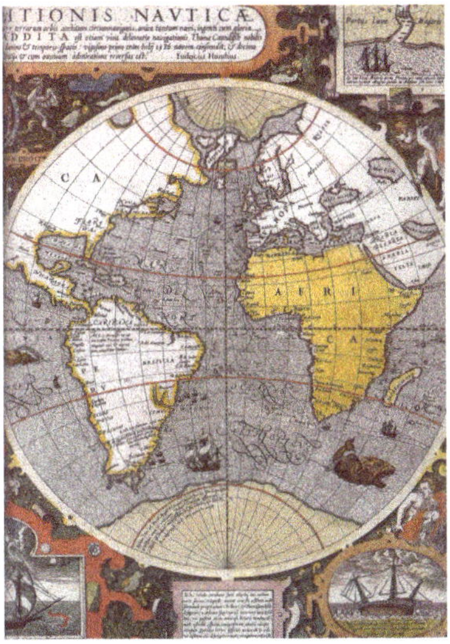

FIG. 3-2 – A portion of the Jodocus Hondius map of the world, 1589 (Robert H. and Margaret C. Power Collection). (From *Christopher Columbus Atlas*, Nebenzahl, 1991.) All rights reserved.

Chapter 2
Geographic characteristics

The problem of measuring longitude was resolved by the arrival of the marine chronometer at the end of the 18th century[185]. Thus, outside of the ice-covered polar regions, the oceanic coasts were well-known at the beginning of the 19th century but little was known about ocean depths, which were still difficult[186] to measure.

In the 19th century, the invention of the telephone brought about the installation of undersea cables between Europe and America and determining depth became essential [See **Section I.3.4.3**], *Challenger* Expedition]. Later, the development of electro-acoustic techniques allowed the collection of much bathymetry data. However, the arrival of multibeam sonar at the end of the 20th century allowed the collection of detailed bathymetry over large regions that were hitherto unknown [**Figure 3-3**].

On the west, the North and South American continents and on the east, the European and African coasts bound the Atlantic Ocean. It is the only ocean in deep communication with the two polar oceans.[187] To the south, it is largely open to the Antarctic Ocean.[188] Oceanographers put the southern limit at the Subtropical Front (the northern boundary of the Antarctic Ocean), located on average about 40°S [**Figure 2-21**]. To the north, coastlines converge in a complex manner. The Atlantic Ocean communicates with the Arctic Ocean across a region composed of the Iceland Sea, the Norwegian Sea, and the Greenland Sea [**Figure 3-3**] from which it is separated by sills, between Greenland, Iceland, and Scotland. The Denmark Strait,[189] located between Greenland and Iceland, is 295 km in width and about 630 m deep. The opening between Iceland and Scotland extends for 960 km. A sill exists at about 480 m between Iceland and the Faeroe Islands. Between the Faeroe Islands and Scotland, a sill reaches down to 840 m in an extension of the Faeroe Bank

[185] John Harrison (1693–1776) self-educated English carpenter and clock maker invented the first marine chronometer in 1759.

[186] The technique of using lead lines covered with tallow took a long time to complete and was imprecise, especially at great depths.

[187] It is 21,000 km between the Bering Strait and the Antarctic continent.

[188] It is 7,400 km between Cape Horn and the Cape of Good Hope.

[189] A strait is an arm of the sea between two landforms, often associated with a sill, namely an elevation of the sea floor between two basins.

Channel. Other more restricted openings to the Arctic Ocean exist across the Canadian Archipelago. Between Greenland and Ellesmere Island, the Nares and Smith Straits allow the Arctic Ocean to communicate with the Baffin Sea. Between the Canadian Basin and the Baffin Sea [**Section III.7.2**] the McClure, Melville, Barrow, and Lancaster Straits are parts of a number of passages across the Canadian Arctic Archipelago. They are blocked most of the time by ice, and are where so many explorers went in search of the Northwest Passage[190] and sometimes lost their lives.

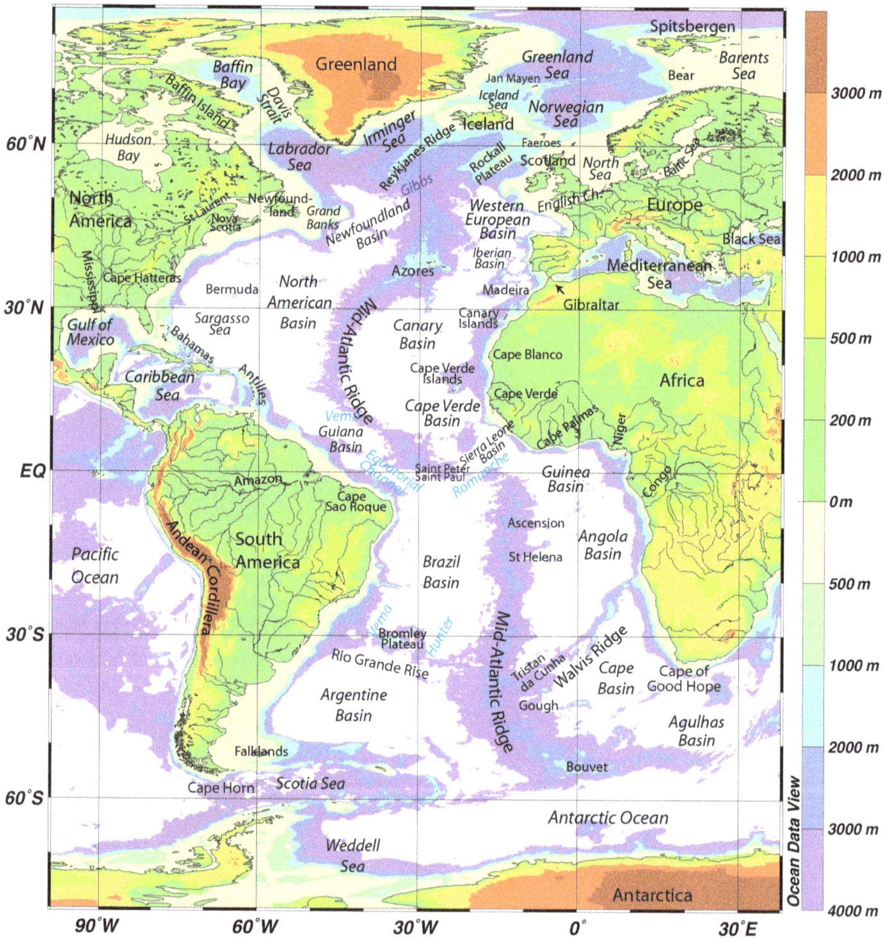

FIG. 3-3 – Atlantic Ocean bathymetry and basins. Depths greater than 4,000 m are in white. (With ODV.)

[190] With a stronger seasonal thaw of the pack ice due to climatic warming, this passage occasionally allows communication between the North Atlantic and the Pacific across northern Canada.

The South Atlantic Ocean presents a rather simple topography and few continental shelves. The North Atlantic, separated by the region of equatorial narrowing between Cape São Roque, Brazil and Cap-Vert, Senegal, has a number of bays and adjacent seas. In the west, the Caribbean Sea and the Gulf of Mexico form the American Mediterranean. Hudson Bay (maximum depth 250 m) is connected to the Labrador Sea. Baffin Bay (maximum depth 2,300 m) communicates with the Labrador Sea *via* the Davis Strait (600 m deep, 270 km wide). In the east, the Mediterranean communicates with the Atlantic through the Strait of Gibraltar (300 m deep, 22 km wide). The Irish Sea, the English Channel, the North Sea, and the Baltic Sea, all shallow, are parts of the continental shelf of western Europe.

The Mid-Atlantic Ridge, whose depth averages 2,000 m, separates the western and eastern basins. It passes by Jan Mayen, then across Iceland and the Reykjanes Ridge. It widens near the Azores, a plateau with tortured relief. It parallels the African coast, joining the equatorial massif, which supports the Saint Peter and Saint Paul Rocks, continuing toward the Antarctic Ocean. The ridge turns eastward and extends as far as California, in the eastern Pacific. All the islands emerging from the ridge are volcanic and are often active,[191] such as Ascension, Tristan da Cunha,[192] Gough, and Bouvet. The ridge is strewn with fracture zones, which shift its axis, sometimes leaving deep passages. The Charlie Gibbs Fracture Zone, around 53°N leaves a passage of 3,000 m, and that of Vema, near 10°N, a passage of about 4,000 m. The Romanche fracture[193] near the equator, with tortured relief, allows a passage of around 4,350 m between the western and eastern basins, including a trench of 7,856 m. In the South Atlantic around 30°S, west of the median ridge, the Rio Grande rise supports the Bromley Plateau of 650 m bordered by two deep passages: the Vema Channel (~ 4,550 m) on the west and the Hunter Channel (~ 4,350 m) on the east. East of the median ridge the Walvis Ridge, between Tristan da Cunha, around 37°S, and Walvis Bay, around 20°S, blocks a large part of the circulation, around 3,000 m. However, there are two passages of more than 4,000 m in the southern part (32°S-36°S) at 2°20W and around 7°W, called the Walvis Passage, and two passages of more than 3,000 m in the northern part around 30°S and 22°S, called the Namib Col.

Various rises, located on either side of the central ridge, delimit a number of basins, as indicated in **Figure 3-3.**[194] Most of these basins have depths

---

[191] The emergence of Surtsey Island, south of Iceland in 1963 is an example.

[192] In 1961 an eruption forced the inhabitants to take refuge in England for several years.

[193] Discovered in 1883 by the vessel *La Romanche*.

[194] West of the ridge are the Argentine Basin, the Brazil Basin, the Guiana Basin, the great North American Basin (including the Sargasso Sea), the Labrador Sea, and the Irminger Sea. East of the ridge are the Cape Basin, the Angola Basin, the Guinea Basin, the Sierra Leone Basin, the Canary Basin, the Iberian Basin, and the Western European Basin.

between 5,000 m and 6,000 m. The greatest depth in the Atlantic Ocean is in the Puerto Rico Trench (~ 8,600 m) along the Antilles Arc.

Between the Subtropical Front to the south, and the Greenland-Iceland-Faeroe-Scotland sills to the north, the Atlantic Ocean extends over an area of 74 million km^2, to which should be added 14 million km^2 for the Arctic Ocean with the Iceland, Greenland, and Norwegian Seas included, and 3 million km^2 for the Mediterranean Sea.

Chapter 3
Climatology

3.1 Pressure and winds

Because of its great meridional extent, most wind and climate regimes can be found over the Atlantic Ocean [**Figure 3-4**]. Continental effects cause modifications, in particular over the Eurasian continental mass and, to a lesser extent, over the North American continent. This produces greater seasonal variability over the North Atlantic than over the South Atlantic, where the less-extensive continents have a weaker influence [**Figure 3-4**]. Thus, the northern subtropical-high belt, centered on 25°N in January, extends from Mexico to the Sahara, while in July, it is displaced north to around 35°N and becomes a large cell centered on the Azores,[195] surrounded by lower pressure over the continents.[196] To the north, subpolar low pressure, the Icelandic Low, is most developed in January, and makes its influence felt from the Labrador Sea to North Cape.[197] To the south, in the austral winter, the high-pressure belt is centered around 30°S. In the austral summer, the center of high pressure, the Saint Helena High, is centered around 10°W and varies little with latitude.

The westerly winds that blow between the subtropical high pressure and the subpolar low pressure are stronger in January,[198] the boreal winter,[199] than in July, the boreal summer. The climate of Western Europe is alternately influenced by the Azores High, or the Icelandic Low, which bring either tropical or polar air.

The NE trade winds[200] blow between the Azores High and equatorial low pressures. The SE trade winds blow between the Saint Helena High and the equatorial lows.[201]

---

[195] From which the cell takes its name, the "Azores High".

[196] Continents warm and cool more rapidly than the ocean and, in consequence, have greater seasonal variations.

[197] North Cape is the extreme northern point of Norway.

[198] In January the pressure gradient is stronger, and many low-pressure cells occur.

[199] The December-March season corresponds to the boreal winter in the northern hemisphere and the austral summer in the southern hemisphere. The June-September season corresponds to the boreal summer in the northern hemisphere and the austral winter in the southern hemisphere.

[200] These winds are moderate and relatively constant, perfect for sailing. Their average speed is about 10 to 12 knots (5 to 6 m s^{-1}).

[201] The average speed of the SE trades is a little greater than that of the NE trades.

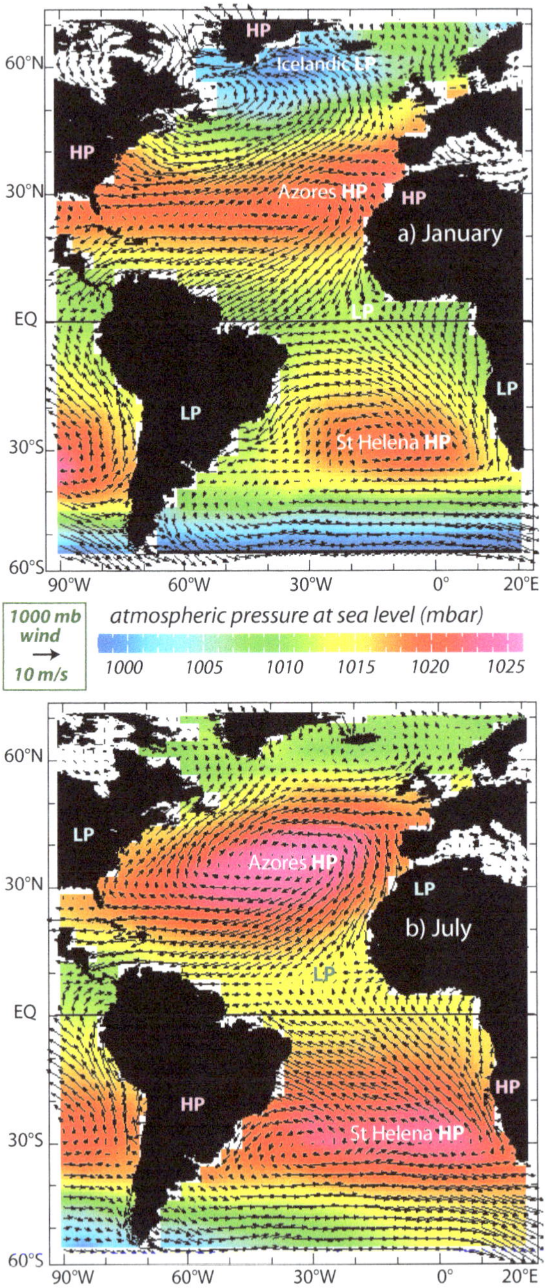

FIG. 3-4 – Mean winds over the Atlantic Ocean superposed over atmospheric pressure (a) in January and (b) in July: HP, high pressure, LP, low pressure. (IRI/LDEO, climatological data, Da Silva, 1994.)

The intertropical convergence zone (ITCZ) [**Figure 3-5**] lies between the NE trades and the SE trades. This is a region of low pressure where winds are generally weak and variable. Because of equatorial narrowing, the extent of the zone of weak winds associated with high surface temperatures varies seasonally.

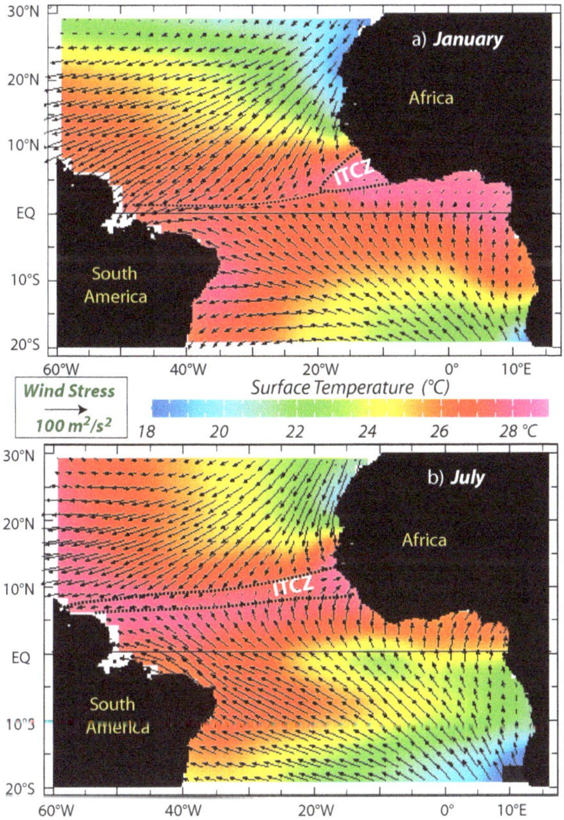

FIG. 3-5 – Mean wind stress superposed on mean sea-surface temperature in the tropical Atlantic Ocean (a) in January, and (b) in July. (IRI/LDEO, FUNCEME data.)

In January-February, the ITCZ shifts southward, is located in the narrowest region of the Atlantic, bounded in a triangle between the equator at 30°W and the African coast from 10°N to 4°30N (Cape Palmas), and does not get to the South American coast. The NE trades extend from 30°N on the east to the equator on the west. This is the season when they are strongest [**Figure 3-4a**] with a maximum intensity in the west. Wind stress, whose magnitude is proportional to the square of the wind speed and which is used in calculating Ekman transport [**Section I.5.2**], delineates the region of maximum wind [**Figure 3-5a**]. The region of weak winds of the ITCZ disappears in the west from 30°W and the NE trades converge with the SE trades. The SE trades extend from 30°S in the east, up to the equator in the west, and as far as 5°N in the Gulf of Guinea, where they acquire a strong meridional component and weaken. Southerly winds weaken along the Angola coast.

In July-August, pressure systems move northward, following the apparent movement of the sun. The region of weak winds of the ITCZ extends north of the equator, along the entire width of the Atlantic from 5°N in the west to 10°N–15°N in the east. The SE trades extend from 30°S in the east to 5°N in the west and up to 10°N in the east [**Figures 3-4b** and **3-5b**]. This is the season when the SE trades are strongest with a maximum in the west near the Brazilian coast. In the Gulf of Guinea, southerly winds are stronger than in January-February. They remain moderate in the Angola basin.

During the summer, around 30° latitude in each hemisphere, the gradients between continental low pressure and oceanic high pressure generate poleward meridional winds. These winds close off the western side of the anticyclonic atmospheric circulation north and south [**Figure 3-4**].

3.2 Precipitation and evaporation

In the ITCZ, surface temperatures are elevated [**Figure 3-5**], winds are weak and variable, atmospheric convection is strong, and stormy precipitation is frequent, amounting to more than 2 m per year [**Figure 1-20**]. Sailors call this region the *doldrums*[202] because ships often stay becalmed under torrential rain without seeing the sun.

In the southwest Caribbean Sea, humid trades arriving after a long ocean trip drop more than 2 m of water per year.

Under westerly winds, between 35°N in the west and 65°N in the east, the passage of frequent depressions, especially in winter in the north, brings precipitation exceeding 1 m to 1.4 m per year. Tropical warm air charged with humidity is often lifted by cold polar air here, bringing about water-vapor condensation: the Polar Front region. In contrast, the anticyclonic zones of the Azores in the north and Saint Helena in the south are subsidence regions where precipitation does not exceed 0.2 m per year.[203] Sometimes they are even less than 0.1 m per year, east of the anticyclonic centers in the regions of Cape Verde and Saint Helena islands. Here winds are weak,

[202] The doldrums is the region where sailors have the blues ("pot au noir" in French).

[203] Known as "horse latitudes" characterized by calm winds, sunny skies, and little or no precipitation. The legend comes from ships carrying horses from Europe, stalled for days and weeks, running short of water and throwing the horses overboard.

the sea is calm, and evaporation [**Figure 1-21**] greatly exceeds precipitation [**Figure 1-20**]. The low-productivity waters[204] are dark blue.

The balance between evaporation and precipitation [**Figure 1-21**] principally reflects the distribution of precipitation except off the North American coasts. In this region, strong westerly winds entrain air masses with dry continental characteristics over the warm waters of the Gulf Stream, intensifying evaporation [**Section I.2.1.5**]. Regions where the balance is negative (precipitation exceeds evaporation) are found in the intertropical convergence zone, in the southwest Caribbean, and in the region of westerlies north of 45°N. The evaporation-precipitation balance represents the freshwater flux at the interface, and directly influences the surface salinity distribution [**Section I.2.6, Figure 1-23**].

In the equatorial region, seasonal variations in the supply of solar energy at the surface [**Figure 1-9**] depend mostly on cloud-cover variations linked to the ITCZ and less on the variations in the inclination of radiation, which are weak. The heat balance [**Figure 1-13**] is largely positive in the equatorial region (the ocean is warmed) and strongly negative in the Gulf Stream region (the ocean is cooled).

[204] Waters are low in nutrients because it is a convergence region [**Section I.5.3.1**, **Figure 1-64**]. Phytoplankton, giving a green color, and zooplankton are not abundant here.

Chapter 4
Surface and subsurface circulation

The surface circulation of the Atlantic Ocean [**Figure 3-6**] consists of an equatorial current system, flanked on the north and south by basin-scale anticyclonic subtropical gyres and on the north by a deformed subpolar gyre. Qualitatively, the circulation brings to mind the theoretical scheme Munk calculated from the winds [**Section I.5.3.3, Figure 1-70**]. In contrast to the

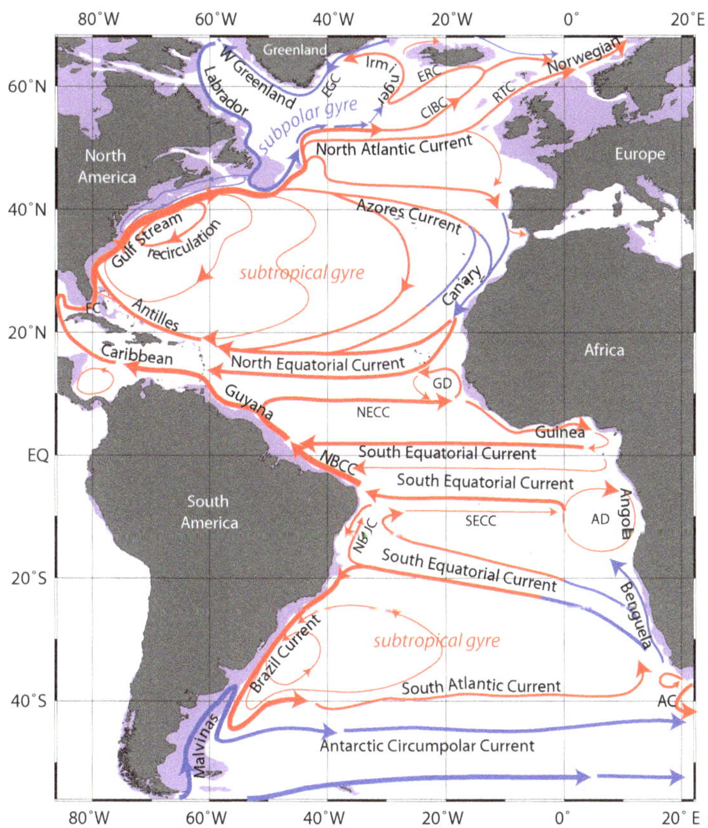

FIG. 3-6 – General surface circulation of the Atlantic Ocean during the northern summer. [Rockall Trench Current (RTC), Subarctic Front Current (SAFC), Central Iceland Basin Current (CIBC), East Reykjanes Current (ERC), East Greenland Current (EGC), Florida Current (FC), Guinea Dome (GD), North Equatorial Countercurrent (NECC), North Brazil Coastal Current (NBCC), North Brazil Undercurrent (NBUC), South Equatorial Countercurrent (SECC), Angola Dome (AD), Agulhas Current (AC).] (After Peterson and Stramma, 1991; Schmitz and McCartney, 1993; Brambilla and Talley, 2008.)

atmosphere, the continental boundaries impose continuity among the zonal currents, through intense western boundary currents, and slow and diffuse currents of the eastern boundary. Even though there is a certain qualitative resemblance, the order of magnitude of the transports in Munk's theory, which does not consider eddies, is much lower than the actual estimates.

4.1 The equatorial current system

A characteristic of the equatorial circulation is that Coriolis force becomes zero at the equator. The Coriolis force begins to have an effect from about 0.5° on either side of the equator. Consequently, transport due to wind (Ekman transport, **Section I.5.2**) is in the direction of the wind at the equator and is perpendicular to the wind, to the right in the northern hemisphere and to the left in the southern hemisphere. The equator acts as a kind of *dynamic barrier* and transfers from one hemisphere to the other take place mostly along the western boundary.

The equatorial circulation is principally zonal and wind-dependent. At the surface, it consists [**Figure 3-7**] of the North Equatorial Current, the North Equatorial Countercurrent, the South Equatorial Current, and the South Equatorial Countercurrent. Below the surface, the Equatorial Undercurrent is the major element, bounded by the North and South Equatorial Undercurrents.

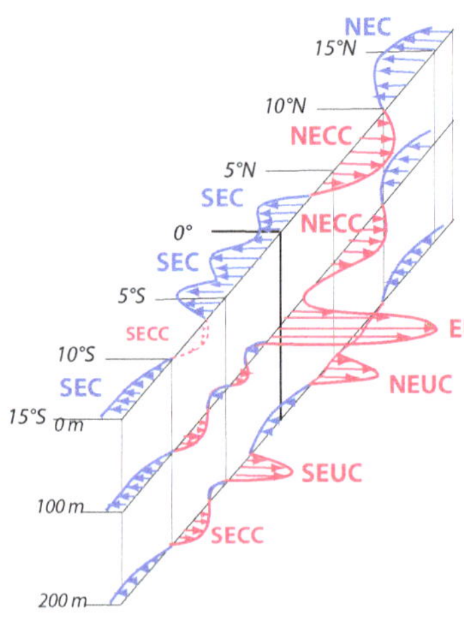

FIG. 3-7 – Schematic of surface and subsurface equatorial currents from 15°N to 15°S during the northern summer. Westward currents are shown in blue, eastward in red. [North Equatorial Current (NEC), South Equatorial Current (SEC), North Equatorial Countercurrent (NECC), South Equatorial Countercurrent (SECC, rarely present at the surface), Equatorial Undercurrent (EUC), North Equatorial Undercurrent (NEUC), South Equatorial Undercurrent (SEUC)]. (After Peterson and Stramma, 1991.)

The offset northward of the current systems is associated with the offset of the ITCZ [**Figure 3-5**] and the trade winds that border it. The narrowness of the ocean in the equatorial zone limits the expansion of the ITCZ in winter and brings about strong seasonal variations in the equatorial circulation.

4.1.1 The North Equatorial Current

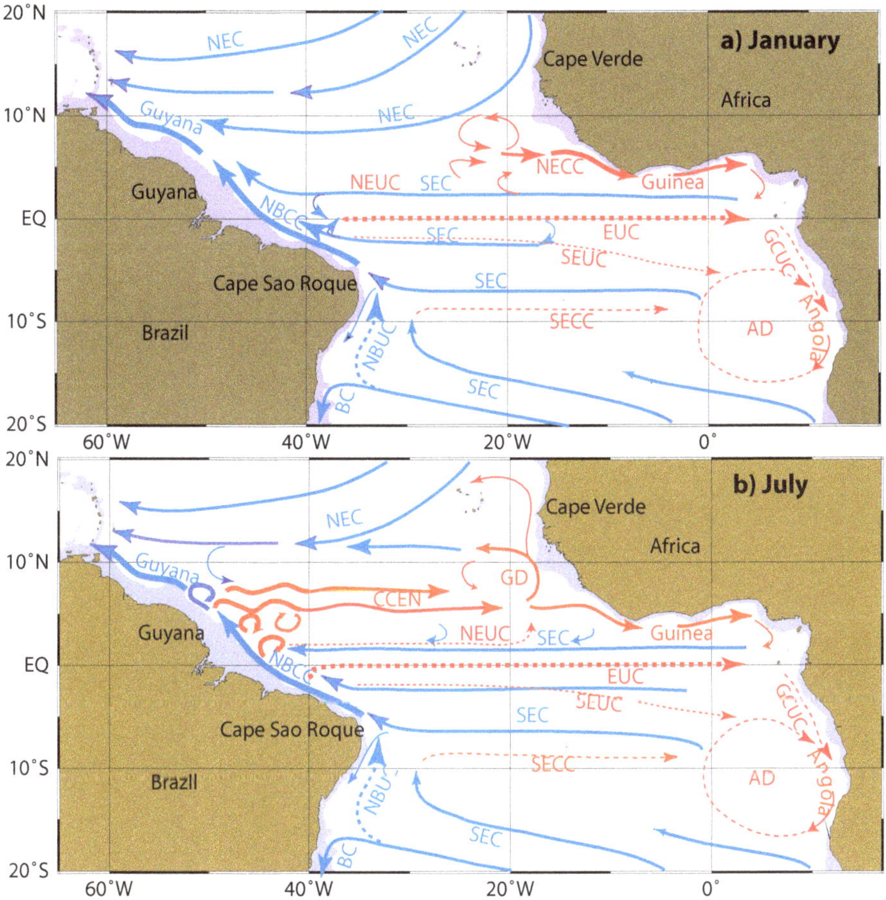

FIG. 3-8 – Schematic of equatorial-zone surface currents (undercurrents as dotted lines) (a) in January, (b) in July [North Equatorial Current (NEC), South Equatorial Current (SEC), North Equatorial Countercurrent (NECC), South Equatorial Countercurrent (SECC), Equatorial Undercurrent (EUC), South Equatorial Undercurrent (SEUC), North Equatorial Undercurrent (NEUC), North Brazil Coastal Current (NBCC), North Brazil Undercurrent (NBUC), Guinea Dome (GD), Angola Dome (AD), Gabon Congo Undercurrent (GCUC), Brazil Current (BC)]. (After Arnault, 1987; Bourlès *et al.*, 1999; Reverdin and McPhaden, 1986; Richardson and Reverdin, 1987; Richardson and Walsh, 1986; Schott *et al.*, 2002; Stramma *et al.*, 2003.)

The North Equatorial Current (NEC) flows in the same direction as the trade winds that generate it. However, it is not directly entrained in the direction of the wind, as one might expect. Ekman transport associated with the NE trades is oriented to the right of the wind, namely to the NW. It comes up against Ekman transport due to the mid-latitude westerlies oriented southward, producing convergence in the center of the anticyclonic gyre (negative Ekman pumping) and oceanic high pressure[205] [**Section I.5.3.1, Figure 1-65**]. The high pressure generates a meridional surface slope, which descends southward [**Figure 1-66**]. Thus there is a meridional pressure gradient which generates the NEC, perpendicular to the gradient [**Section I.4.3, Figure 1-45**], namely in the direction of the wind, which flows westward north of 10°N [**Figures 3-5, 3-7,** and **3-8**]. The NEC begins north of the Cape Verde islands, and covers the region between 25°N and 10°N in July (boreal summer) and between 20°N–25°N and 5°N in January (boreal winter), the season when the NE trades are the most intense and when its transport is maximal.[206] The NEC splits into two branches nearing the Lesser Antilles. One branch penetrates into the Caribbean through inter-island passages [**Section III.4.2.1.1**]. The other branch goes northward, east of the Antilles, to form the Antilles Current.

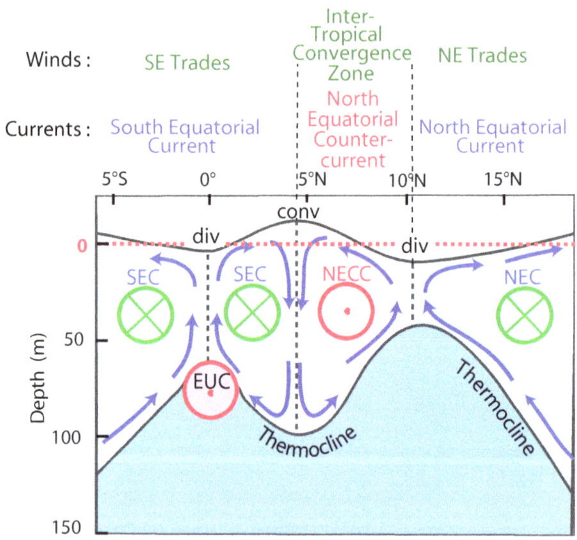

FIG. 3-9 – Schematic of the (highly exaggerated) meridional surface slope and thermocline slope associated with three principal surface equatorial currents and the meridional circulation cells tied to regions of divergence (hollows) and convergence (bumps). The Equatorial Undercurrent (EUC) is located in the thermocline at the equator and flows eastward. Eastward currents are shown in red-pink and westward in green-blue.

205 The convergence produces a hump in the ocean surface.

206 Speeds are of the order of 0.1 m s^{-1} to 0.35 m s^{-1} and the transport is of the order of 16 Sv.

To compensate for the pressure differences due to the surface slope [**Section I.4.3, Figure 1-46**], the thermocline slopes in the inverse sense [**Figures 1-60** and **3-9**]. From the meridional slope of the thermocline, which is easily measured,[207] the position of the various equatorial currents can be determined.

4.1.2 The South Equatorial Current and the North Brazil Current

Driven by the SE trades in the southern hemisphere, the South Equatorial Current (SEC) flows westward between about 20°S–25°S and 3°N–4°N. It extends into the northern hemisphere [**Figures 3-8** and **3-9**]. The SEC consists of intense bands of current around 2°N, 2°S–3°S, 6°S–7°S, and 18°S–25°S. The most southerly band, a continuation of the Benguela Current, splits in the west to feed the Brazil Current to the south and the North Brazil Undercurrent (NBUC) to the north. The NBUC, a western boundary current, extends over the 0–1,000 m layer. It is called undercurrent because its maximum intensity (up to 70 cm s^{-1}) is found around 200 m. Its transport is of the order of 25 $\pm$ 4 Sv. It is weak at the surface and sometimes reverses southward, fed then by a branch of the SEC (around 6°S–7°S), and rejoins the Brazil Current. At Cape São Roque, the NBUC is joined by the principal branch of the SEC and follows the north coast of Brazil, intensifying to form the North Brazil Coastal Current (NBCC), which is prolonged, according to season, by the Guyana Current. The Guyana Current and the NEC join and continue on to the Caribbean driven by the NE trades. As is the NEC, the SEC is subject to strong seasonal variations tied to variations in the strength of the SE trades [**Figure 3-5**].

In July (austral winter) when the system moves northward (and when the SE trades are strongest), the northern branch of the SEC[208] and the NBCC on the western boundary[209] are the most intense. [**Figure 3-10**, at 35°E). Part of the unstable NBCC [**Figure 3-8**] leaves the coast forming meanders and feeds the North Equatorial Countercurrent (see below). In the course of its retroflection to the east, the NBCC casts off strong eddies towards the Guyana Current.

The NBCC is composed of southern hemisphere water and includes the northward transport of warm light water needed to compensate the formation (plunging) of cold dense water in the North Atlantic [**Section I.3.6**]. It is an important element in the transfer of heat from one hemisphere to the other.

---

[207] The ratio of slopes is of the order of 500 to 1,000 [**Figure 1-60**].

[208] With velocities of about 1 knot.

[209] With velocities between 1.5 knots (0.75 m s^{-1}) and 2.5 knots (1.25 m s^{-1}).

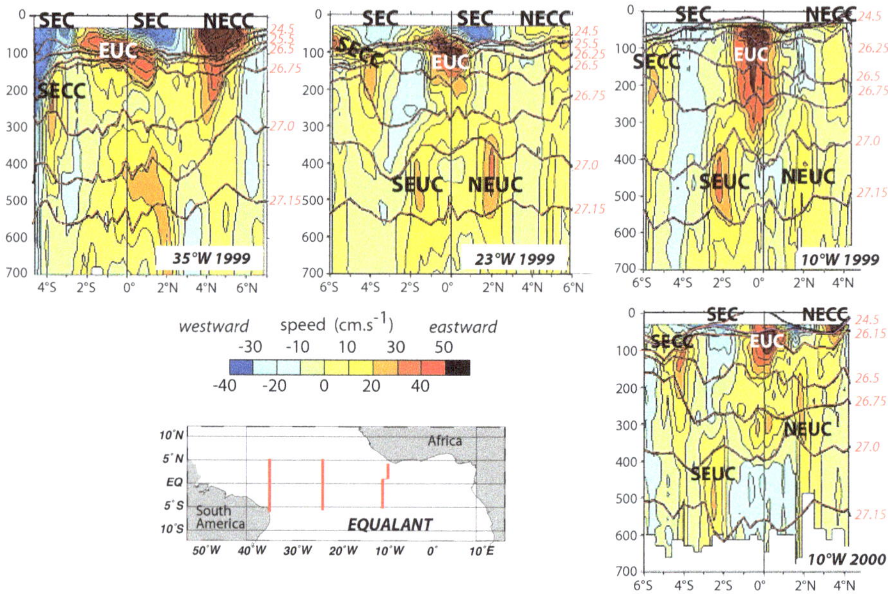

FIG. 3-10 – Meridional current sections measured by a hull-mounted acoustic profiler at 35°W, 23°W, and 10°W during the Equalant expeditions in August 1999 and August 2000 in the equatorial Atlantic. The range of measurements is: 35 m to 600–700 m. Some isopycnals, in red, are superposed. Eastward currents are in yellow and red, and westward in blue and blue-green. The acronyms for currents are shown in the text. (After Bourlès *et al.*, 2002.)

4.1.3 North Equatorial Countercurrent

Globally, in the equatorial region, the trade winds pile up surface water in the west, thereby producing a zonal surface slope and creating a general pressure gradient eastward. At the southern limit of the ITCZ (the northern limit of the SEC), where wind stress is weak, Ekman transport of the SE trades produces convergence, creating a sea-surface ridge [**Figure 1-64**]. At the northern limit of the ITCZ (the southern limit of the NEC), the NE trades produce divergence [**Section I.5.3.1**] creating a sea-surface trough [**Figure 3-9**]. A meridional slope results between the southern edge and the northern edge of the ITCZ, generating an eastward [**Section I.4.3**] geostrophic current [**Figure 3-9**]. Thus the eastward-flowing North Equatorial Countercurrent (NECC) develops seasonally under the ITCZ between the westward-flowing NEC and SEC [**Figures 3-8** and **3-10**].

In July-August, the NECC reaches its full development between the South American coast and the African coast, between 3°N and 7°N [**Figures 3-8** and **3-10**]. West of 50°W, the NEC converges directly with the Guyana

Current and the NECC is principally fed in the west by the retroflection of the NBCC transporting water coming from the southern hemisphere. The NECC is most intense in the west [**Figure 3-10**, section at 35°W]. It is slowed by friction and lateral mixing with the two currents in the opposite direction, that surround it.[210] Arriving at the African coast, which has a zonal orientation at this latitude, the principal branch follows the coast toward the Gulf of Guinea, to form the Guinea Current, which intensifies[211] and then is taken up by the westward circulation at the equator.[212] The other branch flows northward along the coast of Sierra Leone and forms a cyclonic circulation around 22°W, 12°N, called the *Guinea Dome* from where it is entrained westward by the NEC [**Figure 3-11a**]. In a cyclonic circulation, isotherms and isopycnals upwell[213] in its center. For example, the Guinea Dome is marked on a temperature distribution at 50 m by a temperature minimum ($T < 18$ °C) in its center [**Figure 3-11a**]. It is located in the divergence between the NEC and the NECC.

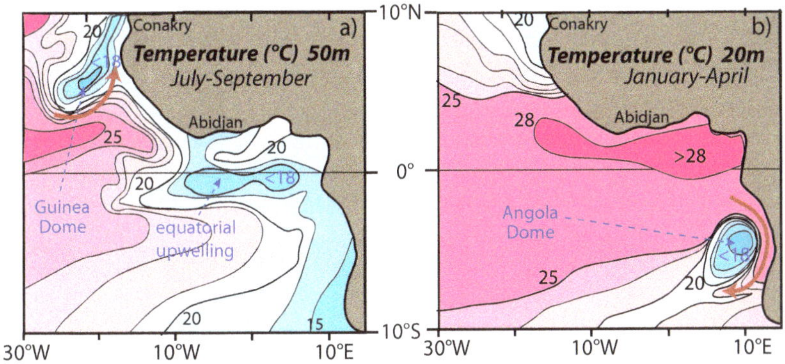

FIG. 3-11 Equatorial Atlantic isotherms (**a**) in July-August-September at 50 m, and (**b**) in January-February-March at 20 m. (After Peterson and Stramma, 1991.)

[210] The average NECC speed is of the order of 0.3 knots (0.15 m s^{-1}) and its transport is of the order of 24 Sv of which 16 Sv comes from the NBCC.

[211] The Guinea Current can reach 1 m s^{-1} near 5°W. It undergoes large seasonal variability. For example, the relaxation of the SE trade winds at the equator strengthens the NECC and, subsequently, the Guinea Current.

[212] Upwelling of cold water appears seasonally along the coasts of the Gulf of Guinea (Côte d'Ivoire, Togo, Bénin, Nigeria) due to a number of factors: the Guinea Current going eastward, wind variations, and the arrival of equatorial Kelvin waves trapped by the coast. The principal season is in July-September, a secondary season occurs in January-February.

[213] The central upwelling is the effect of geostrophic balance.

In January, the reduced ITCZ limits the NECC to the east of 23°W [**Figures 3-5** and **3-8**]. The Guinea Dome is then weakened. In the west, the NECC disappears, and the NEC and the SEC become adjacent.

The convergence between the SEC, going westward, and the NECC, going eastward, produces downwelling [**Section I.5.2.2, Figure 1-63**]. The divergence between the NECC, going eastward, and the NEC, going westward, produces upwelling The result thereby creates meridional-component cells superimposed on the zonal circulation [**Figure 3-9**].

The NECC has low surface salinities [**Figure 3-85b**], because of the strong precipitation under the ITCZ, and elevated temperatures [**Figure 3-85a**].

4.1.4 South Equatorial Countercurrent and the Angola Dome

The South Equatorial Countercurrent (SECC), flowing eastward, is found within the SEC around 4°S–6°S [**Figures 3-7** and **3-10**]. When it surfaces, it is much weaker and much less regular than the NECC. It is most often located beneath the surface, masked by the SEC [**Figure 3-10**] with a velocity maximum[214] at around 100 m to 200 m depth. The SECC may be observed at the surface, for example during El Niño years [**Section V.3.4**]. It turns southward along the Angola coast and takes part in the cyclonic circulation called the *Angola Dome*, especially in the austral summer [**Figure 3-11b**]. Isotherms and isopycnals that upwell[215] in its center characterize the Angola Dome [**Figure 1-66**]. It can be seen on the temperature distribution map at 20 m ($\theta < 18$ °C) during the period from January to April [**Figure 3-11b**]. The meeting of the warm Angola Current, joined by the Gabon-Congo Undercurrent, and of the cold Benguela Current in the opposite direction, results in the Angola-Benguela Front around 15°S, as deep as 200 m. The Angola Current most often flows southward. However, when the SE trades strengthen, the surface layer could reverse northward transporting cold waters [**Figure 3-11a**].

Each dome is more evident during the period of weak trades (February for the southern hemisphere) during which the equatorial undercurrent of the respective hemisphere is the strongest.

In the season when the SE trades are the strongest (July-September), equatorial upwelling can be seen on the temperature map in the east with a temperature minimum (< 18 °C) at 50 m [**Figures 3-11a, 3-5,** and **3-36**].

[214] Of the order of 0.2 knots (0.10 m s^{-1}).

[215] The central upwelling is the effect of geostrophic balance.

4.1.5 The Equatorial Undercurrent

In parallel with the zonal surface currents, the equatorial current system includes a number of subsurface currents, in particular a powerful east-flowing return current centered on the equator: the Equatorial Undercurrent (EUC).

In 1873, on board the *HMS Challenger*[216], J.Y. Buchanan discovered a subsurface current on the equator opposed to the surface current. He returned in 1886, on board the *Buccaneer* and identified the EUC, but this discovery was completely forgotten until 1952 when, in the equatorial Pacific, Cromwell, Montgomery, and Stroup discovered a similar current [**Section V.4.2.5**]. A search aboard the *Lomonosov* rediscovered the EUC of the Atlantic Ocean in 1959, 86 years later!

Meridional sections of direct current measurements [**Figure 3-10**] reveal a velocity maximum toward the east in the thermocline at depths that vary with longitude. At 23°W, the core of the EUC is located between 60 m and 120 m. At 10°W, the maximum velocity (> 0.80 m s^{-1}) nears the surface (at 50 m in August 1999 and at 60 m in August 2000). Its vertical extent is variable from 100 m to 300 m. Its latitudinal extent varies between 200 km and 400 km. It is a thin ribbon of strong undercurrent that crosses the Atlantic.

One of the processes involved in the existence of the EUC, the most intense of the Atlantic equatorial currents, is connected with the presence of easterly winds (the trades) at the equator. The wind effect is transmitted by turbulent viscosity in the mixed layer above the thermocline [**Section I.5.2.2**]. At low latitudes, surface warming by solar radiation is intense and varies little. The thickness of the mixed layer depends mainly on wind action since there is no winter cooling to destroy the stratification. The accumulation of warm and light surface water in the west by wind stress produces a surface slope and an opposed thermocline slope [**Figures 1-62** and **3-12**]. The surface slope produces a pressure gradient to the east. In the mixed layer, the pressure gradient is directly balanced by the stress of the easterly trades. Under the mixed layer, whose extension is blocked by the thermocline, the pressure gradient is not balanced by wind stress and thereby the EUC, centered on the equator, is generated in the direction of the gradient (because the Coriolis force is zero at the equator). Velocity equilibrium is reached when the forces of lateral friction balance the pressure gradient.

[216] HMS: Her Majesty's Ship.

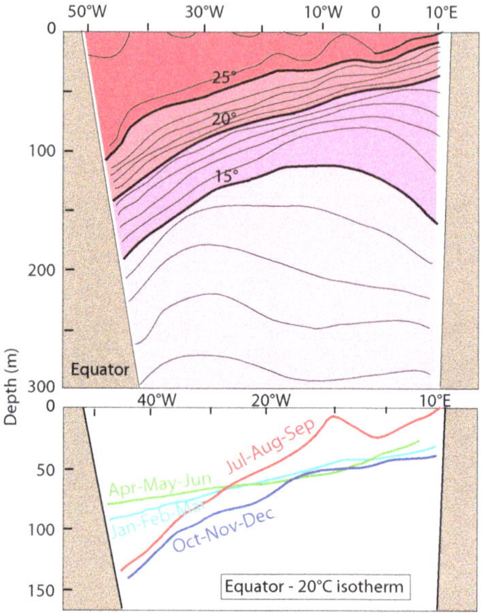

FIG. 3-12 – Temperature section along the equator in the Atlantic and seasonal variability of the 20 °C isotherm located in the core of the thermocline. (After Merle, 1980.)

The slope of the thermocline [**Figure 3-12**] responds to surface-slope variations, which in turn depend on seasonal wind variations [**Figure 3-5**].

Current shear[217] produces a characteristic spreading of the isotherms [**Figure 5-17**] corresponding to a lessening of the vertical temperature gradient. Located in the center of the thermocline, the undercurrent is deeper in the west than in the east, where it sometimes reaches the surface, depending on wind strength. Its slowing in the east can be attributed to wind as it reaches the surface layer and to lateral friction.

Though it is below the surface, the EUC is the most rapid of the equatorial currents, with velocities ranging from 0.60 m s^{-1} to 1.5 m s^{-1} [**Figures 3-10** and **3-13**] and a transport which varies between 4 Sv and more than 35 Sv, taking into account the annual and interannual spatial variability.

The seasonal variation of the westward zonal wind anomaly [**Figure 3-14a**] with respect to the annual average over ten years of measurements from *PIRATA*[218] buoys installed along the equator at 35°W, 23°W, 10°W and 0° longitude shows an increase in intensity of the trades in May-June-July, beginning in mid-April at 10°W, at the four longitudes. After weakening in September-October at 0°, 10°W, and 23°W, a second increase then appears in November in the center and east of the basin, while trades are stable to the west at 35°W.

[217] And the adjustment of the density field to the current (see the dynamics references).

[218] PIRATA: **Pi**lot **R**esearch moored **A**rray in the **T**ropical **A**tlantic.

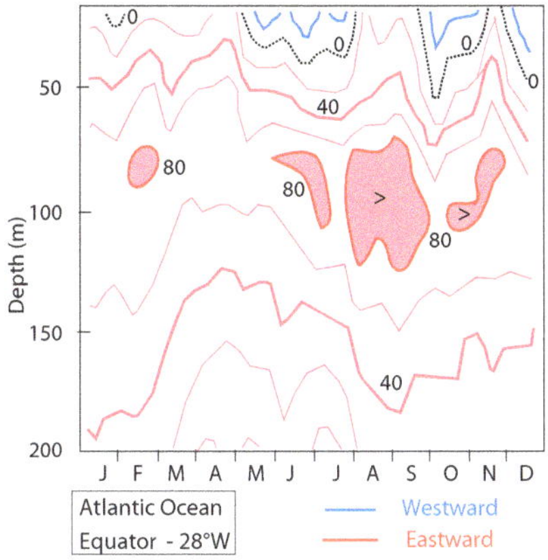

FIG. 3-13 – Fifteen-day averaged zonal current component (in cm s^{-1}) recorded at 28°W showing the annual variability of the Equatorial Undercurrent during 1984. (After Halpern and Weisberg, 1989.)

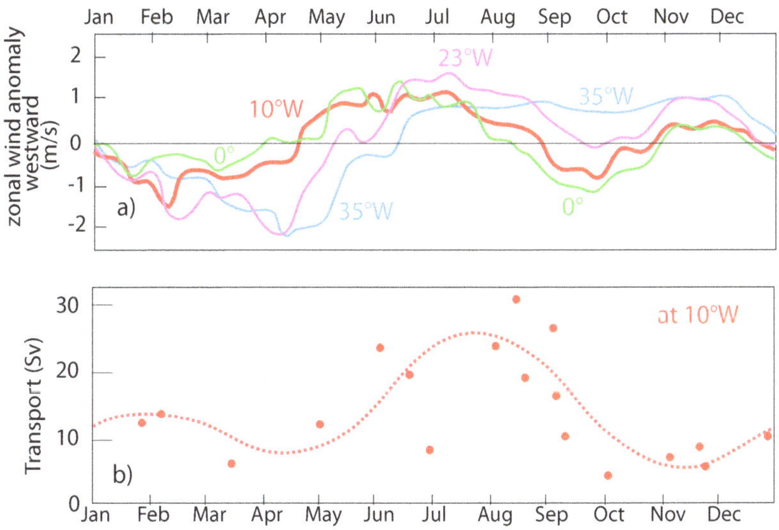

FIG. 3-14 – **(a)** Annual variability of the zonal wind anomaly (positive westward) with respect to the 1997–2007 mean. The wind was measured on PIRATA equatorial buoys at 35°W, 23°W, 10°W, and 0°. **(b)** Equatorial Undercurrent transport at 10°W in the equatorial Atlantic calculated from S-ADCP current measurements during 17 oceanographic cruises between 1997 and 2007. The dotted line is a semi-annual harmonic function fitted to the data. (After Kolodziejczyk *et al.*, 2009.)

The transport of the EUC during the same period [**Figure 3-14b**] at 10°W shows a strong variability, which may be described as a semi-annual component (dotted line) with a summer maximum and a secondary winter maximum. The summer maximum follows the maximum reinforcement of the trades in July by a few weeks [**Figure 3-14a**]. Associated with it is a deep extension of the EUC and sometimes an outcropping at the surface under the effect of an increase in upwelling [**Figure 3-10**]. The winter maximum is due to the reversal eastward of the surface current, tied to the weakening (even a reversal) of the trades which no longer maintain the surface flux opposed to the zonal pressure gradient, thus leading to the relaxation of the flux eastward above the EUC.

The variability of the EUC depends strongly on that of the zonal circulation and thus of the seasonal wind cycle. Superposed on this variability are intraseasonal phenomena, such as local wind events, equatorial waves, non-linear instabilities, and climatic anomalies like ENSO.[219]

The existence of the EUC seems also to be connected to mid-latitude circulation, driven by extra-equatorial winds, through conservation of potential vorticity along lines of current.[220]

The EUC is fed by the North Brazil Undercurrent, which transports the high-salinity Subtropical Water of the South Atlantic [**Figure 3-15**]. As a result, salinity profiles at the equator show a maximum at the depth of the EUC, which nears the surface going eastward [**Figure 3-16**]. In the east, it partly feeds the Gabon-Congo Undercurrent.

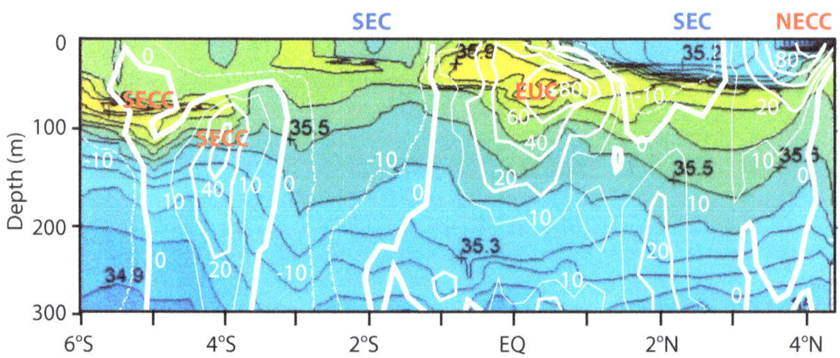

FIG. 3-15 – Salinity section at 10°W during the Equalant 2000 cruise. ADCP velocities (in white) are superposed. They indicate surface currents (SEC, NECC) and subsurface currents (SECC, EUC), negative westward (names in blue), positive eastward (names in red). (After Bourlès *et al.*, 2002.)

[219] El Niño Southern Oscillation.

[220] See the works on fluid dynamics indicated in the references (Gill, 1982).

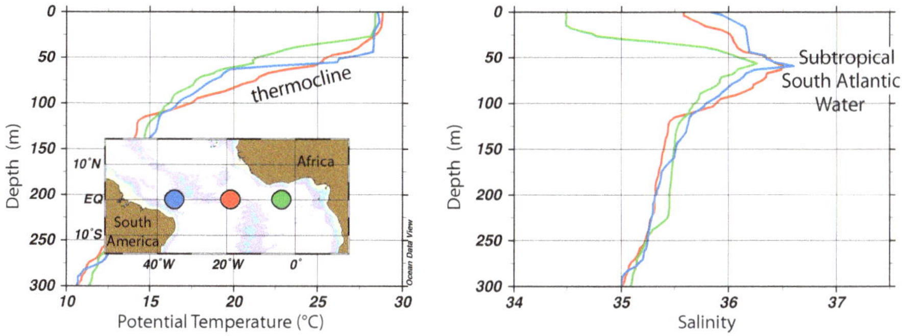

FIG. 3-16 – Salinity and temperature profiles at the equator in the Atlantic. (With ODV and World Ocean Database.)

4.1.6 North and South Equatorial undercurrents

Two eastward-flowing undercurrents are symmetric about the equator but weaker than the Equatorial Undercurrent: the North Equatorial Undercurrent (NEUC) and the South Equatorial Undercurrent (SEUC). They are located around 200 m to 600 m depth and are centered around 2°N–3°N and 2°S–3°S respectively [**Figures 3-7** and **3-10**]. They are narrow, with speeds of about 0.2 m s^{-1}. They intensify during the season when the trades in their hemisphere are weakest and occur further from the equator as they flow eastward, reaching 10°N and 10°S near the African continent. They are deeper and located closer to the equator than the South Equatorial Countercurrent and the North Equatorial Countercurrent.

Arriving at the African coast, the South Equatorial Undercurrent turns southward, joins with the South Equatorial Countercurrent in the Angola Current, whose subsurface speeds reach 0.5 m s^{-1}, and forms the Angola Dome in the southern hemisphere summer.

Because of the form of the African coast at the latitude of the North Equatorial Undercurrent, the part of the current that enters the Gulf of Guinea is not distinct and dissipates rapidly. Between July and September, the major part of the North Equatorial Undercurrent turns northward and joins the deep part of the North Equatorial Countercurrent to reinforce the Guinea Dome.

The SEUC and the NEUC carry Central Water of the South Atlantic (11 °C $< T <$ 15 °C). They are marked by a relative oxygen maximum. Their transport can vary from 5 Sv to more than 20 Sv.

4.1.7 Equatorial jets

At the equator, the current structure is highly stratified. Under the EUC, there are a number of stacked zonal jets. They are rapid, have alternating velocities, and are centered on the equator [**Figure 3-17**]. By their spatial and temporal variability, they play a role in mixing between the water masses that they entrain.

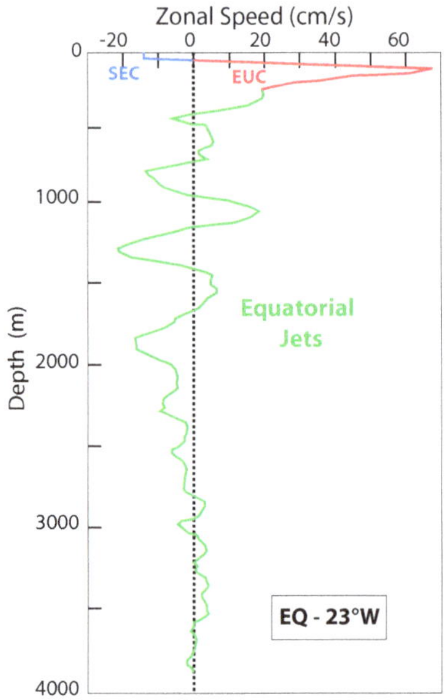

FIG. 3-17 – Zonal velocity component profile on the equator at 23°W during the Equalant 1999 cruise showing multiple equatorial jets at depth. (After Gouriou *et al.*, 2001.)

4.2 The North subtropical gyre

The North Equatorial Current is the equatorial branch of the northern subtropical gyre [**Figure 3-6**], composed of the Gulf Stream system (the western branch), the North Atlantic Drift, the Azores Current and the Canary Current.

4.2.1 The Gulf Stream system

After the discovery of America by Christopher Columbus in 1492, exchanges with Europe began. Ponce de Leon[221] was the first to describe the Florida Current in 1513. Before long, captains who crossed the Atlantic grew

[221] Juan Ponce de Leon, Spanish conqueror, landed in Florida in 1513 looking for the "fountain of youth". He had a lot of trouble returning south against the current of which he was the first to provide a description.

to know the Gulf Stream. Going westward, they used the trade winds and the North Equatorial Current. Returning, they passed through the Florida Straits and were carried by the Gulf Stream as far as 35°N (Cape Hatteras) where they found westerly winds to rejoin Spain. Navigators looking to use or avoid the currents began to try to map them. The first chart showing the Gulf Stream, published by Kircher in 1665, sited its origin as the Gulf of Mexico overall, hence its name. In 1735 Hoxton was the first to describe the Gulf Stream as a rapid narrow current that leaves the coast at Cape Hatteras and turns eastward around 38°N. In 1772, the Inspector General of Florida, De Brahm, published a chart on which the Gulf Stream was represented by a simple thick line beside a detailed map of Florida.[222]

At the same period, the Customs Service in Boston complained to London authorities about the two-week delay in the crossing of packet ships between England and New York compared with merchant ships crossing between London and Rhode Island. Benjamin Franklin, at the time Postmaster General, spoke with Timothy Folger, a Nantucket whaler, who knew the current was delimited by a front[223] between coastal waters and the warm waters of the current, near which whales were numerous. Franklin asked him to draw the position of the current. Thus, a Gulf Stream chart was published in London in 1769–70 and in Paris in 1778 [**Figure 3-18**] showing the North American coast and the Grand Banks, the domain of cod fishermen. He had the idea of using temperature to define the current front, having sometimes observed it himself during passages of his ships. In the 19th century the English oceanographer James Rennell studied the numerous observations compiled by the British Admiralty from which he drew a chart of the Gulf Stream in 1832. Somewhat later, the American Matthew Fontaine Maury undertook a careful compilation of all current information[224] from which emerged the first *pilot charts*[225] (1859), so useful for navigation.

The Gulf Stream system includes the Florida Current, the Gulf Stream in the strict sense, between the northern Straits of Florida and the Grand Banks of Newfoundland, and the interior recirculation in the Sargasso Sea [**Figure 3-19**]. The origin of the Florida Current lies in the American Mediterranean (the Gulf of Mexico and the Caribbean Sea).

[222] The coastline north of Florida did not appear on the map.

[223] The front was observed and recounted for the first time in 1606 by Marc Lescarbot, a travelling lawyer.

[224] Maury had a chart of currents furnished to all captains by the Hydrographic Service in exchange for regular observations of weather, sea state, and currents. He succeeded in convening a meeting of delegates from maritime nations in Brussels in 1853 to define, for the first time, a code of standard observations.

[225] Pilot charts, monthly charts of currents, wind statistics, and sea state, are still maintained and used.

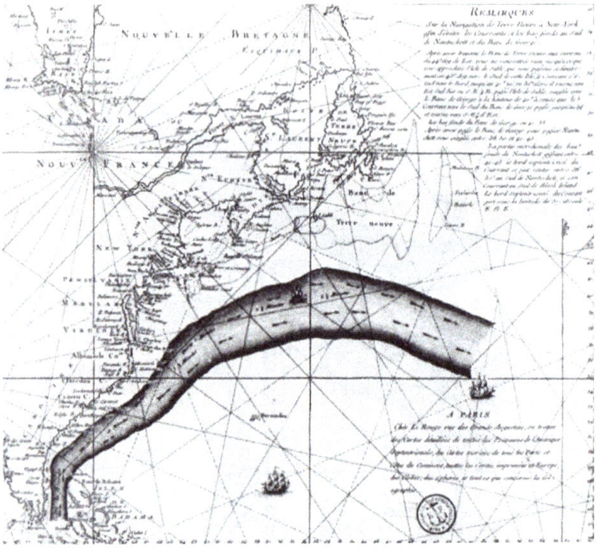

FIG. 3-18 – A portion of the French version of the Gulf Stream chart drawn by Benjamin Franklin and Timothy Folger, published by Le Rouge in Paris around 1778 (Bibliothèque Nationale de France). (After Richardson, 1980.) All rights reserved.

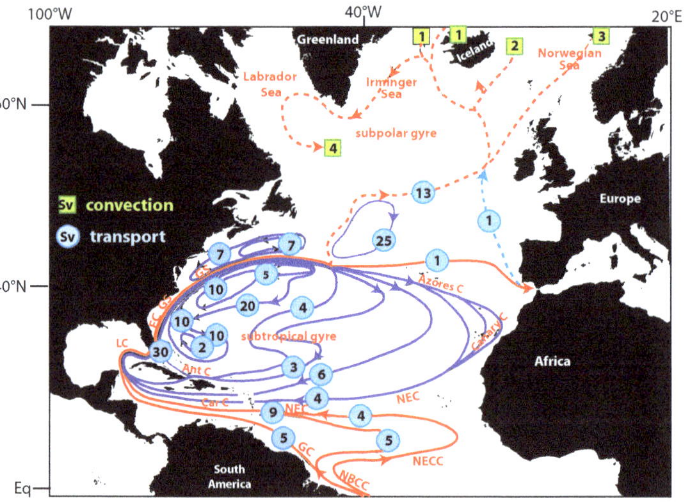

FIG. 3-19 – Transport estimates (in Sv, in blue) of the surface circulation of North Atlantic subpolar gyre branches (for $T > 7$ °C) and convective flows of the subpolar gyre branches (in green). Red lines denote the replacement flow for the meridional cell involving NADW. Dark blue lines characterize the subtropical gyre and recirculation. Dashed blue lines denote the addition of Mediterranean Water to the meridional cell. [Gulf Stream (GS), Florida Current (FC), Loop Current (LC), Caribbean Current (CarC), Antilles Current (AntC), Guyana Current (GC), North Brazil Coastal Current (NBCC), North Equatorial Countercurrent (NECC), North Equatorial Current (NEC).] (After Schmitz and McCartney, 1993; Schmitz, 1996.)

4.2.1.1. Circulation in the American Mediterranean

The term *mediterranean* is used for oceanic regions having limited communication with the main ocean, in the form of a strait or a sill [**Section III.5.1.1**].

The Caribbean Sea (*mer des Antilles* in French) and the Gulf of Mexico form the American Mediterranean. In contrast to classic mediterraneans, it includes a number of linkages with the adjacent ocean and the circulation here is mainly wind-driven. It is the source region for the Florida Current.

The American Mediterranean is a single sea composed of a series of basins with sills, separated from the North Atlantic by the Antilles Arc, a multitude of islands 2,500 km in length between northern South America and southern Florida [**Figure 3-20**]. This arc consists of the Greater Antilles (Cuba, Jamaica, Hispaniola, Puerto Rico) and the Lesser Antilles (Virgin Islands, Guadeloupe, Dominica, Martinique, Saint Lucia, Barbados, Trinidad, etc.) across which there are many passages from 740 m to 1,800 m depth [**Figure 3-20b**]. The deepest passages are the Windward Passage between Cuba and Hispaniola of 1,650 m and the Anegada Passage, extended by the Jungfern Passage, east of Puerto Rico of 1,800 m. Rises delimit the basins of Grenada, Venezuela, Colombia, Cayman,[226] and Yucatan separated by sills of various depths. The Gulf of Mexico is open to the Caribbean through the Yucatan Channel (between Cuba and the Yucatan peninsula) and to the Atlantic through the Florida Straits (800 m) between Cuba and Florida.

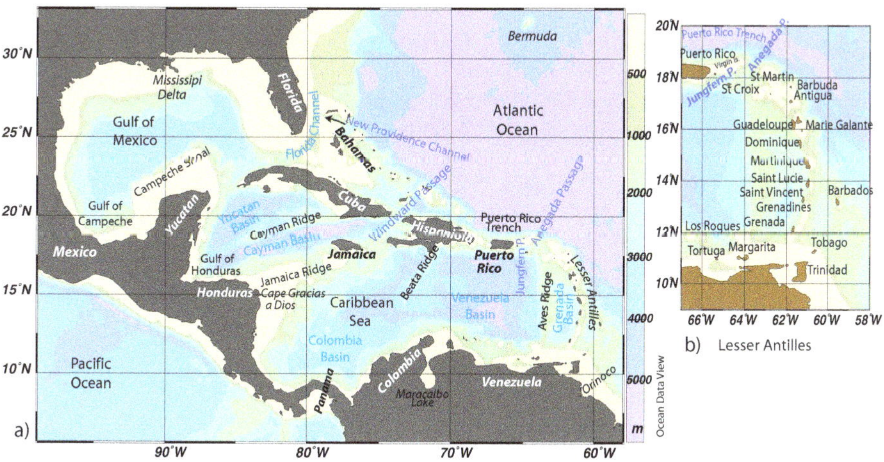

FIG. 3-20 – Bathymetry of the American Mediterranean (the Caribbean Sea and the Gulf of Mexico).

[226] The Cayman Basin includes the Cayman Trough of 7,680 m depth.

The American Mediterranean is thus a basin of many entries and a single exit. A large part of the Guyana Current[227] and of the North Equatorial Current is driven by the NE trades across the numerous passages of the Lesser Antilles to form the Caribbean Current[228] [**Figure 3-19**] which flows through the various basins to the Yucatan Channel (1,885 m) and into the Gulf of Mexico. About 10 Sv enter through passages south of Martinique (St. Lucia, St. Vincent, The Grenadines, and Grenada), about 8 Sv enter between Martinique and the Virgin Islands, and 9 or 10 Sv enter through the Anegada and Windward Passages. The continuation of the Caribbean Current[229] in the eastern Gulf of Mexico generally makes a loop before exiting through the Florida Straits [**Figure 3-21**] from whence it gets the name of Loop Current. It is unstable and forms large meanders from which regularly detach large anticyclonic eddies of 200 to 300 km diameter, which drift toward the southwest Gulf of Mexico.

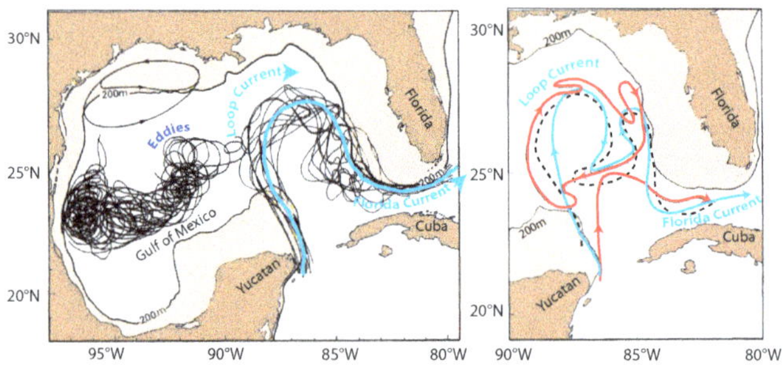

FIG. 3-21 – **(a)** Schematic of Gulf of Mexico circulation from satellite observations and drifting buoy trajectories. The average position of the Loop Current is shown in blue. **(b)** The position of the Loop Current before (blue) and after (red) the detachment of an eddy. The 200 m isobath is shown. (After Vukovich, 1988.)

4.2.1.2. The Florida Current

The accumulation of tropical water by the trades creates a difference in sea level of about 19 cm on opposite sides of Florida. The corresponding pressure gradient generates a current through the Straits of Florida (130 km wide and ~ 800 m deep). The Florida Current first flows eastward, then turns north, channeled by the Great Bahama Bank, where the channel shrinks to 75 km wide and 600 m deep. It fills the channel to the bottom. At the exit from this channel, between Florida and the Little Bahama Bank (150 km wide, 800 m deep), the Florida Current has surface velocities between 1.5 m s^{-1}

---

[227] The Guyana Current is an extension of the northern branch of the South Equatorial Current.

[228] The Caribbean Current forms countercurrents (or recirculation) that are picked up in the east by the main circulation.

[229] The Caribbean Current is called the Yucatan Current between Yucatan and Cuba.

and 3 m s^{-1}. It is a *western boundary current* [**Section I.5.3.3**]. Its dark blue waters form a contrast with the green coastal waters.[230]

In the 19th century, Pillsbury made the first series of direct current measurements[231] along sections across the Florida Current from 1885 to 1890. In 1914, observations of temperature and salinity were taken on the same sections [**Figures 3-22a** and **b**]. The geostrophic current calculated from these observations [**Figure 3-22d**] is close to that of the directly measured current [**Figure 3-22c**]. Even though old, these measurements have become the classic example of the value of the dynamic method [**Section 1.4.4**] for deducing currents based on temperature and salinity distributions. In this example, the reference level, which is habitually a difficulty, was chosen as the sea bottom.

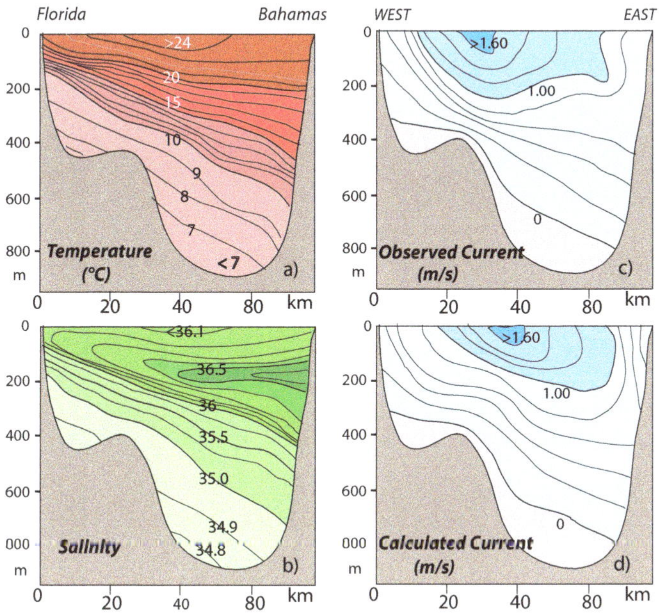

FIG. 3-22 – Sections of (**a**) temperature and (**b**) salinity across the Florida Current, (**c**) speed measured across the Florida Current by Pillsbury, (**d**) speed calculated geostrophically. (After Sverdrup *et al.*, 1942.)

Because the Florida Current is constrained by the Florida Straits, its transport is relatively easier to measure than that of the Gulf Stream. It is of the order of 30 ± 2 Sv, with a maximum in March, a minimum in November, and a seasonal variation of about 4 Sv.

---

[230] The green color of rich waters is due to chlorophyll content in small algae of plankton. Poor waters are blue.

[231] Pillsbury stayed anchored several months on board the *Blake*, and used the first current meter, which he developed himself, with a vane, compass, and rotor.

The Florida Current has two sources [**Figure 3-19**]. One comes from the tropical South Atlantic (by the intermediary of the North Brazil Coastal Current) and the other is part of the wind-driven north subtropical gyre.

The first source represents the warm-water surface flux of the large-scale thermohaline circulation, needed to feed the formation of North Atlantic Deep Water, of about 13 Sv.

The second source corresponds to the western boundary current of the subtropical gyre, which is split in two by the Antilles Arc. The branch that passes through the interior of the Antilles Arc (about 17 Sv) feeds the Florida Current, thus giving a total of 30 ± 2 Sv. At the exit of the Straits of Florida, it receives about 2 Sv through the Bahamas *via* the New Providence Channel [**Figure 3-20**].

4.2.1.3. The Gulf Stream

The Antilles Current passes outside the arc of the Antilles and the Bahamas with a transport of the order of 5–7 Sv. It represents the other branch of the western boundary current. It joins the Florida Current north of the Bahamas [**Figure 3-19**] forming the Gulf Stream, the strong western boundary current of the north subtropical gyre [**Section I.5.3.3**].

Temperature and salinity sections taken across the current show strong isotherm, isohaline, and isopycnal slopes [**Figure 3-23**], in accordance with geostrophic balance.

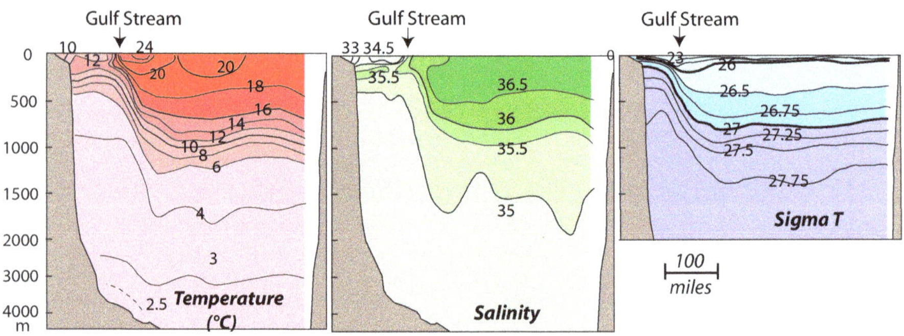

FIG. 3-23 – Temperature, salinity, and density section across the Gulf Stream, from Chesapeake Bay to Bermuda. (After Iselin, 1936.)

The Gulf Stream is a current of tropical origin, transporting warm saline water, well defined on its left side by a sharp temperature drop due to colder and fresher Slope Water.[232] On a 73-m long vessel of the International Ice Patrol,[233] the difference in temperature between the bow, in cold productive water from the north, and the stern, in cobalt-blue, warm, impoverished Gulf Stream water,

[232] Slope Water is found over the slope of the continental shelf.

[233] The Ice Patrol was created after the *Titanic* catastrophe.

reached more than 12 °C! (1.1 °C to 13.3 °C). This intense thermal front[234] is now observed on infrared images taken by satellite [**Figures 3-24** and **3-25**]. Bermuda, bathed in warm water, has the most northern coral isles of the planet.

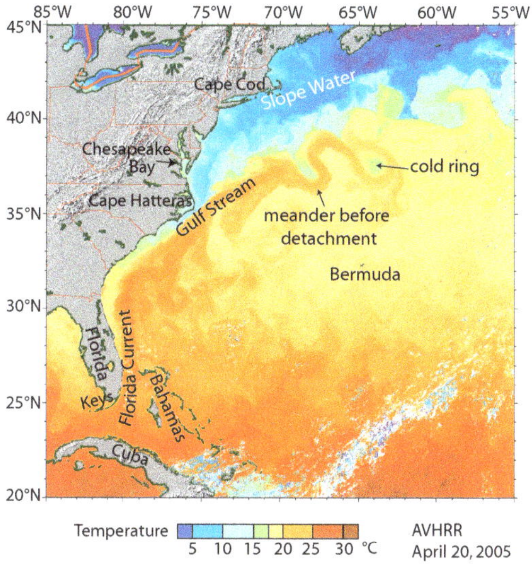

FIG. 3-24 – Satellite image of surface temperature in the western North Atlantic. Warm water is in red, cold in green and blue. (From Johns Hopkins University, 2008.)

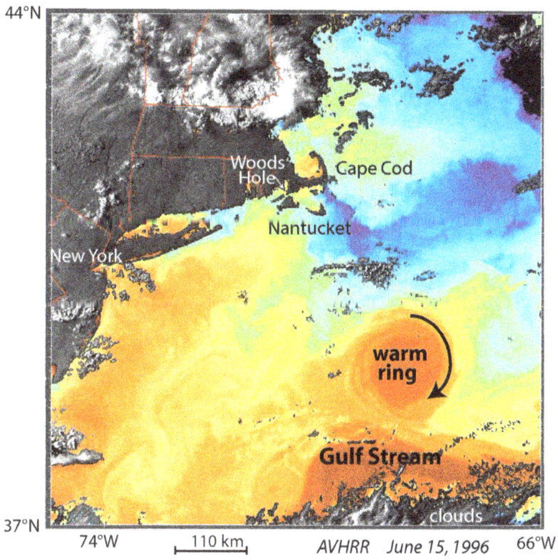

FIG. 3-25 – Close-up of an anticyclonic warm Gulf Stream ring in cold slope water, drifting slowly westward. The warm Gulf Stream is visible at the bottom of the image in red. Grey areas are clouds. (NOAA12 satellite, 15 June 1996.) (From Johns Hopkins University, 2008.)

[234] The front is called the *cold wall*.

As in the Florida Straits, the current extends from surface to bottom (3,000 to 3,500 m) as far as Cape Hatteras **[Figure 3.26]**.

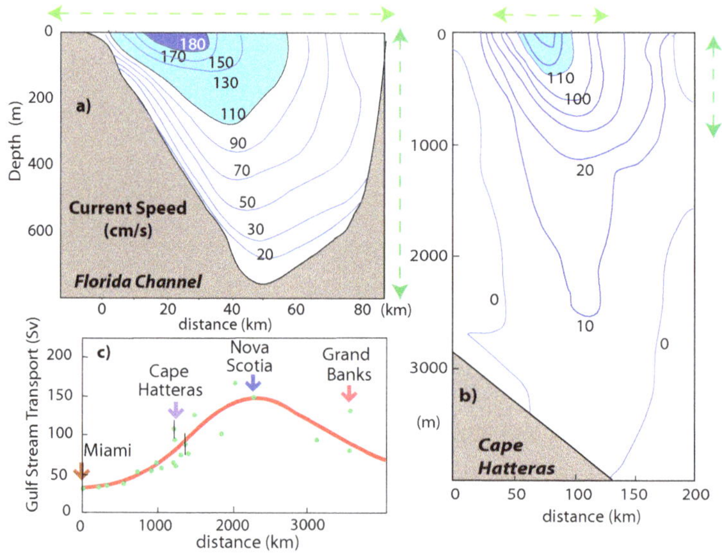

FIG. 3-26 – Current sections (cm s^{-1}) **(a)** across the Florida Current in the Florida Straits, **(b)** across the Gulf Stream, off Cape Hatteras, and **(c)** the variation in transport from southern Florida (Miami) up to the Grand Banks of Newfoundland in Sv (10^6 m^3 s^{-1}). Green points are transport estimates, vertical bars represent uncertainty, and average transport is in red. (After Leaman *et al.*, 1989.)

Indirectly calculated on the basis of numerous temperature and salinity observations, the average height difference between one side and the other of the Gulf Stream is about 1 meter **[Figure 3-27]**. This estimate was later confirmed by direct altimetric sea-surface measurements from satellite **[Figure 3-28]**. Henry Stommel said, "the Gulf Stream is a narrow ribbon of high-velocity water acting as a dynamic barrier which restrains the warm Sargasso Sea water from overflowing cold denser coastal water" **[Figure 3-29]**.

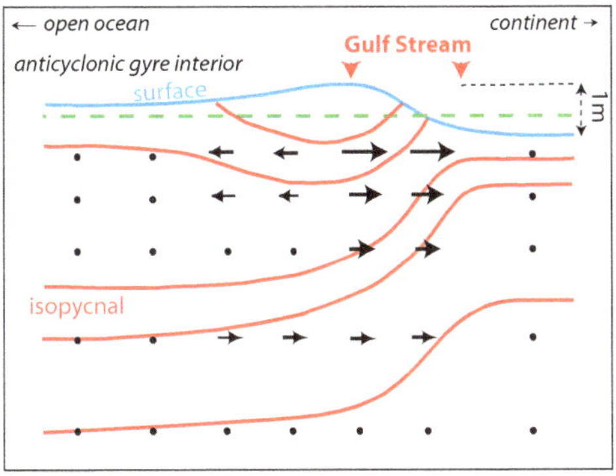

FIG. 3-27 – Schematic section of the surface slope (in blue) and the pressure gradients (black arrows) across the Gulf Stream and its recirculation. Arrows indicate the direction and intensity of the horizontal pressure gradient, points indicate a null gradient. The current is perpendicular to the gradient. Arrows to the left represent currents entering into the page (southward recirculation currents) and arrows to the right for those coming out of the page (northward). Red lines show isopycnals. (After Stommel, *The Gulf Stream*, 1972.)

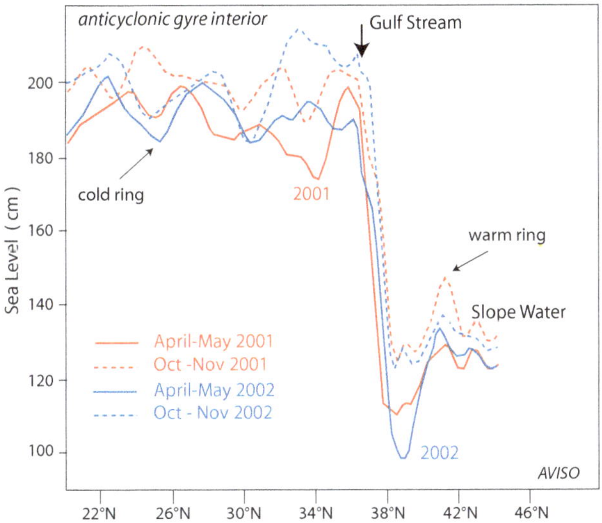

FIG. 3-28 – Sea-level variations (from Topex-Poseidon altimeter) across the Gulf Stream around 70°W in the period 2001–2002. The zone of sharp variation represents the core of the current. The hollows and humps represent passages over meanders, rings, and eddies. (From Aviso LAS, 2005.)

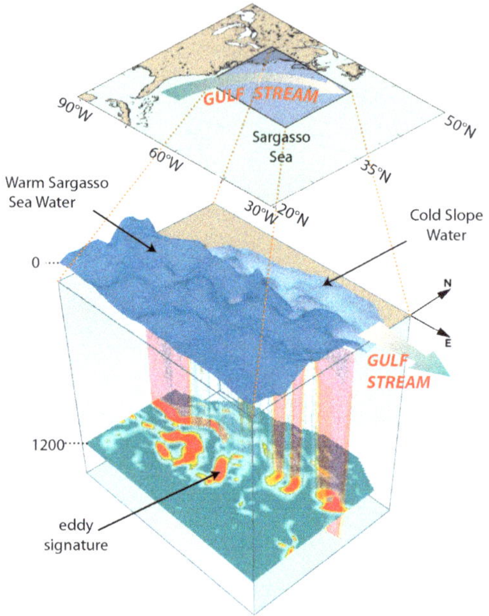

FIG. 3-29 – Block diagram of sea-level height differences measured by satellite between cold coastal water (light blue) and warm water of the Sargasso Sea (dark blue). The Gulf Stream is at the boundary between the two. The pattern of eddies, creating surface humps and hollows, is shown at the 1,200-m level. (From CNES, 1999.)

The Gulf Stream follows the continental slope as far as Cape Hatteras, where it leaves the coast and becomes unstable. It forms meanders enclosing either the warm saline nutrient-poor blue water of the interior of the gyre (Sargasso Sea, 20 °C $< T <$ 24 °C, salinity ~ 36.5) or the cold less-saline richer green water of the coastal zone, originating from the Labrador Current, called Slope Water. The meanders sometimes detach from the main warm current to form cold or warm rings, according to the side of the current from which they detach [**Figure 3-30**]. The warm ring observed on the image of 15 June 1996 was about 120 km in diameter [**Figure 3-25**]. Cold rings are always cyclonic and warm rings are always anticyclonic. Cold rings progress slowly southward and westward and are re-entrained some months later into the main current. In the Sargasso Sea, they propagate in the middle of a turbulent energy field that includes many mesoscale cyclonic and anticyclonic eddies [**Figure 3-31**]. Warm rings, formed north of the current, are limited in their movement by the proximity of the continental slope and are re-entrained more rapidly into the main current [**Figures 3-24** and **3-31**].

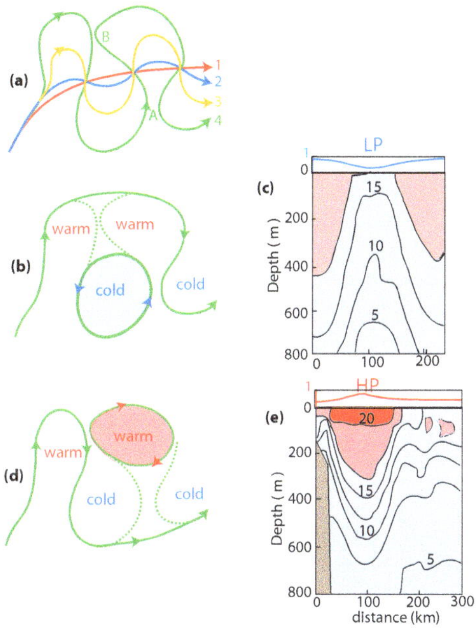

FIG. 3-30 – **(a)** Formation of meanders from the Gulf Stream, **(b)** Separation of a cyclonic ring, with **(c)** the corresponding temperature section (in °C) and associated surface topography (hollow or low pressure) (m), **(d)** separation of an anticyclonic ring, with **(e)** the corresponding temperature section and associated surface topography (hump or high pressure). (After Richardson, 1983.)

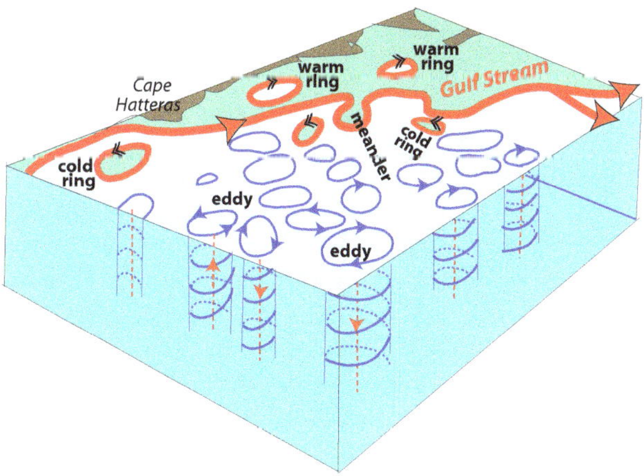

FIG. 3-31 – Schematic snapshot of the Gulf Stream with meanders, warm rings (sky blue) and cold rings (green) on either side of the main current and cyclonic and anticyclonic mesoscale eddies. (After Richardson, 1983.)

Gulf Stream rings have high velocities, comparable to those of the main current from which they emerge. The oceanographic characteristics of ring interiors can remain isolated during several months. As does the Gulf Stream, rings extend several thousand meters in depth, and have a diameter of 100 to 300 km.[235] They have an active role in heat exchange with the atmosphere and with regions on either side of the Gulf Stream. Following meanders and rings by altimetry [**Figure 3-20**] and drifting-buoy trajectories shows the strong turbulent intensity of a Gulf Stream that is much more than the "river in the ocean" of Franklin and Folger [**Figure 3-18**]. This turbulent intensity manifests itself by a maximum in sea-level variability observed by altimetry [**Figure 3-32**].[236]

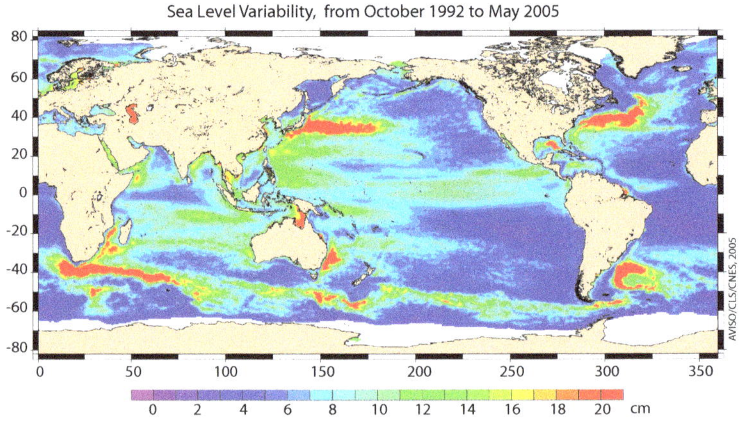

FIG. 3-32 – Sea-level variability (standard deviation with respect to the mean) obtained from combined observations of the altimetric satellites Topex-Poseidon, ERS-1, and ERS-2 from October 1992 to May 2005. (From AVISO/CLS/CNES, 2008.)

The Gulf Stream entrains adjacent water along its path [**Figure 3-19**]. Its transport thus increases from 30 ± 2 Sv off Miami to 90–100 ± 30 Sv off Cape Hatteras [**Figure 3-26c**]. It increases until it reaches a transport of about 150 Sv around 65°W, south of Nova Scotia. Two countercurrents are found, one on either side of the Gulf Stream [**Figure 3-19**]. One is located between the continental slope and the Gulf Stream. It transports cold water rich in nutrients coming from the north. The other is offshore of the Gulf Stream and represents the recirculation southward in the Sargasso

[235] In the early 1970's, the American-Soviet MODE (Mid-Ocean Dynamic Experiment) and POLYMODE programs discovered the turbulent field beyond the Gulf Stream.

[236] Standard deviation measures the dispersion of values from their mean value.

Sea, associated with high variability due to the presence of many eddies [**Figures 3-31** and **3-32**].

The Gulf Stream is a region where the ocean provides great quantities of heat to the atmosphere by evaporation and by conduction. More than 200 W m^{-2} are exchanged on average [**Section I.2.2.5**, **Figure 1-13**]. When cold air masses transported by west winds arrive over the ocean in winter, they form persistent fogs in contact with the warm water, visibly indicating the exchanges with the atmosphere [**Figure 3-33**]. For a short period between May and August, warm and humid air from the south can lightly reverse the tendency.

FIG. 3-33 – The Gulf Stream Front (between warm Gulf Stream Water and cold Slope Water) as seen from the research vessel *Atlantis I*. (Photo C. Neumann.)

4.2.2 The North Atlantic Drift and eastern boundary currents

The Gulf Stream is prolonged by the North Atlantic Drift, whose transport diminishes eastward [**Figure 3-26c**], as it divides into several branches east of the Grand Banks of Newfoundland [**Figure 3-19**]. One branch detaches southward to form the Azores Current, with transport of the order of 15 Sv between 35°N and 40°N, which joins the Canary Current. Another branch, further east, closes the anticyclonic gyre with the Portugal Current, of which about 1 Sv penetrates into the Mediterranean, and joins the Canary Current. The Canary Current, an eastern boundary current, is weak and diffuse in contrast to western boundary currents. Around 15°N the Canary Current comes into confluence with the northward cyclonic current of the Guinea Dome and turns westward. Under the NE trades, it is continued by the North Equatorial Current and closes the North Atlantic subtropical anticyclonic gyre *via* the Antilles and Florida Currents.

To the north, a branch departs from the North Atlantic Drift and follows the continental slope of the Grand Banks to form the southern part of the subpolar gyre complex.

4.3 The subpolar cyclonic gyre

North of the subtropical gyre, under the Icelandic low, a cyclonic gyre [**Figure 3-6**] is associated with the cyclonic wind circulation [**Figures 3-4** and **1-65**]. Greenland, Iceland, and topography contribute to make the subpolar gyre complex. The northern branch of the North Atlantic Drift (or North Atlantic Current) represents the southern branch of this gyre. It splits into several branches that compose the gyre pattern shown in **Figure 3-34**. After skirting the Grand Banks northward and crossing the mid-Atlantic Ridge, a first branch breaks off as the Central Iceland Basin Current (CIBC). It follows topography northward, then turns southwest as the East Reykjanes Current (ERC), which borders the Reykjanes Ridge southward, crosses the ridge toward the Irminger Sea and is continued on by the Irminger Current (IrmC) (8 to 11 Sv transport). The largest part of the Irminger Current turns westward south of Iceland and passes under the surface East Greenland Current (EGC), which goes around Cape Farewell, the southern point of Greenland, and becomes the West Greenland Current (WGC) in the Labrador Sea. North of the Labrador Sea it is joined by a current coming from the Arctic Ocean across the open straits of the Canadian Archipelago, the Baffin Sea, and the Davis Strait, to form the Labrador Current which hugs the Canadian coastline in turning past Newfoundland. The Labrador Current transports cold water, rich in nutrients.[237] Its average transport is of the order of 34 Sv with seasonal variations of 6 Sv and a maximum in February. As with the East Greenland Current, the Labrador Current is a cyclonic western boundary current, extending to 3,000 m depth. It enters into confluence with the Gulf Stream and the North Atlantic Drift to produce the Subarctic Front (SAF), thus closing the subpolar cyclonic circuit.

A filament of the Irminger Current detaches northward toward the Iceland Sea passing through the Denmark Strait [**Figure 3-34**] west of Iceland.

Another branch of the North Atlantic Drift continues northeast, splitting into several filaments, steered by bottom topography, and crosses the sills between Iceland and Scotland to join the Nordic Seas (Iceland Sea, Norwegian Sea, and Greenland Sea). A current filament passes west of the Faeroe Islands and forms the Subarctic Front Current (SAFC) in joining the East Iceland Current in the Norwegian Sea. East of the Faeroes the Rockall Passage Current, the deepest, joins the Subarctic Front Current to form the Norwegian Current (NC) [**Section III.5.2.4**].

[237] The Grand Banks off Newfoundland are a rich region for fisheries.

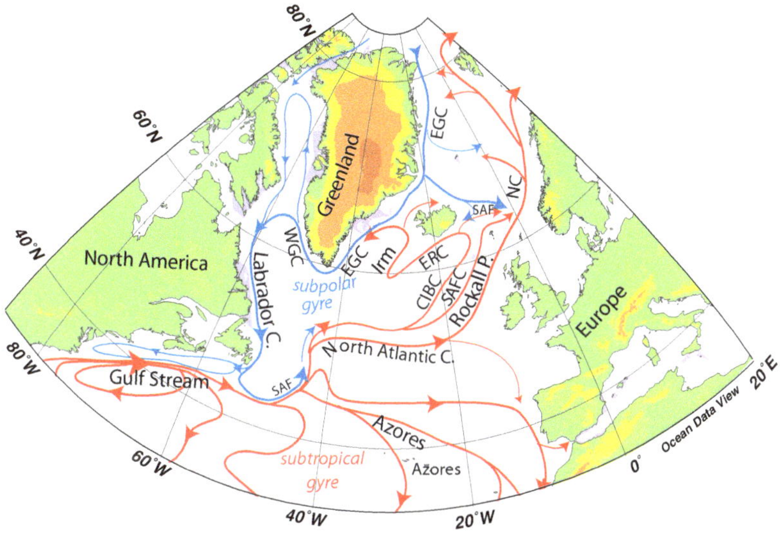

FIG. 3-34 – Schematic of the northern subtropical anticyclonic gyre, the subpolar gyre, and the Nordic-Sea circulation of the Atlantic Ocean. [Norwegian Current (NC), East Greenland Current (EGC), Subarctic Front Current (SAFC), Central Iceland Basin Current (CIBC), East Reykjanes Current (ERC), Irminger Current (Irm), West Greenland Current (WGC), Subarctic Front (SAF)]. (After Osterhaus and Gammelsrod, 1999; Schmitz and Mc Cartney, 1993; Schott *et al.*, 2004; Brambilla and Talley, 2008.)

4.4 The south subtropical gyre

In the southern hemisphere, as in the northern, an anticyclonic subtropical gyre is associated with anticyclonic winds rotating around subtropical high pressure [**Figures 3-4**, **3-6**, and **1-65**]. The drivers of this circulation are SE trade winds in the north and westerly circumpolar winds in the south. Various branches of the South Equatorial Current (SEC) [**Section III.4.1.2**, **Figures 3-6** and **3-8**] constitute the northern part of the gyre. Around the northeastern point of Brazil, Cape São Roque, a small part of the SEC turns southward along Brazilian coast and joins the southernmost branch of the SEC to form the Brazil Current, which transports warm water to the region of confluence with the Malvinas Current. The Brazil Current is shallow. Its axis follows the 200 m isobath, at the edge of the continental shelf. Around 10°S its transport of 4 Sv (1,500 m reference level) is relatively weak. The transport increases by entrainment of adjacent water, reaching 19 to 22 Sv south of the Rio Grande Rise around 38°S.

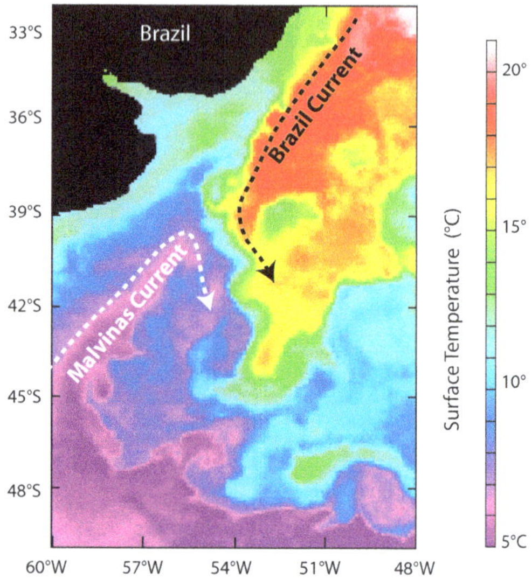

FIG. 3-35 – GHRSST hyper-frequency satellite image (4 July 2008) showing the front between Brazil Current warm water (red and yellow) and Malvinas Current cold water (violet and blue). (From POET/PO.DAAC NOAA/NASA.)

Between 33°S and 38°S, it enters in confluence with the cold Malvinas Current, which penetrates along the coast and creates a temperature front that is visible on satellite images [**Figure 3-35**]. In leaving the coast in the region of confluence, the Brazil Current becomes unstable and forms large meanders and eddies. The latitude at which it leaves the coast varies seasonally. It is further south in summer (December-February) than in winter, following the latitudinal movements of the wind system [**Section III.3.1**]. Warm water transported by detached eddies reaches 38°S to 46°S according to the season. Most eddies are entrained eastward into the South Atlantic Current. The Brazil Current is a western boundary current like the Gulf Stream. However, the transport of the Brazil Current is noticeably less than that of the Gulf Stream because of the topography of the South American continent and because the draw of water northward to compensate the formation of dense water in the North Atlantic adds to the Gulf Stream and subtracts from the Brazil Current.

The Antarctic Circumpolar Current, driven by Antarctic Ocean westerly winds is distinct from the South Atlantic Current [**Figure 3-6**]. The two currents contain strong turbulent energy and are easily distinguished on the map of sea-level variability [**Figure 3-32**].

To the east, the Benguela current closes the anticyclonic gyre of the southern hemisphere. It is a cold current because it comes from colder regions and because coastal upwelling cools the surface water [**Section III.4.5.3**].

The Benguela Current is the eastern boundary current of the south anticyclonic gyre [**Figure 3-6**], whose transport is of the order of 20–25 Sv (2,000 m reference level). It gradually leaves the coast between 30°S and 25°S, while the Benguela Coastal Current, linked to upwelling, extends as far as Cape Fria[238] at 18°S with a transport of about 7 Sv.

4.5 Upwelling zones

4.5.1 Equatorial upwelling

Easterly winds at the equator produce equatorial divergence compensated by upwelling of colder underlying water [**Section I.5.2.2**, **Figure 1-61**]. Equatorial cooling is strongest in the east, where the thermocline is nearer the surface [**Figure 3-12** and **3-36**], where the Equatorial Undercurrent sometimes reaches the surface, and in the austral winter (July-September) when the SE trades are the most intense. Equatorial waves [**Section V.4.2.9**] are visible on the satellite image of July 2008 between the high temperature of the North Equatorial Countercurrent and the low temperatures linked to upwelling. Equatorial upwelled waters are rich in nutrients.

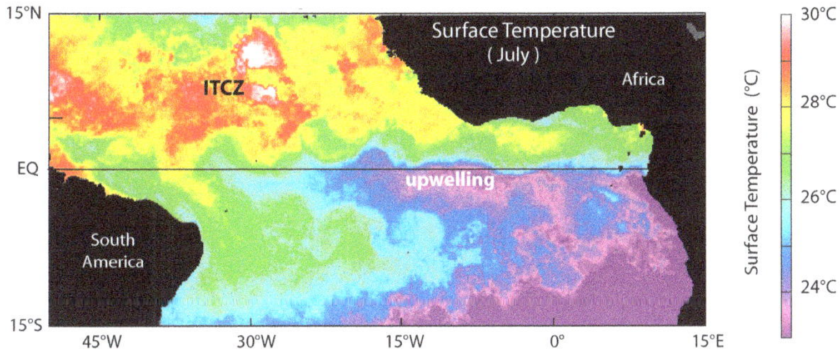

FIG. 3-36 – Sea-surface temperature in the Atlantic equatorial region. (Satellite image of 28 July 2008.) (From PO.DAAC NOAA/NASA.)

4.5.2 Canary upwelling

The coastal part of the Canary Current is cold and full of fish. Along the coasts of Morocco and Mauritania, the NE trades blow parallel to the coast [**Figure 3-4**]. The associated Ekman transport drives surface water offshore [**Section I.5.2.1, Figure 1-59**] generating upwelling of colder subsurface

238 The name "Cape Fria" was inspired by the cold upwelled water.

water near the coast [**Figure 3-37a** and **b**]. It is rich in nutrients, making the region biologically productive.[239]

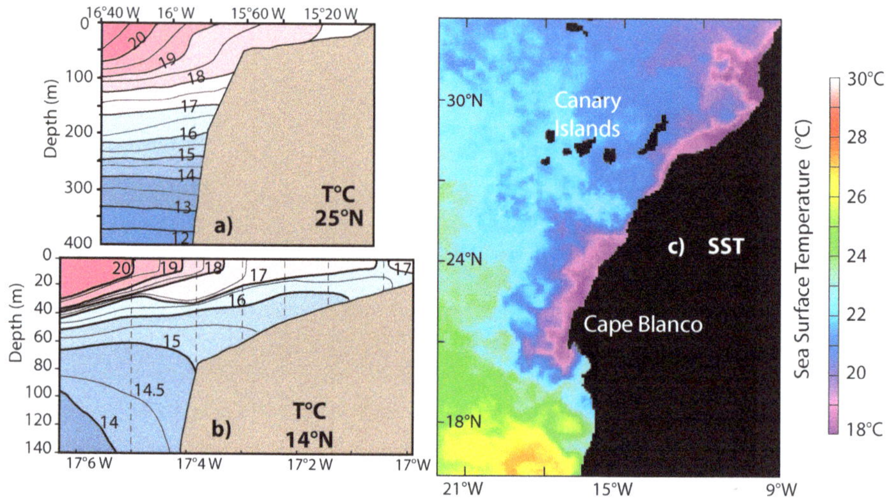

FIG. 3-37 – The Canary upwelling. **(a)** Temperature section at 25°N in November 2007. (After Makaoui *et al.*, 2008.) **(b)** Temperature section at 14°N across the southern upwelling part off the Senegal coast in February when the upwelling migrates southward. (After Capet *et al.*, 2016.) **(c)** Sea-surface temperature in the Canary upwelling region. (Satellite image of 21 November 2008.) (From PO.DAAC NOAA/NASA.)

Upwelling extends as far south as 10°N–8°N in February and only to 20°N (Cape Blanco) in August, following the variation in latitude of the trade winds. The upwelling varies depending on wind strength, as does the surface water temperature. An undercurrent going northeast flows under the Canary Current.

There is also upwelling along the Portugal Coast in July-September due to southward winds, which reverse in autumn and winter. Below, a northward undercurrent (the Portugal Coastal Undercurrent) entrains Mediterranean Water.

4.5.3 Benguela upwelling

Off southern Africa, southerly winds parallel to the coast bring coastal upwelling by the same process as that of the Canary upwelling. This decreases surface temperature to 12 °C around 26–27°S [**Figure 3-38**]. As in the Canary upwelling, the extension of upwelling northward varies with the

[239] There are numerous fisheries along the coast.

variation in latitude of the SE trade winds. The coastal cooling is visible on the map of average temperature [**Figure 1-17**]. An undercurrent flowing south under the Benguela Current transports water of tropical origin.

In these upwelling regions, precipitation does not exceed 0.1 m per year [**Figure 1-20**]. The cold upwelled water condenses the little humidity that exists in the overlying air masses, sometimes generating fogs, and stabilizes them. Consequently, the coasts bordering these regions are most often deserts (Mauritania Desert, Namib Desert).

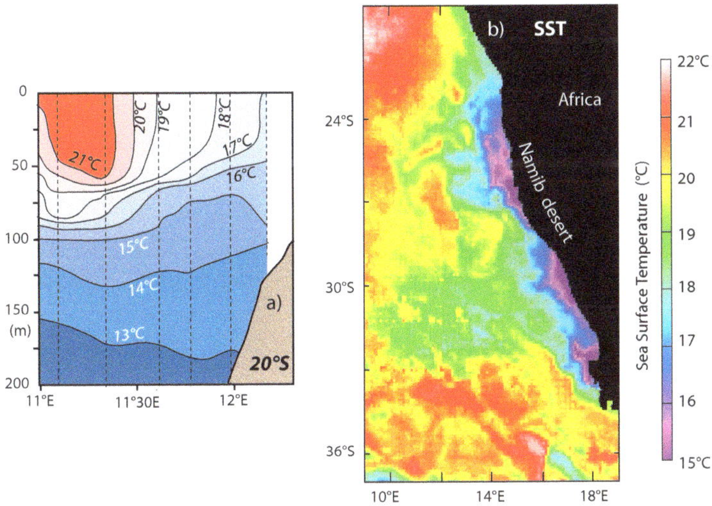

FIG. 3-38 – **(a)** Temperature section at 20°S. (After John *et al.*, 2004.) **(b)** Sea-surface temperature in the Benguela upwelling. (Satellite image of 10 January 2008.) (From PO.DAAC NOAA/NASA.)

Along the eastern coasts close to the equator, coastal upwellings occur seasonally due to equatorial Kelvin waves triggered by the SE trade winds in the western equatorial Atlantic. At the eastern boundary, these waves propagate to the north and to the south trapped as coastal Kelvin waves. Along the Guinea coast, the major upwelling season is in July-September when the ITCZ has its northern most position well north of the equator.

Cold water in upwelling zones is a major process in the regulation of climate. Upwelling regions are efficient "heat pumps" [**Figure 1-13**].

Chapter 5
Adjacent seas

Many water masses penetrate into the Atlantic Ocean and are transformed there. Among them are water masses formed in the Mediterranean Sea and the Nordic Seas, seas adjacent to the Atlantic Ocean.

5.1 The (Eurafrican) Mediterranean

5.1.1 Introduction

Many civilizations have developed around the Mediterranean Sea. The Romans called it "Mare Nostrum", and the Greeks baptized the Strait of Gibraltar, the "Pillars of Hercules". The map of the world drawn following the writings of Herodotus, 450 BCE, represents it rather well, especially the latitude [**Figure 3-39**].

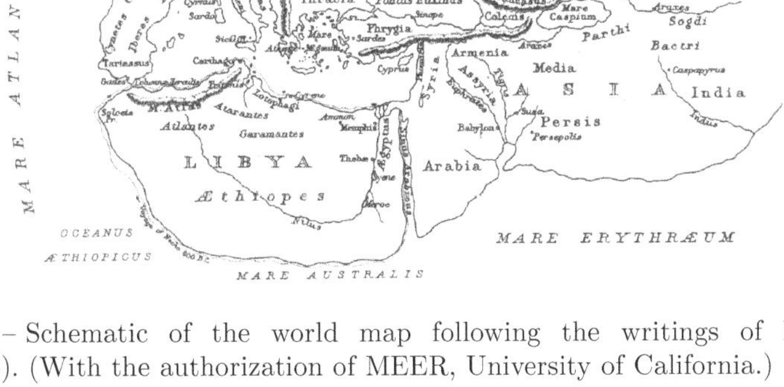

FIG. 3-39 – Schematic of the world map following the writings of Herodotus (450 BCE). (With the authorization of MEER, University of California.)

This extent of water provided a means of communication that was much easier than were the surrounding lands, which are either deserts or mountains. Since the Phoenicians, colonization began by establishing trading posts on the coasts. In this closed sea traders navigated most of the time along the coasts and did not need to know their absolute position. Astronomical navigation allowed them to estimate their latitude, but they avoided navigation in winter, the season of bad weather or overcast skies.

In the Middle Ages, the church represented the Mediterranean in a symbolic form having nothing to do with reality. However, at the end of the Middle Ages maps based on Ptolemy's rediscovered geography represented it rather well [**Figure 3-1**]. The "Carta Pisana" chart, drawn on a pigskin, the oldest Portolan map of western origin, dates from 1296 [**Figure 3-40**]. It represented quite well not only the Mediterranean coasts but also those beyond Gibraltar.

FIG. 3-40 – *Carta Pisana*, the oldest Mediterranean map, drawn on pigskin, dating from the 13th century, found at Pisa and kept at the Archives Nationales in Paris. (Courtesy of François Bellec.)

5.1.2 Topography

The Mediterranean, squeezed between the mountainous continental masses of northern Africa, western Asia, and southern Europe, is a deep interior sea containing few continental shelves (Gulf of Lions, northern Adriatic Sea, Gulf of Gabes, and northern Black Sea) [**Figure 3-41**]. It extends over 4,000 km from east to west and 800 km from south to north. Ranging between 30°N and 45°N, it has an area, including the Black Sea, of 3 million km^2. It is connected to the Atlantic Ocean by narrows of 22 km width, the Strait of Gibraltar,[240] whose sill at the west of the strait is about 300 m below the surface. The Mediterranean is composed of a series of basins with sills that may be subdivided into three principal basins: the Western Basin, the Eastern Basin, and the Black Sea.

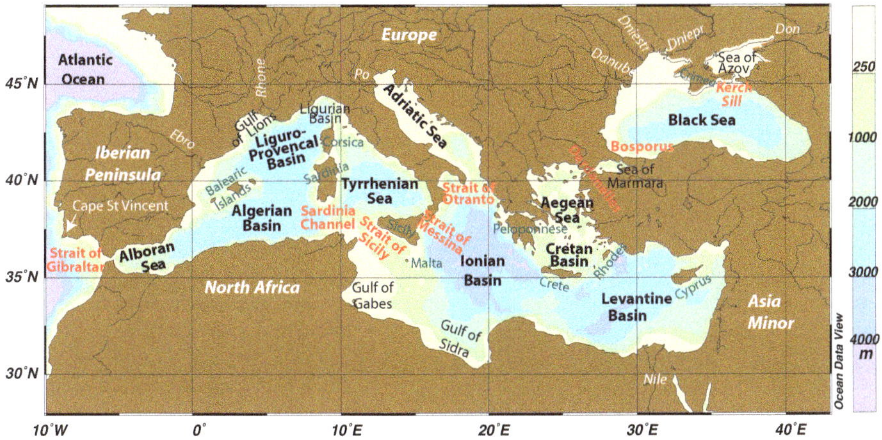

FIG. 3-41 – Topography of the Mediterranean Sea.

The Western Basin includes the Alboran Sea, at the opening of the Strait of Gibraltar, the Algerian Basin, the Liguro-Provencal Basin (2,800 m) composed of the Ligurian (sub-) Basin, the Provencal (sub-) Basin and the Gulf of Lions, and the Tyrrhenian Sea (3,300 m), separated by Corsica and Sardinia. The Western Basin communicates with the Eastern Basin through the Strait of Sicily, between Sicily and Tunisia, with a sill at ~ 430 m. A second narrow connection exists between continental Italy and Sicily through the Strait of Messina whose sill is at ~ 20 m. It is of little importance compared with the Strait of Sicily, but because of relatively violent tidal currents (up to 2 to 3 m s^{-1}) that generate strong eddies and of the channeling of wind, the passage has had a reputation for being dangerous for millennia. In the tale of the *Odyssey*, Homer placed the monsters of Charybdis and Scylla[241] here.

The Eastern Basin includes the Ionian (sub-) Basin, and the Levantine (sub-) Basin separated by a sill ~ 2,000 m between Crete and Cyrenaica [**Figure 3-41**]. The Ionian Basin (5,100 m in the Mattapan Trench southwest of the Peloponnese) is extended northwest by the Adriatic Sea (maximum depth ~ 1,270 m) which communicates with the Ionian Basin across the

[240] The name "Gibraltar" comes from *Djebel Tariq*, the mountain of Tariq Ibn Ziyad, a Berber chief who conquered Spain by crossing the strait in the 8th century.

[241] When escaping one of them, one falls into the reach of the other, even more dangerous, whence the expression "between Scylla and Charybdis".

Otranto Sill around 740 m deep. The Aegean Sea is north of the Eastern Basin from which it is separated by the Cretan Arc,[242] with sills around 600–800 m depth, between the Peloponnese and Turkey. The Aegean Sea [**Figure 3-42**] with complex relief is itself subdivided into three regions. In the north, the Mount Athos (sub-) Basin is separated from the Chios (sub-) Basin by a sill around 400 m southwest of the island of Limnos. The Chios (sub-) Basin is separated from the Cretan (sub-) Basin (the Sea of Crete) by sills around 200 m between the Cyclades and the Dodecanese. The Sea of Crete reaches 2,500 m depth.

The Black Sea is north of the Aegean Sea with an area of 420,000 km^2. A continental shelf and the Crimean peninsula are found in the north [**Figure 3-41**]. East of Crimea, the shallow Sea of Azov represents 9% of the surface and only 0.5% of the volume of the Black Sea. The Kerch Sill (about 5 m) separates it from the Black Sea. The region south of Crimea is occupied by a central trench ~ 2,245 m deep. The Black Sea is linked to the Aegean Sea through the Bosporus (35 m deep, 760 m wide, and 31 km long), the Sea of Marmara (1,000 m deep), and the Dardanelles[243] (70 km long).

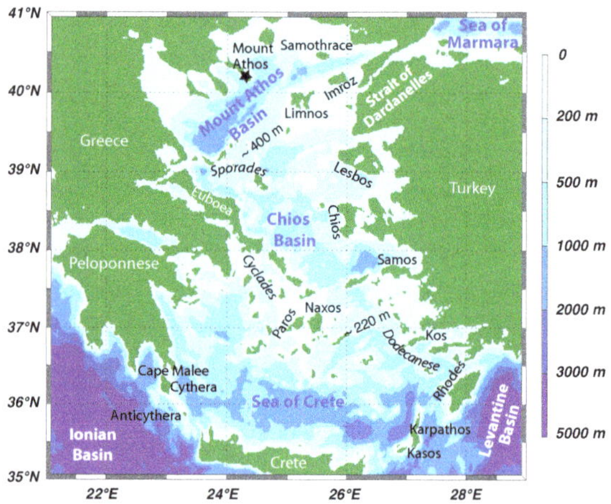

FIG. 3-42 – Topography of the Aegean Sea.

5.1.3 Climatology

The Mediterranean is a temperate region, with a climate that becomes more arid eastward. Summers are warm and dry and winters are relatively mild and humid.

242 The Cretan Arc is composed of the Islands: Cythera, Anticythera, Crete, Kasos, Karpathos, and Rhodes.

243 The *Hellespont* of the ancients.

The dominant winds are from the northwest but the northern regions are exposed in winter to northeast continental winds transporting dry and cold continental polar air masses [**Figure 3-43**]. The mountainous topography north of the Mediterranean channels winds, which, when depressions from the Atlantic affect the Mediterranean, flow out over the Mediterranean through four paths. The Mistral and the Tramontane blow over the Gulf of Lions and are felt as far as the Sicily Channel. The first follows the valley of the Rhone River (between the Alps and the French Massif Central) and the second crosses the Sill of Naurouze (between the Massif Central and the Pyrenees). The Bora[244] is channeled by the relief of the southern Alps, coming out over the northern Adriatic Sea. Regions of the Aegean Sea and of Rhodes-Cyprus receive northern continental winds.[245] In these four regions the frequency of windstorms is highest. They produce significant heat losses in winter, leading to the formation of water masses.

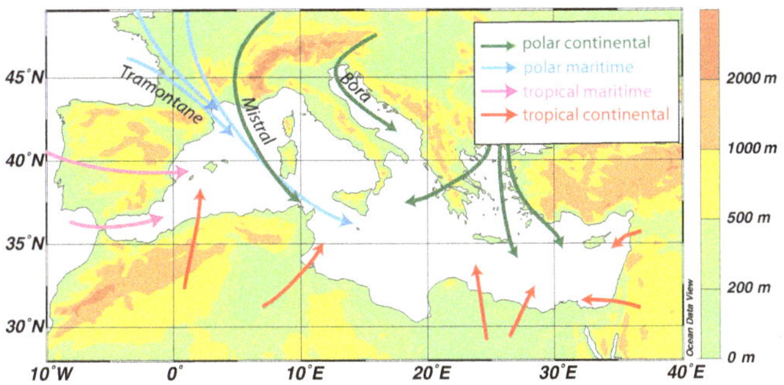

FIG. 3-43 – Topography and air-mass trajectories over the Mediterranean Basin.

North winds and dry southern winds drive strong evaporation all year long. In February, evaporation is maximal in the Aegean Sea and strong in the Levantine Basin, the Adriatic Sea, and the Gulf of Lions. This can be seen on the latent heat-flux distribution map calculated from simulations [**Figure 3-44**].

Precipitation is weak, of the order of 0.2 m/year in the east and 0.6 m/year in the west.

The water balance between evaporation losses and gains by precipitation and inflow results in a deficit of about 0.5 m to 1 m per year over the entire surface of the Mediterranean.

The Mediterranean functions as a concentration basin connected to the main ocean by a single strait.

244 The Bora can be katabatic (a high-density cold air mass) violently descending the alpine slopes.

245 In summer the Meltem (or Etesian) winds blow over the Aegean Sea and set up preconditions for the formation of dense water in winter.

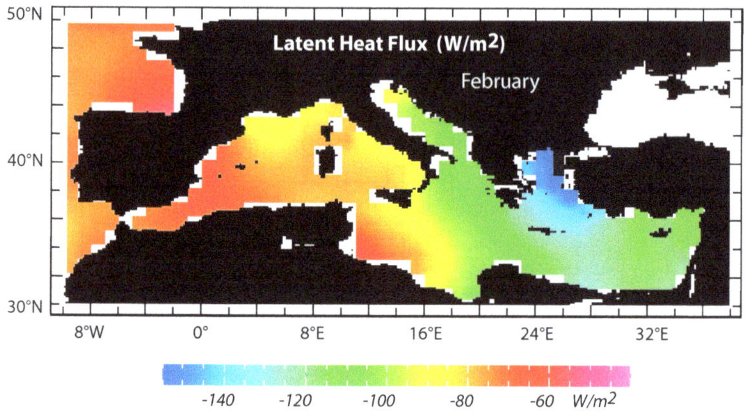

FIG. 3-44 – Latent heat flux calculated for February. Values are in W m^{-2}, flux is negative for heat loss by the sea. (From IRI/LDEO, climatology SOC GASC97.)

Depending on the fresh-water balance, mediterraneans may be basins of concentration (*e.g.*, the Eurafrican Mediterranean or the Red Sea) or of dilution (*e.g.*, the Black Sea, the Baltic Sea, or the Arctic Ocean) [**Figure 3-45**].

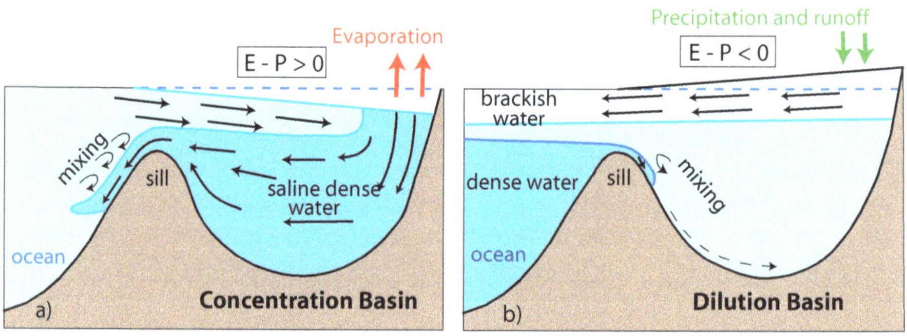

FIG. 3-45 – Schematic of **(a)** a concentration basin and **(b)** a dilution basin. Surface slopes are greatly exaggerated.

In the case of a dilution basin, inputs of fresh water are more important than losses by evaporation so that surface salinity is low, stratification is high, and the level is highest in the basin, which drives a flow of low-salinity surface water out through the strait. Renewal of deep water then occurs sporadically.

In the case of a concentration basin, evaporation holds sway over precipitation and inflow, and the salinity of surface waters increases continually after their arrival into the basin. Fresh-water loss in the interior of the basin lowers the level of the sea, generating the entry of more water into the Mediterranean through the strait. The increase in winter density at the surface (due as much to evaporation as to losses of latent and sensible heat) generates convective motion of varying depth (in forming so-called intermediate or deep waters) that drives a renewal and an evacuation of dense water called Mediterranean Water. The most saline and dense[246] Mediterranean Water exits over the sill under the lighter and less saline ocean water. This is the case of the Eurafrican Mediterranean [**Figure 3-46**]. The entry of water at the strait is induced by the losses of fresh water over the Mediterranean.

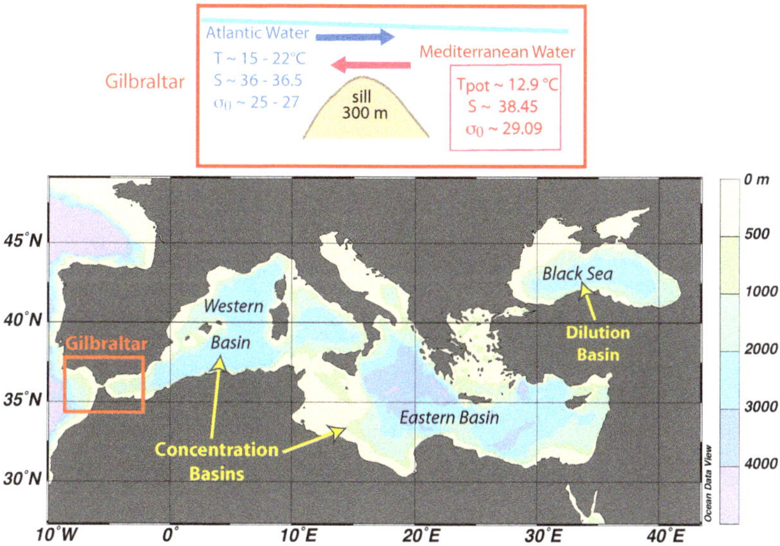

FIG. 3-46 – Schematic of the Mediterranean system.

5.1.4 Surface circulation

Under the effect of the Coriolis force, the general circulation of the Mediterranean is characterized by cyclonic circuits in each basin including strong mesoscale activity and seasonal variations due to the seasonal variation of winds constrained by topography [**Figure 3-47**]. South coastal currents are especially unstable and form numerous anticyclonic eddies.

At the entry, the narrowing of the Strait of Gibraltar intensifies speeds (through hydraulic action) that can exceed 1 m s^{-1} and creates a jet of Atlantic Water, channeled by topography.[247] In the Alboran Sea, the jet

[246] The pressure gradient at depth between the more saline and dense Mediterranean Water and the oceanic water generates an exit over the sill. At the surface, the gradient is directed toward the Mediterranean.

[247] It has always been delicate for sailing vessels to leave the Mediterranean by the Strait of Gibraltar since the current, and frequently the wind, are constant, strong and against them here. The ancients sometimes stayed for weeks, blocked by contrary winds, but some already knew how to use the deep current by dropping a ballasted sail at depth. During the Second World War, submarines positioned themselves in the deep current to leave the Mediterranean incognito.

forms one or two anticyclonic eddies of about 150 km diameter. Associated with these eddies, the thermal front between the arriving Atlantic Water and the already modified Atlantic Water (the Almeria-Oran Front) is visible by satellite [**Figure 3-48**]. Atlantic Water continues eastward along the continental slope of North Africa in the Algerian Current,[248] which is unstable and forms meanders and large anticyclonic eddies of 150 to 250 km diameter.

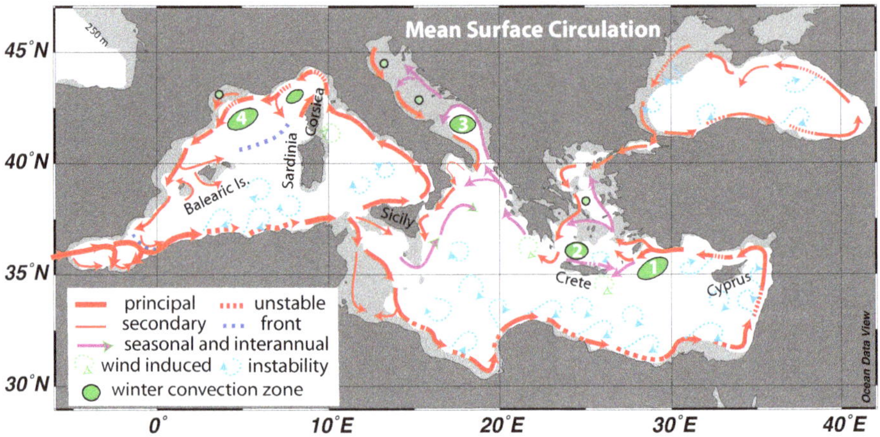

FIG. 3-47 – Mean Mediterranean surface circulation showing principal and secondary circulation, seasonal and interannual circulation (in the western Ionian basin), current instabilities (meanders and eddies), wind-induced eddies, and regions of wintertime convection: principal (1: Rhodes-Cyprus, 2: Cretan Sea, 3: Adriatic Sea, 4: Gulf of Lions) and secondary. The isobath is at 250 m. (After Millot and Taupier-Letage, 2005.)

Arriving at the Sicily Channel, a branch diverges toward the Tyrrhenian Sea (or sub-basin). There it feeds regions of dense-water formation in the north of the basin. The main branch enters into the Eastern Basin [**Figure 3-47**] participating in the cyclonic circulation to the east by following the Tunisian, Libyan, and Egyptian continental slopes (where the current is unstable, forming meanders and anticyclonic eddies) and turns northward along the coast of Asia Minor. Near Rhodes the current most often splits in two. One part goes southwest to feed the region of winter dense-water formation between Rhodes, Crete, and Cyprus. The other part enters the Aegean Sea and after transformation there in winter emerges into the Ionian Basin west of Crete. Following the western Peloponnese coasts, one branch of current enters the Adriatic Sea by the Strait of Otranto and follows a cyclonic circulation with several branches. The other branch goes off south of Italy and Sicily.

---

248 The Algerian Current has speeds of 0.4 m s^{-1} to 0.8 m s^{-1} and transport of the order of 0.5 Sv.

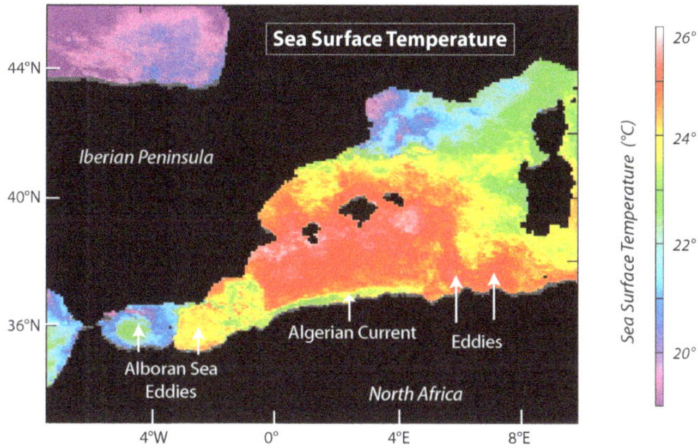

FIG. 3-48 – Alboran Sea and North African eddies seen by satellite. (POET/NASA/GHRSST data.)

In the Black Sea, the general circulation is cyclonic and includes meanders and eddies [**Figure 3-47**]. The average surface topography over the 1992–2008 period [**Figure 3-49**] obtained by satellite altimetry shows the principal cyclonic circulation in each basin (as a surface depression in dark blue) and numerous meanders and eddies.

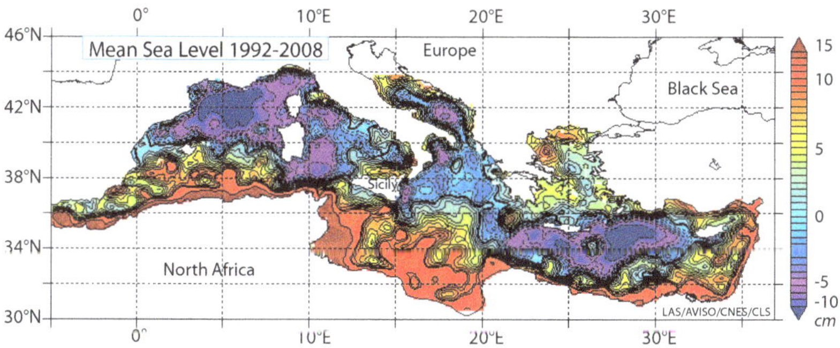

FIG. 3-49 – Sea-surface topography from satellite altimetry (in cm) averaged from 14 October 1992 to 23 January 2008. (From LAS/AVISO/CNES/CLS.)

5.1.5 Mediterranean oceanography

5.1.5.1. Atlantic Water

Atlantic Water enters through the Strait of Gibraltar in a surface layer 100–200 m deep[249] [**Figures 3-50** and **3-53**]. Under the effect of evaporation, the upper layer of Atlantic Water becomes more saline as it travels eastward

[249] The depth is modulated by variations in atmospheric pressure and by tides.

but is warm enough in summer to stay above the lower layer of Atlantic Water characterized by a salinity minimum.

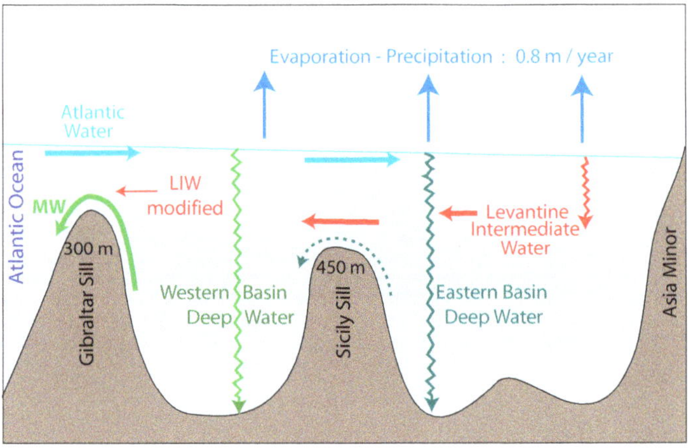

FIG. 3-50 – Schematic of the Mediterranean system. The zigzags represent wintertime convection. In the Eastern Basin, the deep convection occurs in the Adriatic Sea. (MW: Mediterranean Water, LIW: Levantine Intermediate Water).

The extension of minimum-salinity Atlantic Water may be followed as far as the Levantine Basin where it is no longer discernible [**Figure 3-51**]. The minimum value goes from ~ 36.3 at Gibraltar to more than 39 in the east through mixing with water above that has been modified by climatic effects and more saline water underneath. In the Levantine Basin surface water reaches a salinity of more than 39.30 but in summer its high temperature (24 °C, σ_θ = 27.0) allows it to stay on the surface.

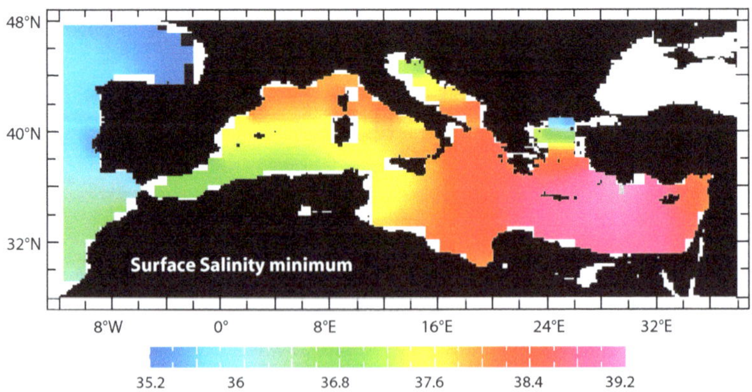

FIG. 3-51 – Evolution of the salinity minimum of surface Atlantic Water. (IRI/LDEO with Levitus data.)

5.1.5.2. The Black Sea

Northeast of the Mediterranean, the Black Sea ("Pontus Euximus" of the Romans) is connected to the Sea of Marmara through the Bosporus [**Figures 3-41** and **3-52**]. It drains a vast extent of Europe from the Swiss Alps to the Russian steppes, receiving many rivers such as the Danube, the Dniester, the Dnieper, and the Don. Evaporation does not compensate precipitation and especially not the river inputs. In contrast to the Mediterranean overall, the fresh-water balance of the Black Sea is positive. It is a dilution basin [**Figure 3-45b**].

With the accumulation of low-salinity ($S < 18$) low-density surface water, sea level is higher than in the Sea of Marmara and the Aegean Sea. It establishes a surface pressure gradient that drives the outflow of low-salinity surface water of 27 m to 30 m thickness over the Bosporus Sill. In the Sea of Marmara low-salinity ($S = 18.2$) surface layer 20 m to 25 m deep flows above Mediterranean water ($\theta \sim 14.5$ °C, $S \sim 38.5$, $\sigma_\theta \sim 28.79$) [**Figure 3-52**] and extends as far as the northern Aegean Sea [**Figure 3-51**].

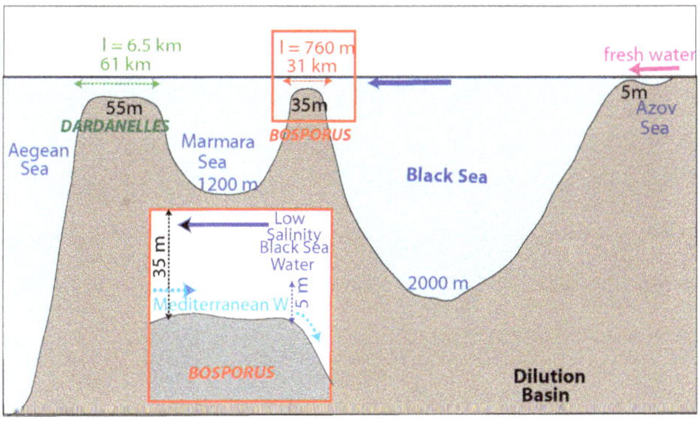

FIG. 3-52 – Schematic of the Black Sea system.

Since the surface water is always less dense ($\sigma_\theta \sim 14$) than Black Sea Deep Water, even in winter, no deep convective motion is possible. Deep Water ($\theta \sim 9$ °C, $S \sim 22.3$, $\sigma_\theta \sim 17.2$) is isolated from the surface layer by a sharp pycnocline whose depth varies seasonally. In summer, surface waters warm and form a shallow thermocline of a few tens of meters [**Figure 3-53a**]. During winter cooling, a low-salinity mixed layer of about 50 m depth is formed. At this depth, the oxygen level decreases strongly forming an "oxycline". Below 200 m, oxygen is completely consumed because of the lack of ventilation and is replaced by hydrogen sulfide (H_2S).

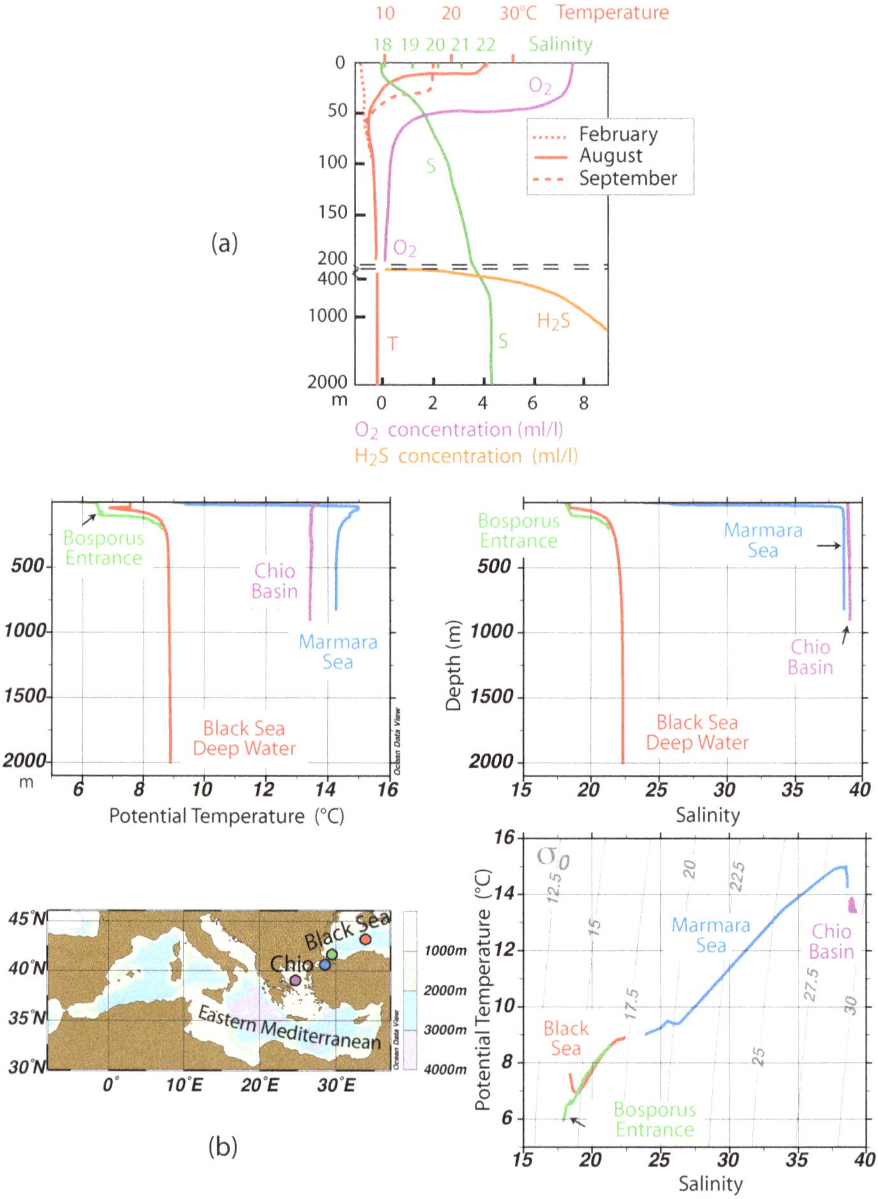

FIG. 3-53 – **(a)** Profiles of temperature (by season), salinity, dissolved oxygen, and hydrogen sulfide (H_2S) in the Black Sea. (1 ml/l = 44 μmol kg^{-1} for $\sigma_\theta = 14$). The scale changes at 200 m (After Tolmazin, 1985.) **(b)** Evolution of profiles of potential temperature (°C), and salinity and θ-S diagrams from the Black Sea to the northern Aegean Sea in winter. (With ODV and World Ocean Database.)

The inflow of Mediterranean Water (θ ~ 14.5 °C, S ~ 38.5, σ_θ ~ 28.79) [**Figure 3-53b**, Sea of Marmara Station] over the bottom of the sill takes place over a thickness of only 5 to 8 m, and is sometimes impeded by perturbations which modify the level of the interface and block its entry. Renewal of deep water is slow with respect to the volume of the Black Sea. Current speed at the bottom of the Bosporus is less than 0.1 m s^{-1}. Mediterranean Water, more saline, warmer, denser, and more oxygenated, descends along the Anatolian continental shelf in a cyclonic movement, mixing with older deep water [**Figure 3-53b**]. Between 300 m and the bottom (about 2,160 m), waters are homogeneous in salinity and temperature and the age of this Deep Water (estimated by Carbon-14) varies between 600 and 2,200 years.

The low-salinity water of the Black sea mixes rapidly in the northern Aegean Sea and has little impact on Mediterranean oceanography. In **Figure 3-53b**, the station in the Chios Basin is homogeneous in winter without a trace of the low-salinity water.

5.1.5.3. Formation processes of Mediterranean water masses

a – Offshore convection: The cyclonic circulation of each basin implies a weaker stratification in the center of the circuit [**Figures 1-66, 3-47** and **3-49**]. Lighter water is ejected at the periphery of the cyclonic circuit leaving denser water in the center. This is the phase of preconditioning the water for convection. In regions where cold and dry winds blow in winter and where the surface salinity is high enough, cooling and evaporation (increasing the density) drive convective motion that homogenizes a deep layer whose depth varies according to region and winter severity.

b – Continental shelf convection: In shallow regions, the effects of cooling and evaporation are intensified when applied to a small volume of water. In winter, cold and dense water is formed. It flows along the continental slope where it mixes with the intermediate water that it encounters. The result descends to a depth that depends on the density and volume created.

5.1.5.4. Intermediate Levantine Water

The region between Rhodes, Crete, and Cyprus is subject to north winds and is an offshore-convection region [**Figures 3-43** and **3-47**]. A homogeneous layer 100–300 m deep forms in winter, varying according to the severity of the winter[250], called Levantine Intermediate Water (LIW) ($14.70 < \theta < 16.40$, $38.90 < S < 39.20$, $28.89 < \sigma_\theta < 29.05$) [**Figure 3-55**, Station 4]. Then LIW subducts and spreads along isopycnal surfaces. In summer the surface layer warms again but the Levantine Intermediate Water keeps its winter characteristics at depth and extends throughout the Mediterranean Basin in the

[250] During some winters, as in 1995, deep convective chimneys were observed down to 900 m, forming Levantine Deep Water.

form of a salinity maximum (between 150 m and 500 m). Its core is located around 250 m and it deepens to 300 m in the Ionian (sub-)Basin [**Figure 3-54**]. The salinity maximum of Levantine Intermediate Water crosses the Sicily Sill (~ 430 m) and continues to deepen to around 400 m in the Western Basin, where it is still characterized in winter by a salinity and temperature maximum[251] [**Figure 3-55**, Stations 1 and 2]. Mixing with less-saline water above and below decreases its salinity maximum as it extends from its formation region [**Figures 3-54** and **3-55**]. Its signature can be followed as far as the Strait of Gibraltar.

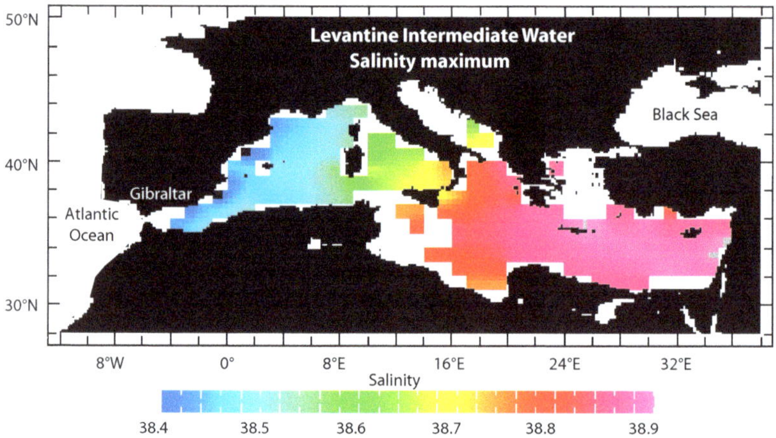

FIG. 3-54 – Evolution of the salinity maximum of Levantine Intermediate Water. (IRI/LDEO with Levitus data 1994.)

When winter conditions are severe it can happen, as in the 1990's, that convective motion can be much deeper. Levantine Deep Water can then be formed, whose salinity reaches 38.87 (σ_θ ~ 29.16). During this climatic episode Levantine Deep Water, along with water formed in the Adriatic and Cretan Seas, participate in Eastern Basin Deep Water formation.

5.1.5.5. Aegean Sea

Deep Water of the Cretan Sea (in the southern Aegean Sea, **Figure 3-55**, Station 5) $38.88 < S < 38.94$, $14.15\ °C < \theta < 14.24$, $29.16 < \sigma_\theta < 29.18$, normally arises from a mixture of Eastern Basin Deep Water flowing across the passages of the Cretan Arc and Cretan Intermediate Water (S ~ 38.90; θ ~ 14.80 °C; σ_θ ~ 29.03), formed by winter convection under the action of cold and dry wind events.

[251] Levantine Intermediate Water represents a reservoir of heat and salt for the Mediterranean Basin.

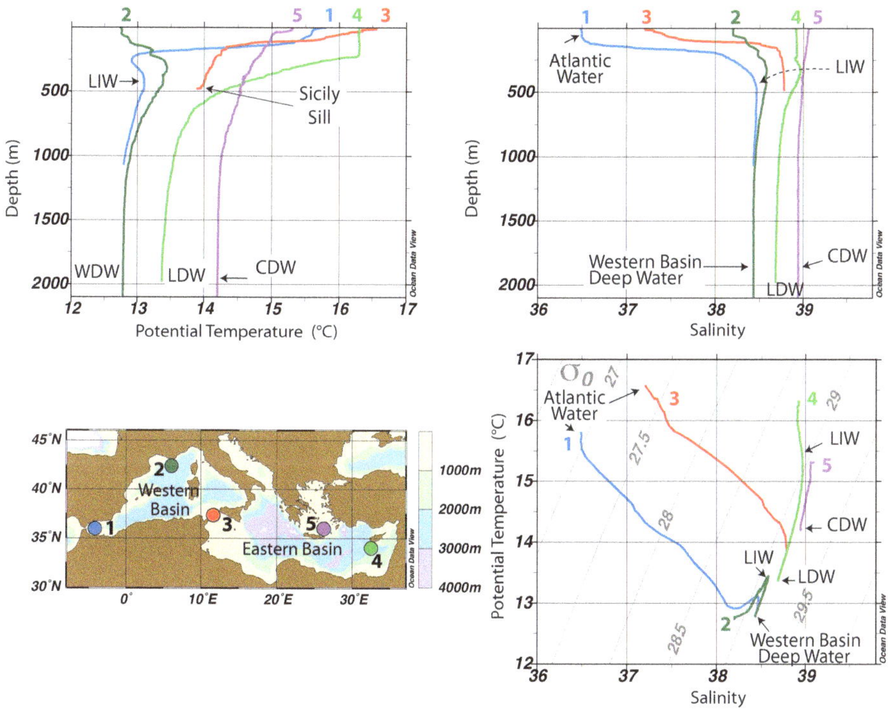

FIG. 3-55 – Wintertime temperature and salinity profiles and θ-S diagrams showing the evolution of Mediterranean water masses. [Levantine Intermediate Water (LIW), Cretan Deep Water (CDW), Levantine Basin Deep Water (LDW), Western Basin Deep Water (WDW).] (With ODV and World Ocean Database.)

During the 1990s, winter conditions were significantly more severe and a new water mass was formed in the Cretan Sea by offshore convection that was more saline, colder, and especially denser ($39.05 < S < 39.10$, $13.90 < \theta < 14.00$, $29.34 < \sigma_\theta < 29.36$). The new denser Cretan Deep Water extended southward across the deep passages of the Cretan Arc. It passed under the Deep Water of the Ionian Basin normally formed in the Adriatic Sea and modified the characteristics of the Deep Water in the Ionian and Levantine Basins. It provides an example of climatic variability that can affect deep oceanographic conditions.

5.1.5.6. Adriatic Sea and Eastern Basin Deep Water

Under Surface Water of Atlantic origin between 200 m and 500 m, Levantine Intermediate Water, somewhat diluted by mixing ($\theta \sim 14$ °C; $S \sim 38.70$; $\sigma_\theta \sim 29.05$) and Intermediate Water formed in the Cretan Sea (CIW) enter the Adriatic [**Figures 3-56** and **3-57**, Station 4]. To the north, the wide continental shelf receives a number of rivers (such as the Po) that drain northern Italy. The water balance between evaporation loss and inputs from precipitation and rivers is slightly positive. The Adriatic Sea is on average a basin of dilution. Surface Water, lightly desalinated by fresh water inputs ($S \sim 38.25$–38.5), exits on the surface through the Strait of Otranto.

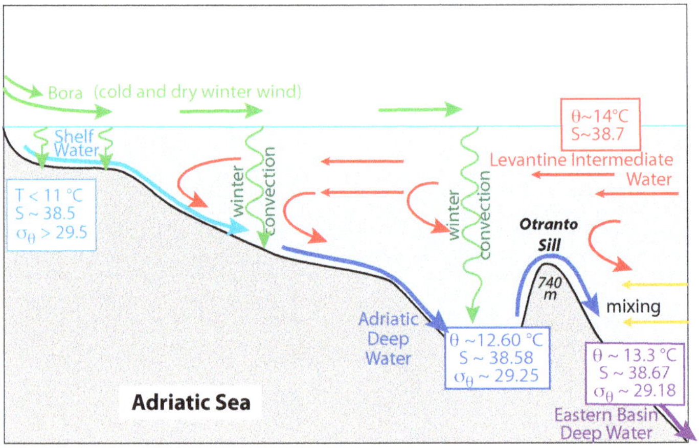

FIG. 3-56 – Schematic of the Adriatic Sea system.

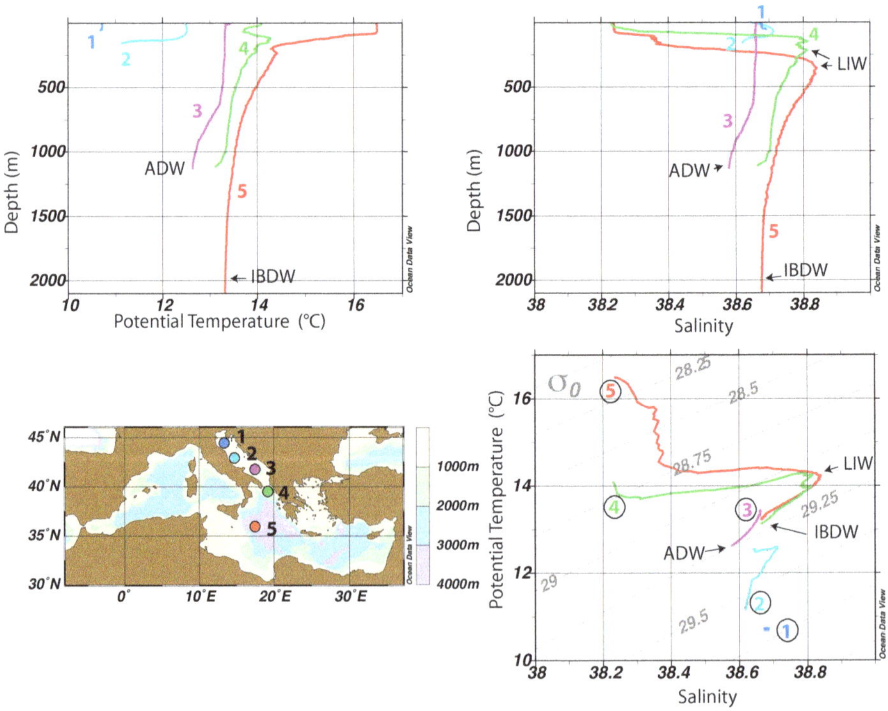

FIG. 3-57 – Temperature and salinity profiles and θ-S diagrams in the Adriatic Sea and Ionian Basin in winter [Levantine Intermediate Water (LIW), Ionian Basin Deep Water (IBDW), Adriatic Sea Deep Water (ADW)]. (With ODV and World Ocean Database.)

Nevertheless, two dense-water formation processes coexist in the Adriatic Sea. In winter, under the effect of the Bora [**Figure 3-43**], a cold dry wind, cooling and evaporation [**Figure 3-44**] intensify over the entire Adriatic Sea. On the wide continental shelf to the north, temperature drops to less than 11 °C [**Figure 3-57**, Station 1] triggering cold-water formation, relatively low in salinity (~ 38.5) and dense (> 29.5), which descends along the continental slope mixing with Levantine Intermediate Water and Cretan Intermediate Water. Some years they join and mix with dense water which forms in the central (Station 2) and southern [**Figures 3-47** and **3-57**, Station 3] Adriatic Sea.

In the center of the southern cyclonic circuit where stratification is weaker, northerly winds in winter generate convective motion over more than 800 m depth leading to the formation of South Adriatic Deep Water ($\theta \sim 12.60$ °C, $S \sim 38.58$, $\sigma_\theta \sim 29.25$) [**Figure 3-57**, Station 3]. It exits over the Otranto Sill (740 m) into the Ionian Basin [**Figures 3-56** and **3-57**, Station 4] and forms Eastern Basin Deep Water (Station 5) ($\theta \sim 13.3$ °C, $S \sim 38.67$, $\sigma_\theta \sim 29.18$). Since stratification is relatively weak between intermediate waters and deep waters, during the most severe winters in the Levantine Basin south of Rhodes convection can reach the bottom and can form Levantine Basin Bottom Water which mixes with that of the Ionian Basin.

The Sicily Sill depth allows Levantine Intermediate Water to freely penetrate into the Western Basin, while Eastern Basin Deep Water has difficulty exiting [**Figure 3-55**]. It passes alongside the Sicily Sill, mixes with Western Basin Deep Water and extends out around 2,000 m depth to form Tyrrhenian Sea Deep Water.[252]

5.1.5.7. Western Basin Deep Water

The final region subject to cold dry winds in winter is the Gulf of Lions. The Mistral (channeled by the Rhone valley) and the Tramontane[253] (channeled by the Naurouze Sill between the Pyrenees and the Massif Central) blow there [**Figure 3-43**]. The circulation in the Liguro-Provençal Basin is cyclonic. Light surface water is expelled at the periphery of the circuit leaving denser water in the center where stratification is weaker. These are the preconditioning conditions required for deep convection.

[252] Recent studies have even shown traces of this water mass on the Gibraltar Sill.

[253] The Tramontane is called *Cers* in the Languedoc region of France.

The oceanographic vessel *Jean Charcot* in the western Mediterranean in 1969.

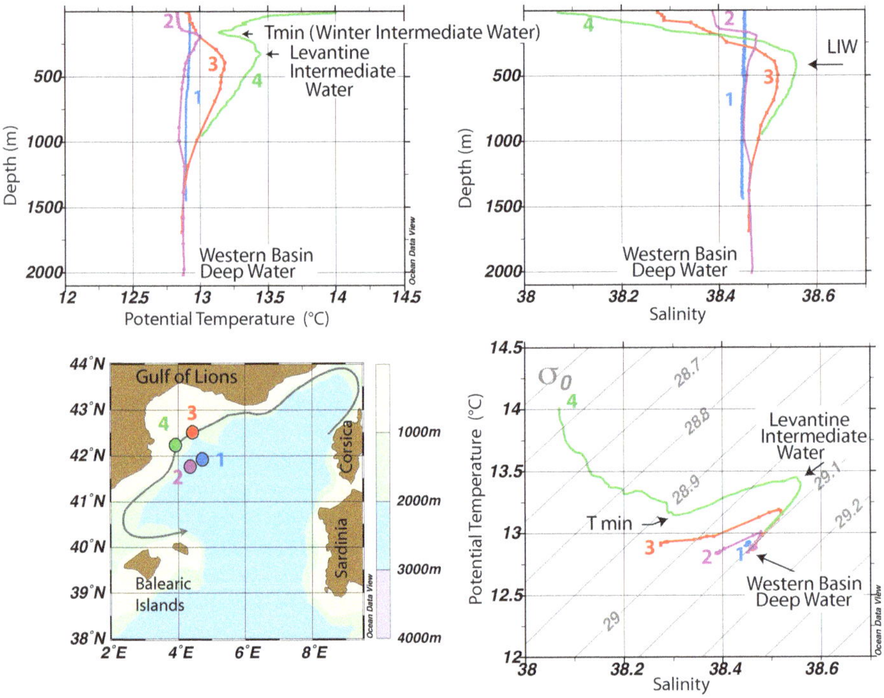

FIG. 3-58 – Potential temperature and salinity profiles and θ-S diagrams in the Gulf of Lions at the end of winter in the center (Stations 1 and 2) and at the periphery of the cyclonic circuit (Stations 3 and 4). (With ODV and World Ocean Database.)

In summer the oceanographic structure schematically includes four layers [**Figure 3-58**]:

- Warm surface water (vestiges of Atlantic Water transformed by climatic effects) 150–200 m deep.
- A temperature-minimum layer representing the remains of the winter mixed layer or a trace of dense water formed on the continental

shelf, also now often called Winter Intermediate Water[254] whose depth varies by region ($\theta \sim 13$–13.5, $S \sim 38.0$–38.3, $\sigma_\theta \sim 28.9$).
- Levantine Intermediate Water between 300 m and 600 m, characterized by a salinity and temperature maximum around 400 m ($\theta \sim 13.3$–13.5, $S \sim 38.5$, $\sigma_\theta \sim 29.05$) [**Figure 3-58**, Station 4].
- Western Basin Deep Water whose characteristics vary year-by-year ($\theta = 12.7$–12.8, $S = 38.405$–38.43, $\sigma_\theta \sim 29.10$–29.11) and are homogeneous below 1,200 m.

Exchanges are intensified in winter under the effect of the cold and dry Mistral. Surface temperature diminishes and instabilities appear, provoking increasingly deep convective motion in the course of the winter. In 1969, the international MEDOC experiment[255] followed the formation of Western Basin Deep Water for the first time[256] [**Figures 3-59** and **3-60**]. In January Levantine Intermediate Water layer, characterized by salinities above 38.50, temperatures above 13 °C, and oxygen levels less than 186 μmol kg^{-1} was present.[257] In the beginning of February in the cyclonic circulation center (around 42°N) under the cold and dry Mistral, convective motion had already mixed the surface layer and the upper part of Levantine Intermediate Water. Then the Intermediate Water layer disappeared through convective mixing and from February 15 to 21, homogenization was completed from the surface down to more than 2,000 m in all characteristics.

When convective motion reached the bottom (2,400 m) density reached $\sigma_\theta = 29.11$ over the entire water column.

Formation is not continuous, either in space or time. During convection, which takes place in columns several kilometers in diameter ("chimneys"), both descending surface water and upwelling deep water are present. Vertical velocities measured in the columns reach high values of 0.10 m s^{-1}.

---

[254] Winter Intermediate Water forms in winter in the northern Western Basin, above Levantine Intermediate Water, from Atlantic Water by cooling, mixing, and shallow convection. Covered in summer by Atlantic Water, it is characterized by a subsurface temperature minimum [**Figure 3-58**, Station 4]. Recent studies have been able to distinguish this water mass on the Gibraltar Sill.

[255] The *MEDiterranée OCcidentale* experiment was an international program combining ships from France (*Jean Charcot* and *Origny*), Britain (*Discovery*), United States (*Atlantis*), and Italy (*Bannock*).

[256] Before this experiment the characteristics of Deep Water were known without having observed its formation since the sea state in a violent Mistral forces small ships to seek shelter. With the construction of more stable vessels such as the *Jean Charcot*, measurements could continue during the strong windstorms that trigger convection.

[257] Levantine Intermediate Water is the oldest in the Western Basin and thus has an oxygen-level minimum.

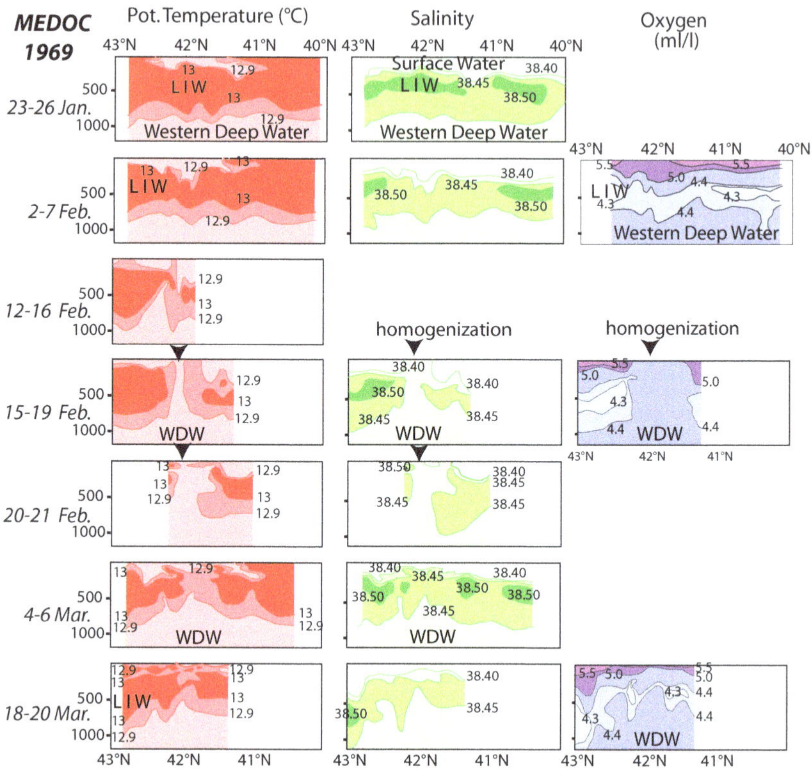

FIG. 3-59 – Evolution of the water structure (θ, S, O_2) along 6°E during the formation of Western Basin Deep Water (WDW) in 1969. Black arrows indicate convection chimneys. [LIW, Levantine Intermediate Water]. (After Tchernia and Fieux, 1971.)

Paul Tchernia and Henry Stommel on board the RV *Jean Charcot* during MEDOC 1970, (photo Michèle Fieux).

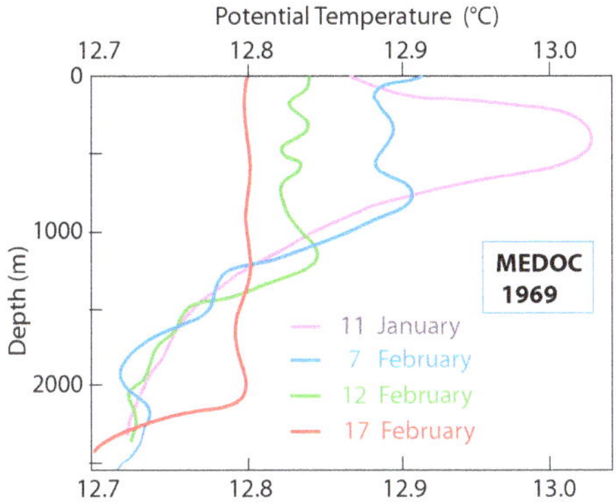

FIG. 3-60 – Potential temperature profiles in the Gulf of Lions around 42°N-5°E (MEDOC 1969) showing the deepening of the mixed layer in one month. (After Sankey, 1973.)

On the continental shelf coastal zone of the Gulf of Lions, cooling involves a less thick layer (minimal thermal inertia) and can result in the formation of water whose density in some years reaches and exceeds the density of Deep Water (29.10–29.11). It descends along the slope and in the canyons cut into the side of the shelf, where it becomes diluted through mixing with the layer of Levantine Intermediate Water. Depending on the quantity formed and the density reached, the mixing can influence Deep Water.

The Buoy Laboratory *BORHA II* (BOuée de Recherche HAbitée) anchored in the Gulf of Lions in 2,000 m of water, during the COFRASOV 76 expedition.

Deep Water (θ = 12.70 °C, S = 38.41, σ_θ = 29.10, O_2 = 4.6 ml l^{-1} or 200 µmol kg^{-1}) expands from the formation region and blankets the bottom of the Western Basin (from 1,000 m to the bottom) [**Figures 3-55** and **3-58**]. It is more saline, warmer, more oxygenated, and denser than water located at the same depth in the Atlantic. Approaching the Gibraltar Sill, the dynamics of the strait make Deep Water rise for several hundreds of meters on the African side[258] [**Figure 3-61**, Stations 1 and 2]. Deep Water, characterized on the sill by a temperature minimum ($\theta \sim 12.90$ °C), exits preferentially on the African side, and Levantine Intermediate Water, characterized by a salinity maximum ($S \sim 38.44$), on the Iberian side. They mix with North Atlantic Central Water on exiting the sill [**Figure 3-61**, Stations 3 and 4] and influence the entire Atlantic Ocean [**Section III.6.5**].

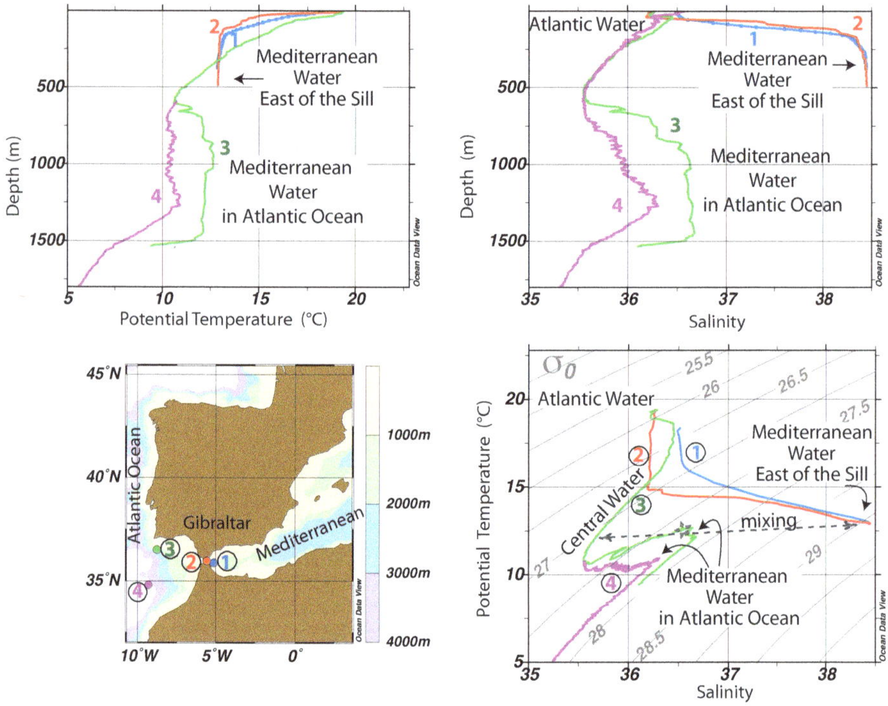

FIG. 3-61 – Potential temperature and salinity profiles and θ-S diagrams of stations located east and west of the Gibraltar Sill. (Station 1 at the east, Station 2 on the sill, Station 3 in the vein of Mediterranean Water, Station 4 SW of Gibraltar.) (With ODV and MEDAtlas data.)

Because of water loss ($E - P > 1$) over the entire Mediterranean Sea, Mediterranean Water transport out is slightly less than that of Atlantic Water transport in, which is of the order of 0.8–1 Sv.[259]

[258] Recent studies propose that only the formation of new dense water "obliges" the old to leave the basin.

[259] Fluxes are strongly modulated by tidal waves on the sill, by internal waves at the interface between Atlantic Water and Mediterranean Water.

5.1.5.8. Mediterranean characteristics

The Mediterranean is like a reduced-scale model of the ocean. Here one finds most oceanic processes, even the formation of bottom water, which is relatively easier to study in more clement weather[260] than in the Antarctic. Located in a region where evaporation dominates largely over precipitation and inflow, the Mediterranean is a concentration basin transforming surface Atlantic Water into Mediterranean Water[261], which influences the oceanography of the entire Atlantic Ocean by transporting salt to intermediate depths.

5.2 The Arctic Ocean, the Norwegian Sea, and the Greenland Sea (or Arctic Mediterranean)

5.2.1 Introduction

The Phoenicians and Greeks knew of a sea north of Norway [**Figure 3-1**]. Following the discoveries of Spanish and Portuguese explorers looking for a route to the "Indies" in the 15th century, the Pope divided the world in two by the treaty of Tordesillas (1492), which fixed the line separating the future colonies "along a meridian 370 leagues west of the Cape Verde islands". Discoveries west of this meridian should go to the Spanish and those to the east to the Portuguese. That left other European sovereigns looking for a maritime commerce route to China to the north. Two possibilities existed: either north of North America, called the "Northwest Passage" or north of the Siberian coast, the "Northeast Passage". Many expeditions risked their lives in the struggle against the cold and the ice trying to find these passages.

In 1594, a Dutch pilot, Willem Barents, aboard the *Messenger*, first reached the NE extremity of Novaya Zemblya[262] while his two other ships passed southward and eastward to the Kara Sea by the strait of Vaigats as far as the mouth of the Ob [**Figures 3-62** and **3-63**]. Blocked the following year, Barents set off again in 1596 and discovered Spitsbergen, then became stuck in the ice north of Novaya Zemblya. He was forced to winter there and died. Following his description (reported by his shipmates) of thousands of animals around Spitsbergen, Englishmen and Dutchmen, aided by

---

[260] Evaporation by cold dry winds replaces the formation of ice to extract fresh water and augment salinity.

[261] Mean renewal time has been estimated to be 60-100 years for Eastern Mediterranean Deep Water and to be 20-40 years for Western Mediterranean Deep Water.

[262] An island northwest of Siberia, 50°E–60°E, 71°N–77°N.

Basque harpooners, began a whale hunt in the region. Later, fishermen from Kamchatka in the north Pacific found harpoons in their catches with the initials of Spitsbergen hunters, proving that the whales knew the passage! Further east, Russian merchants using ships built for ice[263], did business between the mouths of the Ob and Yenissei and the White Sea. With these boats, in 1648, Semyon Dezhnev, a Cossack searching for furs, was able to go beyond the mouth of the Kolyma (in the eastern Siberian Sea) as far as the Anadyr (in the Bering Sea) thereby passing between the NE point of Siberia and the NW point of America[264]. In 1728, the Dane Vitus Bering, sent from the Pacific by Tsar Peter the Great, showed the existence of the strait that now bears his name. Then Tsarina Catherine II sent off the "Great Arctic Expedition" from 1730 to 1740 that explored the entire north coast of Siberia in several expeditions.

FIG. 3-62 – Arctic Ocean map by W. Barents in 1598, elaborated by Hondius in 1625. (From « *Le grand défi des pôles* », Imbert et Lorius, 2006.)

Many explorers lost their lives searching for these maritime passages to the Pacific Ocean. Finally, in September 1879, after having overwintered, the Swede Adolf Erik Nordenskjold managed to lead his ship *Vega* through the Northeast Passage to Yokohama in Japan, having left Sweden in July 1878 [**Figure 3-63**]. Twenty-five years later, it took the Norwegian Roald Amundsen, on board the *Gjoa*, two years (from 1903 to 1905) to master the Northwest Passage by overwintering with the Inuit.

[263] The *kochis* were reinforced and light enough to be dragged onto the ice when stuck in pack ice.

[264] Dezhnev passed through the passage later called the Bering Strait.

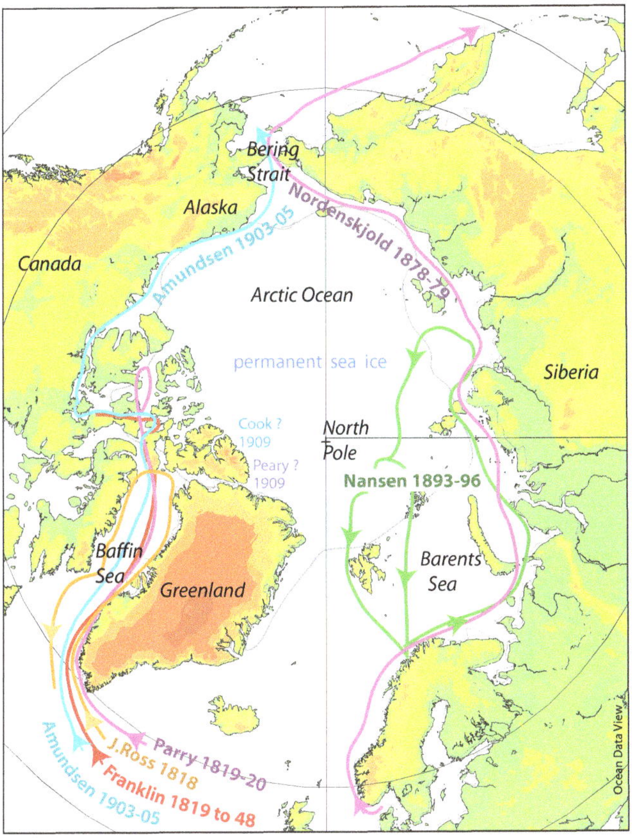

FIG. 3-63 – Tracks of the early expeditions searching for the Northwest and Northeast Passages and for the North Pole. The sky blue tint shows the average position of the ice pack in summer. (After *Le grand défi des pôles*, Imbort and Lorius, 2006.)

5.2.2 Topography

The Arctic Ocean (or North Polar Basin) can be termed "mediterranean" since it is surrounded by land and connected to two oceans, Pacific and Atlantic, by sills [**Figure 3.64**]. However, it is not a classic mediterranean because it includes multiple communications of which the widest is with the Atlantic Ocean. Exchanges with the Pacific Ocean through the Bering Strait, whose depth does not exceed 55 m, are limited.

The Arctic Ocean is connected to the Atlantic by the Nordic Seas, a sort of "antechamber" composed of the Norwegian Sea, the Greenland Sea, and the Iceland Sea. The Norwegian Sea (3,700 m), including the Lofoten Basin, is the portion located south of the Mohn Ridge, which extends from Jan Mayen in the direction of Bear Island. The Greenland Sea (3,200 m) is the portion located north of this ridge. The Iceland Sea, less deep, is located north of Iceland. Oceanographers often call the region included between the Greenland-Iceland-Scotland sills and the Greenland-Svalbard-Norway sills the "GIN Sea" (for Greenland, Iceland, Norway).

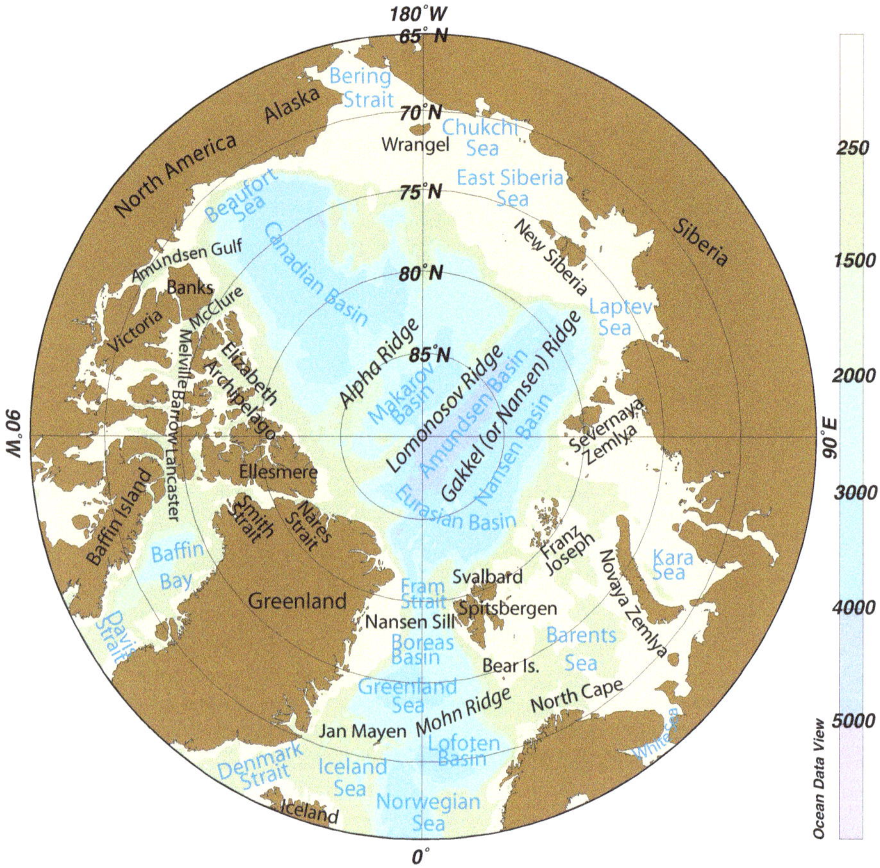

FIG. 3-64 – Topography of the Arctic Basin.

The Nordic seas are wide open to the Atlantic Ocean over 1,700 km between Greenland and Scotland, but exchanges are limited in depth by sills. Between Greenland and Iceland the sill of the Denmark Strait is at a depth of 615 m. Between Iceland and the Faeroe Islands, the sill is only at 400 m, and the sill between the Faeroe Islands and northern Scotland is the deepest at around 850 m. To the north, the Fram Strait (450 km wide), located between Greenland and Spitsbergen, is the main passage between the Greenland Sea and the Arctic Ocean. Cut by the Lena Trough, of about 3,400 m, its sill at about 2,600 m (the Nansen Sill) allows deep communication with the Arctic Basin. Between Spitsbergen (Svalbard Archipelago) and the northern extremity of Norway at North Cape, the shallow continental shelf of the Barents Sea limits exchanges to the upper layers.

The configuration of the Arctic Ocean [**Figure 3-64**] is the inverse of that of the Antarctic Ocean. The Antarctic Ocean surrounds the extremely cold Antarctic continent covered with a high ice sheet [**Figure 2-4**]. In contrast, the Arctic Ocean is in a central position, nearly entirely surrounded by low-lying lands except for Greenland, whose ice sheet reaches 3,730 m. The area of the Arctic Ocean is 14×10^6 km^2. The continental shelf is extensive, especially north of Siberia where it may reach 800 to 1,000 km width. Its depths vary from 20 to 60 m in the Chukchi Sea and in the eastern Siberian Sea, 10 to 40 m in the Laptev Sea, 50 to 200 m in the Kara Sea, and 100 to 350 m in the Barents Sea (located between the Franz Joseph Archipelago, Novaya Zemblya, and the Svalbard Archipelago). A number of rivers[265] run into the Arctic, draining an immense surface of the continents of Eurasia and North America [**Figure 3-65**].

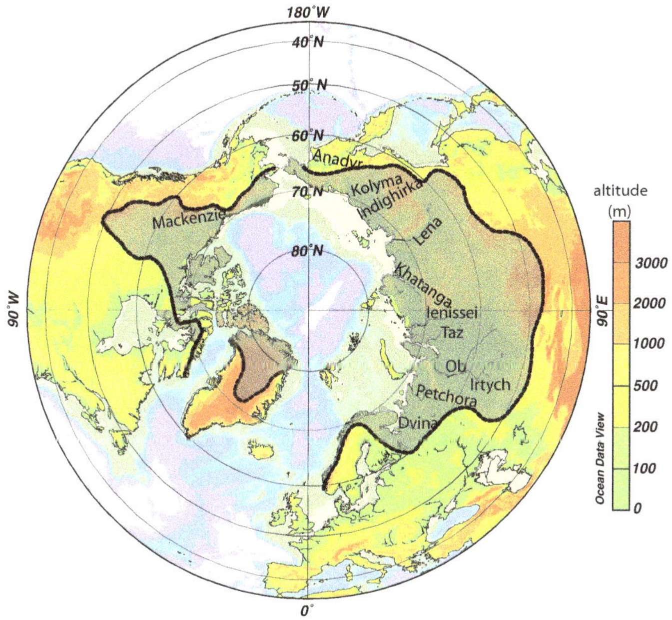

FIG. 3-65 – The Arctic Ocean drainage basin, from the Canadian Rockies to the high Tibetan Plateau.

[265] Northern Dvina, Pechora, Ob, Taz, Yenissei, Kheta, Khatanga, Olenyok, Lena, Indigirka, Kolyma, Mackenzie.

Several ridges divide the deep part of this ocean, having an area of only 5×10^6 km^2. The main ridge is the Lomonosov Ridge between the New Siberian Islands and Ellesmere Island, which divides the basin around 1,500 to 2,000 m depth[266]. It is found between two deeper ridges, the Alpha Mendeleyev Ridge (about 1,500 m deep, allowing passages above 2,000 m) and the Gakkel (or Nansen) Ridge (about 2,500–3,000 m), cut in the middle by a deep rift. These ridges subdivide the Arctic polar basin into four basins: the Canadian Basin (3,600–3,800 m) between the Canadian coast and the Alpha Ridge; the small Makarov Basin (4,000 m maximum) between the Alpha-Mendeleyev Ridge and the Lomonosov Ridge; the deep Amundsen Basin (4,450 m) between the Lomonosov Ridge and the Gakkel Ridge; and the Nansen Basin (3,900 m) between the Gakkel Ridge and the Eurasian continental shelf [**Figure 3-64**]. The latter two basins are often grouped as the Eurasian Basin. The extensive continental shelf gives the Arctic Ocean an average depth of only 1,200 m.

5.2.3 Climatology

5.2.3.1. Pressure and winds

Linked to cold air, a high-pressure center located near the North Pole, on the Canadian Basin side, dominates the wind system [**Figure 3-66**]. The pressure gradient between the polar high and the low pressure that surrounds it varies seasonally depending on the pressures over nearby land. The distribution of pressure over the Arctic Ocean generates a wind system from the east in winter, the season when the gradients are strongest. The Nordic Seas zone is dominated by Icelandic low pressure, which generates a cyclonic circulation with winds from the southwest in the east and winds from the northeast in the west, particularly in winter. The openings of the polar basin on the Pacific Ocean and on the Atlantic Ocean allow the arrival of maritime air from the south in summer, which can produce strong perturbations in the dominant regime. North of Siberia, winds are from the west in winter and from the east in summer.

[266] In the mid 20th century, because of ice cover, Arctic sea-floor topography was still poorly known. By noting differences in deep temperatures off Canada, and off the Eurasian continental shelf, oceanographers concluded the existence of a ridge blocking communication beyond 1500 m, which was later verified.

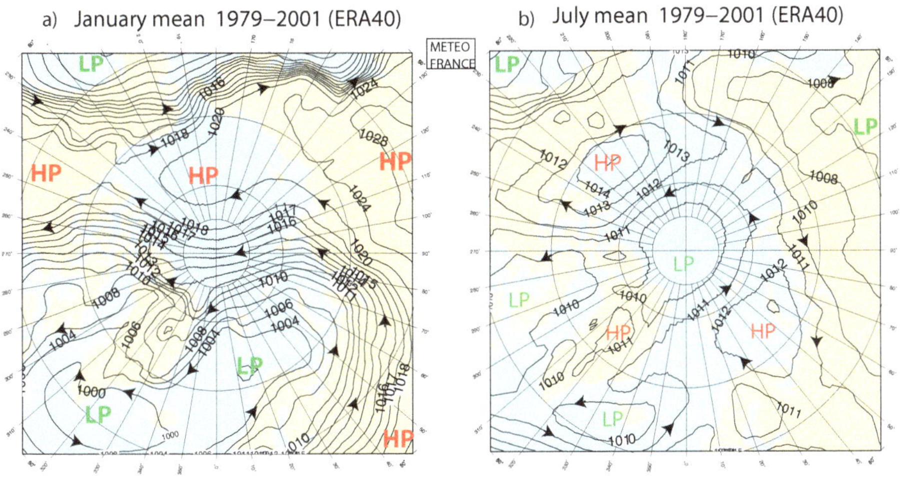

FIG. 3-66 – Pressure systems and average Arctic winds. (Courtesy of M. Hontarrède, using the ERA40 EMCWF/Meteo France model.)

5.2.3.2. Precipitation and heat flux

Precipitation is poorly known over the polar basin. It is estimated at 100 to 250 mm/year. Precipitation falls on the pack ice, mostly in the form of snow. The Norwegian Sea experiences a number of storms that bring strong precipitation (averaging 1 m per year) half of which is in the form of snow.

The coldest air temperatures are found near the North Pole, but they do not reach the extreme values found at the continental South Pole (located at 2,800 m altitude) since, at the North Pole, the thermal reservoir provided by the ocean is only separated from the atmosphere by the thickness of the pack ice. Three climatic regions may be defined:

- the oceanic polar region (*e.g.*, the Svalbard Archipelago, including Spitsbergen) where temperatures vary between –13 °C in winter and +2 °C in summer;
- the continental polar region (*e.g.*, Siberia) between –20 °C and +7 °C;
- the glacial polar region (*e.g.*, the center of the polar basin and the center of Greenland) between –25 °C and –3 °C.

The central part of the Arctic polar basin is covered by year-round pack ice [**Figure 3-67**] whose thickness varies between 2 m and 4 m. It isolates the ocean from the atmosphere. The coverage is not always continuous, however. Movements of ice sheets due to winds and currents can produce breaks in the pack ice resulting in the piling up of broken ice sheets (hummocks) that can reach tens of meters in thickness, as well as opening channels allowing short-term heat exchanges with the atmosphere. In summer, the pack ice melts only over a part of the periphery of the Arctic Ocean and reforms the following winter. In 1980, the ice coverage was estimated at 8×10^6 km^2 in summer and 15×10^6 km^2 in winter (of which about 14% was free water in the form of channels and polynyas). Observations in 2015 showed a decrease by nearly half in summer pack ice over 35 years, linked to climatic warming, which had opened the famous "passages" in summer [**Figure 3-67**]. On the borders of the basin, the pack ice melts and re-forms every year while in the center of the polar basin, the pack ice is many years old. It is subject to currents and accumulates snowfall on the surface.

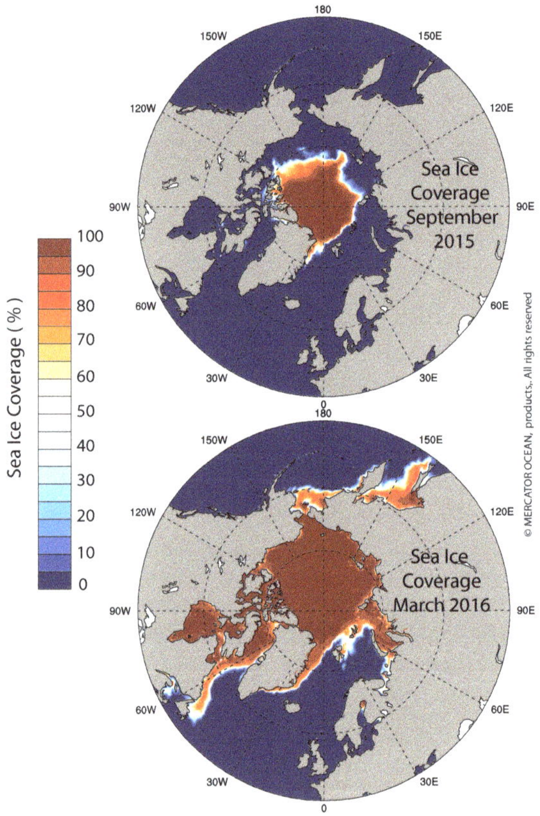

FIG. 3-67 – Seasonal variability of ice-pack cover over the Arctic Ocean in September 2015 and March 2016. (Image generated using E.U. Copernicus Marine Service Information and interpreted by G. Garric, Mercator Océan. All rights reserved).

A strong asymmetry exists in the region between Greenland and Norway. The Norwegian coasts are ice-free as far as North Cape (71°N) while the eastern coasts of Greenland are ice-bound all year.

Evaporation is reduced over the Arctic Ocean because of the great extent of pack ice. On the other hand, the significant inflow of fresh water from the rivers of Siberia and Canada, of the order of 0.2×10^6 km^3, results in a positive fresh-water balance and a reduction in surface salinity. The Arctic Ocean is a dilution basin [**Figure 3-45**] that exports desalinated water and sea ice.

In spite of its ice cover, the Arctic Ocean transfers heat to the atmosphere, especially in winter, at the time of pack-ice formation on the periphery and in channels and polynyas, where the ocean is in direct contact with the atmosphere.

Arctic icebergs on the west coast of Greenland. Arctic icebergs come from Greenland glaciers, descending most often on steep slopes and breaking off when arriving at the sea. They are smaller and have contorted forms much different from the great tabular icebergs from the Antarctic Ocean. (Photos Jacques Darchen.)

5.2.4 Surface currents of the Arctic Ocean

Ice cover made the Arctic Ocean impenetrable for a long time and slowed its study, which is difficult even today. It was first thought that the central Arctic pack ice did not move, that it only melted near coasts and refroze the following winter. The demonstration that pack ice moved was made in 1884 when the Inuits found, off the southeast coast of Greenland, the remains of the little steamboat *Jeannette*,[267] which had been trapped by the pack ice in the Chukchi Sea north of the Bering Strait in 1879, and was crushed by the ice in June, 1881 [**Figure 3-68**].

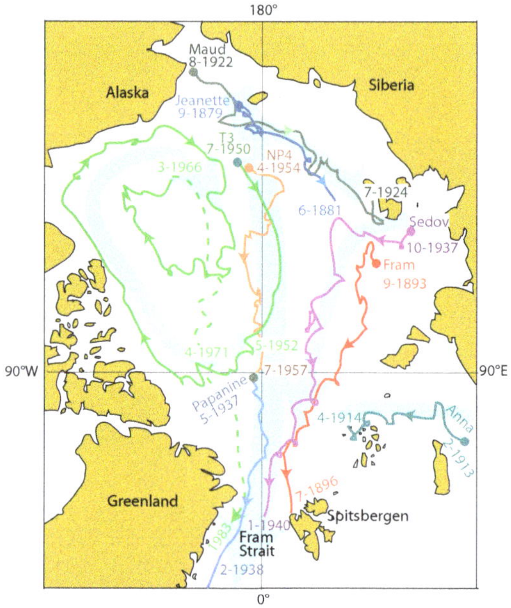

FIG. 3-68 – The drift of ships caught in the ice pack, either voluntarily or accidentally, and of stations installed on the ice pack. (After Tchernia, 1978.)

Fridtjof Nansen then deliberately had his ship frozen into the ice at the beginning of winter, to have it serve as a tracer of polar circulation, with the secret dream of passing the North Pole. He had the *Fram* specially constructed to resist the pressure of the ice without being crushed. He assembled a team of scientists who were ready for anything, to explore the world of ice where the night lasts six months, without knowing when they would return. Leaving Norway, they set course eastward along the coast of Siberia, and were stuck in ice in the Laptev Sea, north of the New Siberian Islands in November 1893. The *Fram* drifted slowly, entrained westward by the Transpolar Drift, passing between the Pole and the Franz Josef archipelago, to finish on the northwest coast of Spitsbergen... three years later![268]

[267] It was sent, following the success of Nordenskjöld by the publisher of the *New York Herald* to report on the "mystery of the North Pole".

[268] In October, 1896. On the *Fram*, Nansen observed for the first time that ice did not drift in the direction of the wind [**Section I.5.2**].

The drift of the *Fram* had extended the route mapped out by the sailors of the *Jeannette* until it was crushed. In August 1922, the ship *Maud* became ice-bound north of the Bering Strait and drifted along the path of the *Jeannette* to finish north of the New Siberian Islands in August 1924. The Soviet ship *Sedov*, beset in the ice in October 1937, drifted toward the north and east, and then was taken westward, as was the *Fram*, 44 years earlier so to emerge in January 1940, between Greenland and Spitsbergen, thereafter known as the Fram Strait [**Figure 3-68**]. The trajectories of stations installed on vast pieces of pack ice and supplied by air and later buoys placed on the ice and followed by satellite,[269] all showed the same "exit" between Greenland and Spitsbergen. They permitted a sketch of the ice circulation to be drawn [**Figure 3-69**]. Today, some people still embark on this grand adventure, such as the crew of the *Damocles* experiment on the schooner *Tara* in 2007.

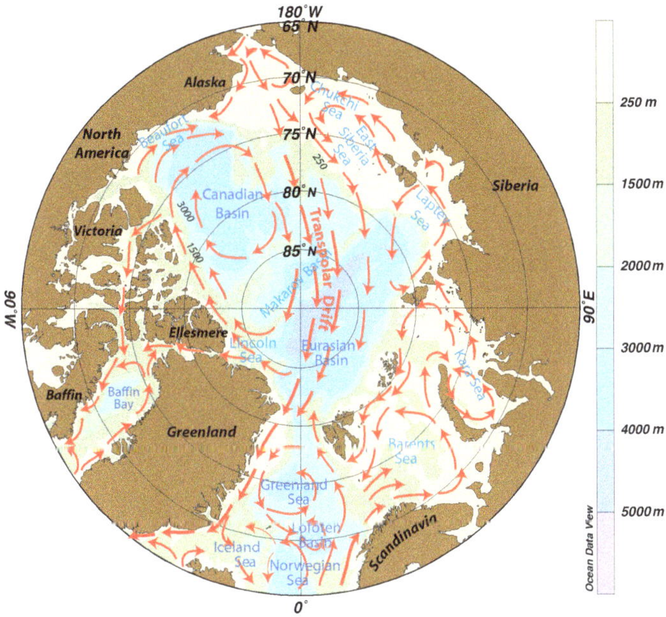

FIG. 3-69 – Average ice drift in the Arctic Ocean obtained from ships drift, from hydro stations and from buoys on the ice pack. (After Aagaard *et al.*, 1985; Jones *et al.*, 1995; Rigor and Wallace, 2004; Rudels *et al.*, 1989.)

---

[269] Such as: the NP1 station, which began at the North Pole in May 1937 and was recovered on the east coast of Greenland in February 1938, the T3 station (or Fletcher's Ice Island, a flat iceberg discovered in 1947 by Colonel Joseph Fletcher) followed the Canadian Basin anticyclonic circuit, intermittently occupied from 1952 to 1974, until it finally exited the Fram Strait in 1983, or the NP4 station, installed in April 1954, which drifted to the Pole, reaching it in July 1957.

Ice cover in the polar circuit strongly reduces the action of wind on currents, which are of the order of a few cm s^{-1}.[270] Recent estimates of geostrophic currents from temperature and salinity measurements from icebreakers, pack-ice stations, and military submarines[271] have helped to define the major components of the average surface circulation.

The dominant current is the Transarctic Current (or Transpolar Drift). It crosses the ocean from north of the Bering Strait, passes north of Wrangel Island, the New Siberian Islands, Severnaya Zemlya, and the Franz Josef Archipelago, to emerge along the eastern coasts of Greenland as the East Greenland Current, bringing pack ice and icebergs detached from the Greenland glaciers. In the Canadian Basin, under the dominant effect of anticyclonic winds [**Figure 3-66**], the circulation of the ice is anticyclonic [**Figure 3-69**] contrary to the circulation below the Ekman layer, which is cyclonic. On the Siberian continental shelf an average eastward current, driven by westerly winds, is perturbed by the presence of numerous islands and peninsulas, to form cyclonic circuits in the bordering seas. A current bearing eastward follows the Alaska coast, prolonging the current of the Bering Strait. There are several exits in the Canadian archipelago through narrow and shallow passages [**Figure 3-69**]. All end up in the Baffin Sea, either by the Straits of Nares and Smith whose sills are less than 250 m (between Ellesmere and Greenland) or through the straits of McClure, Melville, Barrow, and Lancaster (north of Victoria and Baffin islands) then into the Labrador Sea through the Davis Strait.[272]

A cyclonic circulation with many branches exists in the Nordic Seas [**Figure 3-70**].

The Norwegian Current, whose transport is of the order of 30 Sv, extends the North Atlantic Drift of tropical origin. (Its water is relatively warm and saline for the region.) It divides into several branches. One branch goes westward north of the Norwegian Sea south of the Lofoten Basin and another goes westward north of the Greenland Sea. The principal branch enters the Arctic Ocean by the West Spitsbergen Current following the edge of the continental slope. Another branch of the Norwegian Current breaks off eastward to enter the Barents Sea by the North Cape Current, and then joins the Eurasian Basin passing north of Novaya Zemlya. Westward recirculation occurs a number of times along its path, especially in the Fram Strait.

---

[270] Corresponding to a few kilometers per day (1 cm s^{-1} = 0.86 km/day). This is weak but is enough to counteract efforts of explorers to proceed poleward beyond Greenland.

[271] When their observations are publicly available.

[272] The Foxe Basin and Hudson Bay, of a few tens of meters depth, communicate with the Labrador Sea.

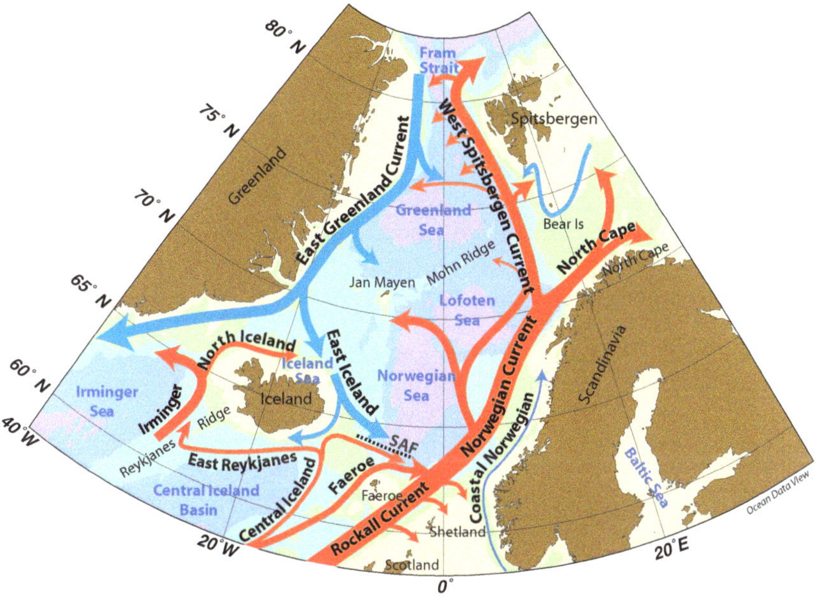

FIG. 3-70 – Schematic surface circulation north of the Atlantic Ocean in the Greenland, Iceland, and Norwegian Seas. Cold currents are in blue, warm currents in red. (After Osterhaus and Gammelsrod 1999.)

The East Greenland Current (cold, with low salinity) is the western branch of the cyclonic circuit. It is fed by the Transarctic Current and hugs the eastern coast of Greenland as it flows toward the Irminger Sea as far as the southern tip of Greenland at Cape Farewell. Secondary branches break off south of the Boreas Basin, south of the Greenland Sea, and north of Iceland. This latter branch, the East Iceland Current, rejoins a branch of the warmer Norwegian Current, to form the thermal and saline Subarctic Front (SAF) [**Figure 3-70**]. In the center of this region, a number of eddies take part in mixing between the two principal currents having inverse directions and contrasting characteristics.[273] The Norwegian Current is separated from the coast south of Norway by a low-salinity coastal current coming from the Baltic Sea [**Section III.7.2.2**].

Thanks to the Norwegian Current, the Norwegian coasts are ice-free year round, even as far as the northernmost North Cape. On the other hand, the current coming out of the Arctic Basin along the eastern coast of Greenland carries pieces of pack ice from the Arctic Basin and icebergs broken off from the Greenland glaciers far southward. These icebergs, entrained by the

[273] The velocities found here are of the order of 0.2 m s^{-1} but can reach 0.5 m s^{-1} within eddies.

Labrador Current, can sometimes reach 40°N,[274] such as the one that sank the *Titanic*[275] in 1912.

5.2.5 Surface temperature

Because of ice cover, significant seasonal sea-surface temperature variations are observed only on the periphery of the Arctic Ocean [**Figure 3-71**]. Their distribution is strongly influenced by circulation [**Figure 3-70**]. Thus there is strong asymmetry between the Norwegian coast and the Greenland coast. The maximum seasonal amplitude occurs in the Baltic Sea.

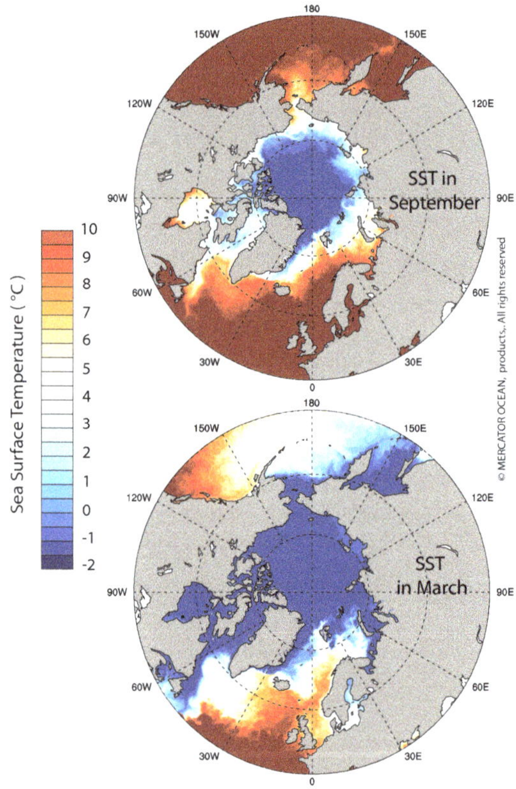

FIG. 3-71 – Sea-surface temperature in September 2015 (end of summer) and in March 2016 (end of winter). Very low temperatures correspond to the presence of sea ice (see **Figure 3-67**). (Image generated using E.U. Copernicus Marine Service Information and interpreted by G. Garric, Mercator Océan. All rights reserved.)

---

[274] 40°N is the latitude of southern Italy!

[275] The *Titanic* sank at 41°46′N, 50°14′W. Following this maritime catastrophe, ships of the International Ice Patrol watched iceberg movements to protect navigation and even tried to deflect or destroy them! The International Ice Patrol continues today using aerial surveillance.

5.2.6 Arctic Mediterranean Oceanography

The Arctic Ocean and the Nordic seas together comprise a "mediterranean" whose internal processes are much more complex than those of a simple dilution basin. It consists of a series of wide and deep single basins, having varying characteristics depending on their topography and climate, where intense convective motions take place. Principally connected to the Atlantic Ocean, the Arctic Mediterranean produces low-density water, which exits at the surface and high-density water, which exits at depth. The Coriolis force "constrains" the exiting current to the right along the Greenland coast and the entering current to the right north of Scotland.

5.2.6.1. Atlantic Water

The principal water mass feeding the Arctic Ocean at its "antechamber" is Atlantic Water transported by the Norwegian Current, as an extension of the North Atlantic Drift [**Figure 3-72**]. Surface temperature and salinity indicate the path of North Atlantic Water that is warm ($T > 10$ °C) and saline ($S > 35$) east of the Norwegian Sea [**Figure 3-70**]. Off North Cape (71°N), well beyond the Arctic Circle (66°33′N), temperatures are still high for this latitude, varying seasonally between 3.5 °C and 8 °C (salinity 34.2–34.4).

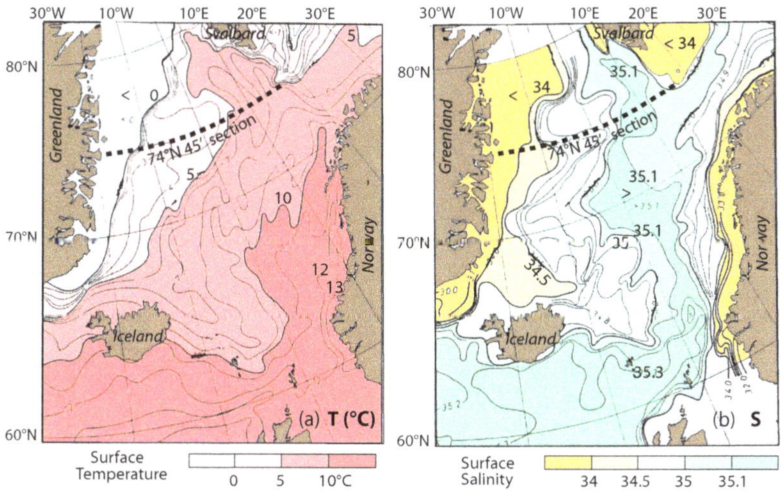

FIG. 3-72 – (a) Temperature and (b) salinity in the Norwegian and Greenland Seas in summer. The position of the section in **Figure 3-73** is shown (After Dietrich, 1969.)

Atlantic Water still retains more than 500 m vertical extent in the Greenland Sea [**Figure 3-73**]. It cools through heat exchange with the atmosphere [**Figure 1-15**] and its salinity decreases progressively northward through mixing with underlying water and with adjacent coastal water on the west (Arctic Surface Water, very cold with low salinity) and on the east (low-salinity Baltic Sea Water, $S < 33$). Its surface-layer oxygen level is a bit lower since its temperature is higher.

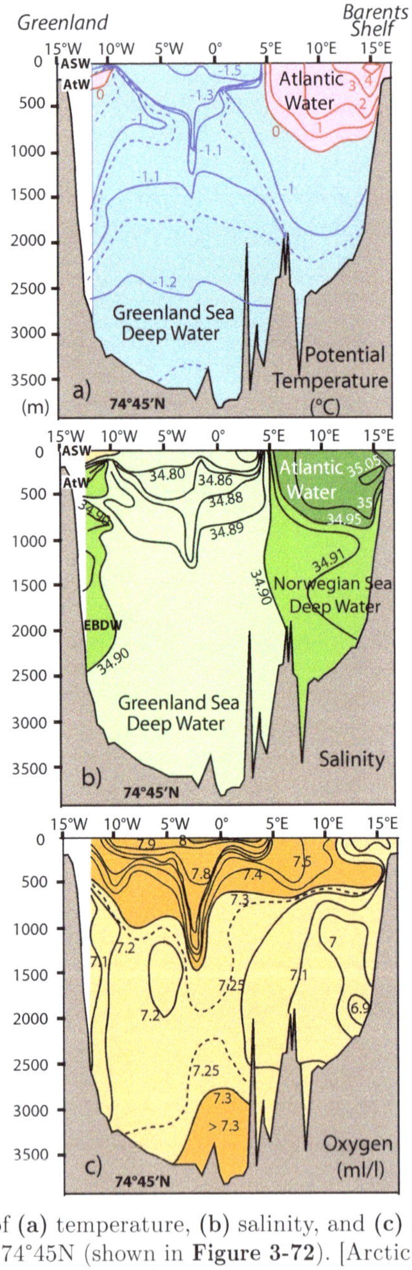

FIG. 3-73 – Sections of **(a)** temperature, **(b)** salinity, and **(c)** dissolved oxygen across the Greenland Sea at 74°45N (shown in **Figure 3-72**). [Arctic Surface Water (ASW), Eurasian Basin Deep Water (EBDW), water of Atlantic origin (AtW).] (After Rudels *et al.*, 1989.)

In the Greenland Sea, part of the Atlantic Water, driven by cyclonic circulation, passes under lighter cold and low-salinity water coming out of the Arctic Basin. The main branch extends northward under cold and low-salinity Arctic Surface Water west of Spitsbergen and into the Barents Sea, producing an intermediate-temperature maximum throughout the Arctic Ocean aside from the shallow continental shelf where the waters are mixed. The Atlantic Water temperature maximum [**Figures 3-74a** and **3-74b**] located at 100 m depth west of Spitsbergen, follows a cyclonic circulation along the continental slope and ends up at 500 m in the Canadian Basin, with recirculation along the Lomonosov Ridge and the Alpha Ridge. The maximum intermediate temperature of the Atlantic Water decreases through mixing along its slow trajectory (0.02 m s^{-1}). It is no greater than 0.4 °C north of the Canadian Archipelago.

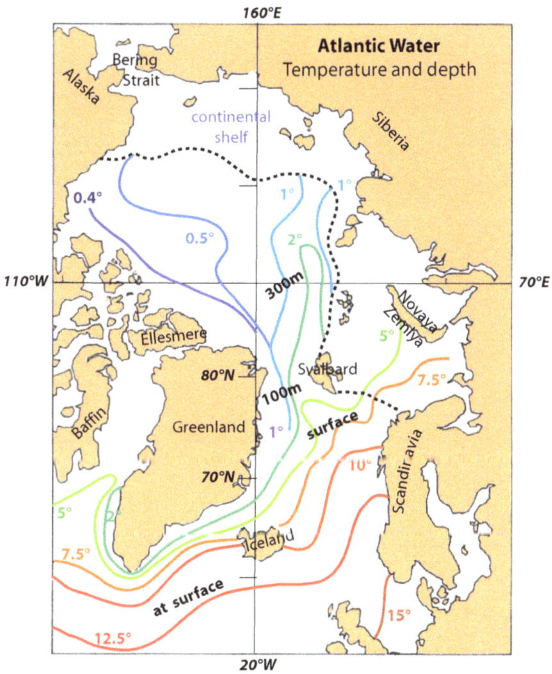

FIG. 3-74A – Temperature maximum due to the influence of Atlantic Water in summer. The maximum is at the surface as far as Svalbard, after which the depth of the maximum is shown in meters. Dashed black line: continental shelf limit. (After Aagaard *et al.*, 1985.)

Under the surface layer, the circulation, not influenced by wind, is cyclonic (the current turns to the right) while at the surface it is anticyclonic (as are the winds).

The branch entering directly north of North Cape allows a large part of the Barents Sea to be ice-free year-round [**Figure 3-67**]. Outside of certain areas of the continental shelf, where vertical mixing is active, the general oceanographic structure of the Arctic Ocean consists of three layers (just as in the Antarctic Ocean south of the Subantarctic Front): a cold and low-salinity surface layer, a warmer and more saline intermediate layer, influenced by Atlantic Water, and a deep layer of intermediate temperature and salinity.

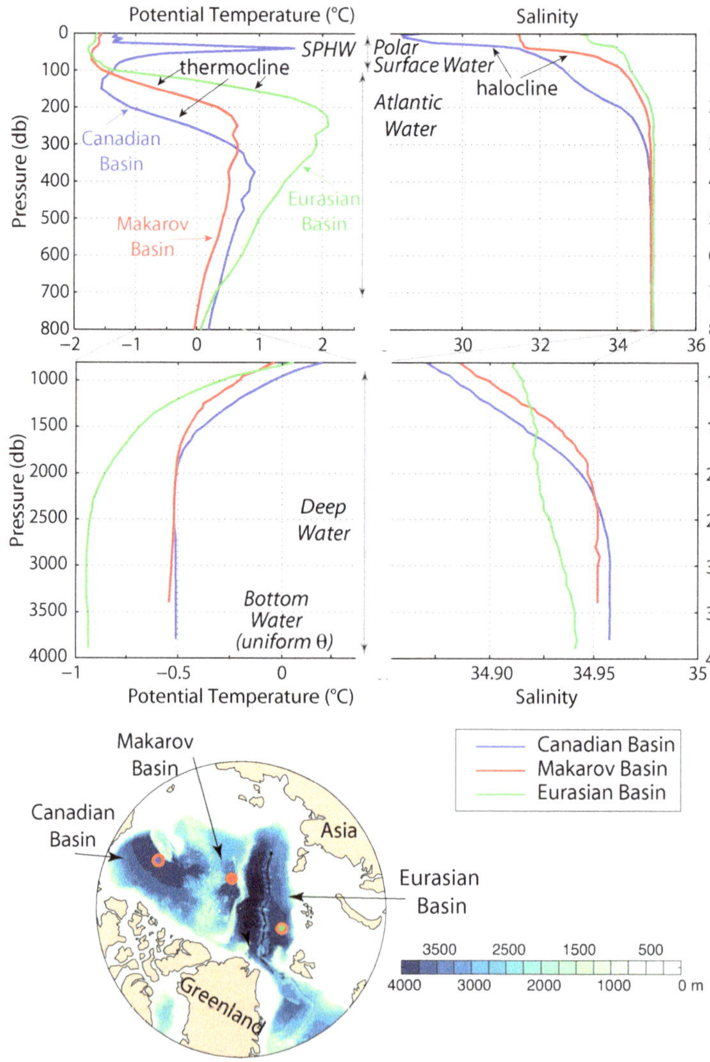

FIG. 3-74B – Potential temperature and salinity profiles in the Canadian Basin, the Makarov Basin and the Eurasian Basin (SPHW: Summer Pacific Halocline Water). (Courtesy of Christophe Herbaut, LOCEAN/Paris VI University, World Ocean Data Base 2013.)

5.2.6.2. Arctic Surface Water

Arctic Surface Water, located just under the pack ice, is cold (–1.9 °C $< \theta <$ –1.5 °C) and near the freezing point [**Figures 3-71, 3-72, 3-73, 3-74b** and **3-75**]. It is low-salinity ($S < 32$) because of fresh water inputs by numerous rivers. While temperature varies little from the surface to 150–200 m (for example Station 5 in blue, **Figure 3-75**), the rapid variation of salinity between Arctic Surface Water and Intermediate Water of Atlantic influence creates a halocline (a region of a strong vertical salinity gradient) that separates the two layers. In the surface layer, up to 25–50 m depth, salinity varies between 28 and 33.5 depending on melting or formation of pack ice and on the proximity of fresh-water inputs. In the subsurface layer, extending from 25–50 m to 150–200 m, salinity increases rapidly and forms a "barrier" layer (see **Section V.4.2.5**), which limits heat exchange between the surface layer and the intermediate layer.

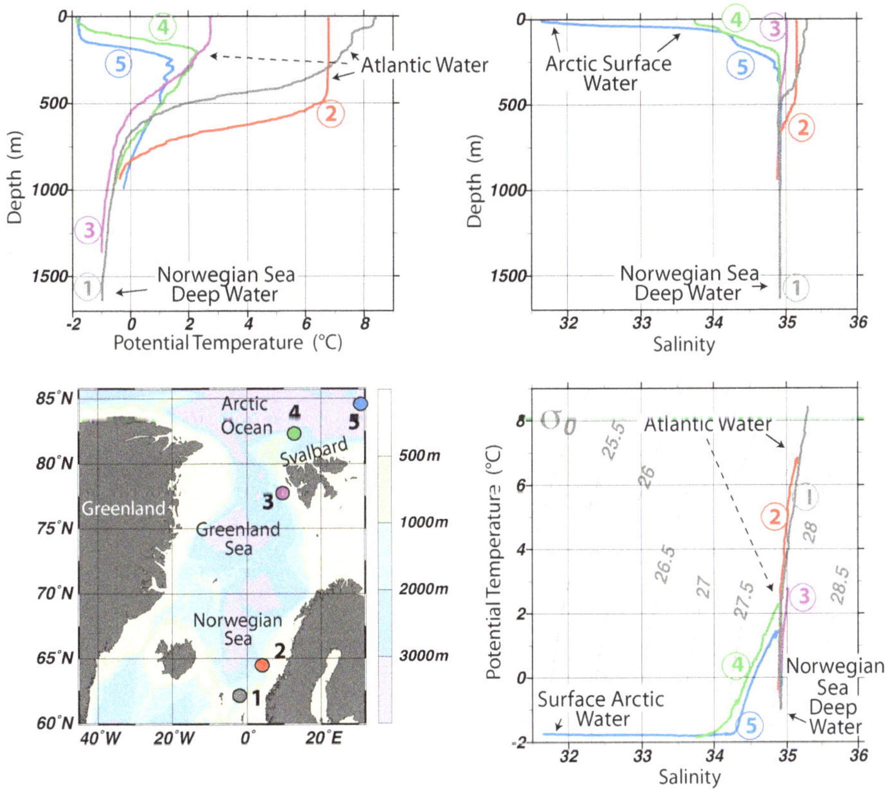

FIG. 3-75 – Potential temperature and salinity profiles and θ-S diagrams showing the evolution of Atlantic Water from its entry into the Norwegian Sea to the Arctic Ocean. (With ODV and World Ocean Database.)

In the Canadian Basin, the Polar Surface Water includes water coming from the Bering Strait (including Alaskan Coastal Water). It makes its influence felt at the depth of the subsurface layer below the Polar Mixed Layer. In summer, it is warmer than the surface layer and produces a temperature maximum between 50 m and 100 m in the halocline and is called Summer Pacific Halocline Water [**Figure 3-74b**]. Around 125–150 m, a temperature minimum is the remnant of winter Bering Strait Water, close to the freezing point. Its signature may be found in the eastern part of the Canadian Basin.

5.2.6.3. Deep Water of the Greenland Sea and of the Norwegian Sea

The Greenland Sea receives Canadian Basin Deep Water (CBDW) on its western side through the Fram Strait. It passes over the Alpha and Lomonosov Ridges following the Greenland continental slope. A salinity maximum > 34.91 is found around 1,400 m with temperatures between $-0.8\ °C < \theta < -0.5\ °C$. The flow joins Eurasian Basin Deep Water (EBDW), which is slightly colder ($-1.1\ °C < \theta < -0.8\ °C$) and has comparable salinity [**Figures 3-73, 3-76,** and **3-77**]. Above these two water masses, water of Atlantic origin comes *via* the west Spitsbergen branch. It is characterized by a temperature and salinity maximum. At the surface, very cold ($\theta < -1\ °C$) and low salinity ($30 < S < 33$) Arctic Surface Water comes from the Arctic Ocean *via* the East Greenland Current.

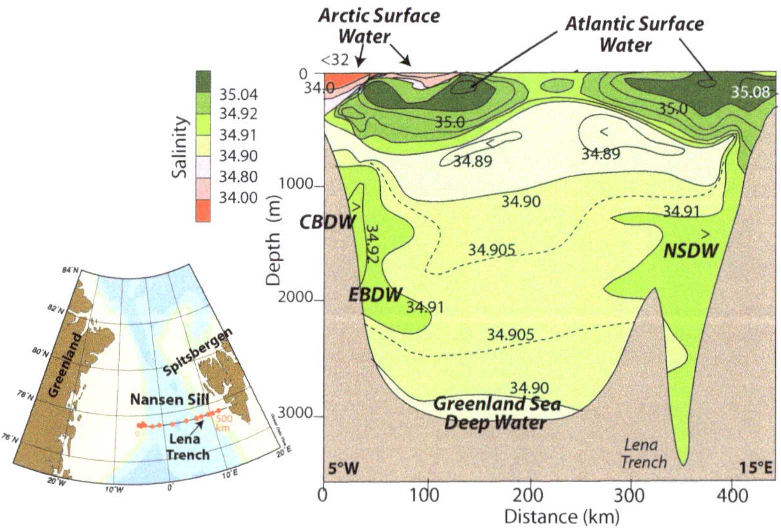

FIG. 3-76 – Salinity section in the Fram Strait, slightly south of the Nansen Sill. [Canadian Basin Deep Water (CBDW), Eurasian Basin Deep Water (EBDW), Norwegian Sea Deep Water (NSDW).] (After Smethie *et al.*, 1988.)

Greenland Sea Deep Water forms sporadically in winter in the center[276] of the cyclonic circuit of the Greenland Sea [**Figure 3-69**] in limited regions (of a few kilometers) during frequent storms that cross the area. The Greenland Sea is never completely covered with pack ice thanks to the heat input provided by Atlantic Water. Cooling down to freezing temperature and ice formation on the edge of winter pack ice increase density enough so that surface water becomes unstable and mixes with warmer and more saline underlying water of Atlantic origin. The section of **Figure 3-73** shows signs of recent convection down to 1,300 m depth around 2°W in the center of the Greenland Sea with associated high oxygen concentration values down to 1,300 m.

During these convective processes, warmer underlying water rises to the surface to compensate for the sinking of surface water. This warmer water melts ice, lowering salinity, and convection ceases until the next storm. The mixing induced by these convective motions renders the stability of the water column progressively weaker during the course of the winter. Each storm advances deep convective motions generating homogeneous columns in midwinter all the way to the bottom and oxygenating the Deep Water (7.3 ml l^{-1} or 320 μmol kg^{-1} at depth) [**Figure 3-73c**].

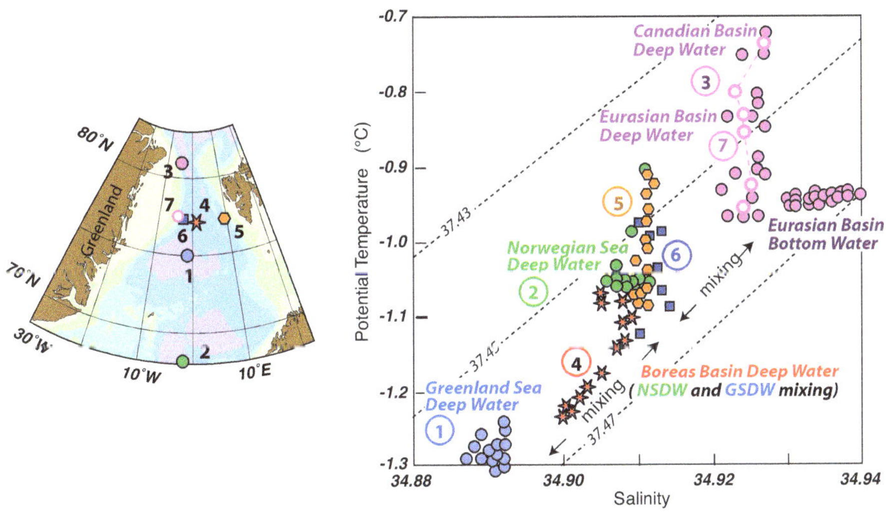

FIG. 3-77 – θ-S diagrams in the Fram Strait, the Boreas Basin, the Greenland Sea and the Norwegian Sea showing mixing of Deep Waters. The isopycnals are with respect to 2,000 db, σ_{2000}. (Polarstern Arktis II/3 1984 and TTO/NAS expeditions, after Smethie *et al.*, 1988.)

[276] Stratification is weaker in the center than on the periphery [**Figure 1-66**].

The θ-S diagram **[Figure 3-77]** shows the characteristics of the various Deep Waters:

- Greenland Sea Deep Water (Station 1, blue) is the coldest (–1.3 °C < θ < –1.25 °C) and the least saline (S = 34.89) since it contains a little Arctic Surface Water;
- Boreas Basin Deep Water (Station 4, red stars) **[Figure 3-77]** located between the Nansen Sill and the Greenland Sea **[Figure 3-64]** is a mixture of Eurasian Basin Deep Water (Station 3, violet) penetrating over the Nansen Sill and Greenland Sea Deep Water (Station 1, blue). Below the Nansen Sill level (about 2,600 m), the deeper the water, the more its characteristics approach those of Greenland Sea Deep Water;
- Deep Water formed in the Greenland Sea circulates toward the Norwegian Sea **[Figure 3-78]** where it mixes with Arctic Basin Deep water (CBDW and EBDW, slightly more saline, Station 3, violet), which follows the topography to form Norwegian Sea Deep Water (θ = –1.05 °C, S = 34.91) (Station 2, green, **Figure 3-77**). The mixing takes place in the Norwegian Sea, then in the Lofoten Basin and exits toward the Eurasian Basin by the Nansen Sill, following the West Spitsbergen Rise **[Figure 3-76]**.

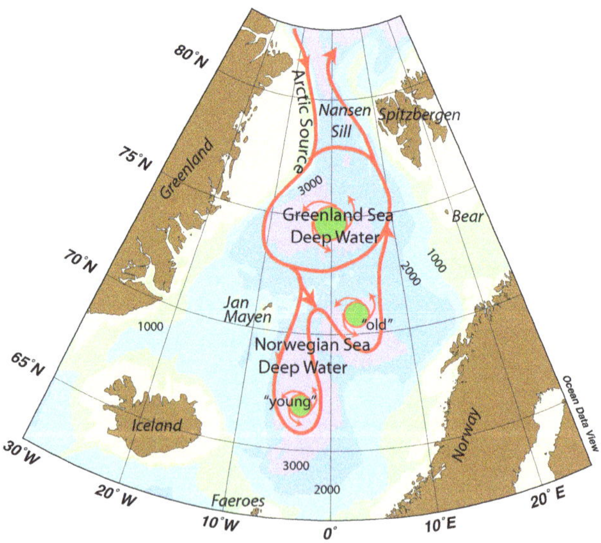

FIG. 3-78 – Stages of Nordic Seas deep-water formation showing deep circulation, deep convection zones, and mixing (in green). The arctic source is Deep Water of the Eurasian Basin and Deep Water of the Canadian Basin modified by dense waters formed on the arctic continental shelf in winter. (After Smethie *et al.*, 1988.)

Profiles and θ-S diagrams of two Norwegian Sea stations [**Figure 3-79**] show Atlantic Water (red station) in the east and weaker stratification in the center (blue station). In the center of the cyclonic circuit, Norwegian Sea Deep Water extends from 800 m to the bottom. The Iceland Sea station (green station) is influenced by low-salinity Arctic Surface Water coming from the Arctic Ocean *via* the East Greenland Current and the East Iceland Current [**Figure 3-70**]. The station in the center of the Greenland Sea (violet station) taken at the beginning of March (at winter's end) shows high density ($\sigma_\theta > 28$) over the whole water column and a nearly homogeneous column from 1,000 m to 3,500 m, the result of winter convection. The slight temperature maximum around 500 m corresponds to traces of Atlantic Water.

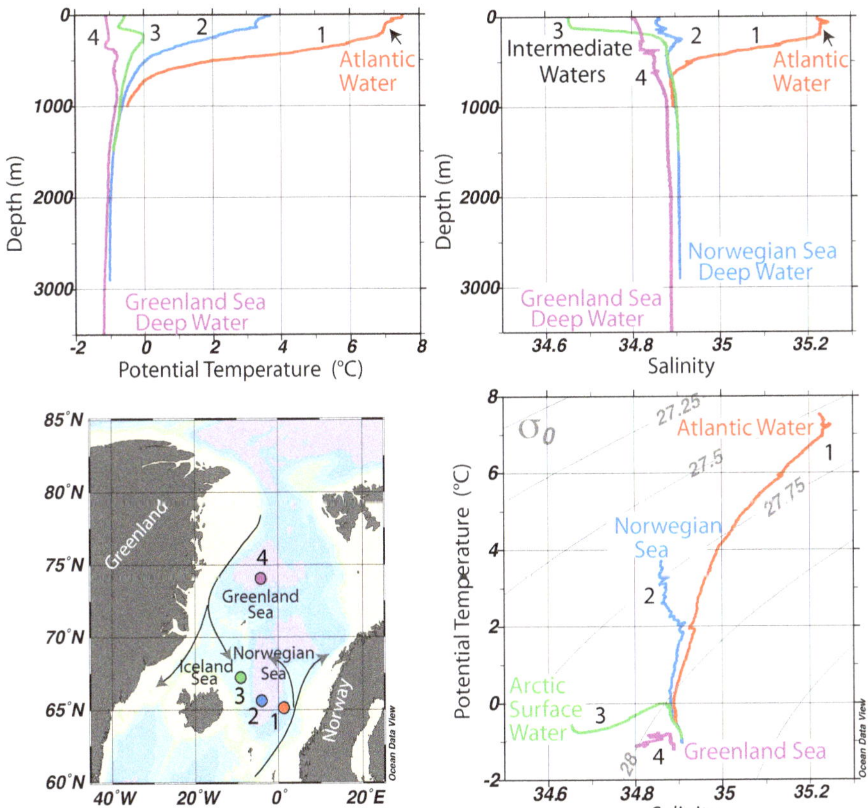

FIG. 3-79 – Profiles of potential temperature and salinity, and θ-S diagrams from stations in the Greenland Sea, Iceland Sea, and Norwegian Sea. (With ODV and WOA05 data.)

5.2.6.4. Arctic Deep Water

Lacking enough measurements in the Arctic Ocean for which access by classic methods is so difficult, and because of the presence of a low-salinity surface layer, it had long been thought that Arctic Deep Water formed entirely in the Nordic Seas (Greenland Sea and Norwegian Sea) then spread to the Arctic Ocean and the Atlantic Ocean. Access to observations, beginning with military submarines, icebreakers, or stations on the pack ice, has improved our knowledge. Even though surface-layer salinity is low in summer, in some Arctic continental shelf areas in winter, pack-ice formation increases surface water salinity to more than 35.[277] Since its temperature is near freezing, this shelf water is then at a density great enough to sink to the bottom of the Arctic Ocean basins and takes part in modifying Arctic Deep and Bottom Waters.

Arctic Deep Water (ADW) formation involves water formed in the Nordic Seas that enters the Arctic Basin east of the Nansen Sill (Fram Strait) [**Figure 3-78**] and dense water formed on the Arctic continental shelf. Eurasian Basin Bottom Water located below the Nansen Sill level, influenced by dense and saline water formed on the Siberian continental shelf, is a bit more saline ($34.93 < S < 34.94$, $\theta = -0.95$ °C) [**Figure 3-74b**] than Nordic Sea Deep Water and Arctic Deep Water. Canadian Basin Bottom Water, under the influence of dense and saline water descending from the Canadian shelf [**Figures 3-74b, 3-80,** and **3-81**], has salinity slightly higher ($S \sim 34.95$) than in the Eurasian Basin. The Lomonosov and Alpha Ridges block colder water formed in the Nordic Seas. Thus bottom temperature is −0.96 °C in the Eurasian Basin (Amundsen and Nansen) and only −0.53 °C in the Canadian Basin.[278]. These water masses are close in density and thus mix easily [**Figure 3-77**].

Deep circulation in each basin is cyclonic [**Figure 3-81**]. Slightly more saline Canadian Basin Deep Water passes above the Lomonosov Ridge to form a weak deep-salinity maximum west of the Amundsen Basin. Then it follows the slope of northern Greenland to enter the Greenland Sea by the Fram Strait.

---

[277] This process is observed in the Chukchi Sea, the Barents Sea, or the fjords of the Svalbard Archipelago.

[278] Tritium and carbon-14 measurements allow the determination of the renewal times: 30 years for all water down to 1500 m depth and for Arctic Bottom Water in the Amundsen and Nansen Basins; 700 years in the Makarov and Canadian Basins at depths below 1500 m, the depth of the Lomonosov Ridge sill.

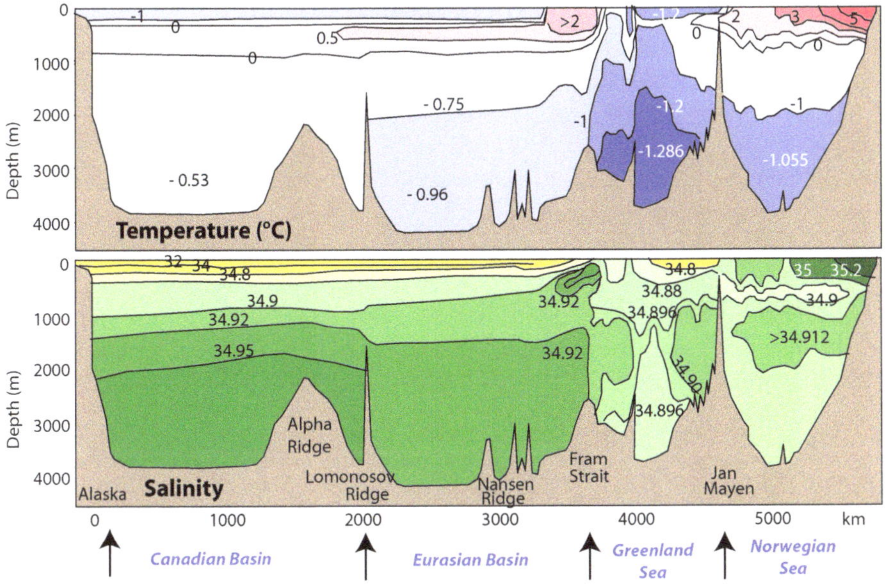

FIG. 3-80A – Potential temperature and salinity section across the Arctic Ocean and the Nordic Seas, from the Canadian Basin to the Norwegian Sea. (See the location on **Figure 3-80b** and the density section, **Figure 1-40**.) (After Aagaard *et al.*, 1985.)

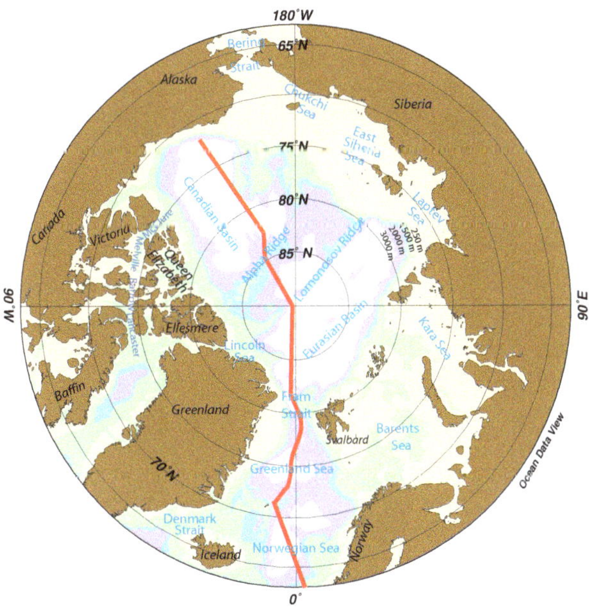

FIG. 3-80B – Location of the Sections shown in **Figure 3-80a** and **Figure 1-40**.

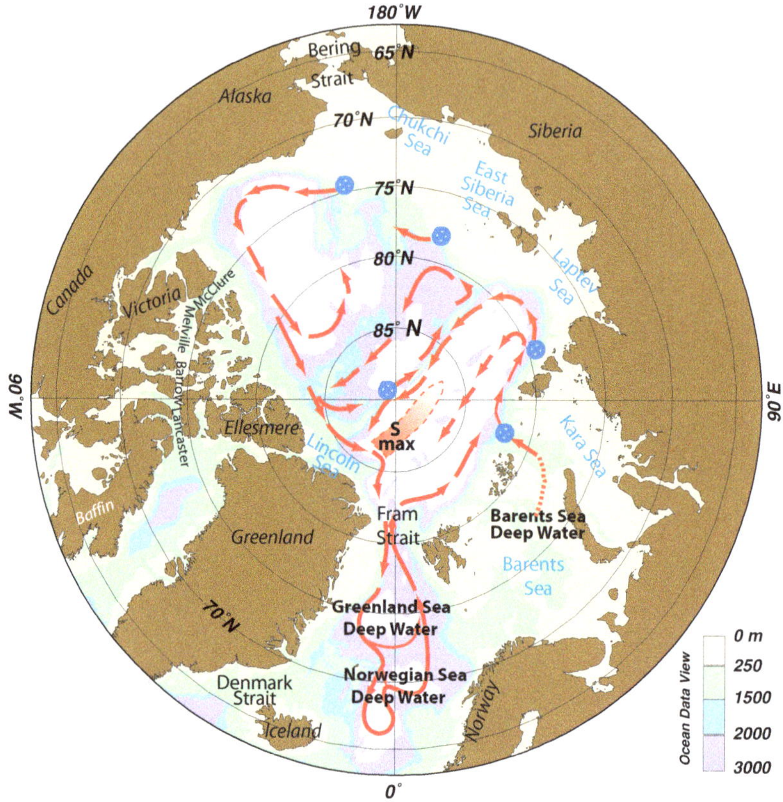

FIG. 3-81 – Deep circulation in the Arctic Ocean basins. Blue circled crosses represent descending dense water from the shelf. The shaded area in diminishing pink shows the influence of more saline water coming from the Canadian Basin into the Amundsen Basin. (After Jones *et al.*, 1995.)

5.2.6.5. Arctic Intermediate Waters

We include Arctic Intermediate Waters here even though they are not formed in the Arctic Ocean, strictly speaking. There are a number of varieties of Arctic Intermediate Waters ranging from the Labrador Sea over to the Norwegian and Greenland Seas. Winter convection processes are not reproduced exactly from one winter to the next. Interannual climatic variations modulate convection depth and the characteristics of Intermediate Water masses formed during these processes. All are high in oxygen and in CFCs. Examples of Intermediate Waters are Greenland Sea Intermediate Water ($\theta \approx 0.7$ °C and $S \approx 34.88$) and Iceland Sea Intermediate Water ($\theta \approx -0.4$ °C and $S \approx 34.84$). The latter was observed in 1999 and could be followed into the Norwegian Sea, where it formed a salinity minimum. North Iceland Winter Water (2 °C $< \theta <$ 3 °C and $34.85 < S < 34.90$) is found south of the Norwegian Sea. It is formed north of Iceland from Atlantic Water modified by the addition of fresh water. Flow into the Atlantic Ocean involves these intermediate waters as well as Intermediate Water resulting from the mixture of surface water and Norwegian Sea Deep Water. Crossing the sills southward, they mix with more saline water influenced by Mediterranean Water and rapidly lose their characteristics south of 60°N.

5.2.6.6. Overflows from the Nordic Seas

Norwegian Sea Deep Water, colder, more saline, and thus denser ($\sigma_\theta = 28.10$) in its place of formation than Antarctic Bottom Water ($\sigma_\theta = 27.88$), easily flows northward into the Eurasian Basin, through the deep Fram Strait [**Figure 3-78**]. On the other hand, shallower sills between Greenland and Scotland hinder the flow of the densest water toward the North Atlantic Ocean [**Figure 3-70**] (which is not the case for Antarctic Bottom Water).

Intermediate Waters mixed with Norwegian Sea Deep Water pour, more or less regularly,[279] over the Denmark sill (615 m) and the sills between Iceland and Scotland. The densest and coldest water flows out on the western sides of the sills under the influence of the Coriolis force [**Figure 3-82**]. The water that pours over each sill has slightly different characteristics between west and east. The dense water flow at the west is called Denmark Strait Water (DSW) (-0.5 °C $< \theta <$ 1 °C, $34.8 < S < 34.9$, $27.80 < \sigma_\theta < 28.05$). Its transport is of the order of 1 to 2 Sv.

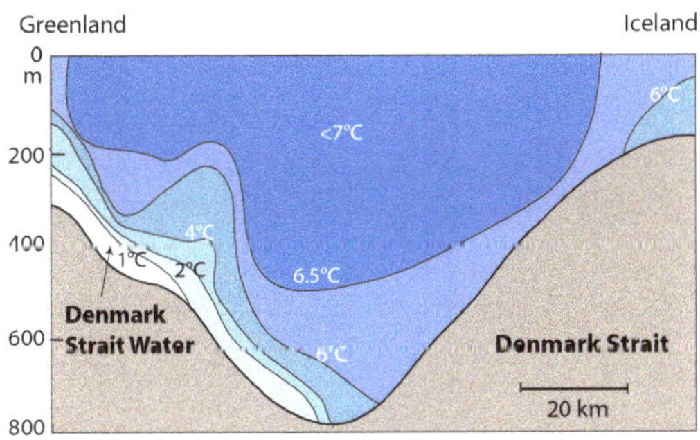

FIG. 3-82 – Temperature section over the Denmark Strait sill. (After Worthington, 1969.)

East of Iceland, on the sill between Iceland and the Faeroe Islands (400 m), a mixture of Iceland Sea Intermediate Water, Norwegian Sea Intermediate Water, and Norwegian Sea Deep Water flows out. The flow over the Faeroe Bank Channel (830 m),[280] the deepest, is a mixture of Norwegian Sea

[279] Flow toward the Atlantic takes place episodically, linked to the passage of perturbations in the area of sills.
[280] East of the Faeroe Islands.

Intermediate Water and Norwegian Sea Deep Water. The water flowing over these sills is called Iceland-Scotland Overflow Water (ISW). It is more saline and warmer than DSW ($\theta < 3$ °C, $S > 34.94$, $\sigma_\theta > 27.8$).

The Arctic Mediterranean receives warm and saline Atlantic Water (the dominant input), low-salinity Bering Sea Water, fresh water from rivers, and some precipitation. It is transformed, in part, as a cold and low-saline water mass exiting at the surface, and in part, as a dense, cold and relatively saline water mass, Norwegian Sea Deep Water (NSDW). NSDW exits at depth toward the north in the Arctic Ocean and toward the south in the Atlantic Ocean. On average, the same amount of water exits as enters. The entrances are found in the surface layers (Atlantic Water, Bering Strait Water, rivers). The exits are found between Greenland and Scotland both at surface and at depth. The order of magnitude of the transports is indicated in **Figure 3-83**. The principal transformation site is found in the Nordic Seas, where exchanges with the atmosphere produce intense convective motions to the bottom. It is one of two descending branches[281] of the immense thermohaline cell that extends as far as the Pacific Ocean by the intermediary of North Atlantic Deep Water [**Figure 1-51**].

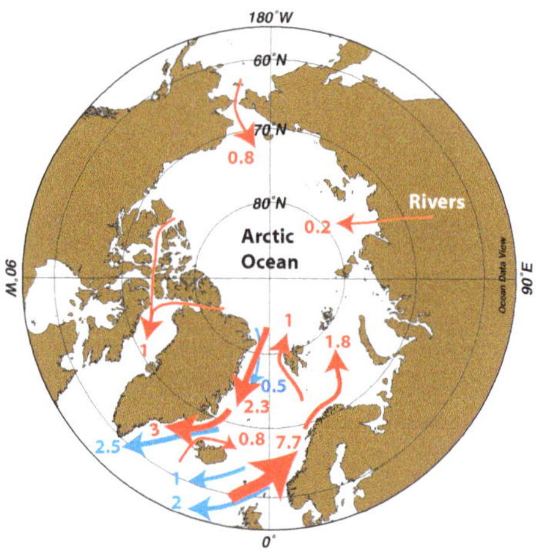

FIG. 3-83 – Transport estimates in the Mediterranean Arctic in Sv (10^6 m^3 s^{-1}). Surface transports are indicated by red arrows. Deep transports are blue (After Jones, 2001 and Rudels *et al.*, 1999.)

As is the case with the Antarctic Ocean, the Arctic Ocean is a supplier of heat to the atmosphere, by warm Atlantic Water, but also by the evacuation of "cold" through the export of ice and very cold Arctic Surface Water.

[281] The second branch is that of the Labrador Sea.

Chapter 6
Water properties

The thermocline separates deep and cold water masses ($\theta < 5$ °C) from warmer water masses between the two subtropical fronts [see the diagram **Figure 1-27**]. Cold water masses may be formed north of the North Subtropical Front or south of the South Subtropical Front. They include Antarctic Bottom Water, North Atlantic Bottom Water, North Atlantic Deep Water, Arctic Intermediate Water, and Antarctic Intermediate Water [**Figures 3-84a** and **3-84b**]. Warmer water masses include Central Waters, Subtropical Waters, and Surface Waters.

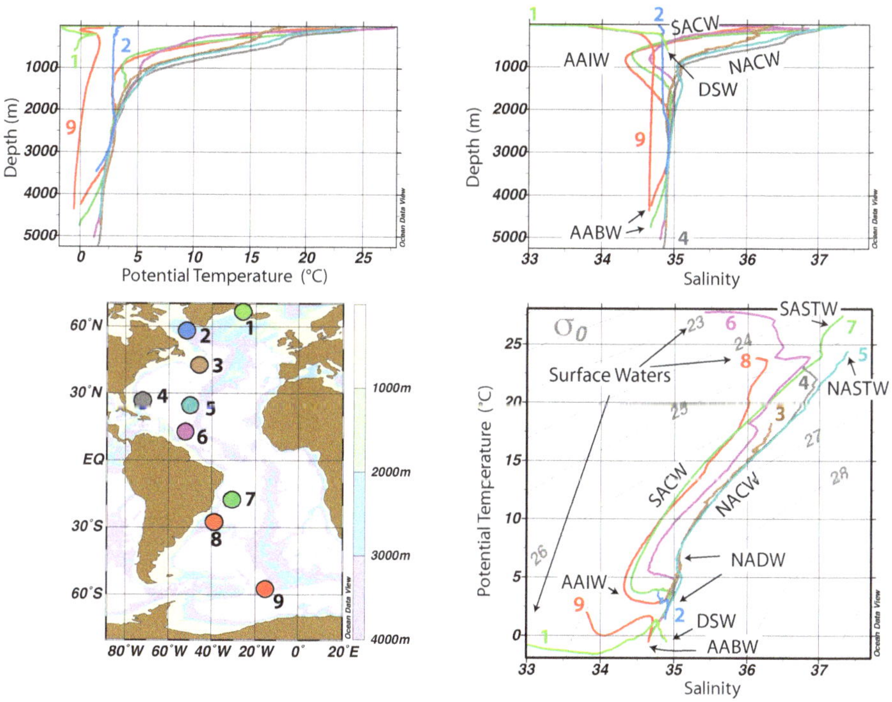

FIG. 3-84A – θ-S diagrams from north to south in the western Atlantic Ocean, showing the various water masses. Station 1 is in the Denmark Strait; Station 9 is in the Antarctic Ocean. [Antarctic Bottom Water (AABW), Denmark Strait Water (DSW), North Atlantic Deep Water (NADW), North Atlantic Central Water (NACW), South Atlantic Central Water (SACW), North Atlantic Subtropical Water (NASTW), South Atlantic Subtropical Water (SASTW), Antarctic Intermediate Water (AAIW).] (With ODV and World Ocean Database.)

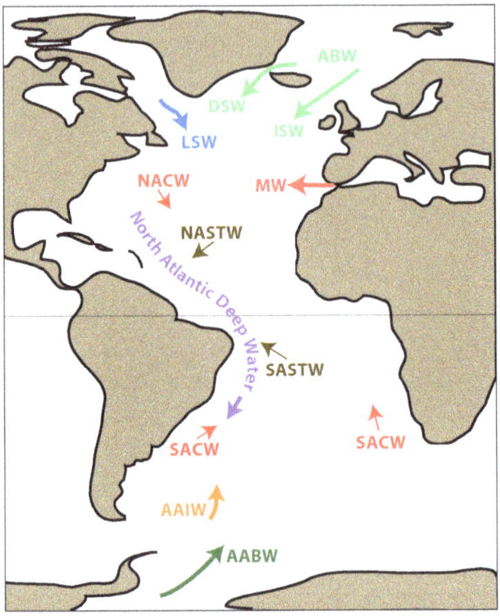

FIG. 3-84B – Schematic of the water masses of the Atlantic Ocean below Surface and Subtropical Waters. [Antarctic Bottom Water (AABW), Arctic Bottom Water (ABW), Denmark Strait Water (DSW), Iceland-Scotland Overflow Water (ISW), Labrador Sea Water (LSW), North Atlantic Deep Water (NADW), Mediterranean Water (MW), Antarctic Intermediate Water (AAIW), North Atlantic Central Water (NACW), South Atlantic Central Water (SACW).]

6.1 Surface water

Surface water characteristics [**Figures 3-85a** and **3-85b**] are directly linked to balances in the exchanges with the atmosphere of heat [**Figure 1-13**] and fresh water [**Figure 1-22**] and thus are climatic indicators. They also depend on surface circulation and oceanic internal circulation [**Section I.2** and **Figure 3-6**]. Warm equatorial water is entrained poleward by western boundary currents of the anticyclonic gyres (Gulf Stream and Brazil Current). In the subpolar cyclonic gyre, the eastern branch transports warm water northward, increasing the spread of isotherms eastward. Vertical motions also have a strong impact on surface temperatures. For example, temperatures decrease in regions of coastal upwelling along the coasts of Mauritania and South Africa. Strong horizontal gradients of temperature and salinity (called fronts) are found in regions of convergence between currents of different origin, at the western boundaries of the subtropical and subpolar gyres such as the Gulf Stream and the Labrador Current in the north, or the Malvinas Current and the Brazil Current in the south.

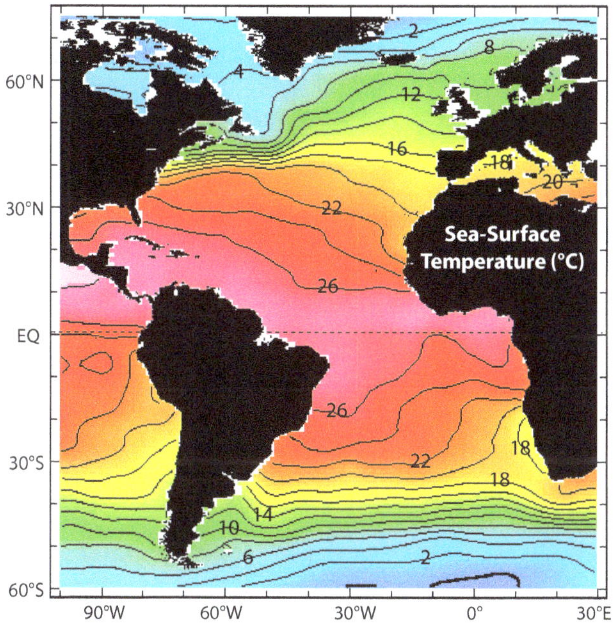

FIG. 3-85A – Atlantic surface temperatures. (IRI/LDEO and WOA05 data.)

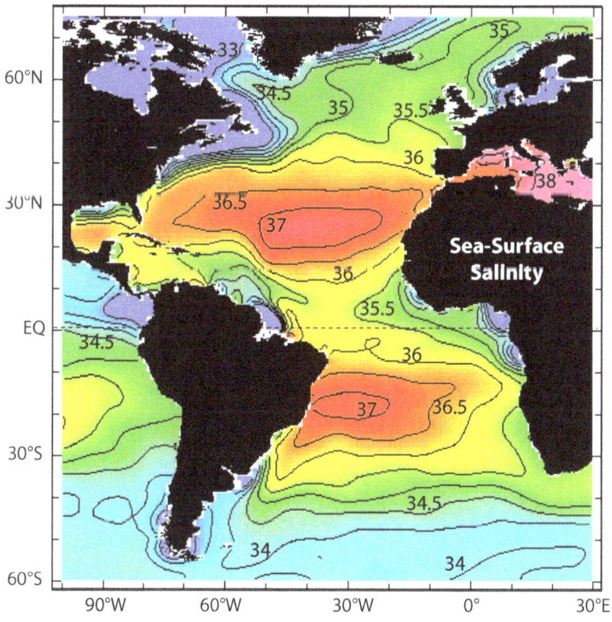

FIG. 3-85B – Atlantic surface salinities. (IRI/LDEO and WOA05 data.)

Salinity maxima occur in the centers of the north and south anticyclonic gyres (where the evaporation-precipitation balance is positive) where values are greater than 37. Between the two maxima, the intertropical convergence zone is a region of lower salinity due to strong precipitation. In the northwest, salinity is low because of the inflow of low-salinity Arctic Surface Water along Greenland, which follows the Canadian coast and converges with saline Gulf Stream water. The Mediterranean, a concentration basin, has high salinity while the Baltic, a dilution basin, has low salinity. Off the mouths of major rivers such as the Congo, the Niger, the Amazon, the Mississippi, and the Saint Lawrence, salinity decreases.

Compared with intermediate and deep waters, surface temperature and salinity show great variability due the variability of exchanges with the atmosphere. This can be seen in that part of θ-S diagrams associated with surface water [**Figure 3-84a**].

Density, the basic parameter in oceanographic structure, depends on temperature and salinity. Regions where deep sinking of surface water occurs are found where surface density is already high in summer. The average surface-density distribution [**Figure 3-86**] shows values greater than 27 in all the northern part of the Atlantic as well as the Mediterranean, making possible the sinking of surface water to great depth in some particular regions.

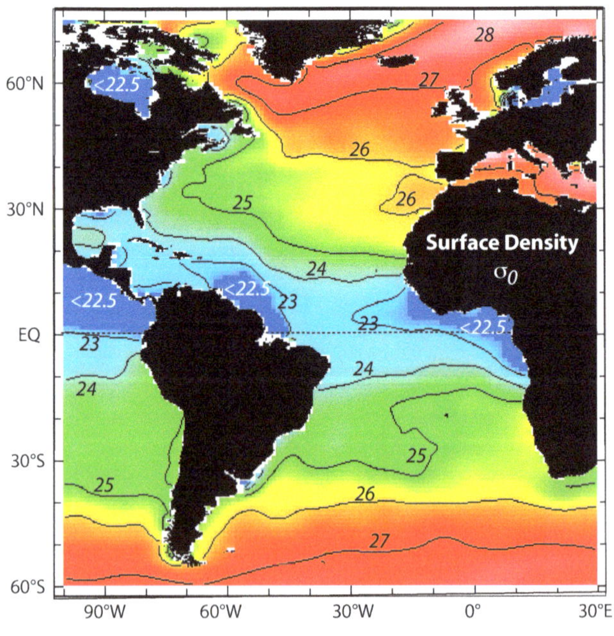

FIG. 3-86 – Atlantic surface densities. (IRI/LDEO and WOA05 data.)

6.2 Subtropical Waters, Central Waters, Subtropical Mode Waters

6.2.1 North and South Subtropical Waters

Subtropical Waters form in regions of low precipitation and strong evaporation, located in the centers of the north and south anticyclonic gyres, where surface salinity [**Figure 3-85b**] is maximal [**Figure 3-87,** Station 2 to the south and Station 6 to the north]. They are subducted under the effect of negative curl τ/f in anticyclonic regions toward the equator under lighter warmer and less saline water of the equatorial zone thus forming a subsurface salinity maximum which represents the less-dense extremity of Central Water. North Atlantic Subtropical Water (NASTW) is present at the surface at Station 6 and at the subsurface (**Figure 3-87**] at Stations 5 and 7. South Atlantic Subtropical Water (SASTW) [**Figure 3-87**, Station 2 at the surface and Stations 1 and 3 at the subsurface] is a little less saline than that of the North Atlantic when subducted under the less-saline water with which it mixes.

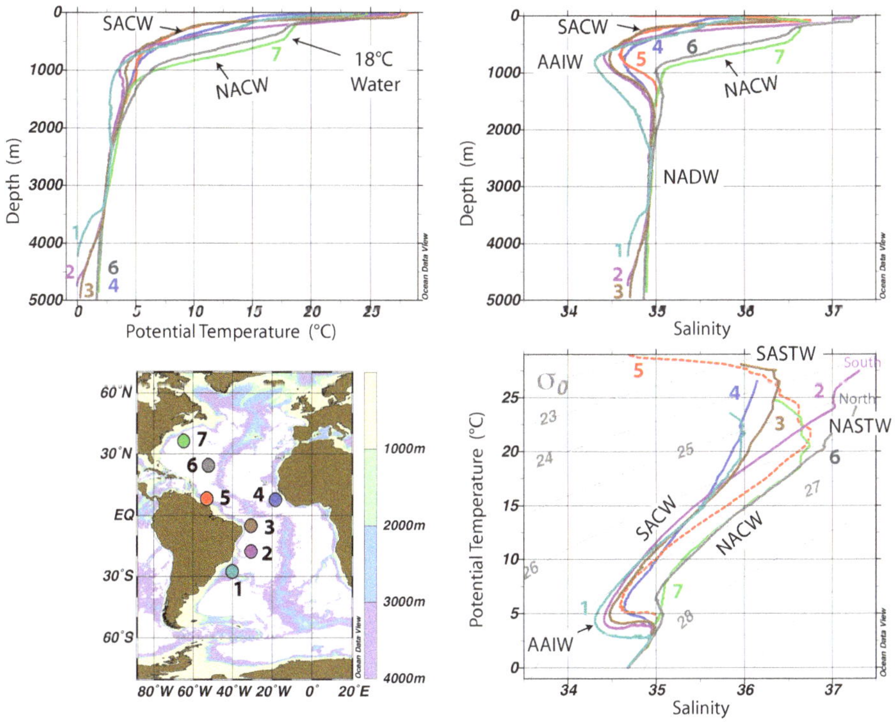

FIG. 3-87 – θ-S diagrams of South and North Atlantic Subtropical Waters. (SASTW and NASTW) and North and South Atlantic Central Waters (NACW and SACW). (With ODV and World Ocean Database.)

6.2.2 North and South Central Waters

Central Waters represent waters of the thermocline. They comprise an important volume of Subtropical Mode Waters formed by winter convection, followed by subduction in the convergence zone (curl $\tau/f < 0$). (The formation process is described in **Section I.3.2**, **Figures 1-31** and **1-34**.) Mode Water[282] is characterized on temperature sections by a spreading of the isotherms, denoting a minimum in the vertical temperature gradient. Of varying characteristics, Central Waters are located on θ-S diagrams [**Figure 3-87**, right] between the salinity minimum of Antarctic Intermediate Water and the salinity maximum of Subtropical Water.

The most important Subtropical Mode Water of North Atlantic Central Water (NACW) forms in the subtropical convergence zone offshore of the Gulf Stream. It is affected by intense winter cooling [**Figure 1-13**] creating convection that is notably deeper than in the South Atlantic. The mixed layer can reach 600 m to 700 m. It includes the formation of Subtropical Mode Water "eighteen-degree water" in the Sargasso Sea, represented in summer by a homogeneous layer between 300 m and 500 m [**Figures 3-87** and **3-88a**] and of Madeira Subtropical Mode Water, formed north of Madeira, which in summer results in a homogenous layer between 80 and 180 m [**Figure 3-88b**]. These two Mode Waters comprise about half the volume of North Atlantic Central Water.

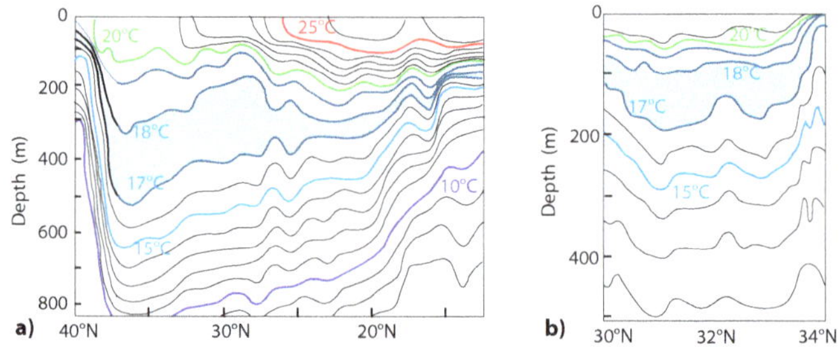

FIG. 3-88 – Temperature section in **(a)** the Sargasso Sea in summer. Spreading of the 17 °C and 18 °C isotherms indicates the presence of 18 °C Subtropical Mode Water and **(b)** the eastern Atlantic Basin, showing the presence of Madeira Subtropical Mode Water. (After McCartney, 1982.)

In the southwest Atlantic, in the convergence zone between the Malvinas Current and the Brazil Current, Southwest Central Water forms (SWCW) [**Figure 3-89**] that includes several varieties of Subtropical Mode Water of slightly different densities. One variety has a temperature of 17°–18 °C similar to that of the North Atlantic but is less saline. Southwest Central Water is limited to the anticyclonic circulation of the South Atlantic [**Figure 3-6**].

[282] Mode Waters have a weak vertical density gradient, corresponding to a minimum of potential vorticity $(f/\rho)(\partial\rho/\partial z)$.

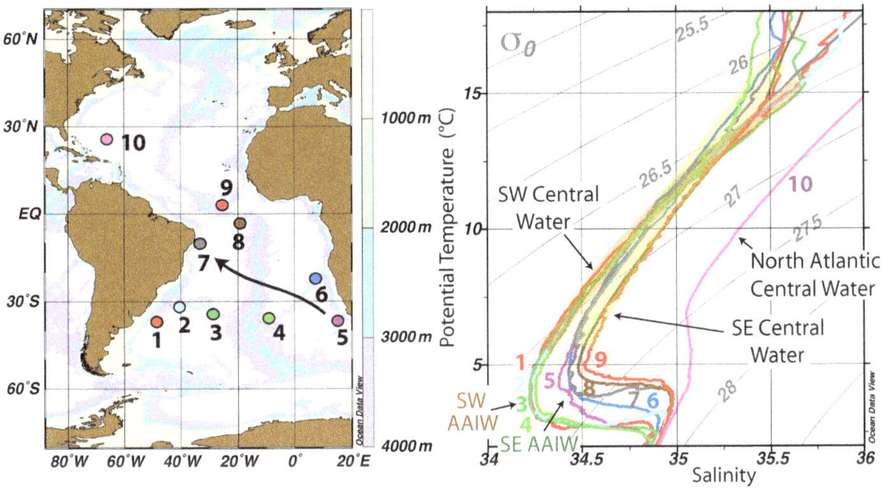

FIG. 3-89 – θ-S diagrams in the South Atlantic showing the difference between SW Central Water, limited to the south anticyclonic circuit (Stations 1, 2, 3, 4) and SE Central Water, which can be followed into the equatorial region (Stations 5, 6, 7, 8, 9). Station 10 (pink) represents the more saline North Atlantic Central Water. SW Antarctic Intermediate Water (SW AAIW) may be distinguished from the more saline SE Antarctic Intermediate Water (SE AAIW). (With ODV and World Ocean Database.)

In the southeast, the Atlantic Ocean receives Central Water from the Indian Ocean (SE Central Water) *via* turbulent eddies of the Agulhas Current that may be tracked along the Brazil coast, entrained by the Benguela Current and the North Brazil Coastal Current [**Figure 3-89** Stations 5, 6, 7, 8] to north of the equator [**Figure 3-89**, Station 9]. In the interior of this water mass, Subtropical Mode Water of "13 °C" can be distinguished, among others.

The two South Atlantic Central Waters have slightly different θ-S slopes [**Figure 3-89**]. Southwest Central Water is less saline than that of the southeast at temperatures less than 12 °C, and more saline at temperatures greater than 12 °C. From the North Brazil Coastal Current, Southeast Central Water is transported eastward by the North Equatorial Countercurrent and meets North Atlantic Central Water in the eastern equatorial region [**Figure 3-87**]. The transition between North Atlantic Central Water and South Atlantic Central Water forms a front around 15°N, near the African coast, which extends under the mixed layer to the bottom of the thermocline. The front is located at the confluence of the Canary Current and the cyclonic branch of the Guinea Dome and thus varies in latitude with the season, as does the Canary Current. In the region of the front, the two central waters have comparable densities, so that mixing is active and forms intrusions of one into the other giving perturbed profiles. The mixture of North Atlantic Central Water and South Atlantic Central Water travels westward in the North

Equatorial Current, then in the Guyana Current, which entrains it toward the North Atlantic where it undergoes transformations leading in the north to the formation of North Atlantic Deep Water. In **Figure 3-87**, Station 5 (dotted red) located around 10°N in the Guyana Current is composed of South Atlantic Central Water at depth (7 °C–12 °C) and of a mixture with North Atlantic Central Water above (12 °C–20 °C). The salinity maximum found above corresponds to North Atlantic Subtropical Water transported by the North Equatorial Current, and the surface salinity minimum reflects the influence of the Amazon.

South Atlantic Central Waters are less saline (by about 0.5) than North Atlantic Central Water, influenced by Mediterranean Water [**Figures 3-87** and **3-89**].

6.3 Antarctic Intermediate Water

Antarctic Intermediate Water (AAIW), the densest of the Subantarctic Mode Waters [**Section I.3.2**], is formed in the Subantarctic Front region, principally in the southeast Pacific Ocean [**Section II.6.3.2**]. Associated with the Subantarctic Mode Water, it arrives mainly through the Drake Passage and the Malvinas Current (SW AAIW) and extends northward between 800 m and 1,000 m depth [**Figure 3-90**] above North Atlantic Deep Water and Upper Circumpolar Deep Water and below South Atlantic Central Water. A second input of Antarctic Intermediate Water (SE AAIW) arrives south of Africa from the Indian Ocean sector. The water from the southeast is somewhat more saline than that from the southwest [**Figure 3-89**].

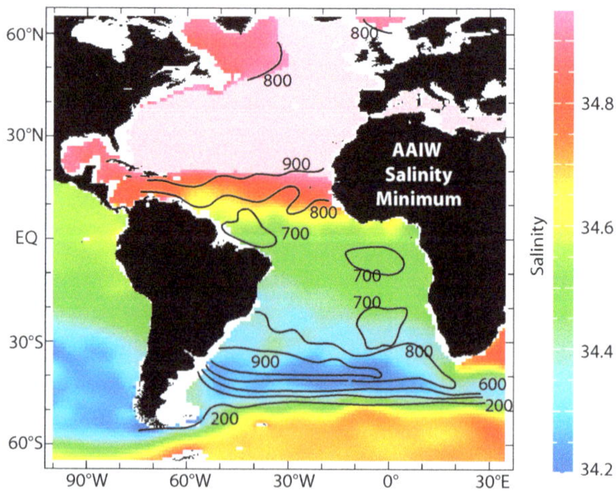

FIG. 3-90 – Isopleths corresponding to the salinity minimum of Antarctic Intermediate Water and Arctic Intermediate Water. The minimum disappears in the pink region. Some depths of the minimum are shown. (IRI/LDEO, and Levitus data.)

Antarctic Intermediate Water (AAIW) is characterized by a salinity minimum ($S \sim 34.2$, $\theta \sim 3$–5 °C), that is intensified toward the western boundary. The salinity minimum erodes during the course of its path through mixing with more saline water above and below [**Figure 3-90**]. Its signature disappears around 20°N, when the AAIW meets high-salinity Mediterranean Water found at the same depth (having the same density) [**Figures 3-90** and **3-91**].

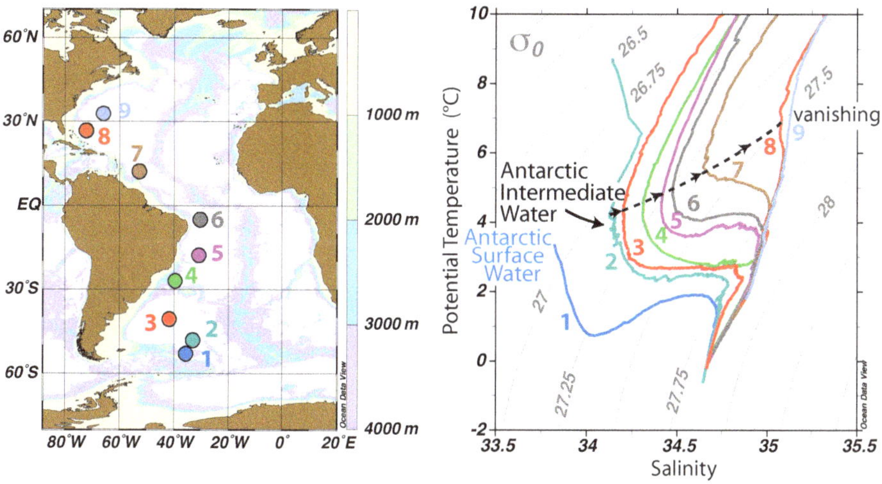

FIG. 3-91 – Evolution of the salinity minimum of Antarctic Intermediate Water from the Subantarctic Front until its disappearance around 20°N in the Atlantic. (With ODV and World Ocean Database.)

6.4 Arctic Intermediate Waters and Subpolar Mode Waters

In addition to the varieties of Arctic Intermediate Waters of the Nordic Seas [**Section III.5.2.6.5**], there is also an Arctic Intermediate Water of the SW Labrador Sea ($\theta \sim 3$ °C, $S \sim 34.5$). It forms by winter convection and subduction along the North Atlantic Subarctic Front, at the boundary between the cold Labrador Current and the warm extension of the Gulf Stream. It propagates eastward, transported by the various branches of the North Atlantic Drift and mixes with the lower part of North Atlantic Central Water. Its salinity minimum is no longer discernible south of 40°N.

The influence of the many varieties of Arctic Intermediate Water is limited because they lose themselves through mixing with water influenced by more saline Mediterranean Water, and are swallowed up in North Atlantic Deep Water.

The northward extension of the North Atlantic Drift exposes homogeneous layers formed by winter convection to cooling that is more intense and longer than in the North Pacific. Subpolar Mode Waters (SPMW) of various densities ($27.3 < \sigma_\theta < 27.5$) form in the NE Atlantic, along the many branches of the North Atlantic Drift [**Figure 3-34**]. Denser waters are entrained toward the Irminger Sea and less dense waters toward the Norwegian Sea. Their density increases along their paths. Subpolar Mode Waters have an important role in the water transfer from the North Atlantic subtropical gyre to the Nordic Seas.

6.5 Mediterranean Water

At the outlet of the Strait of Gibraltar, Mediterranean Water [**Section III.5.1**], of much higher density than the water blanketing the Atlantic depths, descends rapidly along the continental slope and mixes with North Atlantic Central Water [**Figures 3-59** and **3-92**]. Modifications occur so that the mixture does not reach the sea floor. Rather, it finds its equilibrium level around 800–1,200 m. At the sill exit, filaments of Mediterranean Water turn to the right and follow the slope south of Spain and Portugal. (For example, in **Figure 3-59** the influence of Mediterranean Water is more evident at Station 3 than at Station 4.) Off Cape Saint Vincent, the southwestern point of Portugal, Mediterranean Water eddies form and detach from the principal filament and migrate westward [**Figure 3-93**]. These lenses of Mediterranean Water, located around 800–1,200 m, are called "meddies",[283] traces of which can be found just offshore of the Gulf Stream.

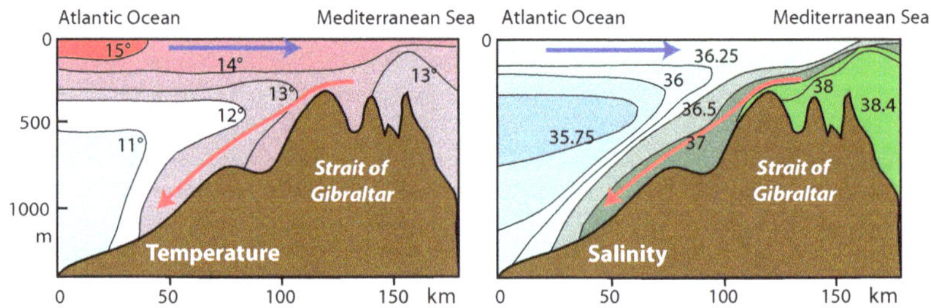

FIG. 3-92 – East-west temperature and salinity sections showing the entrance of Atlantic Water at the surface and the flow and mixing of Mediterranean Water at depth over the Gibraltar Sill. (After Tchernia, 1978.)

Mediterranean Water creates a salinity maximum around 1,000 m [**Figure 3-94**], which extends northward as far as the sills between Iceland and Scotland, where it influences the mixture of Iceland-Scotland Overflow Water (ISW) spreading into the Northeast Atlantic basin.

[283] For *Mediterranean eddies*.

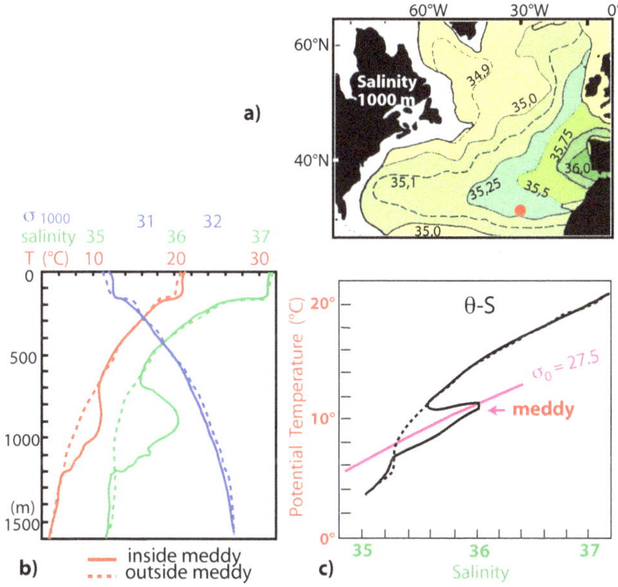

FIG. 3-93 – **(a)** Salinity at 1,000 m in the North Atlantic and station position (red), **(b)** profiles of temperature, salinity, and density (σ_{1000}) in a "meddy" (solid lines) and outside a "meddy" (dotted lines) and **(c)** corresponding θ-S diagrams. (After Armi and Stommel, 1983.)

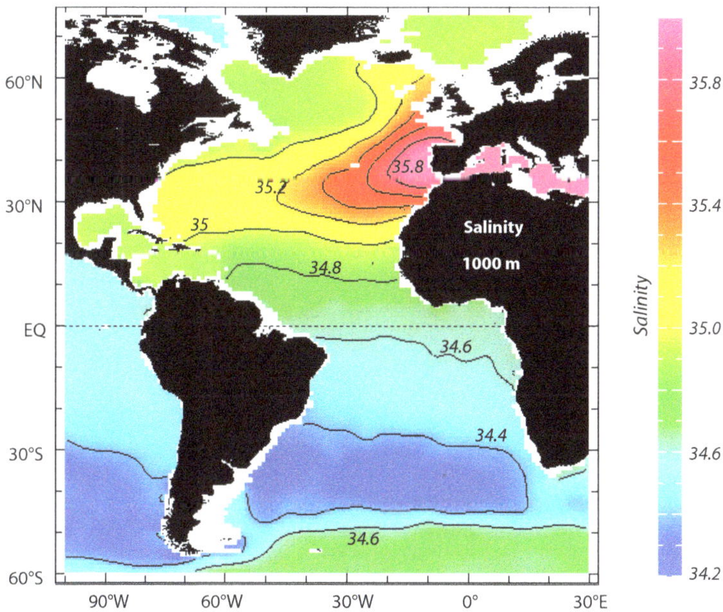

FIG. 3-94 – Salinity at 1,000 m showing the influence of Mediterranean Water. (IRI/LDEO, and 1994 Levitus data.)

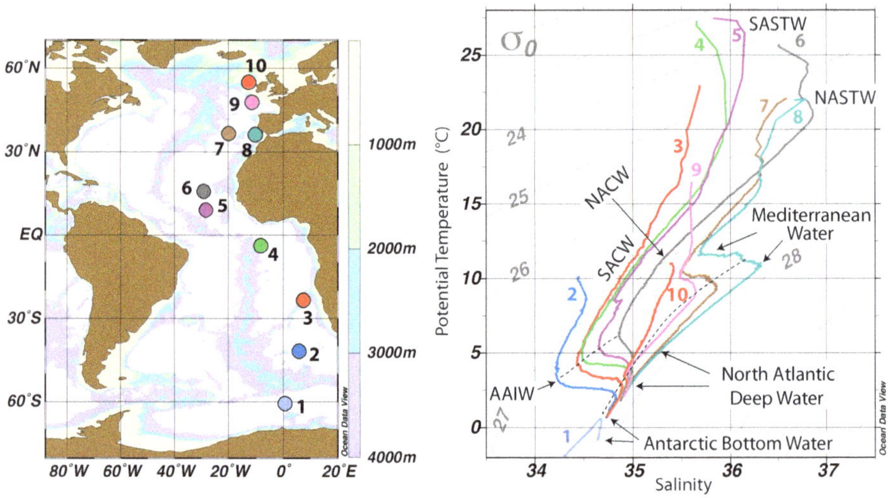

FIG. 3-95 – θ-S diagrams in the Eastern Atlantic Basin. [North Atlantic Subtropical Water (NASTW), North Atlantic Central Water (NACW), South Atlantic Subtropical Water (SASTW), South Atlantic Central Water (SACW), Antarctic Intermediate Water (AAIW).] (With ODV and World Ocean Database.)

Mediterranean Water is responsible for the weak extension of Arctic Intermediate Waters and makes up the saline component of North Atlantic Deep Water [**Figure 3-95**]. Its salinity maximum (quite noticeable at Stations 7 and 8 near Gibraltar) has the same density as the Antarctic Intermediate Water salinity minimum so that the trace of Antarctic Intermediate Water disappears around 20°N through mixing [**Figure 3-90**].

6.6 Antarctic Bottom Water

Antarctic Bottom Water (AABW) formed in the Weddell Sea [**Section II.6.3.4**, **Figure 2-43** and **Figure 3-95**, Station 1] extends northward and eastward, mixing with the lower part of Circumpolar Deep Water. It penetrates into the Atlantic through two basins delimited by the Mid-Atlantic Ridge: the Argentine Basin on the west, and the Cape Basin on the east [**Figure 3-3**]. In the west, it follows the base of the Scotia Arc, extends into the Argentine Basin, forming a cyclonic loop around the Zapiola Rise [see **Figure 3-114**], passes through the deep channels of the Rio Grande Rise (the Vema Channel ~ 4,550 m and the Hunter Channel ~ 4,350 m), into the Brazil Basin. The shallowest Antarctic Bottom Water flows through the equatorial channel, between the Brazilian continental slope and the Mid-Atlantic Ridge, toward the North American Basin. Its characteristics alter through mixing along its path as far as 45°N–50°N in the western basin. Potential temperature at 4,000 m shows warming through mixing with overlying warmer water, accentuated at sill passages [**Figure 3-96**]. Temperature and salinity minima at the bottom indicate the presence of Antarctic Bottom Water west of the Mid-Atlantic Ridge (*e.g.*, on the section along 11°S) [**Figure 3-97**]. Its upper part, made up of Lower Circumpolar Deep Water, flows toward the eastern basin through the Romanche Fracture Zone at the equator and through the Vema Fracture Zone at 10°N.

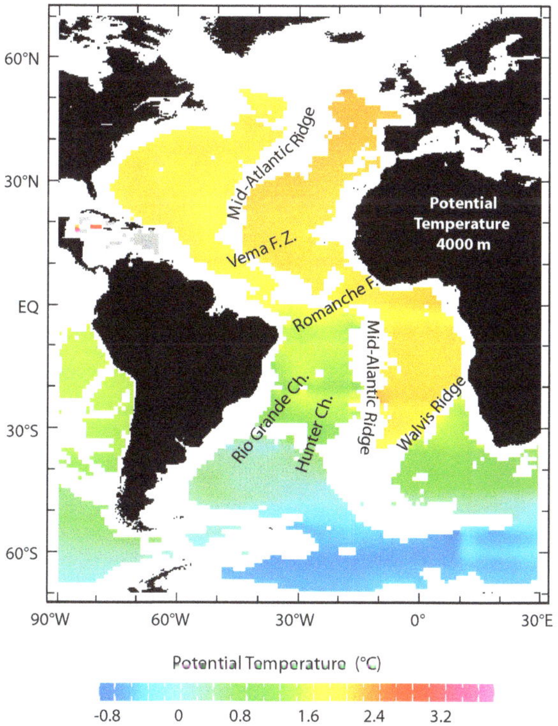

FIG. 3-96 – Potential temperature at 4,000 m. (IRI/LDEO, WOA05 data.)

To the east, Antarctic Bottom Water penetrates into the Cape Basin through a depression of the ridge south of Africa (at about 4,300 m) where its characteristics are by now attenuated (1 °C $< \theta <$ 1.3 °C, 34.72 $< S <$ 34.78) through mixing with warmer and more saline Lower Circumpolar Deep Water. The Walvis Ridge, which is nearly continuous at depths greater than 3,000 m [**Section III.2, Figure 3-3**], limits its northward extension. At the equator, the Romanche Fracture Zone (4,350 m) blocks the coldest water, thus creating a jump in temperature between the western and eastern sides of the ridge [**Figure 3-97**]. This semi-blockage of Bottom Water by topography, which occurs both south and west of the Angola Basin, results in a marked temperature difference (1.2 °C) as well as a salinity difference (0.045) between the southern and northern sides of the Walvis Ridge [**Figures 3-98** and **3-99**].

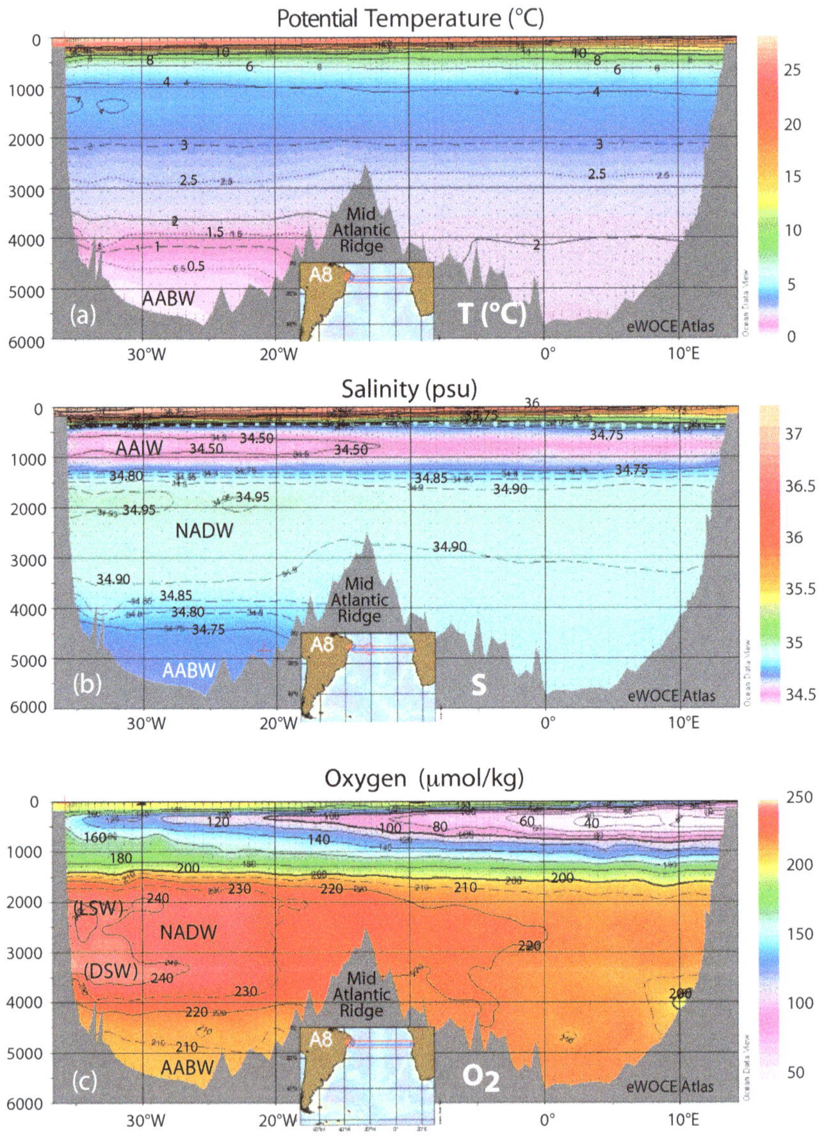

FIG. 3-97 – Sections at 11°S, between South America and Africa: (**a**) potential temperature, (**b**) salinity, (**c**) dissolved oxygen (μmol kg^{-1}). [Labrador Sea Water (LSW), Denmark Strait Water (DSW), Antarctic Intermediate Water (AAIW), North Atlantic Deep Water (NADW), Antarctic Bottom Water (AABW).] (Section A8, *eWOCE Atlas*, Schlitzer, 2002.)

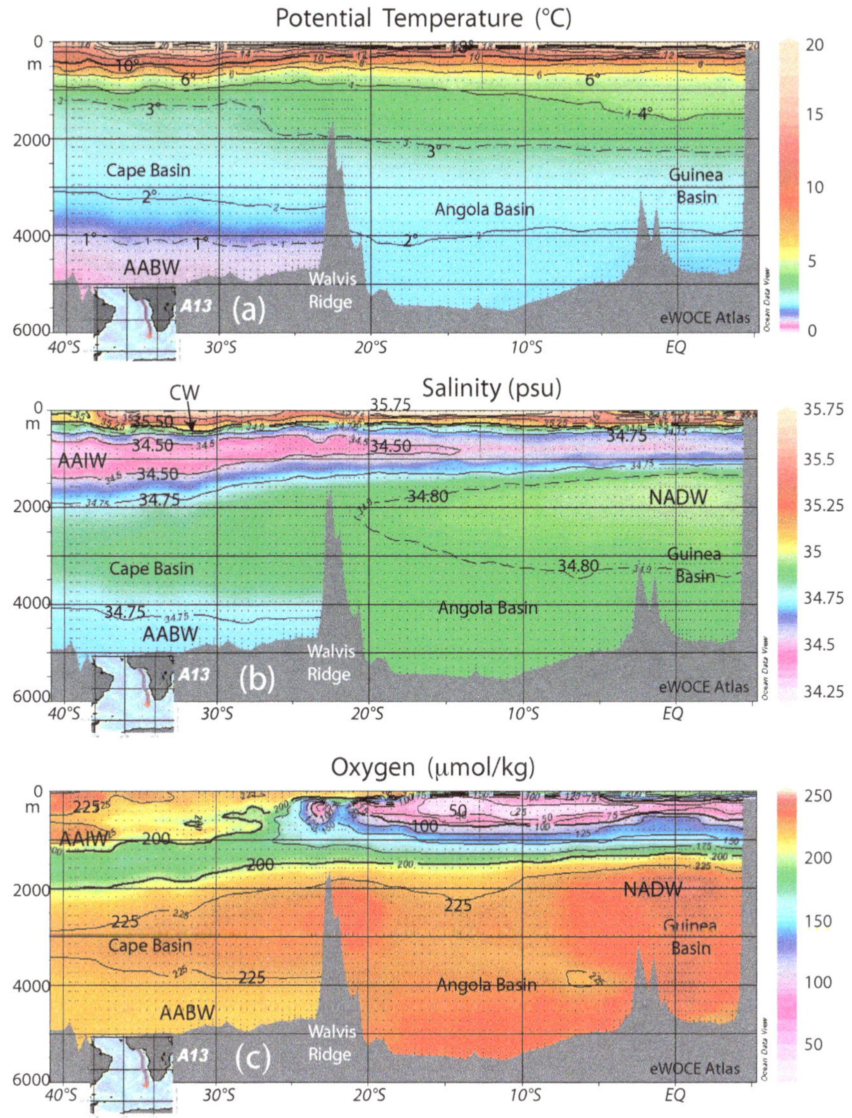

FIG. 3-98 – Sections in the South Atlantic Eastern Basin from the Gulf of Guinea to 40°S: (a) potential temperature, (b) salinity, (c) dissolved oxygen. (Section A13 *eWOCE Atlas*, Schlitzer, 2002.)

North of the Walvis Ridge, the floor of the Eastern Basin is filled with a mixture of the upper part of greatly altered Antarctic Bottom Water and of the lower part of North Atlantic Deep Water coming mainly through the Romanche Fracture Zone on the equator [**Figure 3-96**]. Bottom water in the Angola Basin north of the ridge is more saline, warmer, and richer in oxygen (as is North Atlantic Deep Water at this latitude) [**Figure 3-98**].

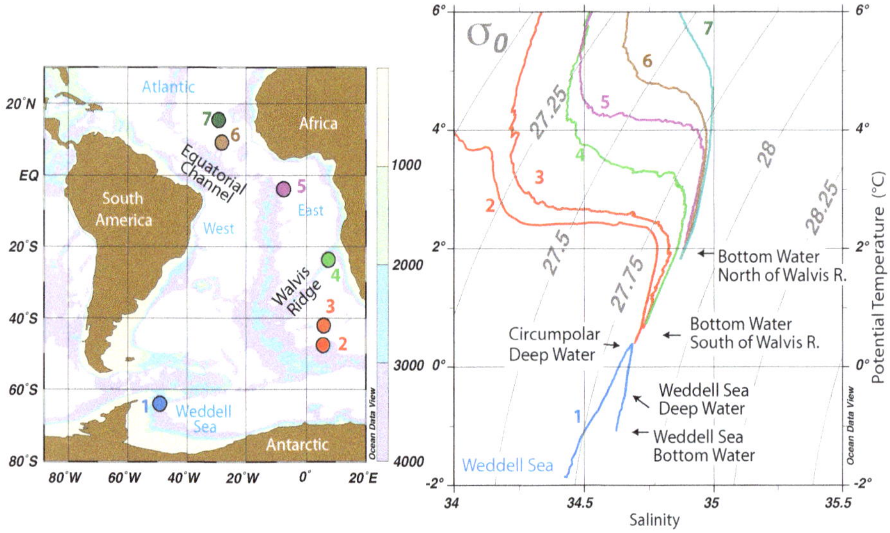

FIG. 3-99 – θ-S diagrams showing the variation of bottom-water characteristics from the Weddell Sea to the equator. Station 1 is in the Weddell Sea, the principal site of Antarctic Bottom Water formation. (ODV and World Ocean Database.)

6.7 Dense waters exiting over the Greenland-Scotland sills

Dense waters exiting from the Nordic Seas over the sills between Greenland and Scotland [**Section III.5.2.6.6**] are channeled by topography. They undergo mixing with water encountered while crossing the sills to reach the deep Atlantic Basin, which alters their characteristics and in particular their density.

6.7.1 Sill Water between Iceland and Scotland

Dense water exiting across the sills between Iceland and Scotland (Iceland-Scotland Overflow Water, ISW) mixes with water that is heavily influenced by saline Mediterranean Water. Held against the Mid-Atlantic Ridge by Coriolis force, one branch finds a passage to the Irminger Sea through the Charlie Gibbs Fracture Zone (3,600 m, 51°N), entraining nearly 3 Sv of adjacent waters [**Figure 3-100**].

The other denser branch ($\sigma_\theta > 27.85$), located below the level of the Gibbs passage, follows the eastern slope of the ridge southward [**Figures 3-100** and **3-101**] mixing with Lower Circumpolar Deep Water that has penetrated through the Vema and Romanche Fracture Zones, expanding northward along the eastern side of the basin to form Bottom Water of the Northeast Atlantic, more saline and warmer than that of the Northwest Atlantic.

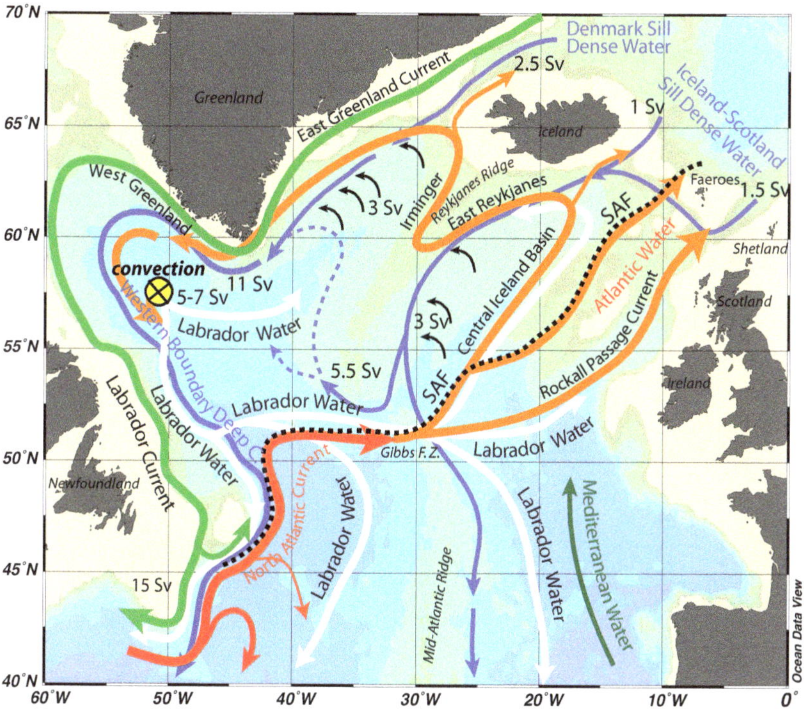

FIG. 3-100 – Surface and deep circulation south of the Greenland-Scotland sills: (1) General surface circulation. The North Atlantic Drift is red and orange, cold currents from the Arctic including the East Greenland Current, the West Greenland Current, and the Labrador Current are green. (2) Deep paths and transports of Denmark Strait Water, and Iceland-Scotland Overflow Water are blue. (3) The zone of formation of Labrador Sea Water is a yellow cross and its paths (around 1,500 m) are white. (4) Mediterranean Water, around 1,000 m, is dark green. (5) The Subarctic Front (SAF) is a dotted black line. (After Schott *et al.*, 2004; Brambilla and Talley, 2008.)

6.7.2 Denmark Water Strait

Denmark Strait Overflow Water (DSW) is dense, cold [**Figure 3-82**], and ventilated (rich in tritium, CFCs, and oxygen). Descending the slope of the Denmark Sill [**Figure 3-102**], it mixes with Irminger Sea Water, Labrador Sea Water ($\theta \approx 3.25$ °C, $S \approx 34.90$, $\sigma_\theta \approx 27.80$) and Iceland-Scotland Overflow Water (ISW) that is more saline ($S > 34.95$), warmer, and lower in oxygen to form Bottom Water of the Northwest Atlantic (1 °C $< \theta <$ 2.2 °C, $34.83 < S < 34.90$, $\sigma_\theta \approx 27.90$) [**Figure 3-101**]. The distribution of tritium concentration [**Figure 3-102**] injected into the atmosphere[284] during the 1960's tracks the descent of Denmark Strait Water rich in tritium at the

[284] By nuclear tests in the atmosphere.

outlet of the strait. Measurements repeated nine years later show its progression southward. Denmark Strait Water entrains adjacent waters in the course of its passage to reach a transport of about 10–11 Sv south of Greenland [**Figure 3-100**] and penetrates into the Labrador Sea in a cyclonic circulation.

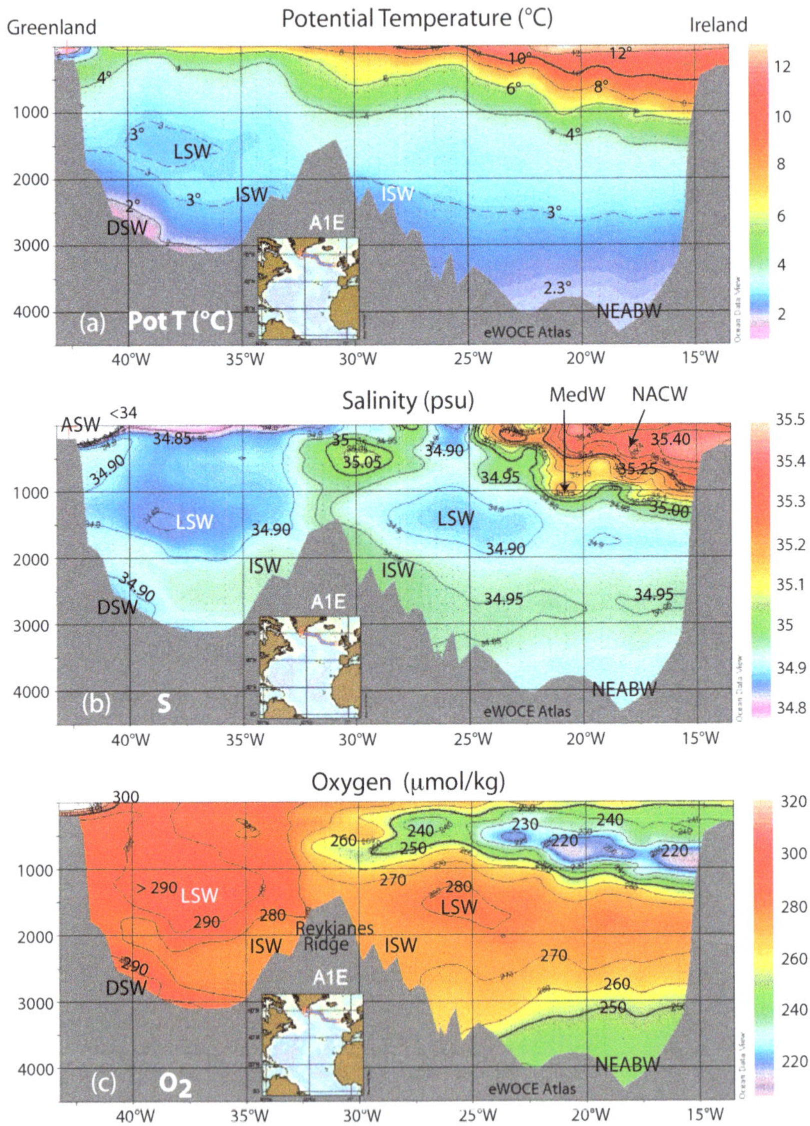

FIG. 3-101 – Sections from south of Greenland to Ireland, around 60°N–55°N: **(a)** potential temperature, **(b)** salinity, **(c)** dissolved oxygen. [Denmark Strait Water (DSW), Iceland-Scotland Overflow Water (ISW), Labrador Sea Water (LSW), North Atlantic Central Water (NACW), Mediterranean Water (MedW), Arctic Surface Water (ASW), Northeast Atlantic Bottom Water (NEABW).] (Section A1E, *eWOCE Atlas*, Schlitzer, 2002.)

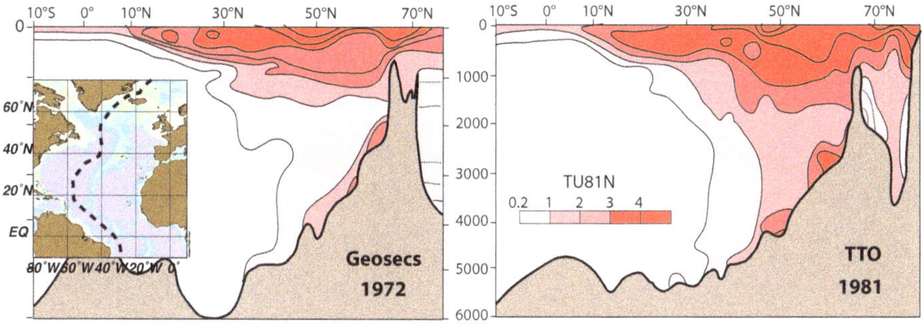

FIG. 3-102 – Tritium concentration in the Atlantic western basin along a section between 10°S and 75°N carried out in 1972 and repeated in 1981. A tritium unit TU81N corresponds to a ratio of 10^{-18} between the number of tritium atoms (^{3}H) and the number of hydrogen atoms (^{2}H). The ratio is normalized to 1981. (After Ostlund and Grall, 1987.)

The branch of Iceland-Scotland Overflow Water (ISW) coming from the east through the Gibbs Fracture Zone ($S > 34.95$) [**Figure 3-101**] may be tracked by its salinity maximum and its oxygen minimum around 2,500 m on the east and on the west of the Reykjanes Ridge. It is located above Denmark Strait Water, which is denser, colder, and richer in oxygen and CFCs [**Figure 3-103a**]. Denmark Strait Water, Iceland-Scotland Sill Water, and Labrador Sea Water are entrained in a cyclonic recirculation between the Irminger Sea and the Labrador Sea [**Figure 3-100**].

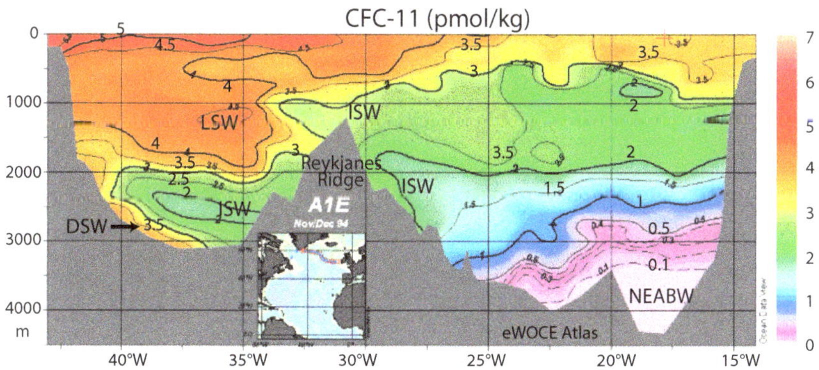

FIG. 3-103A – CFC-11 concentration along the section between Greenland and Ireland. (Section A1E, *eWOCE Atlas*, Schlitzer, 2002.)

A strong asymmetry exists east and west of the Reykjanes Ridge [**Figures 3-101** and **3-103a**]. On the west, in the Irminger Sea, young water rich in CFC-11 characterizes recently formed Labrador Sea Water down to 2,000 m (orange), water formed by winter convection in the Irminger Sea down to 600 m,[285] and water exiting the Denmark Strait on the bottom, against the Greenland Slope. Water from the Iceland-Scotland Overflow *via* the Gibbs Fracture Zone is less rich in CFCs (green), because it mixes with other waters, attenuating its characteristics. Below 3,500 m, in the Northeast Atlantic, water low in CFC (pink) indicates a slow renewal of Deep and Bottom Water.

6.8 Labrador Sea Water

Located between the western coast of Greenland and the Labrador Coast, open to the Atlantic Ocean [**Figure 3-3**], the Labrador Sea is a region of active mixing between water of arctic origin and water from the North Atlantic Drift. It is fed at the surface and at depth by water from the Arctic Mediterranean [**Figure 3-100**]. On the bottom it receives Denmark Strait Water mixed with altered Iceland-Scotland Sill Water and highly altered Antarctic Bottom Water. At the surface it is fed by the cold and brackish West Greenland Current (Arctic Ocean outflow), which flows along the Greenland coast under which it is joined by subsurface water from the warmer and more saline Irminger Current, a branch of the North Atlantic Drift [**Figure 3-103b**].

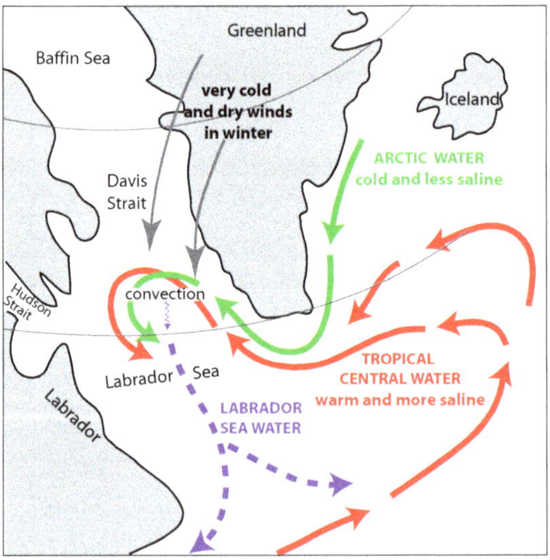

FIG. 3-103B Schematic of Labrador Sea Water formation. Labrador Sea Water is formed in winter in the center of the cyclonic circuit under cold and dry winds from Greenland.

[285] Depending on the climatic characteristics of each winter, the depth is variable and may reach 1000 m.

The cyclonic circulation [**Figure 3-100**] of the Labrador Sea results in reduced stability in its center [**Figures 1-66** and **3-104**], providing a precondition to winter convection. In winter, under the influence of cold northerly winds from the Canadian arctic and the high icy plateau of Greenland, the Labrador Sea undergoes an intense winter cooling (5 °C to 10 °C difference between winter and summer), increasing surface density. In the center of the cyclonic circuit, the increase is enough to develop convective motion, which homogenizes the existing water masses down to 1,500–2,000 m, though it does not reach the bottom [**Figure 3-105**]. The characteristics of Labrador Sea Water vary between 2.8 °C $< \theta <$ 3.5 °C, 34.84 $< S <$ 34.94, and $O_2 >$ 6.3 ml l^{-1}, or 273 μmole kg^{-1}.

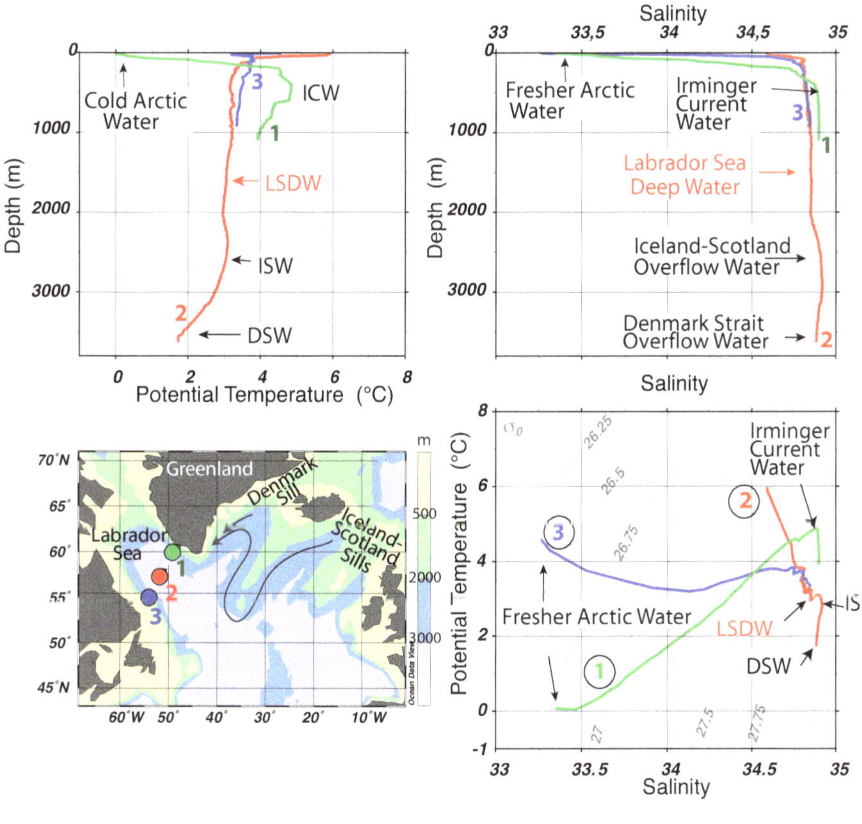

FIG. 3-104 – θ-S diagrams in the Labrador Sea in June 1998. (With ODV and World Ocean Database.)

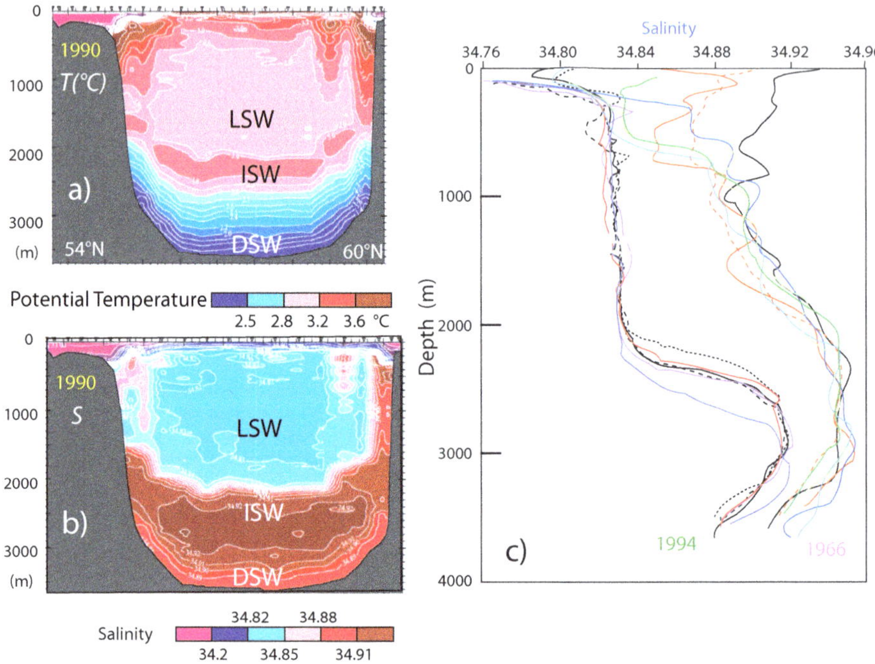

FIG. 3-105 – **(a)** Temperature and **(b)** salinity sections in July 1990 between Greenland and Labrador (location in **Figure 3-104**), **(c)** Salinity profiles in the Labrador Sea during two series of measurements in 1966 and 1994. [Labrador Sea Water (LSW), Iceland-Scotland Overflow Water (ISW), Denmark Strait Overflow Water (DSW).] (After Yashayaev *et al.*, 2007 and World Ocean Database.)

In a section made in 1990 across the Labrador Sea [**Figure 3-105a** and **b**], Labrador Sea water is the homogeneous water mass down to 2,000 m with a temperature near 3 °C and salinity near 34.84. The warmer and more saline layer around 2,500 m represents a trace of water coming over the Iceland-Scotland sills (ISW) through the Gibbs Fracture Zone that has recirculated in the cyclonic circuit of the Labrador Sea and the Irminger Sea. Colder and denser Denmark Strait Water is found at the bottom.

At the time of formation, the water is homogeneous from the surface down to the depth reached by the convection. This depth and the characteristics of the water that is formed vary year-by-year depending on the meteorological conditions that cool the surface and the circulation conditions that bring different "ingredients" to the mixture. When the winter is too mild there can be no water formation. A comparison of two series of salinity profiles in the convection zone shows significant differences between 1966 and 1994 [**Figure 3-105c**]. In 1966, no homogeneous structure was present above 1,500 m and deep water is more saline than in 1994, when salinity profiles were homogeneous from the subsurface to 2,000 m. The situation in 1994 is typical of a phase of convection and formation of Labrador Sea Water. In contrast, 1966 corresponds to a "warm" year with a thin core of Labrador Sea Water, while the strong cooling observed in 1994 corresponds to the formation of a thick layer of Labrador Sea Water [**Figure 3-105c**].

This type of variability is associated with climatic anomalies of the North Atlantic Oscillation (NAO).[286] Changes in salinity and tracer concentrations by deep convection in the 1990's have been tracked as far as the tropical Atlantic in the first decade of the 21st century.

Labrador Sea Water is a mode water characterized, as are all mode waters, by a minimum of potential vorticity [**Section I.3.2**]. **Figure 3-106** shows the evolution of salinity values in the potential vorticity minimum corresponding to Labrador Sea Water.

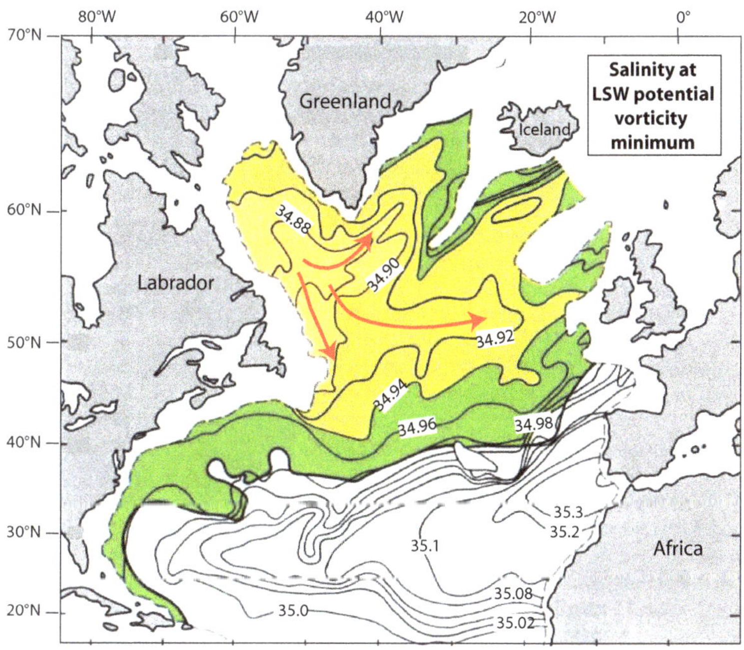

FIG. 3-106 – Salinity at the Labrador Sea Water potential-vorticity minimum. Beyond the colored region, the minimum can no longer be distinguished. Spreading Mediterranean Water can be seen in the isopleths at the same density level. (After Talley and McCartney, 1982.)

286 The NAO index is based on the difference between Icelandic low pressure and Azores high pressure.

On average, about 3 to 7 Sv of Labrador Sea Water is formed. With free communication with the rest of the Atlantic Ocean, it spreads out in three principal directions [**Figure 3-106**]. One branch is taken up in the cyclonic circulation between the Irminger Sea and the Labrador Sea [**Figure 3-100**] where Labrador Sea Water may be identified by a salinity that is slightly less than that of Denmark Strait Water, and by high concentrations of oxygen, between 200 m and 2,000 m [**Figure 3-101**]. One branch extends eastward into the Eastern Basin passing preferentially through the Gibbs Fracture Zone. One branch is entrained southward in the deep western boundary current. It is one of the major components of North Atlantic Deep Water.

6.9 North Atlantic Deep Water

North Atlantic Deep Water is the special case of a water mass that does not form at the surface, as do most water masses but which is the result of a mixture of several water masses, themselves formed at the surface. The θ-S diagram [**Figure 3-107**] indicates the principal water masses involved in the formation of North Atlantic Deep Water.

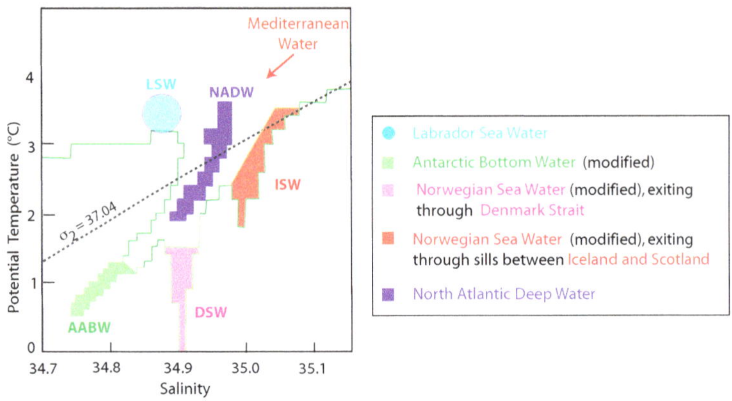

FIG. 3-107 – θ-S diagram of the principal water masses involved in the formation of North Atlantic Deep Water. The isopycnal with respect to 2,000 m (σ_{2000}) is shown. (After Swift, 1984.)

In addition to the classic temperature and salinity tracers, oxygen levels[287], CFCs, tritium, neodymium, etc., allow three slightly different layers to be distinguished in the interior of this water mass. These are Lower

[287] Oxygen and nutritive salts are not conservative, but are often used to trace water masses at depth.

North Atlantic Deep Water (LNADW), Central North Atlantic Deep Water (CNADW) and Upper North Atlantic Deep Water (UNADW).

Lower North Atlantic Deep Water (8 Sv), coldest and most oxygenated, is a mixture of:

- Denmark Strait Water (DSW) (3 Sv) characterized by oxygen and CFC maxima between 3,500 m and 3,900 m [**Figures 3-97, 3-101** and **3-103a**];
- Subpolar Waters (3 Sv) entrained at the Denmark Strait outlet, which are Labrador Sea water and Arctic Intermediate Water formed in the Subarctic Front region south of the Labrador Sea;
- Antarctic Bottom Water (AABW) much altered, coming from the south along the western boundary of the Mid-Atlantic Ridge, characterized by an oxygen minimum (and a maximum of silicates).

Central North Atlantic Deep water (6 Sv) (1,900 m to 3,500 m), characterized by an oxygen maximum around 2,500 m, is a mixture of:

- Water spilling over the sills between Iceland and Scotland (ISW) (2 Sv);
- Subpolar Waters (2 Sv) entrained at the exits of sills, which are the lower part of Labrador Sea Water and Irminger Current Water;
- Eastern Basin Deep Water (2 Sv) coming from the south, composed of the upper part of Antarctic Bottom Water coming through the Romanche and Vema Fracture Zones and of Antarctic Circumpolar Deep Water characterized by a minimum of CFCs and a minimum of oxygen.

Upper North Atlantic Deep Water (4 Sv) (1,200 m to 1,900 m), characterized by a salinity maximum around 1,700 m, is a mixture of:

- Labrador Sea Water (LSW) with a maximum of CFC and a maximum of oxygen;
- Mediterranean Water (MW) with a salinity maximum;
- Irminger Current Water;
- Upper Circumpolar Deep Water (without CFCs) whose influence is felt increasingly southward, its region of origin.

North Atlantic Deep Water is formed through mixing in the Western Basin, south of about 50°N, and is entrained by the deep western boundary current, where its characteristics are most marked. The θ-S diagrams in the Western Basin [**Figure 3-108**] and in the Eastern Basin [**Figure 3-109**] allow its evolution to be followed in each basin. With the appearance of the salinity minimum of Antarctic Intermediate water around 20°N, (coming from the south) North Atlantic Deep Water acquires its distinctive sign of a deep salinity maximum. The Eastern Basin is characterized by the marked influence of Mediterranean Water.

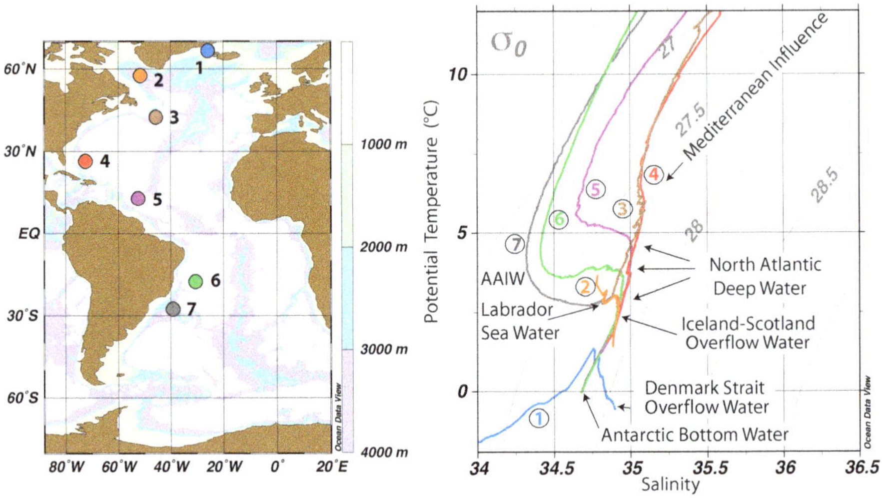

FIG. 3-108 – Deep θ-S diagrams from stations in the Atlantic western basin, from the Denmark Strait, Station 1, to 30°S, Station 7, showing the water masses forming North Atlantic Deep Water and the evolution of its salinity maximum consisting of Mediterranean Water, Labrador Sea Water, Denmark Strait Overflow Water, Iceland-Scotland Overflow Water, and Antarctic Bottom Water. [Antarctic Intermediate Water (AAIW).] (With ODV and World Ocean Database.)

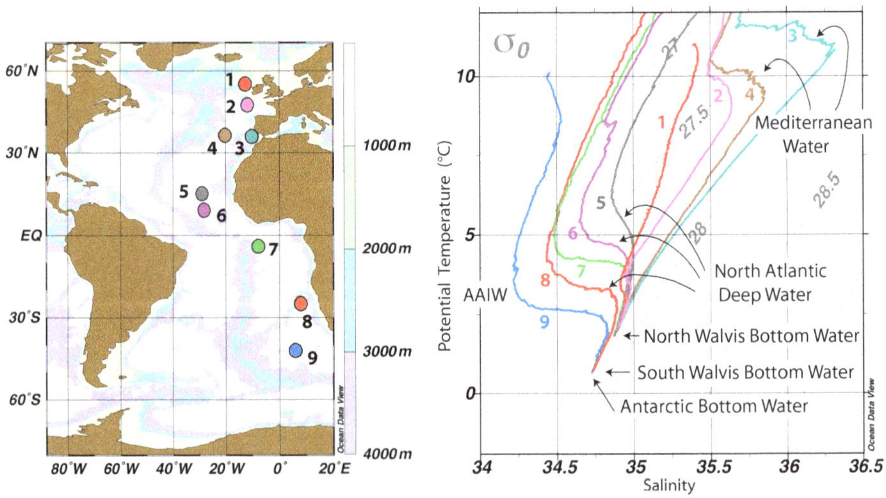

FIG. 3-109 – The evolution of North Atlantic Deep Water in the eastern basin under the large influence of Mediterranean Water. (With ODV and World Ocean Database.)

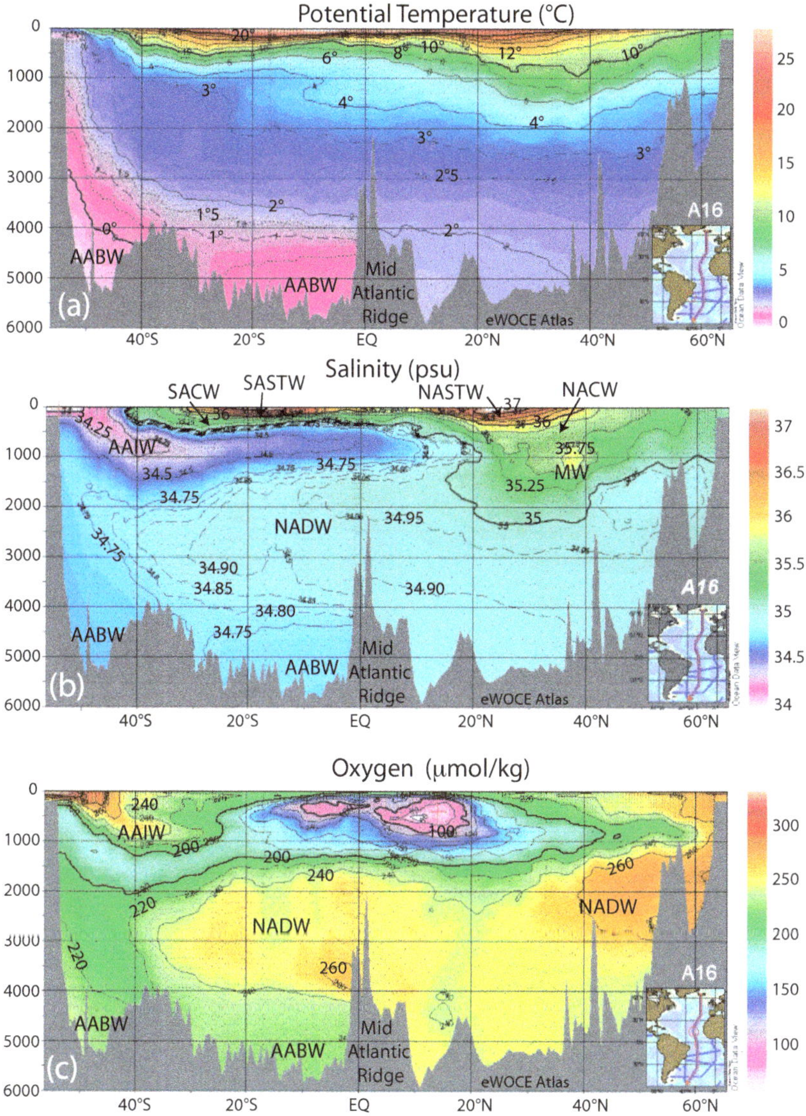

FIG. 3-110 – Atlantic meridional sections from south of Iceland to the equator (in the Eastern Basin) and from the equator to the Antarctic (in the Western Basin). (**a**) temperature, (**b**) salinity (**c**) dissolved oxygen. (Section A16, *eWOCE Atlas*, Schlitzer, 2002.)

Temperature, salinity, and oxygen concentration sections [**Figure 3-110**] show the evolution of characteristics, from south of Iceland to the equator across the Eastern Basin, then in the Western Basin as far as the Antarctic Ocean. To the north, water rich in oxygen, between 1,200 m and 3,000 m, shows the influence of Iceland-Scotland Overflow Water, of Labrador Sea

Water and of Arctic Intermediate water, young waters formed by convection. Around 38°N, the core of high salinity around 1,000 m depth ($S > 35.75$) indicates the influence of Mediterranean Water, which limits the influence of Antarctic Intermediate Water around 20°N, characterized by a salinity minimum at the same depth. South of the equator, in the Western Basin, the presence of Antarctic Bottom Water, whose region of formation near the Antarctic Continent may be seen, is shown by a decrease of temperature, salinity, and oxygen at the bottom. Only the upper part of Antarctic Bottom Water mixed with Lower Circumpolar Deep Water flows toward the Eastern Basin through the Romanche Fracture Zone. The path of North Atlantic Deep Water may be followed, characterized by a salinity maximum between 1,500 m and 3,000 m. Its salinity decreases through mixing with Antarctic Intermediate Water above and Antarctic Bottom Water below. The oxygen level of North Atlantic Deep Water decreases slowly southward but remains greater than that of surrounding waters as far as about 40°S, the region where it mixes strongly with less saline and less oxygenated Antarctic Circumpolar Deep Water [**Section II.6.3.1**].

In the Atlantic sector, Circumpolar Deep Water tends to infiltrate northward into North Atlantic Deep Water at different levels. The table in **Figure 3-111** provides a résumé of the various water masses, intermediate and deep, encountered in the South Atlantic.

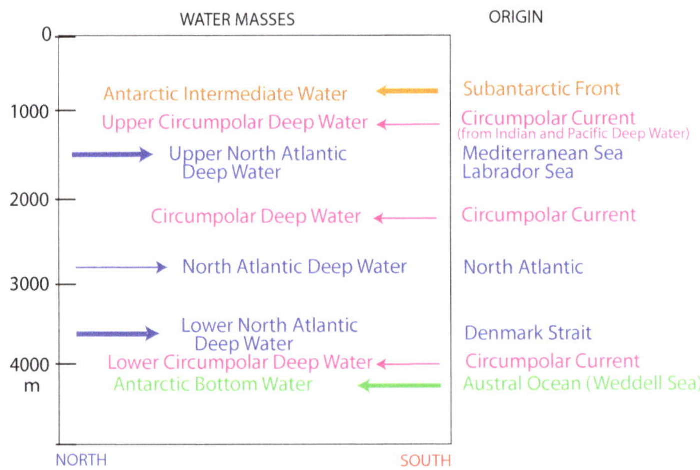

FIG. 3-111 – Summary table, showing directions and origins of South Atlantic water masses between 1,000 m and the bottom.

The deep "western boundary" current which entrains the various components of North Atlantic Deep Water undergoes strong mixing in its upper levels, when it passes under the Gulf Stream around Cape Hatteras.

At 7°30′N the two maxima of CFC, between 1,200 m and 1,900 m and between 3,500 m and 3,900 m in the western part of the section [**Figure 3-112**] represent the influence of Labrador Sea Water, in Upper North Atlantic

Deep Water, and of Denmark Strait Water, in Lower North Atlantic Deep Water. They are younger than the surrounding water masses. The maxima are most marked in the west, since the circulation is intensified on the western boundary of basins.

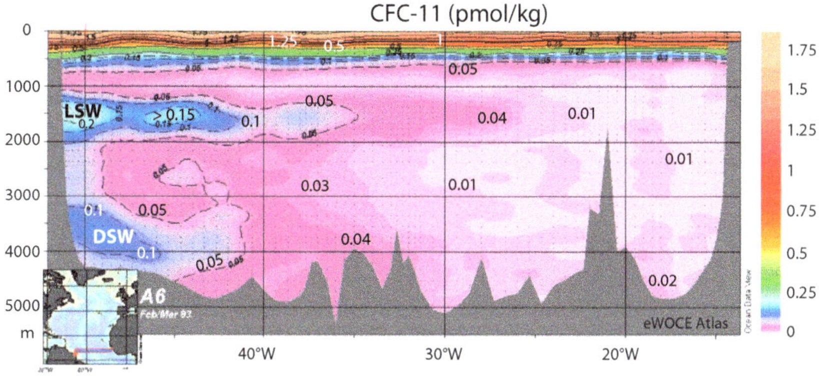

FIG. 3-112 – CFC-11 concentration (in pmol kg^{-1}) at 7°30′N in the Atlantic (After Andrié *et al.*, 1998.) (Section A6, *eWOCE Atlas*, Schlitzer, 2002.)

At the equator, the deep current splits into two branches. One continues southward, along the American continent; the equatorial zonal circulation entrains the other eastward. CFC measurements in the equatorial region [**Figure 3-113**] show an eastward entrainment of the tongue of water rich in CFC. Repeated measurements 10 years later show a progression that allows the velocity of propagation to be estimated at 1–2 cm s^{-1}.

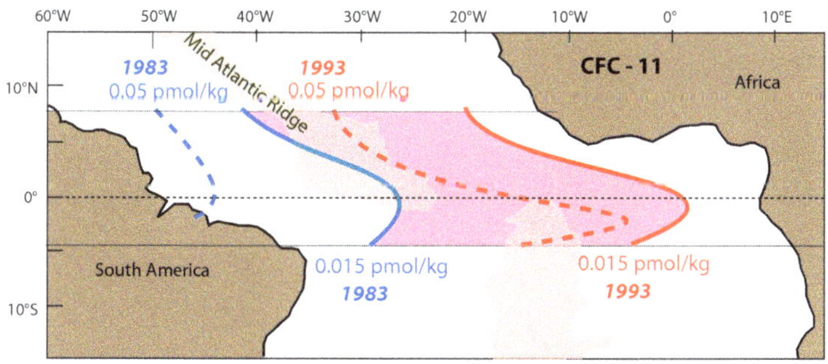

FIG. 3-113 – CFC-11 concentration (0.015 and 0.05 pmol kg^{-1} isopleths) in the 1,600 m to 1,800 m layer in the equatorial Atlantic, measured in 1983 and 1993. (After Andrié, 1996.)

A North Atlantic Deep Water salinity maximum at 11°S, with a width of about 500 km, is found on the western boundary, between 1,600 m and 2,200 m, associated with a slight temperature maximum around 1,400–1,500 m [**Figure 3-97**]. Two richer levels can be seen (2,200 m and 3,500 m) on the oxygen concentration section, corresponding to CFC maxima at 7°30′N representing Labrador Sea Water and Denmark Strait Water influence.

In the South Atlantic, some North Atlantic Deep Water follows the western boundary. From the western boundary, one branch is entrained eastward along the equator and returns westward around 10°S. A second branch turns eastward in the region of the Vitoria-Trindade Ridge around 20°S, passes the Mid-Atlantic Ridge, and rejoins the eastern boundary through three deep passages across the Walvis Ridge, entering the southern Indian Ocean *via* the Cape Basin [**Figure 3-114**]. The branch that follows the western boundary to the confluence with the Malvinas Current turns eastward to join the Antarctic Circumpolar Current to form Circumpolar Deep Water.

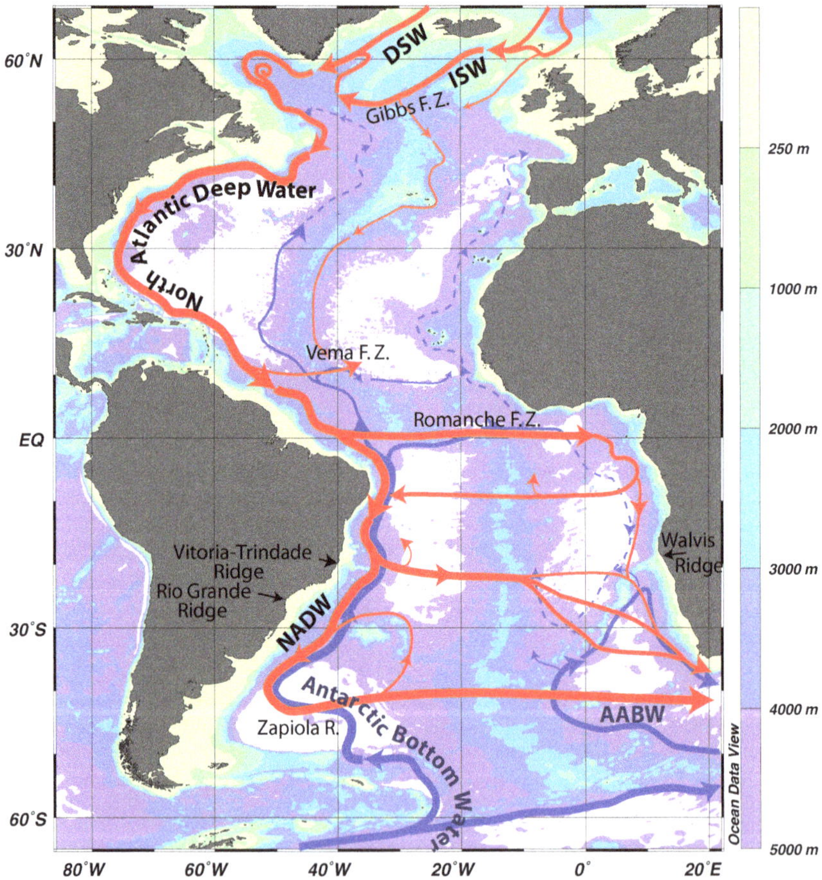

FIG. 3-114 – Schematic Circulation of North Atlantic Deep Water (in red) and Antarctic Bottom Water (in blue). (After Andrié *et al.*, 1998; Arhan *et al.*, 1998, 2003; Bourlès *et al.*, 2003; England, 2001; Gouriou *et al.*, 2001; McCartney, 1993; Mémery *et al.*, 2000; Mercier and Speer, 1998; Peterson and Whitworth, 1989; Reid, 1994; Schmitz and McCartney, 1993; Schott *et al.*, 2003; Speer and McCartney, 1991; Stramma and England, 1999; Stramma and Schott, 1999; Warren, 1981.)

Chapter 7
Water properties of adjacent and epicontinental seas

7.1 American Mediterranean (Caribbean Sea)

Located in the tropical zone, the American Mediterranean is a region where surface temperature is high and varies little. The Caribbean receives an important part of the waters of the Orinoco and Amazon,[288] transported by the Guyana current, and the Gulf of Mexico receives the Mississippi.[289] High surface temperature and the NE trade winds produce evaporation that reaches more than 1.5 m annually [**Figure 1-21**]. The excess of evaporation over fresh-water inputs leads to a slight increase in salinity which, averaged over the 0–200 m level, goes from 36.09 at the southeast entry to 36.19 in the Yucatan Strait and to 36.39 in the Florida Straits. Nevertheless, values are relatively low compared to subtropical salinities in the Atlantic Ocean, which exceed 37.2 [**Figure 3-85b**]. Because of the high surface temperature, stratification is strong year-round and there is no formation of any specific water.

As for the circulation, the oceanographic properties of the American Mediterranean are strongly constrained by basin bathymetry [**Figure 3-20**]. The entry of various water masses from the Atlantic Ocean depends on sill depths. North and South Atlantic Subtropical Water, North Atlantic Central Water (including "eighteen-degree water") mixed with South Atlantic Central Water, and the remainders of Antarctic Intermediate Water are all found under Surface Water. The minimum salinity value of Antarctic Intermediate Water increases from its entry in the southeast (34.70) to the Gulf of Mexico (34.85) through mixing with adjacent waters. The upper part of North Atlantic Deep Water penetrates through the Windward Passage, between Cuba and Hispaniola, and the Anegada-Jungfern Passage, east of

---

288 Which drain northern South America (about 0.2 Sv).

289 Which drains a large part of the United States (about 0.02 Sv).

Puerto Rico, the two deepest passages. The section from Puerto Rico to the Venezuela Basin shows that deep-water temperatures are always greater than 3.7 °C (even below 5,000 m) because of the blockage of colder Atlantic water by sills [**Figure 3-115**]. Below 2,000 m, water is homogeneous. This particular structure depends on the sill depths of the Antilles Arc.

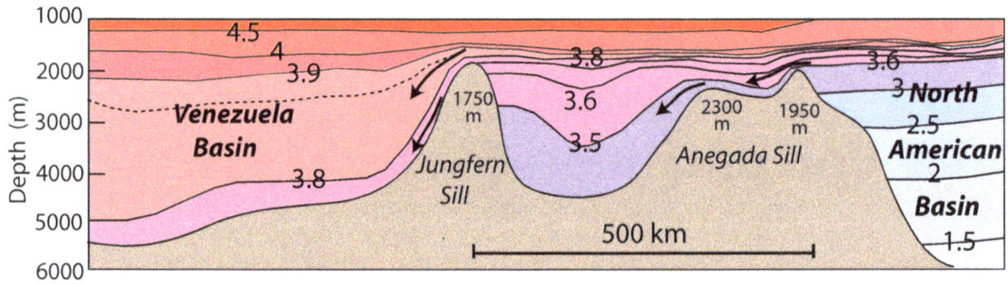

FIG. 3-115 – Temperature section from 1,000 m to 6,000 m between the North American Basin and the Venezuela Basin through the Anegada-Jungfern Passage east of Puerto Rico (see **Figure 3-20**). (After Wüst, 1963.)

7.2 Some epicontinental seas

We will give here only some examples of epicontinental[290] seas having a large extension in the northern Atlantic Ocean [**Figure 3-3**], but relatively little influence on the oceanographic aspects of the Atlantic Ocean. In these shallow seas, currents are mainly due to tides, which generate strong mixing over the entire water column.

7.2.1 The North Sea

The North Sea lies between Great Britain, Norway, and northeastern Europe. Its depths vary between 40 m in the south and 150 m in the north, with the exception of a deep of a few tens of kilometers width along the Norwegian coast, whose depths vary between 300 m and 700 m[291] and a shallow depth of 13 m (Dogger Bank), feared by fishermen during winter storms. The North Sea receives water from the English Channel and especially, *via* its wide opening to the Norwegian Sea, from a branch of the Norwegian Current with relatively saline water [**Figure 3-116**]. It receives lower-salinity water from the Baltic Sea *via* the straits of Skagerrak and Kattegat, which then flows northward, hugging the Norwegian coast [**Figure 3-72**].

[290] Located on the continental shelf.

[291] Submarines hid there during the war.

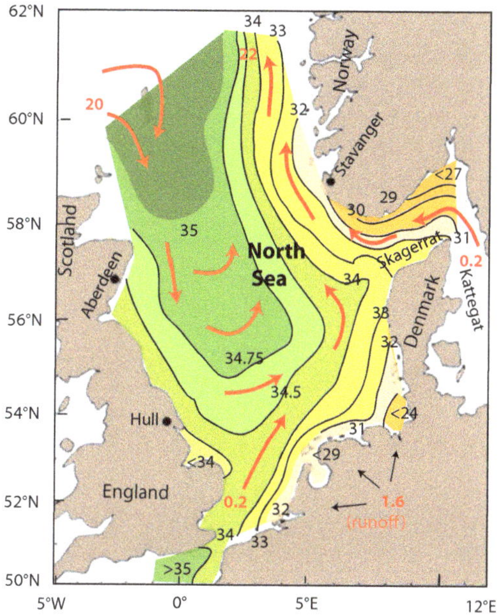

FIG. 3-116 – Surface salinity and average circulation in the North Sea. Transports are in red, in 10^3 km^3 per year, equivalent to 0.03 Sv. (After Goedecke *et al.*, 1967.)

7.2.2 The Baltic Sea

The Baltic Sea is a sea with a shallow sill, which receives much fresh water from run-off since it drains a vast region [**Figure 3-117**].

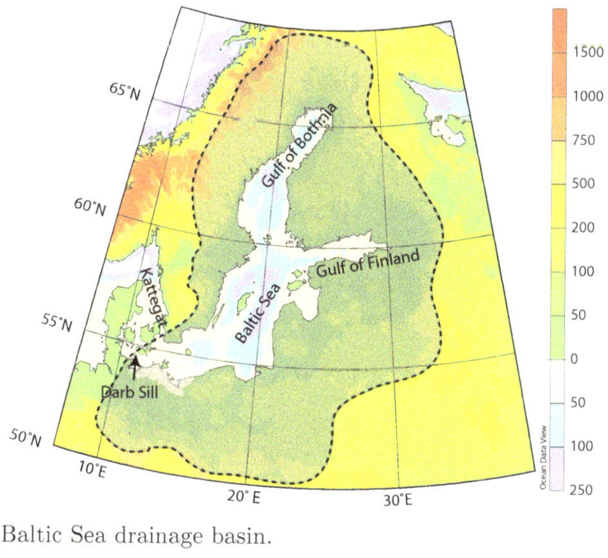

FIG. 3-117 – Baltic Sea drainage basin.

The salinity section over the Darb Sill, connecting the Kattegat to the North Sea, shows an outflow of brackish water and an inflow on the bottom of saline water from the North Sea, rapidly eroded by mixing at the sill [**Figure 3-118**]. Transport across the sill is of the same order as that of the Bosporus but the Baltic Sea has only 4% of the volume of the Black Sea. Renewal by saline and oxygenated water from the North Sea occurs intermittently. When there has not been an inflow[292] for several months or even several years, the water below 120–150 m becomes anoxic[293] but does not remain so continually as in the Black Sea.

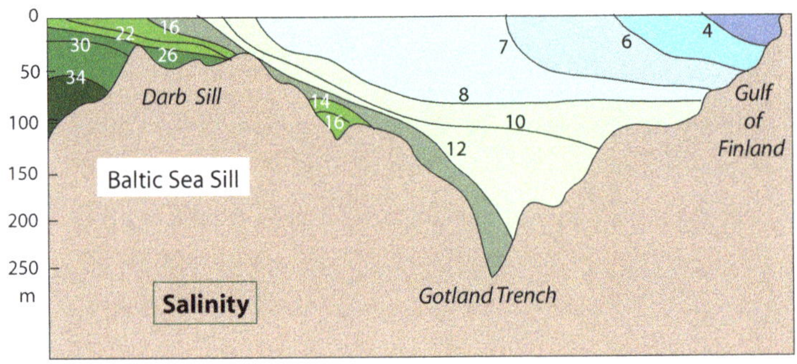

FIG. 3-118 – Salinity variation in the Baltic Sea from the Gulf of Finland to the Darb Sill, south of the Kattegat. (After Matthaüs and Frank, 1990.)

The vertical salinity gradient creates a strong stratification, which limits winter convection to about 70 m. In the brackish surface layer seasonal temperature variations reach more than 20 °C [**Figure 3-71**], leading to frequent winter ice formation. The bottom temperature of about 5.5 °C varies little.

The Baltic Sea is a dilution basin, which supplies fresh water to the North Sea and to the Norwegian Sea along the coast on the surface. Baltic Sea Water dilutes in the Atlantic Water of the Norwegian Current.

7.2.3 Hudson Bay

Hudson Bay has an average depth of only 250 m and receives considerable fresh water from seasonal river inflow. It is ice-covered for much of the year and functions as a large estuary, in the interior of which the circulation is cyclonic [**Figure 3-119**].

292 Intermittent inflow is controlled by winds.

293 In anoxic conditions, oxygen is replaced by hydrogen sulfide.

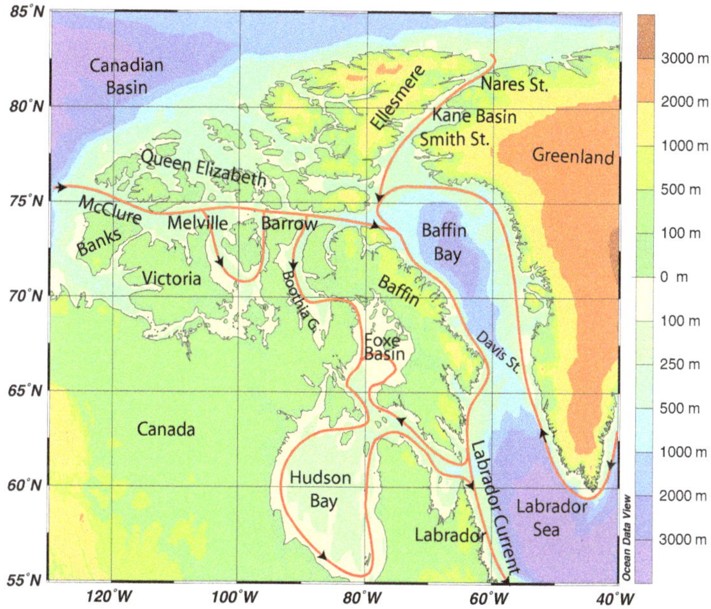

FIG. 3-119 – Bathymetry and general circulation in the Canadian Arctic Archipelago and Baffin Bay. (After Prinsenberg, 1986.)

7.2.4 Baffin Bay

Baffin Bay is a deep basin reaching more than 2,300 m in its center, separated from the Labrador Sea by Davis Strait (600 m). It receives water from the Arctic to the north, through Nares Strait, Kane Basin, and Smith Strait (sill around 200 m), and to the west, through the Straits of McClure, Melville, Barrow, and Lancaster (sills around 150–200 m). The principal source is to the south, from the West Greenland Current, which transports cold and brackish water at the surface and warmer water at the subsurface coming from the Irminger Current [**Figure 3-120**] around 500 m. Deep Water of Baffin Bay below 1,800 m (θ ~ 0.4 °C, S ~ 34.49) has an oxygen concentration of only 3.6 ml l^{-1}, 155 μmol kg^{-1}. It is a dilution basin whose renewal below the level of the sill occurs slowly.

The passages between the Canada Basin and Baffin Bay are ice-bound for most of the year. However, in some regions, violent north winds can sometimes push ice offshore forming polynyas and creating upwelling regions where warmer subsurface water prevents ice from re-forming.[294]

[294] Explorers searching for the Northwest Passage used these ice-free zones.

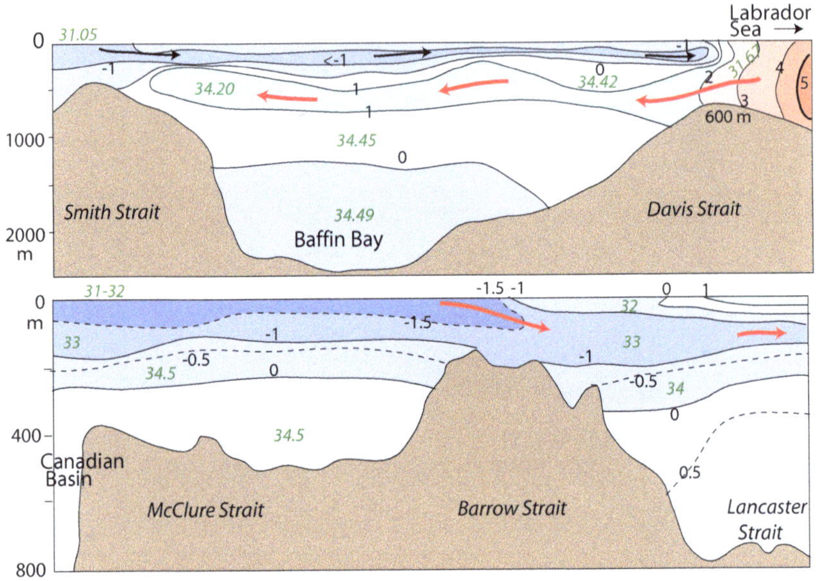

FIG. 3-120 – Temperature section with salinity values (**a**) across Baffin Bay from Smith Strait to the Labrador Sea, (**b**) across a part of the Northwest Passage from the Canadian Basin to Lancaster Strait. Arrows show water-mass movements. (After Collin, 1966.)

Chapter 8 Distinctive features of the Atlantic Ocean

The Atlantic Ocean, of great meridional extent, communicates with the two polar basins, but topography limits its north polar influence. The Mediterranean Sea strengthens salinity throughout the North Atlantic. The Mid-Atlantic Ridge, by blocking the circulation of bottom water, accentuates the asymmetry between east and west due to the effect of Earth's rotation, which intensifies currents on the western boundary of basins. Surface circulation and deep circulation contribute to strong meridional exchanges, induced partly by intense vertical motion in the Labrador Sea and in the Nordic Seas, the principal sources of the world's deep water. The thermocline separates deep and cold water masses ($\theta < 5$ °C) from warmer water masses between the subtropical fronts sketched in **Figure 3-121**. Variations in thermocline depth are linked to variations in Ekman pumping due to wind. After leaving the surface, water masses evolve under the effect of mixing by diffusion (diapycnal or isopycnal mixing), by internal waves, including internal tidal waves, and by turbulence generated at the seafloor.

The favored region for transformation of Surface Water into Deep Water, the Atlantic Ocean is a key element in the thermohaline circulation of the planetary ocean.

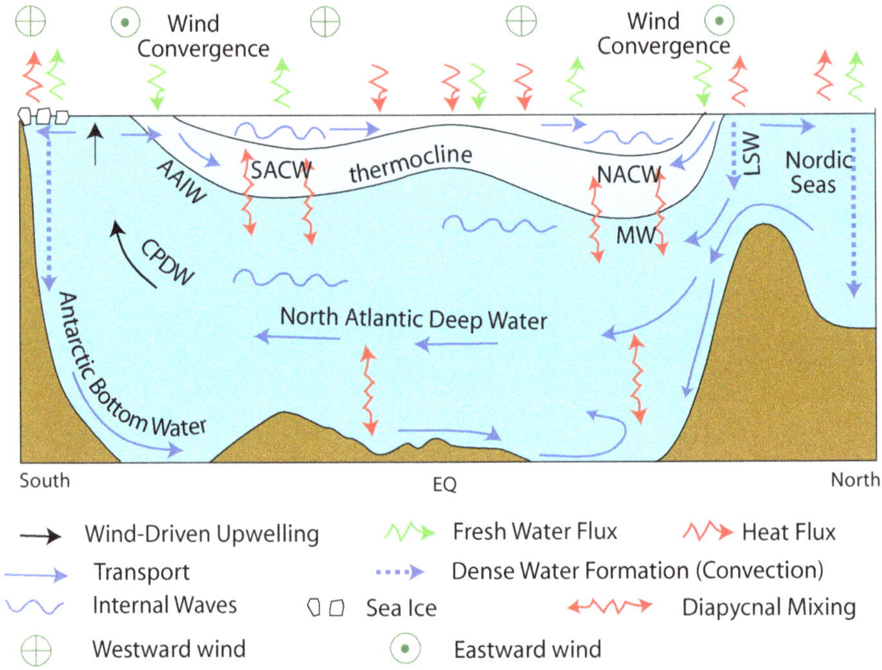

FIG. 3-121 – Thermohaline circulation of the Atlantic Ocean up to Antarctica with fluxes of heat and fresh water according to regions and mixing processes. [South Atlantic Central Water (SACW), North Atlantic Central Water (NACW), Circumpolar Deep Water (CPDW), Antarctic Intermediate Water (AAIW), Labrador Sea Water (LSW), Mediterranean Water (MW).] (After Kuhlbrodt *et al.*, 2007.)

PART IV
The Indian Ocean

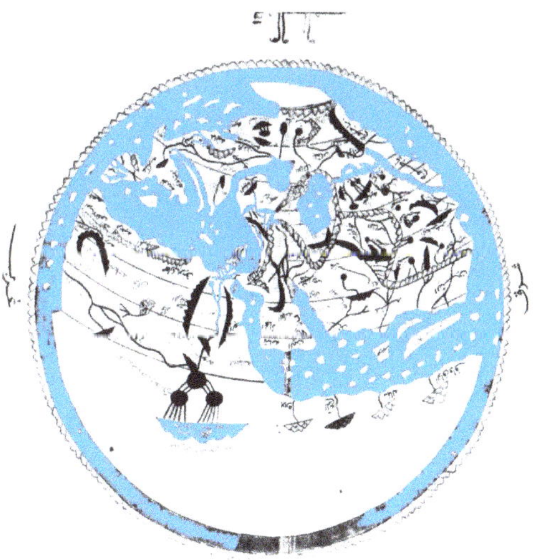

FIG. 4-1 – *Al Idrisi planisphere* of 1154. (After Aleem, 1980.)

Chapter 1
Introduction

The Indian Ocean is special and fascinating for many reasons. It was already known in the time of Egyptian Queen Hatshepsut (15th century BCE), who sent an expedition as far as the Somali[295] Coast. It can be found on maps of the Alexandrian, Claude Ptolemy, in the 2nd century [**Figure 3-1**]. Over the centuries Arab merchants (from Oman, Arabia, Persia...) traversed the Indian Ocean to trade with India, China, the African east coast, and islands such as Zanzibar, Madagascar, Sri Lanka, Sumatra, and Java. Their knowledge was compiled on maps. In 1154, Al Idrisi, an Arab scholar of the court of the Norman King Roger II of Sicily, drew a round Earth [**Figure 4-1**] revealing the Mediterranean as far as Gibraltar, the Black Sea, the Caspian Sea, the great lake at the source of the Nile, the Red Sea, the Persian Gulf, the northern Indian Ocean strewn with islands, China, and the Indonesian islands, as well as oceanic continuity south of Africa and north of Asia. The Chinese made numerous and forceful incursions in the reign of the Mongol emperors. In the 15th century, Admiral Cheng Ho led a fleet of sixty giant ships around the Indian Ocean, to demonstrate Chinese power. Merchant mariners already knew how to profit from the seasonally reversing winds of the monsoon,[296] which dominate a large part of this ocean, giving it its special character. In 1487 Bartolomeu Dias rounded Cape of Good Hope in a storm and was the first to enter the Indian Ocean by the south, opening the way to the Spice Islands.

Far from the first large centers of oceanographic study, the Indian Ocean was still poorly known oceanographically until the 1960s. The veil began to be lifted with the International Indian Ocean Expedition from September 1959 to December 1965, which involved 44 ships from 25 nations. Subsequently, many observations have been made, coordinated in the framework of large international programs (INDEX: Indian Ocean Experiment; TOGA: Tropical Ocean and Global Atmosphere; WOCE: World Ocean Circulation Experiment; CLIVAR: Climate Variability and Predictability; GOOS: Global Ocean Observing System).

---

295 To the Kingdom of Punt.

296 The word *monsoon* comes from the Arab *mawsin* meaning "season".

Chapter 2
Geographic characteristics

The Indian Ocean is closed to the north, around the Tropic of Cancer, by the continental mass of Eurasia [**Figure 4-2**]. It is largely open to the south on the Antarctic Ocean, as are the Atlantic and Pacific Oceans. The Subtropical Front, located around 40°S, acts as a southern limit [**Section II.5.2.1**].

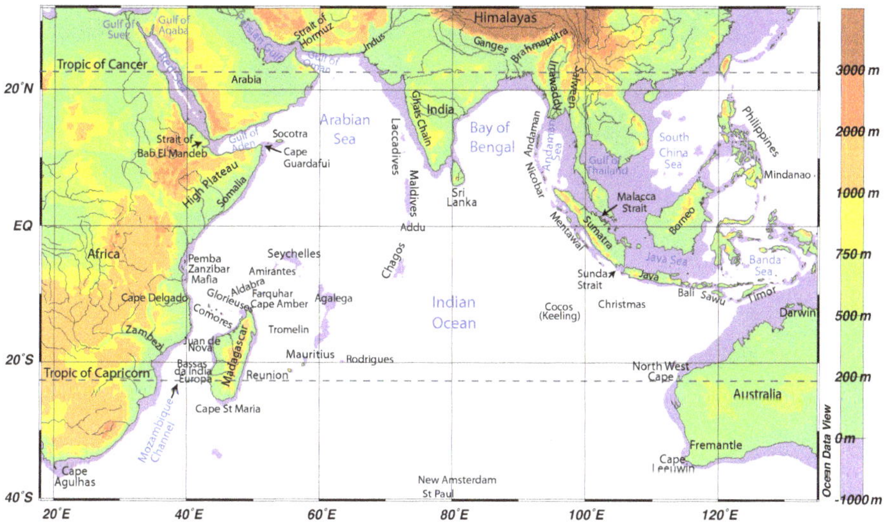

FIG. 4-2 – Map of the Indian Ocean.

The Indian Ocean is bordered on the west by Africa and the Arab peninsula, on the north by the south and southeastern Asian continent, and on the east by the arc of the Indonesian Archipelago and Australia. To the north, the Indian peninsula and the island of Sri Lanka separate the Arabian Sea on the west from the Bay of Bengal to the east. The large island of Madagascar is southwest. In the northwest, two adjacent seas, the Red Sea and the Persian Gulf, connect to the Arabian Sea by, respectively, the Gulf of Aden and the Gulf of Oman. These two seas extend the geographic domain of the Indian Ocean to 30°N.

The Indian Ocean is connected to the equatorial Pacific Ocean through the Indonesian Seas, which contain a number of deep basins, of more than 7,000 m, separated by sills with a maximum depth of free communication of about 1,300 m. It is 6,800 miles from the northern limit (the Bay of Bengal) to the southern limit (40°S), and Africa is 7,800 km distant from Australia at 30°S latitude. It is the smallest of the oceans with a surface of 49×10^6 km^2 out of 361×10^6 km^2 for the world ocean. Only a fifth of its area is in the northern hemisphere.

Topographic features of note are the high plateaus of East Africa, the Arabian peninsula, the Himalayan ranges, and the high Tibetan plateaus, which all play a role in the circulation of air masses at low altitude [**Figure 4-2**].

The average depth of the Indian Ocean is 3,800 m, intermediate between that of the Atlantic (3,300 m) and that of the Pacific (4,200 m) [**Figure 4-3**]. The continental shelf is little developed except in the Persian Gulf, in northwest Australia, and on the Mascarene Plateau. Numerous Ridges divide the Indian Ocean sea floor. The principal ridges, seismically active, are part of a global network. In the southern Atlantic, the Mid-Atlantic Ridge bends eastward in the Antarctic Ocean at Bouvet Island and becomes the Southwest Indian Ridge. It supports the Marion and Prince Edward Islands and extends to a central massif called the Rodrigues Triple Point (20°S, 67°E). From the triple point the Southeast Indian Ridge takes off in the direction of the Pacific across the Antarctic Ocean (here are found the islands of New Amsterdam and St. Paul). The Central Indian Ridge goes northward and then extends northwest as the Carlsberg Ridge which connects, *via* the Gulf of Aden, with the East African continental rift to the south, and *via* the Red Sea, with the Gulf of Aqaba[297] to the northwest. In addition to this system of seismically active ridges there are numerous ridges and rises: the Ninety East Ridge, a rectilinear ridge that was discovered only in the middle of the last century, the Mascarene Plateau, which includes the granitic massif of the Seychelles, the Cargados Carajos Shoals (Saint Brandon Rocks), Agalega Islands and the Saya de Malha Bank, the Chagos-Laccadive Ridge, the South Madagascar Plateau, the Natal or Mozambique Plateau, and the Broken Ridge at 32°S off Australia. There are volcanic islands such as La Réunion (whose volcano is still active), Mauritius, Rodrigues and Comoros, coral islands such as Bassas da India, Europa, Juan de Nova, Glorioso (Glorieuses), Tromelin,[298] Farquhar, and Aldabra, and the island groups of Chagos, Maldives, Lakshadweep (Laccadive), Andaman, Nicobar and Mentawai to the east, as well as Christmas and Cocos Islands.

---

[297] Prolonged on the continent by the Dead Sea and the Lake of Tiberiade.

[298] On each of these five islands a meteorological station is maintained.

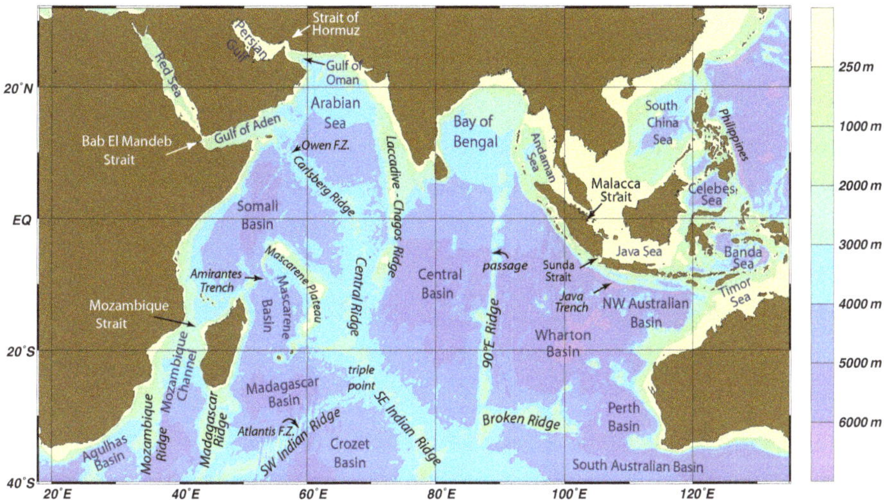

FIG. 4-3 – Indian Ocean topography, showing the separation into a number of deep basins.

This system of ridges, rises, and plateaus breaks up the Indian Ocean into numerous basins whose depths reach 5,000 to 6,000 m and even 6,500 m in the Wharton Basin northwest of Australia [**Figure 4-3**]. The greatest depth in the Indian Ocean is found in the Java Trench, which reaches 7,450 m at the boundary of the Indo-Australian Plate and the Eurasian Plate. Fracture zones across the ridges permit some deep communication, however: between the Arabian Sea and the Somali Basin by the Owen Fracture Zone (around 12°N); between the Indian Central Basin and the Northwest Australian Basin by a passage around 5°S; between the Madagascar Basin and the Crozet Basin by the Atlantis Fracture Zone; and between the Mascarene Basin and the Somali Basin by the Amirante Trench.

Chapter 3
Climatology

Climatology influences oceanic surface characteristics through exchanges of water and heat with the atmosphere, modulated by winds generated by pressure differences, which intensify the exchanges.

3.1 Atmospheric pressure and wind regime

The presence of the Asian continent around the Tropic of Cancer is the cause of an atmospheric circulation regime, north of 10°S, that is very different from that of the Pacific and Atlantic Oceans. The continental effect, namely seasonal heating and cooling of the earth that is more rapid than that of the ocean, generates an alternation of high pressure in winter and low pressure in summer over the Asian continent [**Figure 4-4**]. This contrasts with the relative stability of subtropical high pressure found over the Atlantic and Pacific at the same latitudes.

3.1.1 Pressure and winds over the southern Indian Ocean

The southern part of the Indian Ocean, south of 10°S, is relatively stable under the influence of high subtropical pressure, comparable to the other oceans at the same latitudes.

In the austral summer (January–February) subtropical high pressure centered around 35°S is weaker than in the austral winter since it is affected by low pressure over the continental masses of southern Africa and Australia, linked to summer warming of the land [**Figure 4-4a**]. The limit between the two systems north and south, represented by the intertropical convergence zone (ITCZ, a low-pressure region) is found south of the equator, between 10°S and 15°S, in contrast to the Atlantic and Pacific where the ITCZ is slightly north of the equator. The ITCZ covers a zone located between Africa (20°S), northern Madagascar, and Sumatra and extends over the Indonesian region as far as northern Australia, around 10°S on average, except in the east and west. The southeast trades blow between subtropical high pressures and ITCZ low pressures [**Figure 4-4**]. The westerlies of the Antarctic regions blow between subtropical high pressures and periantarctic low pressures. Low pressure over Australia induces a cyclonic wind circulation around Australia producing a divergence in the trade winds to the west and northwest of Australia where winds are southerly, then southwesterly [**Figure 4-4a**].

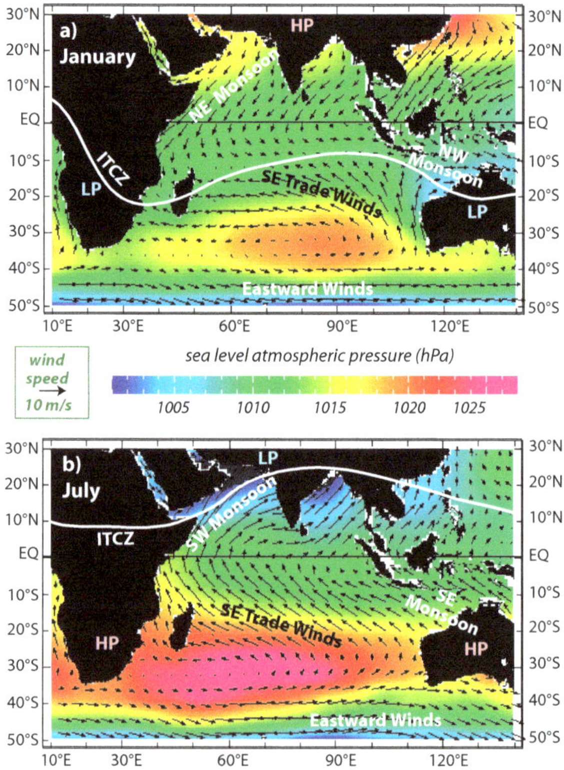

FIG. 4-4 – Average winds superposed on atmospheric pressure (**a**) in January and (**b**) in July, also showing the average position of the Intertropical Convergence Zone (ITCZ). (IRI/LDEO, and Da Silva climatological data 1994.)

In the austral winter (July–August) subtropical high pressure migrates northward (following the apparent solar movement) and strengthens, forming a continuous belt around the globe near 25°S–30°S [**Figure 4-4b**]. The ITCZ is driven far northward onto the Asian continent. When subtropical high pressure is highest in the austral winter, the trades and the periantarctic westerlies [**Figure 2-8**] are most intense.[299] In the western part of the basin, southeast trades blow as far as the equator [**Figure 4-4b**]. North of Australia, the site of high pressure at this season, winds are southeast (the southern hemisphere southeast monsoon) and reinforce the southeast trades.

[299] The southeast trades of the Indian Ocean are the strongest of all southeast trades.

3.1.2 Pressures and winds over the northern Indian Ocean

In the boreal winter (January, February), intense cold over the high Tibetan plateaus and Siberia induces the highest pressures over the globe and the ITCZ is forced south of the equator [**Figure 4-4a**].

In the boreal summer (July, August), warming over southern Asia induces a low-pressure zone centered over Arabia, Pakistan, and northern India [**Figure 4-4b**].

Thus, north of 10°S, the seasonal seesaw of pressure over the continent [**Figure 4-4**] generates a complete reversal of winds: this is the seasonal alternation of the monsoon.[300] The variability of the atmospheric circulation is such that a description requires a division into four periods.

a) December to March

From December to March, pressures are elevated over the Asian continent [**Figure 4-4a**]. Winds blow from the land to the sea, turning around high pressure, giving the northeast monsoon. Continental high pressures are centered north of the Himalayas. Winds are moderate, with a maximum of only 6 to 8 m s^{-1} along the Somali coast. Air masses coming from the continent are dry for India. In contrast, it is the humid monsoon for some islands such as the Seychelles, located around 5°S, since the air masses are charged with humidity over the Arabian Sea and the Somali Basin before reaching them.

At this period the wind system over the Indian Ocean is similar to that of the Atlantic or Pacific. The northeast monsoon is the analog of the northeast trades. However, the intertropical convergence zone is offset from 12° to 13° southward since the anticyclone (high pressure) is much stronger over Asia than over the subtropical ocean. Crossing the equator,[301] the northeast monsoon winds become winds from the northwest, which blow in particular over northern Australia and over southern Indonesia. The meridional component of Ekman transport is oriented northward on both sides of the equator.

b) April to thc bcginning of May

April to the beginning of May is an inter-monsoonal period, between the end of the northeast monsoon and the beginning of the southwest monsoon, during which the winds are weak or non-existent. At the equator, a light gradient of pressure between Africa and Indonesia generates moderate[302] westerly winds.

[300] One may crudely compare the monsoon regime to an immense system of "land breeze-sea breeze", analogous to the response of land to diurnal variations of solar radiation, on the scale of a season and over an oceanic basin under Coriolis force.

[301] The direction of the Coriolis force changes at the equator.

[302] These moderate winds are in contrast to those in the Atlantic and Pacific Oceans, which come from the east at the equator. As the Coriolis force is zero, winds blow in the direction of the pressure gradient.

c) June to September

From June to September the monsoon dominates from the southwest.[303] Winds blow violently and regularly north of the equator, particularly along the Somali coast and in the Arabian Sea [**Figure 4-4b**]. The high plateaus of Ethiopia channel the lower part of the air mass and intensify the atmospheric circulation along the Somali coast, where winds, often referred to as the Findlater Jet, regularly exceed 30 to 40 knots (15 to 20 m s^{-1}) off Cape Guardafui. This phenomenon is similar to the intensification of currents on the western boundaries of oceans, due to the presence of high plateaus and to the latitudinal variation of the Coriolis force. During this season, there is continuity between the southeast trades of the southern Indian Ocean and the southwest monsoon in the western equatorial zone. Wind direction changes from southeast to southwest crossing the equator. The meridional component of Ekman transport is directed southward on both sides of the equator. Air masses entrained by the southwest monsoon take up water vapor along their trajectory over the ocean and release it over India. This is the humid monsoon (the "real" monsoon for India). In the central and eastern equatorial zone winds are much weaker with a principal south-north component [**Figure 4-4b**].

d) October to the beginning of November

October to the beginning of November is the second inter-monsoonal period, during which winds are weak or nonexistent with only moderate westerly winds at the equator, as during the first inter-monsoonal period.

The limit of the influence of the monsoon regime on average is around 10°S. It reaches 20°S in the Mozambique Channel and along northern Australia because of seasonal pressure changes over southern Africa and Australia [**Figure 4-4b**].

Since the 1980s data-based climatic assimilation models using observations obtained by merchant ships and satellites reproduce the wind-stress[304] field and the associated wind curl, permitting studies of the oceanic response to atmospheric forcing. The wind-stress field [**Section I.5.3**] obtained this way shows the difference between the two seasons [**Figure 4-5**]. The southwest monsoon stress is much higher, especially along the Somali coast, than that of the rather moderate northeast monsoon. In the southwest monsoon [**Figure 4-5b**] wind curl is negative (anticyclonic) over the Arabian Sea while it is positive (cyclonic) in the northeast monsoon. The inverse occurs over the Bay of Bengal. Ekman pumping may be calculated by dividing the wind stress by f, the Coriolis parameter, and obtaining the curl of this

[303] The onset of southwest winds along the Somali coast varies from year to year from the beginning of May (early monsoon) to the beginning of June (late monsoon). Onset may occur suddenly or progressively.

[304] Wind stress is proportional to the square of the wind speed [**Section I.5.2**].

term (curl τ/f) [**Section I.5.3.1**]. Values are positive in divergence zones and negative in convergence zones. Along the Somali coast in the southwest monsoon, there is divergence at the coast and convergence offshore.

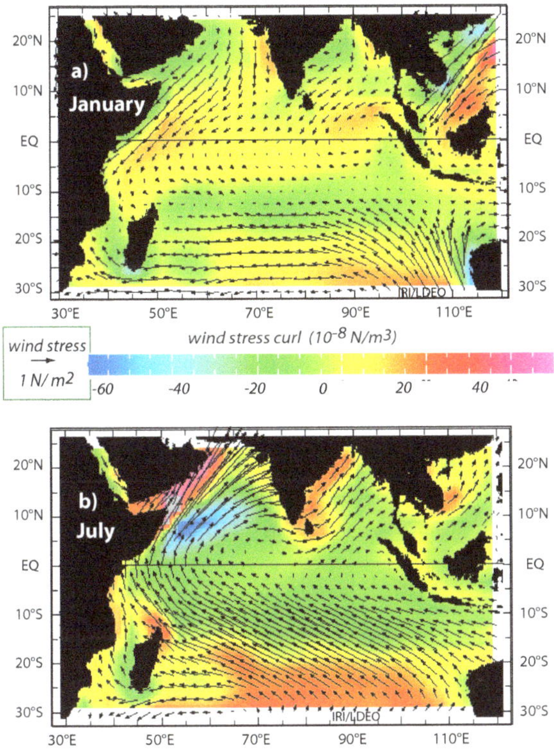

FIG. 4-5 – Mean wind stress superposed on mean wind-stress curl (downward Ekman pumping is negative (blue) in the northern hemisphere and positive (red) in the southern hemisphere, (**a**) during the NE monsoon, (**b**) during the SW monsoon. (IRI/LDEO, and monthly average climatological data, 1980–1989, Trenberth *et al.*, 1989.)

3.2 Precipitation, evaporation, heat exchange

Three quarters of the Indian Ocean are located between the tropics of Cancer and Capricorn[305], that is to say in the tropical and equatorial zone, where the sea receives a maximum of heat by solar radiation. However, the monsoon regime in its northern part imposes a strong asymmetry between the two tropical climates.

[305] Latitudes 23° 26′ N and S. The latitude of the tropics is slowly oscillating back and forth. At the moment, the tropics are moving equatorward at about 15 m per year.

The lowest air temperatures are found at the southern limit of the Indian Ocean, in the temperate zone. They are of the order of 10 °C to 15 °C according to the season. Around 30°S seasonal variations reach 6–7 °C (T_{air} is around 20 °C) and diminish in the equatorial zone where variations are only 2–3 °C (25 °C $<$ T_{air} $<$ 30 °C). Temperatures are highest in the north. They reach a maximum in summer in the two adjacent seas surrounded by deserts: the Red Sea and the Persian Gulf.

The precipitation distribution shows great spatial and seasonal variability, particularly north of 20°S [**Figure 4-6**].

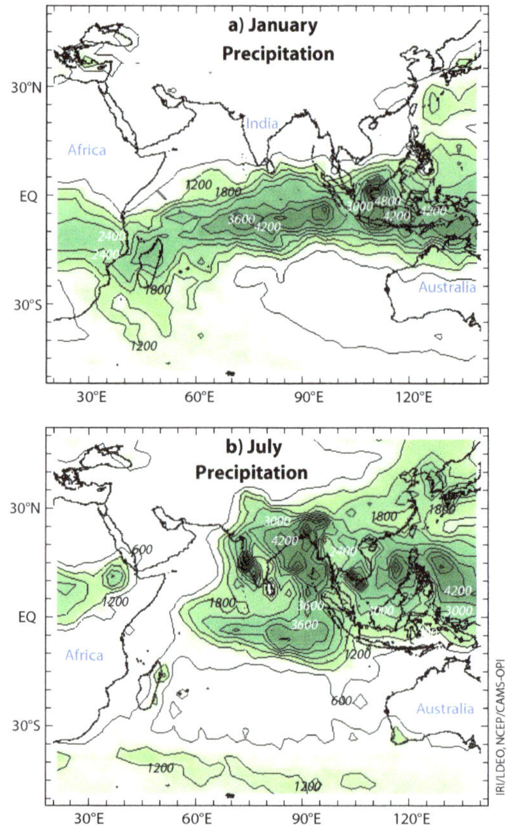

FIG. 4-6 – Precipitation **(a) in January and (b) in July** (in mm/year) averaged over 1979–1995. (IRI/LDEO, and NCEP/CAMS-OPI data.)

The region of strong precipitation follows the Intertropical Convergence Zone. During the northeast monsoon [**Figure 4-6a**] it is located between the Mozambique Channel and Sumatra-Java. The southwest monsoon season is characterized by a strong asymmetry between the NW and NE regions of the Indian Ocean. On an annual average, precipitation varies between

100 mm/year in the west (Somalia, Arabia) and more than 3,000 mm/year near Sumatra and over the Andaman Sea. Seasonal variations due to the monsoon in turn amplify this asymmetry. On the western coast of India and in the western part of the Bay of Bengal, precipitation goes from 1,500 mm/year during the northeast monsoon to 6,000 mm/year during the southwest monsoon. This distribution is the opposite of what is normally found in tropical zones where the trades, whether they are from the northeast or from the southeast, are dry near eastern coasts and humid near western coasts, being charged with humidity along their oceanic path. Strong winds of the southwest monsoon intensify evaporation over the Arabian Sea [**Figure 1-21**], while the Bay of Bengal receives fresh water from the humid summer monsoon and from the rivers[306], which drain the Indian peninsula and the Himalayas.

The classic precipitation pattern for anticyclonic regions is found in the southern part of the Indian Ocean, with less than 50 mm/year on the Australian west coast and 2,000 mm/year on the Madagascar east coast,[307] where the trades drop their humidity. The center (25°S to 30°S) and the east of the south anticyclonic gyre are regions of weak precipitation (200 mm/year) and strong evaporation, especially under the SE trades [**Figure 1-21**]. The fresh-water balance (evaporation-precipitation) here is clearly positive [**Figure 1-22**].

The fresh-water balance pattern correlates well with the pattern of precipitation, whose variability dominates over that of evaporation [**Section I.2.5, Figure 1-22**].

The heat-exchange balance with the atmosphere [**Figure 1-15**] shows that on average the Indian Ocean, of which two thirds are in the tropics, gains heat.

[306] Though the drainage basin of the Indian Ocean is relatively small [**Figure 4-2**], it is subject to strong precipitation from the southwest monsoon which feeds great rivers like the Indus in the north of the Arabian Sea, the Ganges and Brahmaputra in the northern Bay of Bengal, and the Salween and Irrawaddy in the northern Andaman Sea. The Ghats mountain chain located west of the Indian peninsula channels the majority of monsoon rains toward the Bay of Bengal.

[307] Precipitation is amplified by the high relief on the east side of Madagascar [**Figure 4-2**].

Chapter 4
Surface circulation

4.1 General circulation

The relative stability of the wind system in the southern Indian Ocean results in an anticyclonic oceanic circulation that is in turn relatively stable. On the contrary, the great variability of the atmospheric circulation north of 10°S, specific to the Indian Ocean, generates an oceanic response that, too, is specific. **Figures 4-7 and 4-8** sketch the circulation for the extremes of January and July.

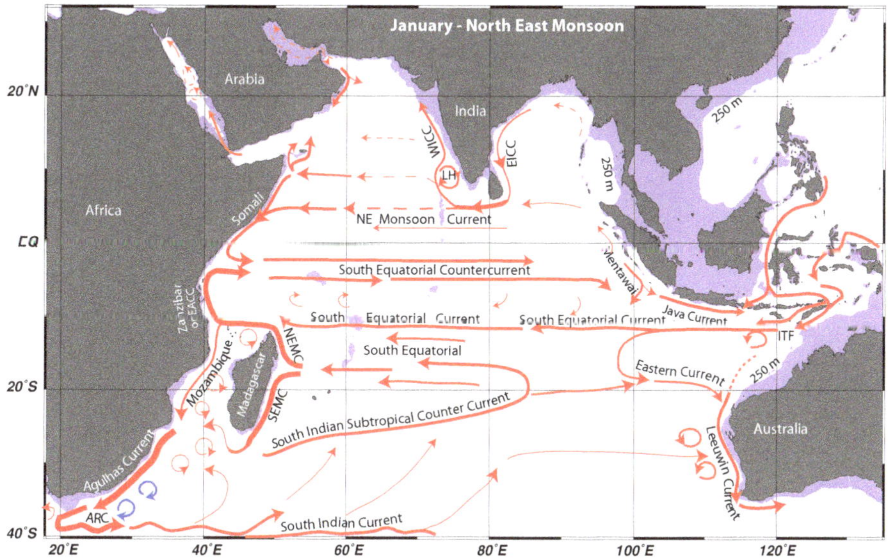

FIG. 4-7 – Indian Ocean surface circulation in January (NE monsoon). Depths less than 250 m are in violet. [Agulhas Return Current (ARC), NE Madagascar Current (NEMC), SE Madagascar Current (SEMC), Zanzibar Current (ZC) or East African Coastal Current (EACC), West Indian Coastal Current (WICC), Laccadive High (LH), East Indian Coastal Current (EICC), Indonesian Throughflow (ITF).] (After Schott and McCreary, 2001; Schott *et al.*, 2009; Siedler *et al.*, 2006.)

4.2 Circulation South of 10°S

In the southern Indian Ocean, the anticyclonic wind system drives an anticyclonic surface circulation, broken up by the presence of the island of Madagascar. Its seasonal variability is weak compared with that of the northern region [**Figures 4-7 and 4-8**].

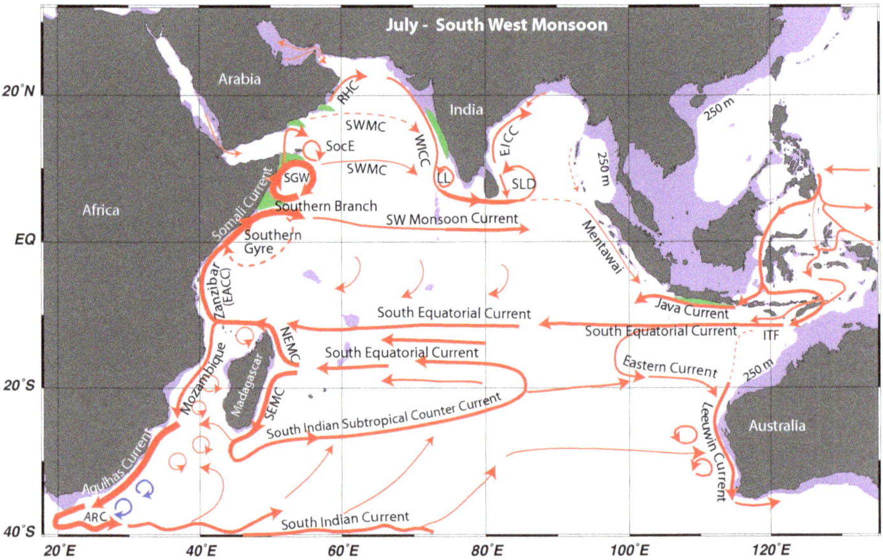

FIG. 4-8 – Indian Ocean surface circulation in July (SE monsoon). Upwelling regions are green, depths less than 250 m are violet. [See also **Figure 4-7**, Somali Great Whirl (SGW), Socotra Eddy (SocE), Ras Al Hadd Current (RHC), SW Monsoon Current (SWMC), Laccadive Low (LL), Sri Lanka Dome (SLD)]. (After Schott and McCreary, 2001; Schott *et al.*, 2009; Siedler *et al.*, 2006.)

4.2.1 South Equatorial Current

The northern branch of the anticyclonic gyre is composed of the South Equatorial Current (SEC) going westward, driven by the SE trades. In the southern hemisphere winter (June to September) when the trades are the most intense, the SEC is strongest. Its transport is of the order of 40 Sv.[308] It extends from about 20°S–25°S to 8°S–10°S. In the western Australian Basin it is joined by the current coming from the Pacific across the Indonesian Seas around 12°S (the Indonesian Throughflow, ITF), which also is at a maximum while the SE monsoon dominates the region from June to September.

[308] 40 Sv is 40×10^6 m^3 s^{-1}.

4.2.2 SE and NE Madagascar Currents

The SEC feeds the currents of the western boundary, intensified along the eastern Madagascar coast [**Figures 4-7 and 4-8**] and along the African coast. Arriving at the eastern coast of Madagascar around 17°S, the SEC divides a first time. The northern branch forms the NE Madagascar Current (NEMC) transporting 27 ± 9 Sv at Cape Amber [**Figure 4-9**]. The southern branch follows the coast southward and forms the SE Madagascar Current (SEMC) transporting 20 ± 6 Sv, with velocities that can reach 1.5 m s^{-1}. The NE Madagascar Current joins the northern part of the SEC north of Madagascar (Cape Amber) and continues westward. The SEC splits a second time on the African coast at Cape Delgado around 11°S. The northern branch forms the Zanzibar Current (ZC or East African Coastal Current, EACC) whose transport is of the order of 20 Sv.[309] The southern branch forms the Mozambique Current (15 ± 5 Sv), which displays many eddies in the channel of the same name. The Zanzibar Current flows northward year-round beyond Cape Delgado, but its northward extension varies seasonally depending on the variability of the Somali Current.

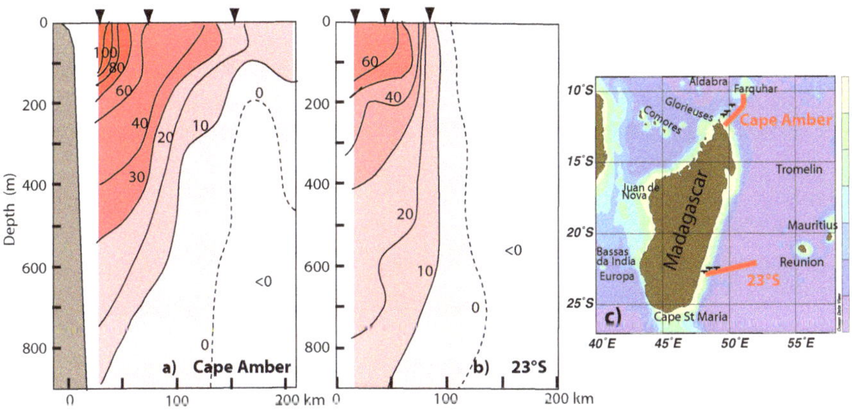

FIG. 4-9 – Geostrophic velocity sections (in cm s^{-1}) calibrated with the aid of direct current measurements (**a**) at Cape Amber, and (**b**) at 23°S. The abscissa is distance from the coast; the black triangles represent mooring positions. (**c**) Position of the two sections and the moorings. (After Swallow *et al.*, 1988.)

At the southern point of Madagascar, the SE Madagascar Current becomes unstable and emits eddies [**Figure 4-10**]. Part of its warm water undergoes a retroflection eastward toward the interior of the anticyclonic gyre, driven by a branch of the anticyclonic gyre going northeast, to form the South Indian Subtropical Countercurrent [**Figures 4-7 and 4-8**]. The other part propagates westward as eddies, which fill the Mozambique Channel.

[309] 16 Sv from 0 m to 300 m and 3.2 Sv from 300 m to 900 m.

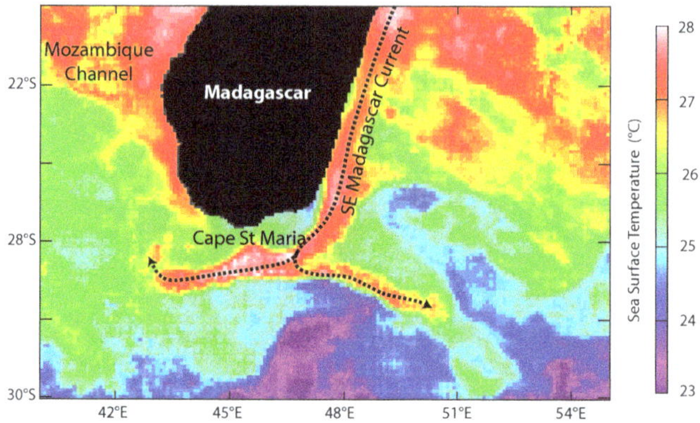

FIG. 4-10 – Satellite image of 17 April 2008 showing the eastward retroflection of the SE Madagascar Current (SEMC) and the extension of warm water (in red and yellow) toward the Mozambique Channel. (POET/NASA and GHRSST data.)

4.2.3 Agulhas Current

The other part of the SE Madagascar Current, which proceeds in the form of eddies toward the Mozambique Channel, joins the Mozambique Current going south along the African coast. The Agulhas Current [**Figures 4-7** and **4-8**] extends the Mozambique Current and intensifies southwards. It is a powerful western boundary current, transporting warm water, whose warmth depends on the season, with a transport that increases from 70 Sv around 31°S to 95 Sv–135 Sv around 35°S, with velocities that can exceed 2.5 m s^{-1}. It extends down to around 3,400 m. Below the Agulhas Current is an undercurrent flowing northward, which entrains North Atlantic Deep Water.

The Agulhas Current, as the SE Madagascar Current south of Madagascar, becomes unstable south of Africa forming meanders and eddies, which intensify the mixing with cold water of the Circumpolar Current [**Figure 4-11**]. Part of its water of tropical origin undergoes a retroflection eastward (Agulhas Return Current), joining the South Indian Current. Another part escapes westward by the ejection of filaments and warm eddies [**Figures 4-11** and **4-12**] toward the Atlantic Ocean.

Observations show on average a transfer of heat from the Indian Ocean to the Atlantic Ocean above 1,500 m in the thermocline layer. The transport of these eddies is estimated at between 5 Sv and 15 Sv. It represents the warm branch of the thermohaline circulation cell on the global scale [**Figure 1-51**, the Conveyor Belt], which feeds the transformation of surface water into North Atlantic Deep Water. Between the coast and the Agulhas Current, a deep counter/undercurrent flowing northward entrains North Atlantic Deep Water at depth into the South Indian Ocean.

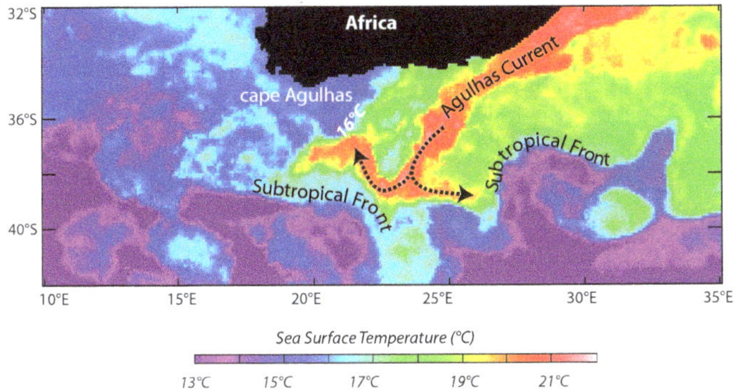

FIG. 4-11 – Satellite image of 30 September 2008 showing surface temperatures off South Africa. (POET/NASA and GHRSST data.)

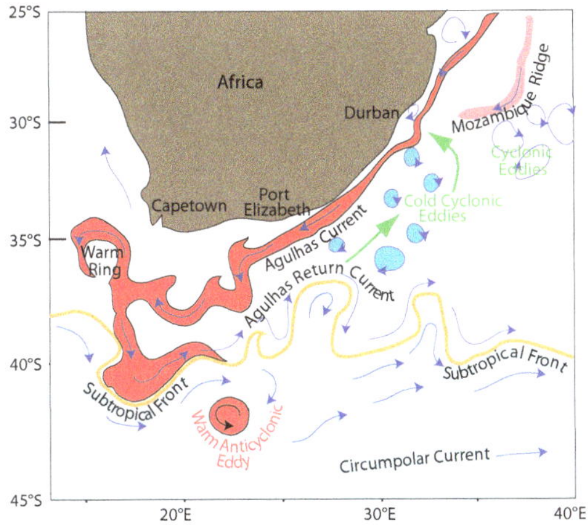

FIG. 4-12 – Circulation schematic south of Africa. (After Boden *et al.*, 1988, and Grundlingh, 1987.)

4.2.4 South Indian Current, West Australian Current and the eastern cyclonic circuit

The South Indian Current, which prolongs the retroflection of the Agulhas Current, travels eastward north of the Subtropical Front [**Figure 4-7** and **4-8**]. It transports about 60 Sv at the longitude of Madagascar, where it is distinct from the Circumpolar Current, which is located more to the south [**Figure 2-18**]. The South Indian Current sends a large part of its water to the interior of the anticyclonic gyre, a convergence zone. It is limited by a current heading northeast that crosses the Indian Ocean from southern Madagascar between 22°S and 28.5°S, called the South Indian Ocean

Subtropical Countercurrent (SICC) and joins the SEC, closing the relatively narrow anticyclonic gyre of the South Indian Ocean.

Associated with the wind divergence NW of Australia in the austral summer, an eastern cyclonic circuit, the East Gyral Current, develops in the northeast part of the subtropical anticyclonic gyre [**Figure 4-7**]. The cyclonic circuit forms by detaching southward from a branch of the SEC. It joins the Eastern Current heading eastward and is unique to the Indian Ocean.

The Eastern Current is fed in part by the SEC, transporting water of the Indonesian Throughflow and in part by the South Indian Subtropical Countercurrent coming from south of Madagascar. Off the NW coast of Australia, the circulation is rather diffuse and includes a number of eddies. A branch coming from the south takes a direction perpendicular to the Australian coast, and feeds the coastal circulation of the Leeuwin Current.

4.2.5 Leeuwin Current

The Leeuwin Current is one of a kind. In contrast with the eastern coasts of the southern Atlantic and Pacific which are bordered by cold north-flowing currents (the Benguela and Peru Currents, respectively) the west coast of Australia is bathed by the warm south-flowing water of the Leeuwin Current (LC)[310] [**Figure 4-13a**].

Surface temperature is 4–6 °C higher than at the same latitude along the Peru and Namibian coasts. As a result, precipitation is twice as strong on the Australian coast, with a maximum in June and a second maximum in February. The Leeuwin Current is a coastal current flowing against the wind. Its existence is in part due to brackish and warm water from the Indonesian Seas, which produces a meridional pressure gradient oriented southward. The meridional gradient generates a geostrophic current coastward, opposite to the offshore Ekman transport due to the southerly winds, and feeds the Leeuwin Current. Most of the time the pressure gradient dominates over the effect of the winds.

Variations in the Leeuwin Current are seasonal and interannual. When the meridional wind component weakens during the austral winter,[311] the current intensifies. When the wind strengthens in the austral summer, the current decreases and the velocity maximum then is below the surface, around 30 m to 40 m. At the surface the pressure gradient current is counteracted by the effect of the wind. Variations in the meridional pressure gradient induce variations in the current, which accelerates when the gradient is strongest and vice versa. The core of the current is trapped at the edge of the continental shelf. Its thickness is from 150 m to 300 m and its average

[310] Warm-water aquatic fauna in this region differ from those found at the same latitude in the eastern Atlantic and Pacific.

[311] During the austral winter Australia is under the influence of high pressure.

transport is about 3.4 Sv, with a June–July maximum (5 Sv). Off Fremantle, sea level varies with current intensity. The level is highest in winter, corresponding to a strong current, and low in summer, corresponding to a weak current. The difference in level between summer and winter reaches 23 cm. The maximum speed of the surface current (55 cm s^{-1}) occurs in April–May, two months before the transport maximum and minimum (25 cm s^{-1}) from November to February.

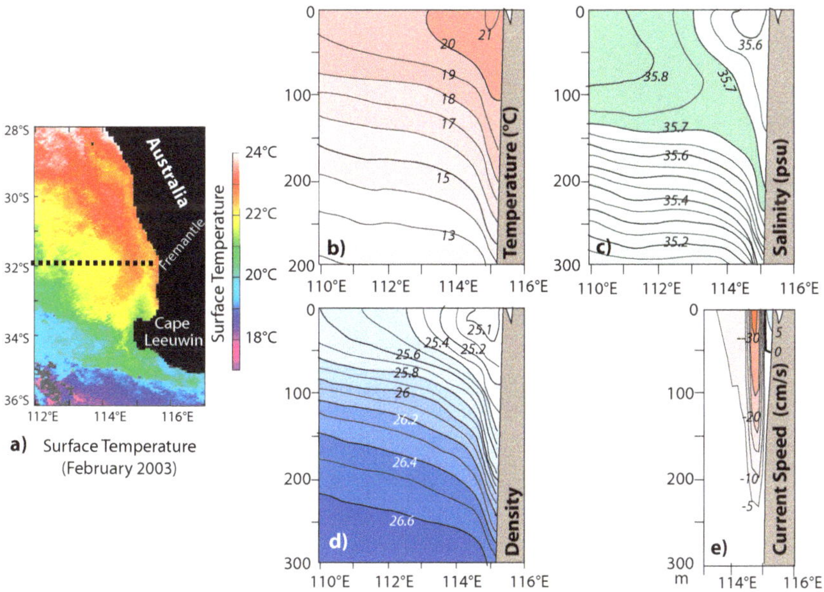

FIG. 4-13 – **(a)** Satellite image of February 2003, showing warm water entrained southward along the coast by the Leeuwin Current west of Australia. (POET/NASA and GHRSST data.) Black dots locate the section off Fremantle: **(b)** potential temperature, **(c)** salinity, **(d)** potential density (in kg m^{-3}), and **(e)** geostrophic velocity. (1950–2000 averages, after Feng *et al.*, 2003.)

Salinity varies with current intensity: the stronger the current, the lower the salinity and vice versa. The Leeuwin Current is in geostrophic equilibrium. Isotherms, isohalines, and isopycnals deepen toward the coast (downwelling) [**Figure 4-13**], drawing the subtropical salinity maximum layer down to 250 m along the slope. The pressure gradient is equatorward under the Leeuwin Current, generating a northward undercurrent between 300 m and 800 m. The undercurrent carries South Indian Central Water, Subantarctic Mode Water, and Antarctic Intermediate Water. Its speed can reach 40 cm s^{-1} and its transport can be 1 to 2 Sv.

Semi-annual and interannual climatic signals propagate poleward from the equatorial Pacific Ocean as waves topographically trapped on the edges of the continental shelf across the Indonesian Seas and northern Australia. The Leeuwin Current thus decreases during El Niño years (3 Sv) and increases during La Niña years (4.2 Sv) [**Section V.3.4**].

These trapped waves on the coast interact with the Leeuwin Current to form instabilities that produce warm eddies [**Figure 4-13a**: white patches around 29°S] which transport heat west-northwestward.

4.2.6 Upper connections with the Pacific and the Atlantic Oceans

Besides the Throughflow which connects the tropical Pacific and Indian Oceans, a linkage south of Australia through one branch of the "super gyre" connects the three subtropical gyres of the southern hemisphere (**Figure 6-9**). The westward branch occurs at intermediate depth south of Australia coming from the East Australian Current, passing south of Tasmania. Southwest of Australia it passes under the eastward flowing Leeuwin Current and crosses the southern Indian Ocean north-westward under the north-eastward surface circulation, then flows southward and rounds Africa to reach the south Atlantic. The eastward branch of the "super gyre" flows at shallower depths along the northern edge of the Antarctic Circumpolar Current and sends filaments northward into the subtropical anticyclonic gyres [**Figures 4-7 and 4-8**].

4.3 Circulation North of 10°S

4.3.1 Circulation in the Arabian Sea and along the Somali coast

Northeast monsoon winds [**Figure 4-4**] are comparable to the northeast trades of the other oceans. They drive the Northeast Monsoon Current, which flows westward south of Sri Lanka and whose transport is about 12 Sv [**Figure 4-7**]. In December–January the Northeast Monsoon Current partly feeds the coastal current, which flows northward west of India, forming the anticyclonic Laccadive Eddy. The general circulation in the Arabian Sea is directed westward. One branch of the Northeast Monsoon Current crosses the Arabian Sea near 5°N and turns southward in the coastal Somali Current (SC). Another branch of the Northeast Monsoon Current arrives on the Somali coast around 10°N, south of Socotra, and divides into three branches. One branch feeds the Somali Current to the south, another turns off into the Gulf of Aden through the passage between Cape Guardafui and Socotra, and the third passes east of Socotra northward. Under the effect of NE winds, the Somali Current crosses the equator in a layer only 150–180 m deep [**Figure 4-14a**] transporting about 5 Sv. Below, a countercurrent flows to the north and seems to prolong the Zanzibar Current between 180 m and 400 m (called the Zanzibar Undercurrent).

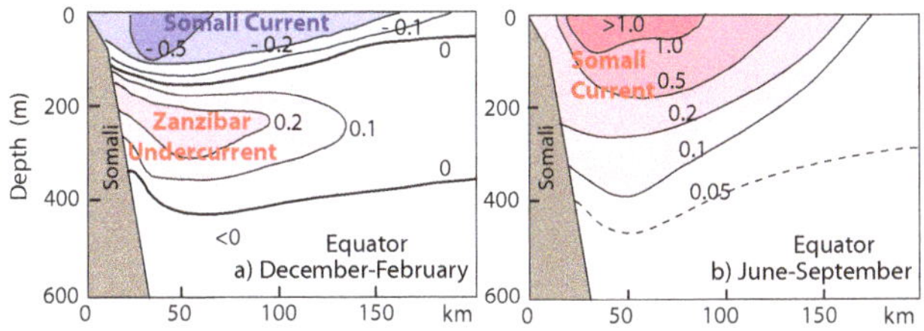

FIG. 4-14 – Sections showing average meridional velocity (in m s^{-1}) in the western boundary current at the equator based on two years of direct measurements (1984–1986). **(a)** From December to February, during the NE monsoon and **(b)** from June to September, during the SW monsoon. (After Schott *et al.*, 1990.)

John Swallow, Michèle Fieux and Fritz Schott on board the Marion Dufresne in 1985 in the Indian Ocean, during SINODE 19 (Photo Sabine Arnault).

At the surface, the Somali Current enters into confluence with the Zanzibar Current, or East African Coastal Current, around 2°S–4°S and forms the South Equatorial Countercurrent (SECC) flowing eastward. **Figure 4-15** shows the seasonal evolution of circulation off Somalia.

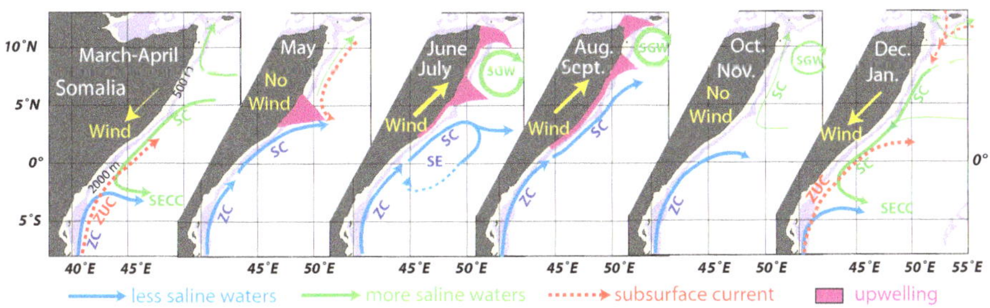

FIG. 4-15 – Evolution of the Somali Current circulation during the year. Weak salinity, originating in the South Equatorial Current is blue, higher salinity, coming from the Arabian Sea, is green. Regions of upwelling are violet, undercurrents are dotted red, and the average wind direction is shown by a yellow arrow. [South Equatorial Countercurrent (SECC), Somali Current (SC), Zanzibar Current (ZC or East African Coastal Current), Somali Great Whirl (SGW), Zanzibar Undercurrent (ZUC).] Isobaths are 500 m and 200 m. (After Fieux, 1987; Schott and McCreary, 2001.)

In March–April, the main flow of the Northeast Monsoon Current arrives on the Somali coast around 6°N–8°N and splits into one branch going north and one going south.

In April–May, under the north-flowing current, north of 4°N, an undercurrent going south[312] between 100 m and 300 m depth develops. It converges with the undercurrent coming from the south and leaves the coast around 4°N.

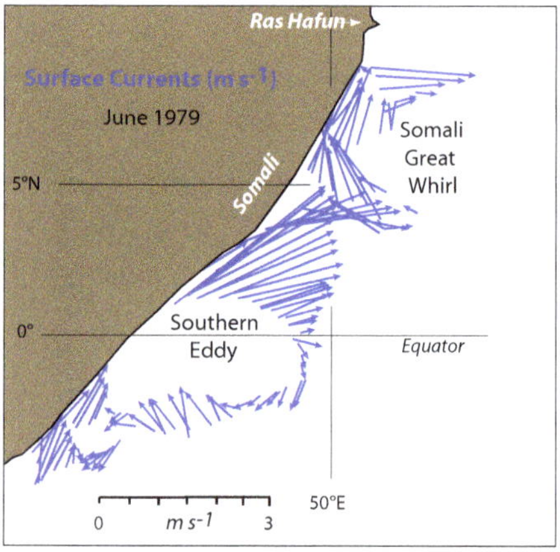

FIG. 4-16 – Surface currents observed by the RRS *Discovery* in the Somali Basin in June 1979 at the beginning of the SW monsoon. The center of each arrow corresponds to the measurement point. (After Swallow and Fieux, 1982.)

Along the Somali coast, the Somali Current responds rapidly[313] to the onset of the southwest monsoon, generally at the end of May[314] [**Figure 4-15**]. After the arrival of southwest winds, there is continuity between the Zanzibar Current (fed by the SEC) and the Somali Current, flowing north along the coast. It is a violent western boundary current sometimes[315] reaching 3.5 m s^{-1} [**Figure 4-16**] whose transport can reach 80 Sv. The Somali Current leaves the coast around 4°N, producing upwelling on the northern edge of the current. One part of the current returns southward then is re-entrained into the Zanzibar Current [**Figures 4-15** and **4-16**] forming a circuit sitting on the equator, called the Southern Gyre (or Southern Eddy), the most energetic part of which is the coastal branch, or southern branch of the Somali Current.

[312] Speeds reaching 0.6 m s^{-1} have been measured in the undercurrent.

[313] The response is rapid because of the proximity of the equator where the Coriolis force is zero.

[314] The onset of the southwest monsoon may occur suddenly in a few days, in two episodes over a month from mid-May (early monsoon) to mid-June (late monsoon), or progressively.

[315] This is an extreme value for the open sea, the equivalent of 7 knots.

In June, when the southwest monsoon is well established, a strong anticyclonic eddy develops under the negative wind-curl maximum [**Figure 4-5**]. Centered on 8°N –9°N, it is called the Great Whirl or Somali Great Whirl (SGW). It generates a second strong upwelling along the coast and north of the northern branch [**Figures 4-15** and **4-17a**].

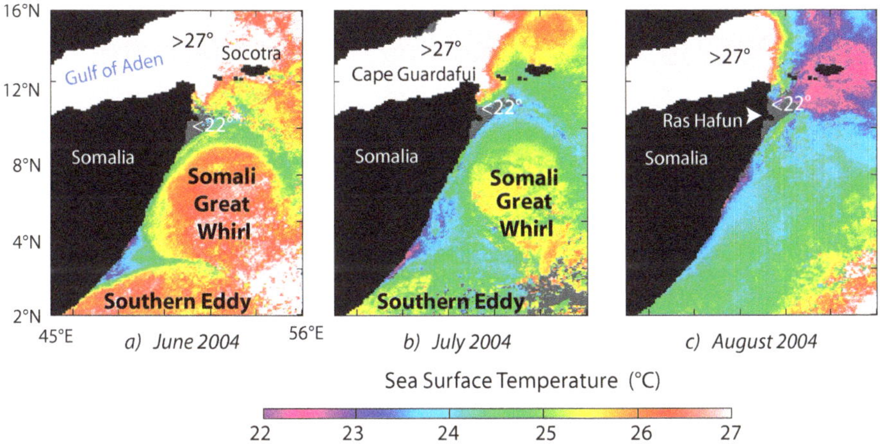

FIG. 4-17 – Surface-temperature evolution along the Somali coast observed by satellite during the SW monsoon **(a)** in June 2004 and **(b)** in July 2004. There are two eddies associated with two stronger upwelling regions. Water temperatures less than 22 °C are in grey. (c) In August, there is upwelling all along the coast and strongest around 10°N. (POET/NASA and GHRSST data.)

In June–July the Somali Great Whirl is clearly separated from the Southern Gyre by a temperature front due to cold upwelling water and by a haline[316] front between the low-salinity water coming from the southern branch of the Somali Current, fed by the SEC, and the more saline water of the Arabian Sea, entrained in the Great Whirl [**Figure 4-15**]. To the northeast of the island of Socotra, another smaller anticyclonic gyre frequently develops, called the Socotra Gyre.

During the southwest monsoon, winds parallel to the coast generate cold-water upwelling [**Section I.5.2.1**] all along the Somali coast [**Figure 4-17a** and **b**], less strong than the two upwelling areas at 4°N and 10°N, intensified by the departure of the current offshore. The cold water of the upwelling systems, rich in nutritive material, allows a rapid planktonic bloom[317] which turns the water green, as opposed to blue water offshore, which is warm

[316] Of salinity.

[317] During a June oceanographic expedition off the Somali coast, a strong fish odor was noticed throughout the ship. The thermosalinograph [**Section I.6.1.5**] recording revealed that we had just entered the upwelling zone!

and impoverished in nutritive material and particles. The thermal front is thus associated with a distinct color front as is the western edge of the Gulf Stream.

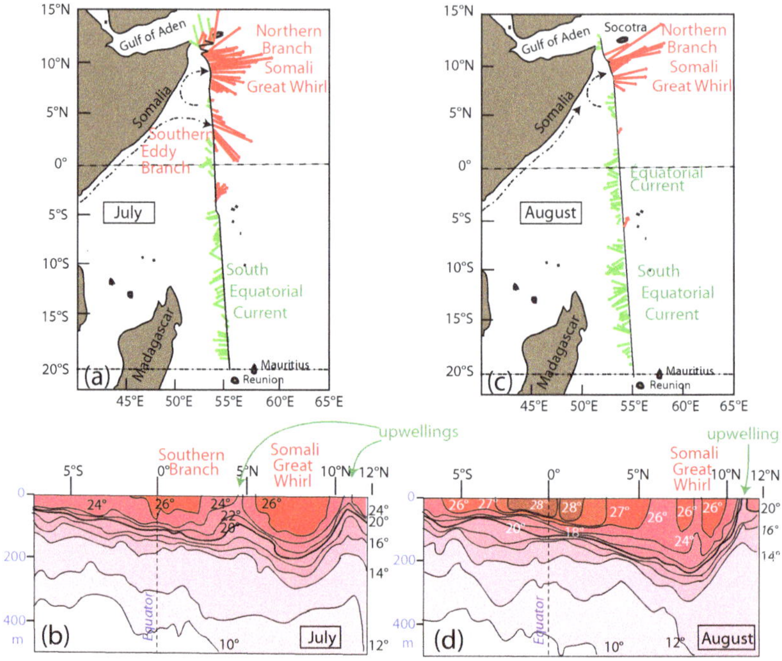

FIG. 4-18 – Surface current maps (red eastward, green westward) along a route between Cape Guardafui and La Réunion **(a)** in July and **(c)** in August. The corresponding temperature sections, taken during the program SINODE[318], are **(b)** in July and **(d)** in August. (After Fieux, 1987.)

At the end of the southwest monsoon (late August–September) the Southern Gyre migrates northward entraining the upwelling at 4°N, which converges with that at 10°N [**Figures 4-15** and **4-17c**]. The current along the Somali coast is then continuous as far as 10°N, and entrains within it low-salinity water from the south toward the Arabian Sea. A temperature section [**Figure 4-18**] between Cape Guardafui and La Réunion, taken in July and repeated in August, shows the presence of two gyres in July and only one in August. The thermal structure of the anticyclonic Somali Great Whirl north of the section shows an accumulation of warm water in the center of the gyre and upwelling of colder water of the thermocline on the periphery. Even though the section angles away from the coast southward, the southern

[318] From the SINODE program: Surface Indian Ocean Dynamics Experiment, on board the *Marion Dufresne*.

branch, associated with eastward currents and upwelling, is clearly marked in July and disappears in August.

Between the end of the southwest monsoon and the beginning of the northeast, the wind ceases in the northern part, but the Somali Great Whirl persists until October–November.

At 5°N, along the Somali coast, below 200 m depth, the undercurrent flows southward year-round [**Figure 4-19**]. The surface layer reverses northward as early as March–April. In the peak of the southwest monsoon, at 5°N, the northward Somali Current only reaches to 200 m depth.

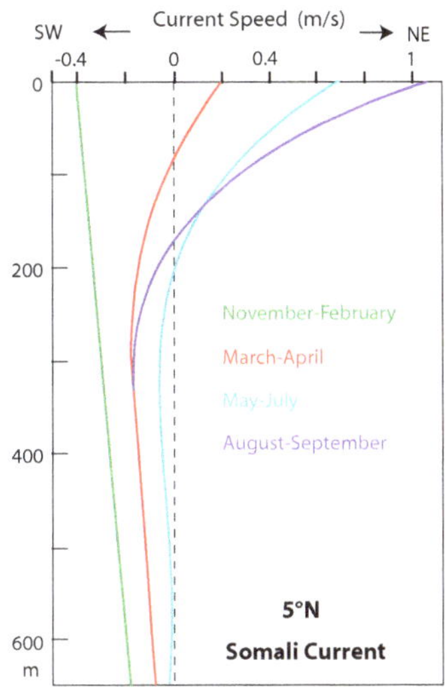

FIG. 4-19 – Seasonal variation of the velocity component parallel to the coast (positive to the NE) in the Somali Current at 5°N, based on 30 months of current measurements. (After Quadfasel and Schott, 1982.)

Along the coast of the Arabian Peninsula in July, southwest winds generate upwelling similar to that along the Somali coast. In the Arabian Sea, under the effect of negative wind curl generated by the Findlater Jet off Cape Guardafui [**Figure 4-5**], the circulation is anticyclonic [**Figure 4-8**]. There is convergence (an accumulation of warm water) in the center of the gyre where the mixed layer is deepest.[319] South of India and Sri Lanka, the Southwest Monsoon Current (SWMC) heads east toward the Bay of Bengal with a transport of about 8 Sv. It is directly fed by a branch of the Somali Current around 10°N and by the West Indian Coastal Current (WICC) coming from the north. Along the West Indian coast, the West

[319] The gyre structure has a bowl shape, as opposed to the dome shape of cyclonic structures.

Indian Coastal Current, associated with a slope of the isotherms, which rise toward the coast, appears in March.[320] It is maximal in July and disappears in October. It generates the cyclonic Laccadive Eddy (with a sea-surface hollow as seen by satellite and an isotherm dome) north of the Laccadive Islands. An undercurrent in the opposite direction occurs between 100 m and 300 m (with a transport of 4.7 Sv northward). It transports fresh water from the Bay of Bengal, which mixes with more saline water from the Arabian Sea. Winds favorable to upwelling along the western Indian coast begin in April and reach a maximum in August, then die down in October. During the NE monsoon, the WICC flows northward. It extends over ~ 400 km at around 10°N and 100 km around 22°N. A southward-flowing undercurrent is associated with it. The WICC maximum is not correlated with the force of the wind, indicating that it is under the influence of coastal waves coming from the Bay of Bengal. The Laccadive Eddy is anticyclonic in this season. The presence of the Laccadive Eddy displaces westward the link between the Northeast Monsoon Current coming from south of Sri Lanka and the WICC. The current reversal south of India and Sri Lanka (see below) and the cyclonic circulation around the Sri Lanka Dome were tracked with a drifting buoy by satellite in 1981–1982 [**Figure 4-20a**].

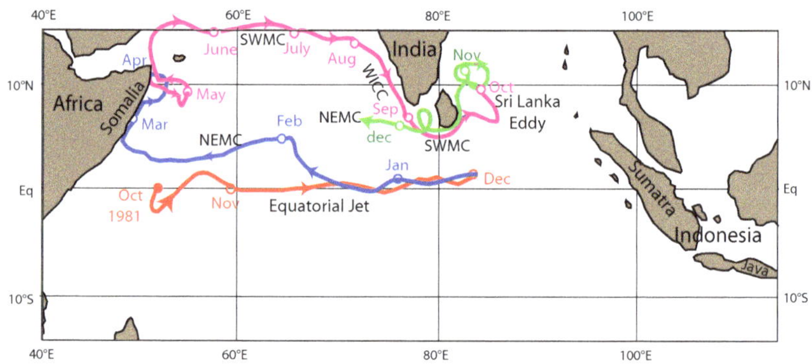

FIG. 4-20 – A drifting buoy trajectory, from October 1981 to December 1982, tracked by satellite during the SINODE Program. The buoy was entrained eastward in the Wyrtki equatorial jet from November to December (red) then the current reversed westward during the NE monsoon (blue) [NE Monsoon Current (NEMC)]. It then was entrained in the anticyclonic circuit of the Arabian Sea [SW Monsoon Current (SWMC), West Indian Coastal Current (WICC).] It passed south of India and Sri Lanka during the SW monsoon, was taken up in the turbulent cyclonic circulation east of Sri Lanka (the Sri Lanka Eddy) in the Bay of Bengal and was re-entrained westward (in green) during the NE Monsoon. (After Fieux, 1987.)

[320] Circulation models show that the West Indian Coastal Current depends primarily on the influence of winds blowing over the Bay of Bengal.

4.3.2 Circulation in the Bay of Bengal and the eastern Indian Ocean

A portion of Homen and Reinel map of the Bay of Bengal, 1519 (Bibliothèque Nationale de France, Paris). (From *Atlas de Christophe Colomb et des grandes découvertes*, Nebenzahl, 1991.)

In the Bay of Bengal during the southwest monsoon, the Southwest Monsoon Current follows the cyclonic circuit of the Sri Lanka Dome whose northern branch splits on arriving at the coast around 10°N, then extends toward the northern Bay of Bengal as the East Indian Coastal Current (EICC) [**Figure 4-8**].

The East Indian Coastal Current flows northward from February to September, with a maximum in March–April. From May to September, only the central part goes north. It flows southward from October to January, with a maximum in November. Between February and May the EICC is the western boundary current of the anticyclonic gyre established in the Bay of Bengal during this period. During the southwest monsoon the circuit splits into several smaller-scale circuits. During the northeast monsoon, it forms a cyclonic circuit of which the East Indian Coastal Current is the western branch, heading southward.

Off Sumatra, the current generally heads southeast along the Mentawai Islands.

From June to September, under the influence of the southeast Australian monsoon, the Java Current flows westward, entraining Indonesian water, and is difficult to distinguish from the SEC. Under the effect of winds parallel to the coast[321] upwelling takes place along the coast of Java (extending

[321] When wind blows parallel to a coast with the coast on the right in the southern hemisphere, Ekman transport is directed offshore.

as far as Sumatra, especially during El Niño periods), lowering surface temperature by several degrees. The Mentawai Current and the Java Current converge near the Sunda Strait.

During the northwest monsoon, the Java Current reverses and heads east. At this season, it prolongs the South Equatorial Countercurrent and the Mentawai Current off Sumatra [**Figure 4-8**]. Northwest winds produce downwelling along the coast of Java and a slight upwelling along the northwest Australian coast.

4.3.3 Equatorial Current, Wyrtki Jet

The equatorial current system in the Indian Ocean is certainly different from that in the other oceans.

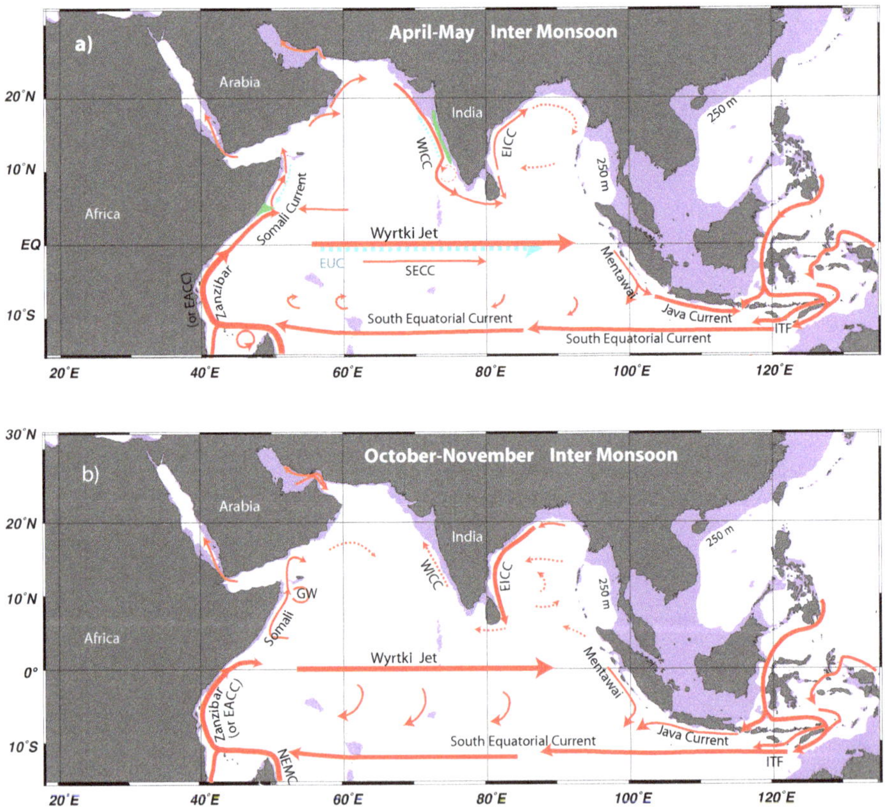

FIG. 4-21 – Surface circulation diagrams during the inter-monsoonal periods: a) in April–May, b) in October–November. Undercurrents are shown in dashed blue lines (Equatorial Undercurrent, EUC). See the legends of **Figures 4-7** and **4-8**. (After Wyrtki, 1973; Fieux, 1987; Reverdin *et al.*, 1983; Schott and McCreary, 2001; Shankar *et al.*, 2002.)

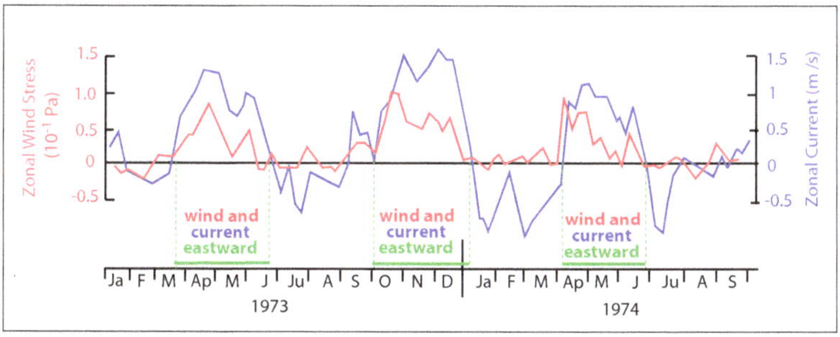

FIG. 4-22 – Variation of the zonal component of surface current and wind stress on the equator at 73°E (Addu Atoll), from January 1973 to October 1974. (After Knox, 1976.)

Winds at the equator are weak during the two monsoons except along the Somali coast and have a dominant meridional component. There are no regular easterly winds as in the other oceans. On the contrary, westerly winds appear on the equator during the inter-monsoonal periods (April–May and October–November). They entrain a strong eastward equatorial current east of 50°E, called the Equatorial Jet or Wyrtki Jet[322] (WJ) [**Figures 4-21b and 4-21c**]. This current is unique to the Indian Ocean. Observations taken from the island of Gan (an atoll at 73°E) in the middle of the basin showed that it was in phase with the appearance of west winds [**Figure 4-22**]. It occurs between April and June and between October and December. Its speed reaches 0.75 m s^{-1} to 1 m s^{-1} in the eastern basin, and up to 1.5 m s^{-1} in November–December. It accumulates warm water eastward in the basin,[323] modifying sea level and mixed-layer thickness above the thermocline [**Figure 4-23**]. Seasonal variations of the 25 °C isotherm depth, located in the thermocline along the equator, indicate a rise to the west and a deepening to the east,[324] twice a year, corresponding to the inter-monsoonal periods [**Figure 4-22**]. Generally, the zonal slope is most accentuated in November–December, and then it weakens after the onset of the following monsoon.

[322] Named after the oceanographer who noticed equatorial eastward drift on pilot charts during inter-monsoonal periods.

[323] This is the inverse of the situation in the equatorial Atlantic and Pacific.

[324] Thermocline slope is the inverse of surface slope [**Figure 1-60**].

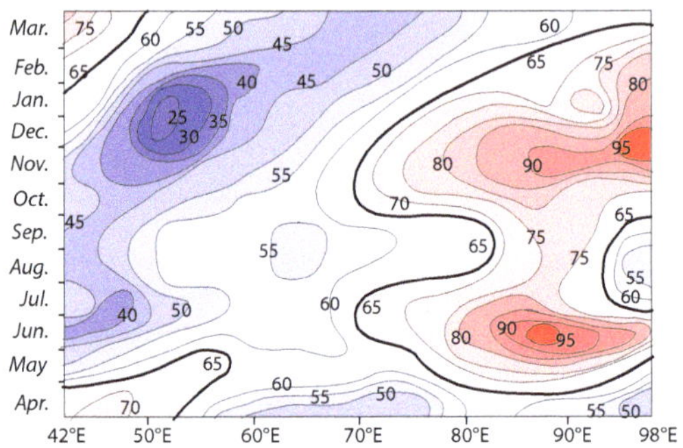

FIG. 4-23 – Seasonal variation of the depth of the 25 °C isotherm (located in the thermocline) at the equator as a function of longitude, under the effect of Wyrtki's semi-annual equatorial jet. The average depth of 65 m is the bold contour, depths less than 65 m are blue; those greater than 65 m are red. (After Reverdin, 1987.)

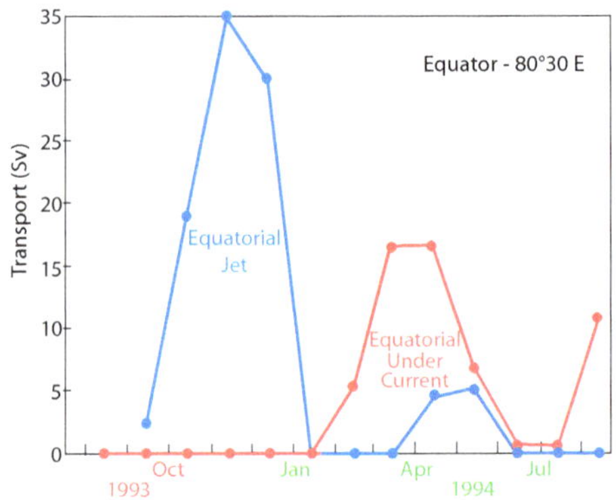

FIG. 4-24 – Variation of the average monthly eastward transport on the equator at 80°30′E, obtained from direct current measurements from September 1993 to September 1994. (After Reppin *et al.*, 1999.)

The Wyrtki Jet is only 80 to 100 m thick and from 400 to 500 km wide. It is centered on the equator. Its transport south of Sri Lanka reaches 35 Sv in November and only 5 Sv in May [**Figure 4-24**] with strong interannual variations. To the east, the jet is extended by the southward current passing the Mentawai Islands off Sumatra, trapped along the slope.

Drifting buoys set in the equatorial jet stay with the current[325] and permit its evolution to be followed [**Figures 4-20** and **4-25a**]. They show that

[325] A current heading eastward at the equator converges as a result of the change of sign of the Coriolis force at the equator [**Section I.4.2**].

the jet ceases suddenly in December[326] when the northeast monsoon starts, under the effect of a wave propagating westward [**Figure 4-20**]. Then the buoy is either picked up by a weak current heading west north of the equator [**Figure 4-20**] or by the South Equatorial Countercurrent [**Figure 4-25a**] south of the equator. During the monsoons, a weak westward current exists at the equator. Some years it happens that the system displaces southward in the midst of the southwest monsoon. Then west winds reach the equator south of India[327] and re-entrain an eastward jet.

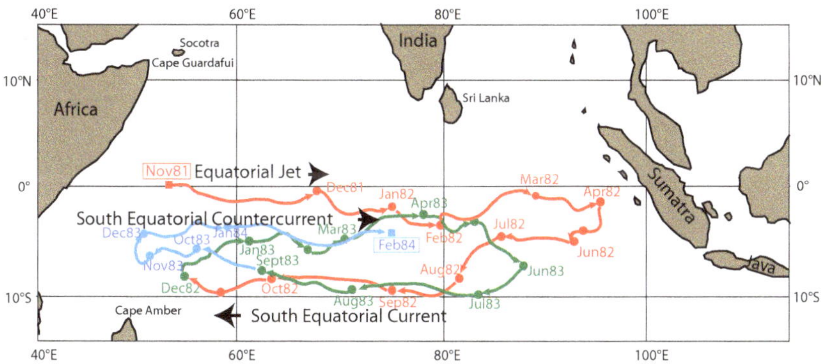

FIG. 4-25A – A drifting buoy trajectory, tracked by satellite, from November 1981 to February 1984 during the SINODE Program. It was entrained eastward by the equatorial jet from November to December 1981 (red), then by the south Equatorial Countercurrent (SECC) until April 1982 (red). Next, it was entrained westward by the South Equatorial Current (SEC) (red), ending up, after a year, in December, 1982 at the season of the South Equatorial Countercurrent (SECC), and was once again entrained eastward (green then blue). It followed nearly the same circuit two and a half times, doing it in nearly one year. (After Fieux, 1987.)

4.3.4 South Equatorial Countercurrent

The South Equatorial Countercurrent (SECC) exists at the surface during the northeast monsoon season when the Somali Current heads south. The convergence along the African coast around 2°S–4°S of the Zanzibar Current and the Somali Current give it birth [**Figure 4-14**]. It flows eastward between 2°S and 6°S–8°S under the zone of weak winds [**Figure 4-4a**].

[326] The jet ceases upon the arrival of an equatorial Rossby Wave, formed by reflection of an eastward-propagating equatorially trapped Kelvin Wave on the eastern boundary [see texts on dynamics]. This explains the rapidity of the response to wind in this region.

[327] Monsoon winds then pass to the south around India and rains cease over the Indian sub-continent (a *monsoon break*).

Buoys placed in the South Equatorial Countercurrent cross the basin and then are generally entrained southward and return westward with the SEC [**Figure 4-25a**].

The South Equatorial Countercurrent is well defined at the surface only during the northeast monsoon [**Figure 4-4**]. In other seasons it is found below the surface under the general westward circulation[328] [**Figure 4-8**].

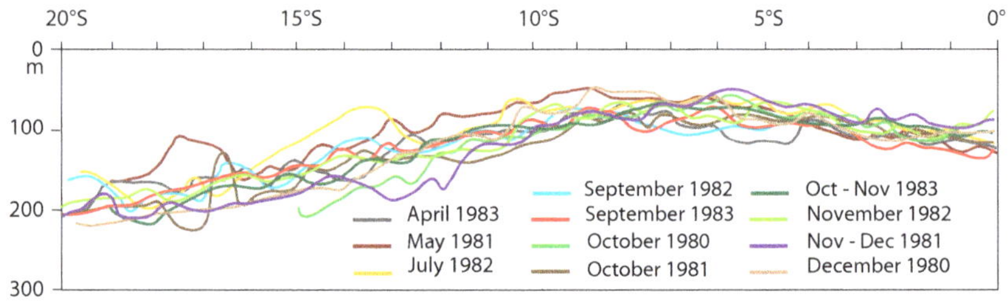

FIG. 4-25B – Depth variability of the 20 °C isotherm between 20°S and the equator around 50°E–60°E. (After Fieux *et al.*, 1985.)

In latitudes 5°S–10°S and longitudes 45°E–90°E, between the westward SEC and the eastward SECC, the thermocline upwells as a response to wind forcing. The Ekman drift transports water southward from the northern edge of the SE trades, which makes the thermocline shallower throughout the year [**Figure 4-25b,** see **Section I.5.3.1**]. In the Atlantic and Pacific Oceans, the upwelling is at the equator under the effect of the SE trades. In the Indian Ocean, the SE trades do not blow over the equator, they are displaced to the south, and the zonal upwelling is also displaced to the south.

4.3.5 Equatorial Undercurrent

In the Indian Ocean, the Equatorial Undercurrent (EUC), heading east, is an episodic phenomenon. It only appears at the end of the northeast monsoon [**Figure 4-24**] from March to June when the winds have accumulated enough water on the western boundary to create an eastward pressure gradient at the thermocline level. Such a situation exists year-round in the other tropical oceans [see **Section III.4.1.5**]. Its core has a depth of around 80–100 m in the thermocline and its speed is of the order of 60 to 90 cm s^{-1} [**Figure 4-26**]. South of Sri Lanka, its transport may reach 17 Sv. The Wyrtki Jet sometimes appears at the end of March [**Figure 4-26**] above the Equatorial Undercurrent.

[328] Wind anomalies may sometimes make the SECC re-appear at the surface.

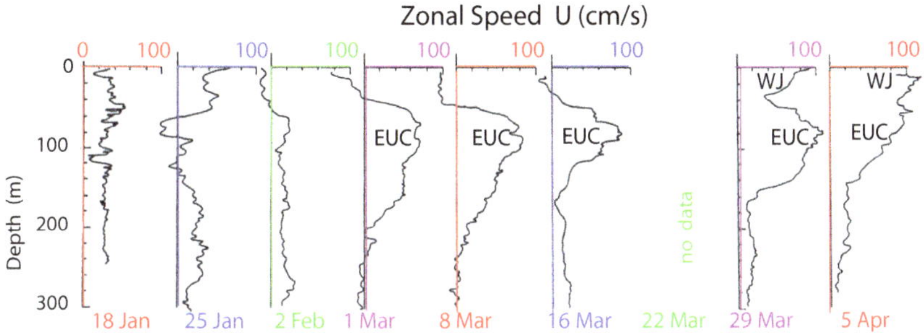

FIG. 4-26 – Profiles, between 0 and 300 m, of zonal component of current velocity on the equator, indicating the presence of the Equatorial Undercurrent (EUC) in the thermocline in March–April, at the end of the NE monsoon and the Wyrtki jet at the end of March, beginning of April at the surface. (After Luyten and Swallow, 1976.)

Current measurements, taken along the equator between 50°E and 62°E show the existence of a suite of zonal currents and countercurrents to great depth, subject to strong semi-annual variability.

Chapter 5
Water properties

5.1 Surface water

5.1.1 Surface temperature

Average annual surface temperatures in the Indian Ocean range from a minimum of about 12–14 °C at the subtropical convergence, around 40°S, to a maximum of 29 °C along the Sumatra coast [**Figure 4-27**].

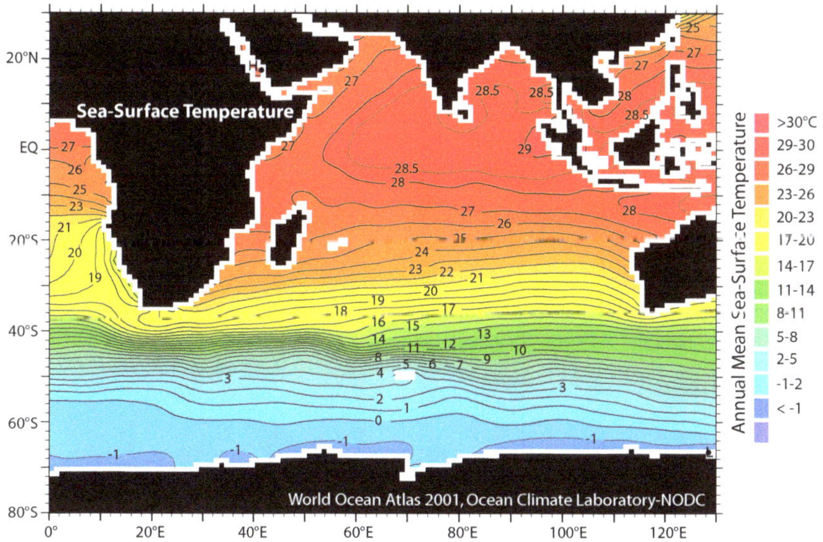

FIG. 4-27 – Annual mean Indian Ocean surface temperature. (WOCE/NODC WOA01 data.)

Surface-temperature variability, due to winds and currents, is such that a four-season division is needed to properly represent the region north of

10°S [**Figure 4-28**]. In January–February, high temperatures are found under the Intertropical Convergence Zone, where winds are weak [**Figure 4-28a**]. In the north, temperatures are not highest in July–August, but when winds are weak during the inter-monsoonal period in April–May [**Figure 4-28b**]. Surface temperature then can exceed 32 °C. In the Red Sea and in the Persian Gulf, temperatures can even reach 35 °C. These temperatures are the maximum found in the ocean. During the season of maximum sunshine, in July, temperatures decrease strongly, particularly along the Somali coast and the Arabian peninsula [**Figure 4-28c**]. This apparent contradiction is due to violent southwest monsoon winds, which mix the surface layer with colder underlying water and diminish surface temperature, forming a mixed layer. Heat is redistributed over a thicker layer. Furthermore, winds drive coastal upwelling that brings up colder water, which is then entrained offshore, often with temperatures less than 18 °C. During the November inter-monsoon, temperatures rise again a little in the west [**Figure 4-28d**]. In January, minimum temperatures in the northern Indian Ocean are found in the northern Persian Gulf and Red Sea, with temperatures from 16 °C to 18 °C, a difference of at least 15 °C from July–August [**Figure 4-28a**]. Isotherms bend southward along the coasts of Australia and Madagascar showing the influence of the Leeuwin Current and the East Madagascar Current transporting warm water southward.

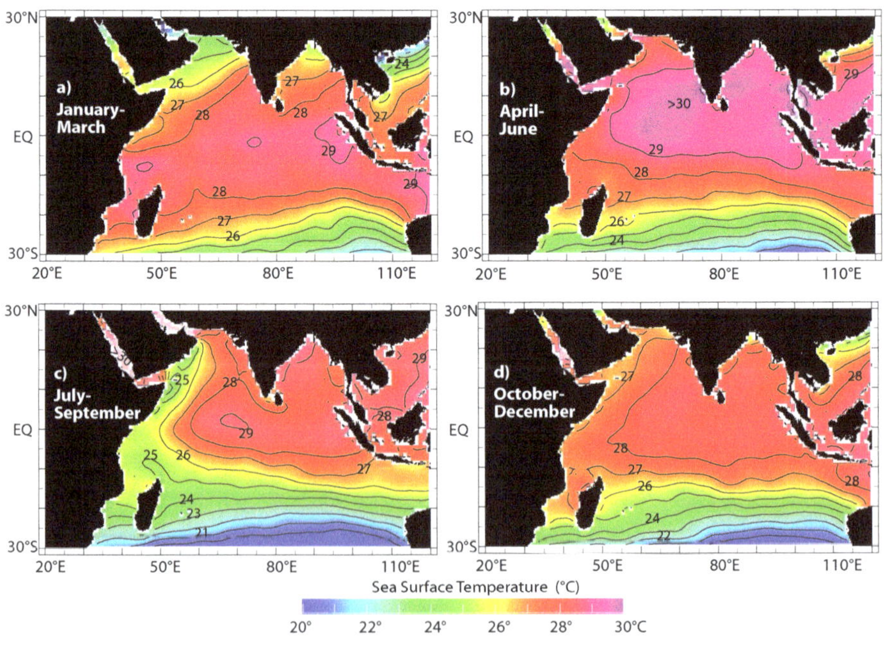

FIG. 4-28 – Seasonal mean surface temperature in the Indian Ocean. (IRI/LDEO, and WOCE NODC/WOA05 data.)

5.1.2 Surface salinity

Surface salinity distribution [**Figure 4-29**] depends on fresh-water balance (*E-P*) [**Figure 1-22**] and circulation [**Figures 4-7** and **4-8**]. Fresh-water inflows from large rivers and precipitation strongly reduce the surface salinity of the Bay of Bengal. Values are less than 33, and even drop to 25 in the Andaman Sea, contrasting with the Arabian Sea where salinities exceed 36.4 under strong evaporation. The *E-P* maximum in the southern anticyclonic zone is manifested by a subtropical salinity maximum greater than 35.8. The input of low-salinity Indonesian water by the South Equatorial Current adds to the effect of strong precipitation under the atmospheric Intertropical Convergence in January. The result is a tongue of low-salinity water from the Indonesian Seas, heading west around 10°S–15°S. Salinities are highest in the northern Persian Gulf and Red Sea due to strong evaporation [**Figures 4-29** and **4-30**]. They can exceed 42 and are the highest in the planetary ocean.

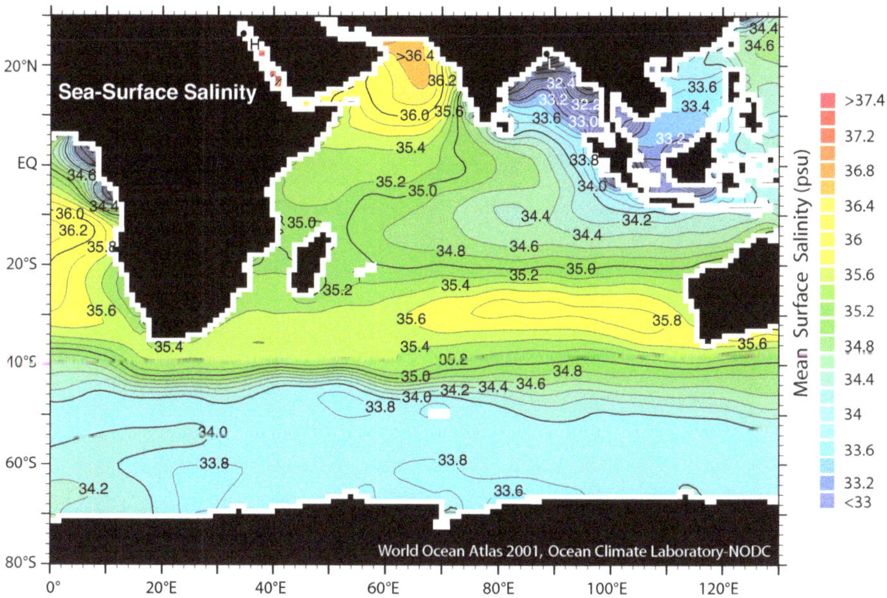

FIG. 4-29 – Annual mean surface salinity in the Indian Ocean. (WOCE/NODC WOA01 data.)

Because of the contrast between Arabian Sea high-salinity water and Bay of Bengal low-salinity water, the current south of India entrains either low-salinity water westward during the northeast monsoon or high-salinity water eastward during the southwest monsoon [**Figure 4-30b**]. Similarly, current reversals along the Somali coast modify the salinity distribution seasonally, as South Equatorial Current water and Zanzibar Current water are less saline than Arabian Sea Water.

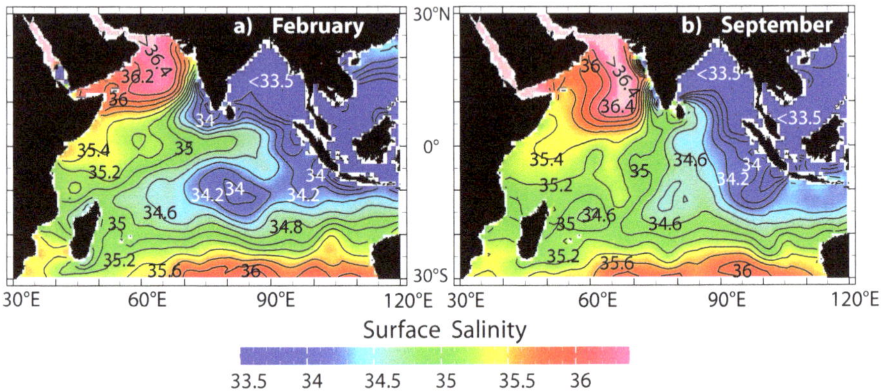

FIG. 4-30 – Seasonal variation of surface salinity in the Indian Ocean. (IRI/LDEO, WOCE NODC/WOA05 data.)

5.2 Adjacent seas

The Indian Ocean has two adjacent seas in the northwest, the Red Sea and the Persian Gulf, and a connection with the Pacific through the Indonesian Seas. Specific water masses formed in these adjacent seas have great influence on the oceanography of the Indian Ocean.

5.2.1 The Red Sea

The Red Sea was known to ancient civilizations (Sumerian, Mesopotamian, Egyptian, Greek, Arab...). The Old and New Testaments refer to it. On old maps, the Red Sea closes off the "Ecumene",[329] the limit of the known world. Most of the time it was represented in red,[330] in contrast with oceanic waters. An example is given on the extract from the world map of Vespucci in the 17th century [**Figure 4-31**]. The Greeks called it the "Erythrean Sea"[331] and the Romans "Mare Rubrum" (Red Sea). Its name came from the color of the shore reddened by the sun and soil rich in iron and lead oxides (minium or red lead).

[329] Ecumene was the world known to the Greeks, *i.e.*, a portion of nature occupied by a human collectivity.
[330] Red symbolizes blood, the heat of the sun, and the color of roses.
[331] According to Herodotus, the Erythrean Sea went as far as the Indian Ocean.

The Red Sea extends over more than 2,000 km, on an NNW-SSE axis. It has an area of 0.45×10^6 km^2. It includes a trench in an axial position, which reaches 2,740 m. However, with a well-developed continental shelf, the mean depth of the Red Sea is only 490 m [**Figure 4-32**]. It is prolonged northward by the continental shelf of the Gulf of Suez, 260 km long and from 50 to 80 m deep, and by the Gulf of Aqaba, 180 km long and 1,800 m deep (with a sill at 300 m). The Red Sea communicates on the south with the Gulf of Aden and the Arabian Sea through a narrow passage, sprinkled with islands,[332] the Strait of Bab-el-Mandeb,[333] 26 km wide. A sill 110 m deep is 135 km to the NNW of the strait near the Hanish Islands.

FIG. 4-31 – Part of the Juan Vespucci's map of the Red Sea, from his world map of 1526 (Hispania Society of America, New York). (From *Atlas de Christophe Colomb et des grandes découvertes*, Nebenzahl, 1991.)

The Red Sea is located in a desert region where temperatures are high and evaporation is intense year round. It is a region where evaporation is the highest in the world ocean, of the order of 2 m per year. This loss of water induces a drop in the level of the basin, which drives the entry of water at the surface ($\theta \approx 29$ °C, $35.2 < S < 36.5$, $\sigma_0 \approx 23$) from the Gulf of Aden through the Strait of Bab-el-Mandeb [**Figure 4-33**]. Very saline water ($S > 40$) exits at depth over the sill. The Red Sea is a concentration basin like the Mediterranean [**Figure 3-45**]. All along its path northward, surface water undergoes an increase in salinity under the effect of evaporation [**Figure 4-33**].

[332] The Perim Islands divide the strait in two: a channel of 3 km, on the Yemen side and a channel of 26 km, south of the Perim Islands.
[333] *Bab-el-Mandeb* means *Gate of Grief.*

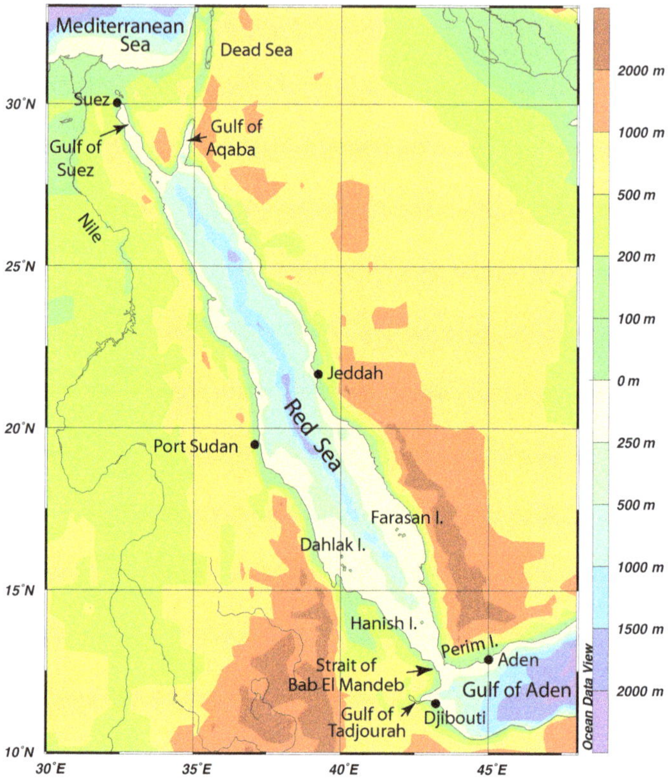

FIG. 4-32 – Red Sea map.

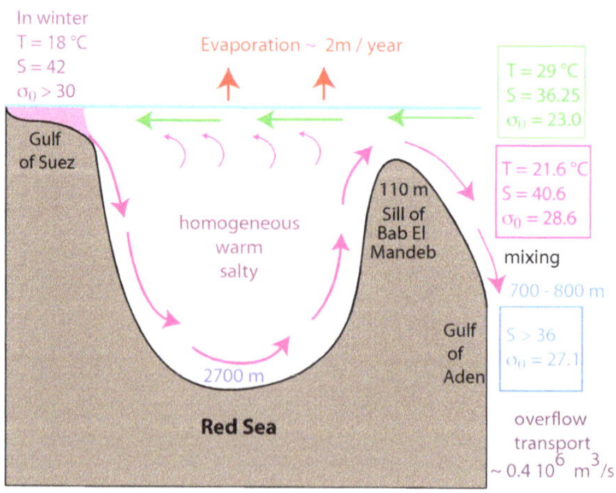

FIG. 4-33 – Schematic of the Red Sea system.

Surface salinity ranges from about 36.2–36.5 at Bab-el-Mandeb to 42 in the Gulf of Suez [**Figures 4-33** and **4-34**] and the density of surface water increases accordingly. In winter, in the Gulf of Suez, under the action of cold and dry north winds, temperatures drop from 31 °C in summer to less than 18 °C. The strong drop in temperature increases surface-water density to more than 30, thus inducing convective motion over the entire depth of the Gulf of Suez [**Figure 4-35**]. Dense water thus formed descends by gravity along the continental slope mixing with intermediate water to form Red Sea Deep Water [**Figure 4-34c**]. Intermediate waters form in winter on the coastal shelves under the effect of cold winds which, by increasing the density of surface water, cause convective motion that does not exceed 200 m in depth. Their characteristics are intermediate between those of surface water, coming from the Gulf of Aden, and those of Red Sea Deep Water.

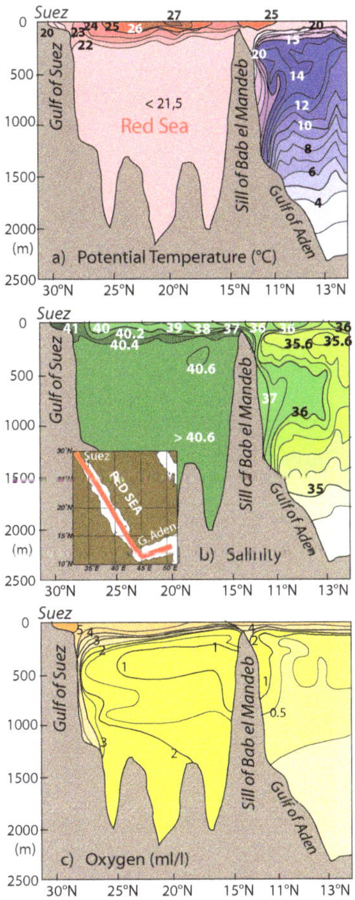

FIG. 4-34 – Longitudinal section of **(a)** potential temperature, **(b)** salinity and **(c)** dissolved oxygen in the Red Sea from the Gulf of Suez to the Gulf of Aden. (After the Wyrtki *Atlas*, 1971.)

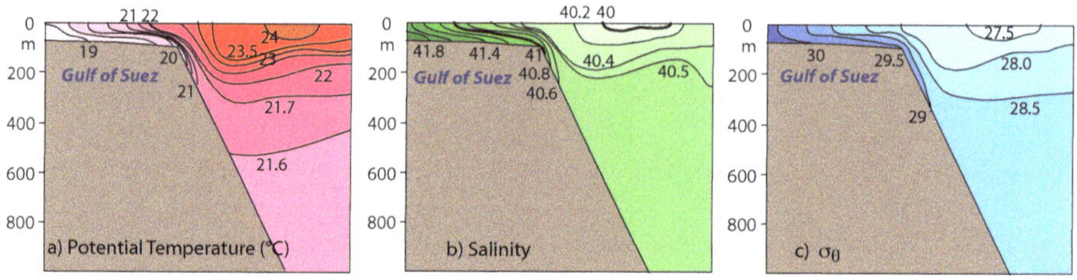

FIG. 4-35 – Wintertime sections of **(a)** potential temperature, **(b)** salinity and **(c)** potential density in the Gulf of Suez. (After Maillard, 1974.)

Red Sea Deep Water, formed each winter, fills the bottom of the basin homogeneously between 100–200 m and the bottom. It is warm (θ = 21.6 °C) but so saline (S = 40.6) that its density reaches 28.60, much greater than that of Antarctic Bottom Water of 27.88.

The Red Sea hydrologic system is shown schematically in **Figure 4-33**. Red Sea Deep Water (RSW) exits over the base of the sill of Bab-el-Mandeb (S > 39.5) into the Gulf of Aden under entering surface water [**Figure 4-36**]. The oxygen level, weak because of high temperatures, decreases with residence time. The oxygen minimum (O_2 < 1 ml l^{-1} or 43 µmol kg^{-1}) is found around 500 m depth [**Figure 4-34c**], the zone of oldest water.

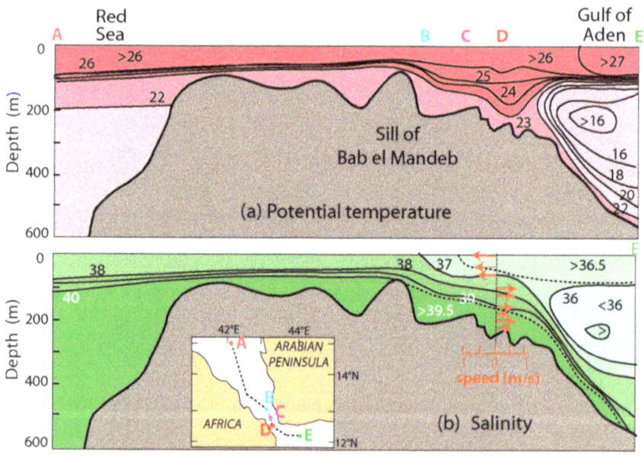

FIG. 4-36 – **(a)** Temperature and **(b)** salinity sections and currents over the Sill of Bab-el-Mandeb. (After Siedler, 1969.)

The Red Sea is the only place in the world, outside of hydrothermal springs, where water temperature exceeds 20 °C[334] at 2,000 m depth. In general, water at this depth forms in regions where surface water is cold enough

334 In the Mediterranean the temperature is greater than 12 °C at 2,000 m.

to reach high density. In the Red Sea as in the Mediterranean, the important increase of surface salinity allows high winter densities.

Atmospheric circulation in the south of the region is influenced by the monsoon regime. North of 20°N, winds are from the NNW [**Figure 4-37**]. From October to May over the southern half of the Red Sea they come from the SSE and reinforce the flow entering at Bab-el-Mandeb.

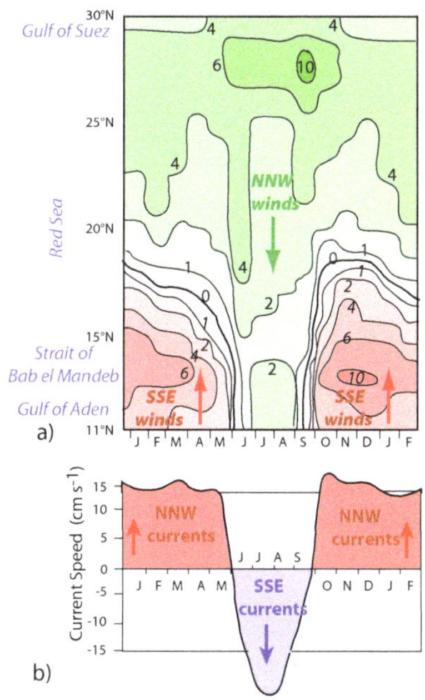

FIG. 4-37 – Seasonal variability of **(a)** winds over the Red Sea (in m s^{-1}) and **(b)** surface currents deduced from ship drifts (in cm s^{-1}) in the Strait of Bab el Mandeb. (After Patzert, 1972.)

In summer, more moderate winds from the NNW blow across the entire basin and drive a current that exits at the surface [**Figure 4-38**]. The flow entering the Red Sea, with temperature and salinity lower than in the Red Sea, is then found in an intermediate level, between Red Sea Deep Water outflow at the base of the sill and surface water driven by the wind toward the Gulf of Aden [**Figure 4-38**]. The outflow of Red Sea Deep Water over the sill is then hindered by water inflow from the Gulf of Aden.

The surface circulation takes place in the form of large meanders generating transversal currents. It intensifies from the Strait of Bab-el-Mandeb (about 0.5 Sv) to the Gulf of Suez, by recirculation of subsurface water (about 1.5 Sv in the center of the basin) entrained by surface water.

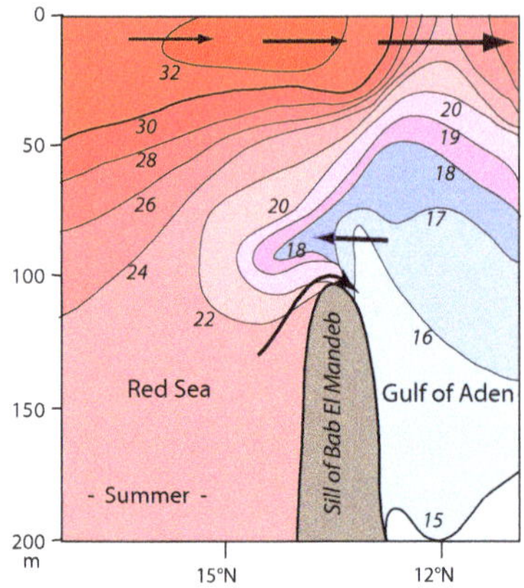

FIG. 4-38 – Temperature section over the sill of Bab-el-Mandeb in summer when the winds blow toward the Gulf of Aden. Three layers can be observed in the circulation over the sill. (After Patzert, 1972.)

As it leaves the Strait of Bab-el-Mandeb, Red Sea Deep Water descends along the continental slope, mixing with colder and less saline water of the Gulf of Aden. Mixing strongly decreases salinity and temperature as well as density, so that the Red Sea Deep Water does not descend to the bottom. The mixture finds a level of equilibrium around 700–800 m in the Gulf of Aden. The high temperature of Red Sea Deep Water associated with high density generates temperature inversions at the outlet of the Strait of Bab-el-Mandeb [**Figure 4-36**] that can be found as far south as the equator [**Figure 4-39**: Station 5]. Temperature inversions are rare in the world ocean. Red Sea Water flows preferentially southward[335] along the Somali coast, deepening and mixing with surrounding water. Profiles at Stations 4 and 5 of **Figure 4-39** show a foliated structure, indicating that mixing is taking place. Red Sea Water is characterized by a salinity maximum around $\sigma_0 = 26.9–27.2$. It can be followed in the Mozambique Channel and as far as 30°S under the Agulhas Current off South Africa, where cores of more saline water, 300 m to 400 m thick, are imbedded in the salinity minimum of Antarctic Intermediate Water that has the same density.

The residence time of water masses in the Red Sea has been estimated at 40 years. Surface water takes nearly 6 years from entry in the south to descent by convection in the north. The outflow at Bab-el-Mandeb of water formed in the Red Sea is about 0.5 Sv, about half the amount of Mediterranean Water exiting at Gibraltar.

Pockets of hot brines have been found in the Red Sea in deep depressions and fissures of the seismically active axis of the basin. Temperatures of more

[335] The flow deviates to the right in the northern hemisphere due to the Coriolis force.

than 56 °C and salinities exceeding 250[336] have been measured. The hot brines cannot be seawater, since their proportion of various salts differs greatly. These brines do not have consequences on the oceanographic characteristics of the Red Sea since, being extremely dense, they do not mix with seawater.

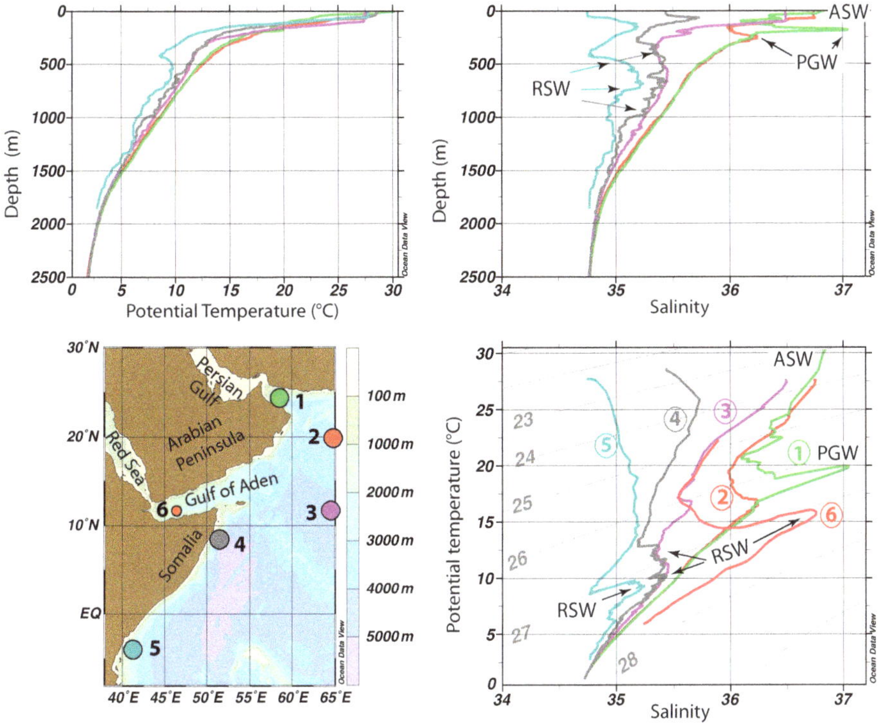

FIG. 4-39 – Temperature and salinity profiles and θ-S diagrams from the NW Indian Ocean showing the saline water formed in the Persian Gulf (PGW), Red Sea (RSW), and the warm and saline surface water of the Arabian Sea (ASW). (With ODV and World Ocean Database.)

5.2.2 The Persian Gulf

In contrast with the Red Sea, the Persian Gulf is a shallow epicontinental sea (average depth 25 m, maximum depth 90 m) 800 km long and about 200 km wide **[Figure 4-40]**[337]. It connects with the Gulf of Oman and the Arabian Sea through the Strait of Hormuz, the limit of the continental shelf, beyond which depth increases rapidly. The Persian Gulf receives runoff water

[336] These temperatures and salinities are beyond the range of conventional instruments.
[337] Its area is about 240,000 km^2.

from the Tigris and Euphrates, which unite to form the Shatt-al-Arab before emptying into the NW Persian Gulf.[338] Climatic conditions are here similar to the Red Sea with an evaporation of 2 m per year, which dominates over fresh-water inputs [**Figure 4-41a**]. The Gulf is thus a concentration basin [**Figure 3-45**] without a sill.

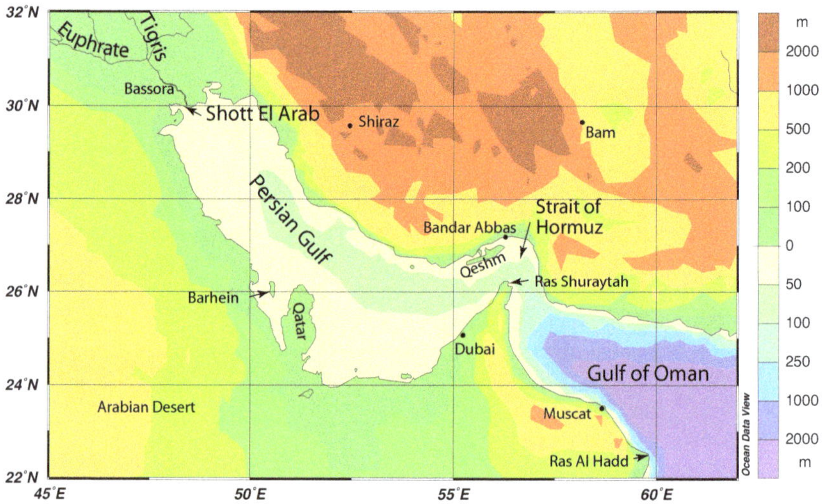

FIG. 4-40 – Topography of the Persian Gulf.

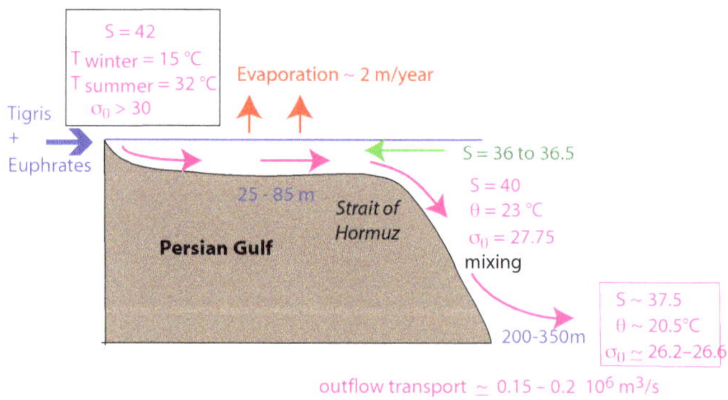

FIG. 4-41A – Schematic of the Persian Gulf system.

Arabian Sea Surface Water enters through the Strait of Hormuz with a salinity between 36 and 36.5. It follows a cyclonic circuit, passing along the Iranian coast northwestward and returning along the Arabian coast

[338] The average flow of the Tigris and Euphrates is 0.0025 Sv.

southeastward (**Figure 4-41b**). It undergoes evaporation throughout the year, which increases surface-water salinity to values of the order of 42, as in the northern Red Sea. In winter cold and dry north winds blow over the Gulf dropping temperatures to 8–10 °C (while they are 30–35 °C in summer). The density of water formed in winter in the northern Gulf reaches values as high as those in the Gulf of Suez ($\sigma_\theta \approx 30$). It flows along the bottom toward the Strait of Hormuz mixing with warmer and less-saline water above. Its characteristics at the Strait of Hormuz are $\theta \approx 23$ °C, $S \approx 40$, and $\sigma_\theta \approx 27.75$. It mixes further as it descends along the continental slope of the Gulf of Oman toward the Arabian Sea, where it reaches a density equilibrium around 200–350 m [**Figure 4-42a**]. Its characteristics are then $\theta \approx 20.5$ °C, $S \approx 37.5$, and $\sigma_\theta \approx 26.6$. The volume of Persian Gulf Water flowing out into the Arabian Sea has been estimated at about 0.2 Sv, around half of that of the volume of Red Sea Water outflow.

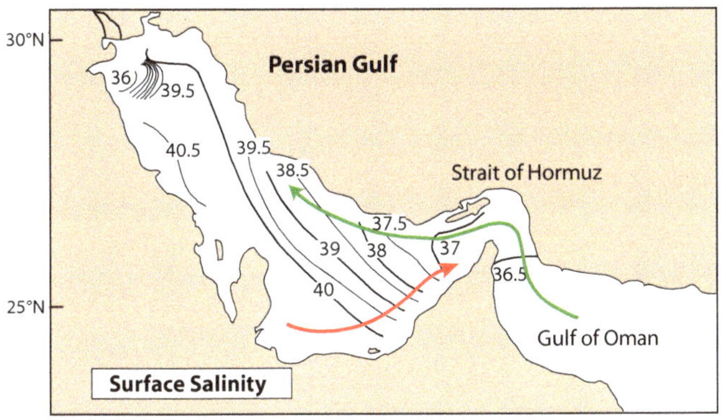

FIG. 4-41B – Distribution of surface salinity in August in the Persian Gulf. (After Brewer and Dyrssen, 1984.)

The Red Sea and Persian Gulf are concentration basins, forming dense and, particularly, saline waters, which influence the oceanographic characteristics of the entire northern Indian Ocean. After leaving their respective basins, they find their density equilibrium at different depths because of differences in flow and mixing. Persian Gulf Water extends in the Gulf of Oman in a thin layer around 200–350 m [**Figure 4-39**]. The farther it travels from its place of formation, the more the salinity maximum is eroded through mixing. Persian Gulf Water is represented on a θ-S diagram by a salinity maximum located around $\sigma_\theta \approx 26.3$–$26.7$. The salinity maximum of Red Sea Water, around $\sigma_\theta \approx 26.7$–$27.3$, lies between 500 m and 1,000 m and deepens to the south [**Figures 4-39** and **4-42b**].

These two water masses, together with surface water of the Arabian Sea [**Section IV.5.3.4**] are responsible for the high salinities of the northern Indian Ocean.

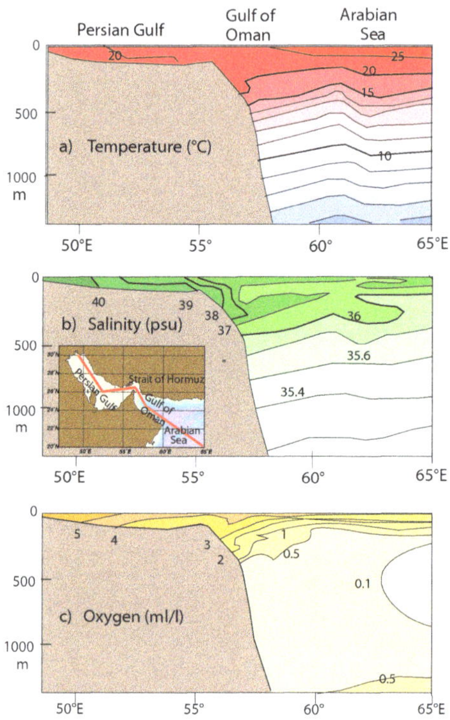

FIG. 4-42A – Sections from the Persian Gulf to the Arabian Sea of **(a)** temperature, **(b)** salinity, and **(c)** dissolved oxygen. (After Wyrtki, 1971.)

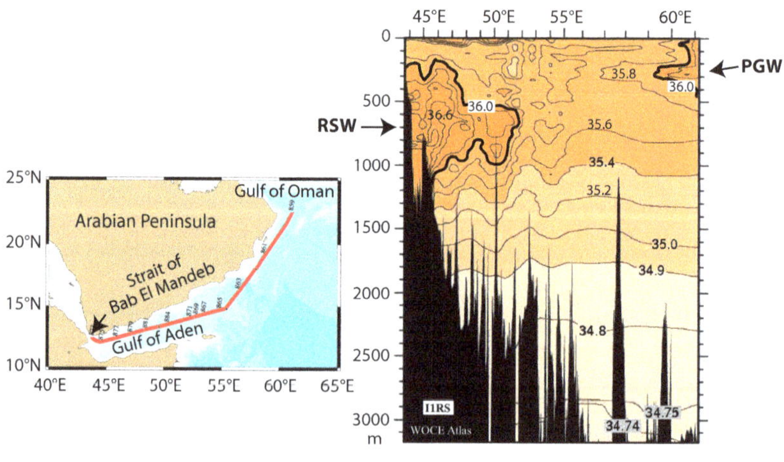

FIG. 4-42B – Salinity section between Gulf of Aden and Gulf of Oman, showing Red Sea Water (RSW) and Persian Gulf Water (PGW) salinity maximums. (From WOCE Indian Ocean Atlas, L.D. Talley, 2011)

5.2.3 Indonesian Seas

The Indian Ocean communicates with the tropical Pacific Ocean [**Figure 4-43**] across the Indonesian Seas whose complicated topography includes a series of basins of varying depths (some with depths greater than 6,000 m), separated by sills of a diversity of depths. The maximum depth of common communication is about 1,300 m. The Torres Strait between Australia and New Guinea [**Figure 4-44**], strewn with islands and shallow (of the order of tens of meters), does not allow significant transport. The Gulf of Thailand, the southern South China Sea, and the Java Sea are shallow regions affected by tidal currents and by alternating monsoon winds. Annual average transports (< 1 Sv) are weak due to the shallow water depth.

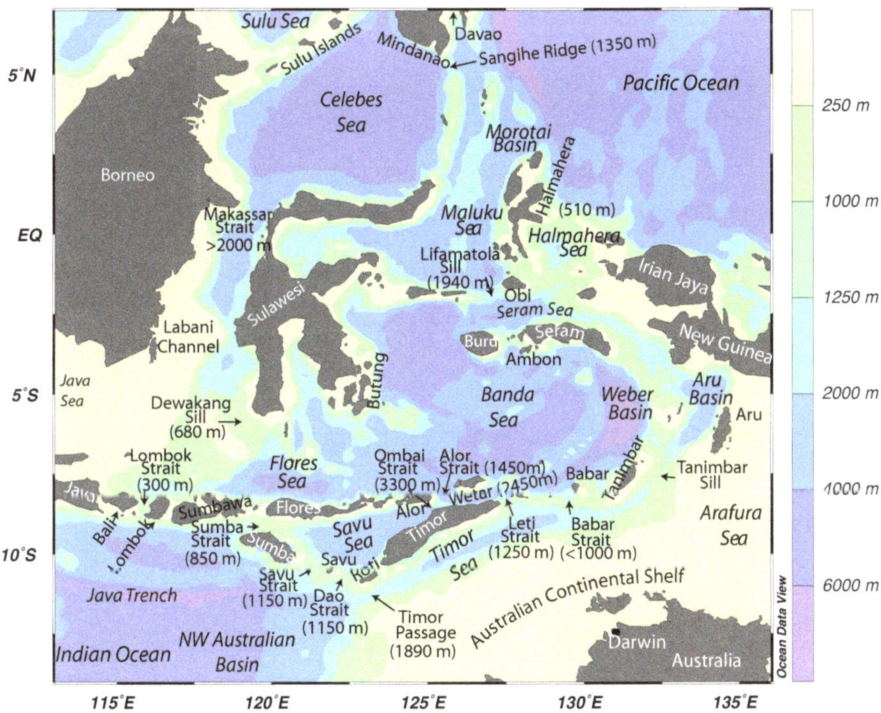

FIG. 4-43 – Detailed topography of the Indonesian Seas.

The distribution of salinity in the thermocline (averaged between the 10 °C and 20 °C isotherms) in the Indian Ocean and western Pacific [**Figure 4-45**] clearly reflects the intrusion of low-salinity water from the North Pacific into the Indian Ocean *via* the Indonesian Seas, around 10°S.

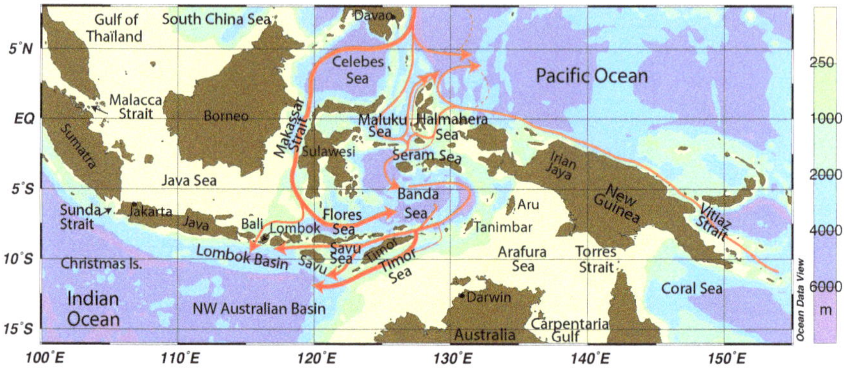

FIG. 4-44 – Topography of the passage between the tropical Pacific Ocean and the Indian Ocean showing the different "Throughflow" paths between the two oceans.

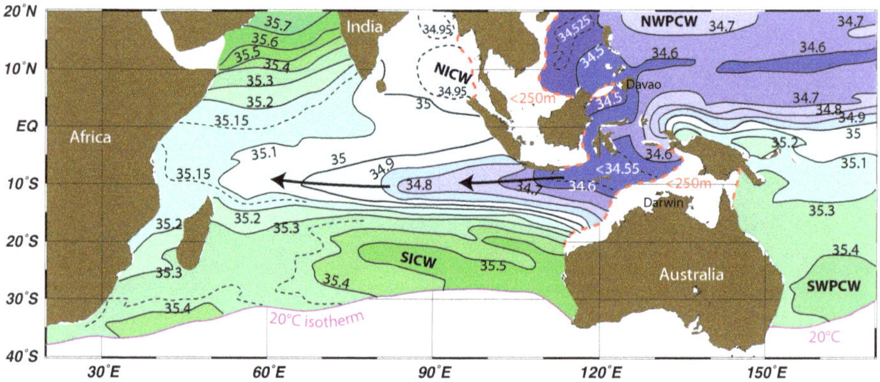

FIG. 4-45 – Average salinity between the 10 °C and 20 °C isotherms in the thermocline. Continuous pink line: 20 °C isotherm at the surface. Dashed pink line: intersection of the 10 °C isotherm with the bottom. [SW Pacific Central Water (SWPCW), NW Pacific Central Water (NWPCW), North Indian Central Water (NICW), South Indian Central Water (SICW).] (Levitus climatological data, after Gordon, 1986.)

Trade winds over the Pacific Ocean accumulate light water on the western boundary of the equatorial Pacific, along the New Guinea coast, the eastern Indonesian islands, and the Philippines [**Section V.3.4**]. As a consequence, a difference in level between the western Pacific and the eastern Indian Ocean (of the order of 16 cm on average between Davao in the Philippines and Darwin in northern Australia) produces a pressure gradient toward the Indian Ocean. The communication between the two oceans through the Indonesian Seas [**Figure 4-44**] allows a water transfer from the Pacific to the Indian Ocean, called the Indonesian "Throughflow" (ITF). The Throughflow is specific to this border region between the two tropical oceans. It includes several points of entry on the Pacific side and several Indian Ocean exits. The passage between the two oceans is complicated by topography [**Figure 4-43**]. The Indonesian Throughflow is composed of different Pacific Ocean water masses [**Figure 4-46**: Stations 1 and 2] that are completely transformed during their passage across the Indonesian Seas. They emerge into the Indian Ocean in the form of Indonesian Intermediate Water [**Figure 4-46**: Stations 3 and 4] where extremes have disappeared, and of seasonally variant Indonesian Surface Water.

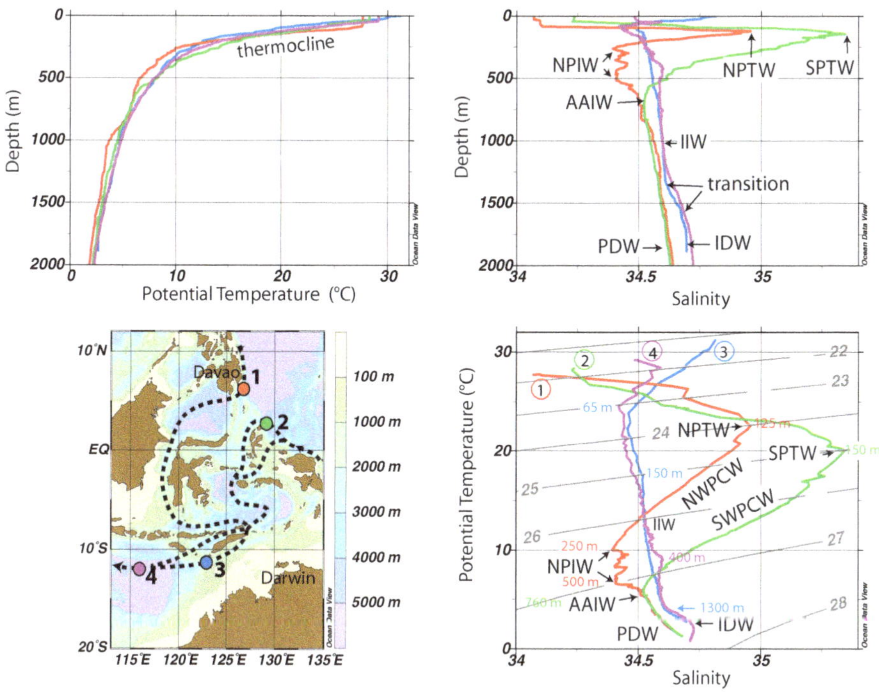

FIG. 4-46 – Profiles of potential temperature, salinity, and θ-S diagrams at the Throughflow entry, south of Mindanao (Station 1), north of Halmahera (Station 2), at the Throughflow exit in the Timor passage (Station 3), and in the NW Australian Basin, north of the front (Station 4). [North Pacific Tropical Water (NPTW), South Pacific Tropical Water (SPTW), NW Pacific Central Water (NWPCW), SW Pacific Central Water (SWPCW), North Pacific Intermediate Water (NPIW), Antarctic Intermediate Water (AAIW), Indian Deep Water (IDW), Pacific Deep Water (PDW).] (With ODV and World Ocean Database.)

The Indonesian Throughflow entry point is between the NW extremity of New Guinea (Irian Jaya) and the southernmost Philippine island of Mindanao. The Throughflow takes different paths to reach the Banda Sea, a deep and extensive Indonesian interior sea, where they mix.

The western, and principal path, fed by the Mindanao Current [**Figure 4-47**: Station 1], penetrates into the Celebes Sea, passing above the Sangihe Ridge (1350 m), then passes through the Makassar Strait, between Borneo (Kalimantan) and Celebes (Sulawesi). This branch is composed of North Pacific waters: North Pacific Surface Water (NPSW), with low salinity; North Pacific Tropical Water (NPTW), characterized by a salinity maximum; North Pacific Central Water (NPCW), corresponding to thermocline water; and North Pacific Intermediate Water (NPIW), characterized by a salinity minimum. Mixing among these water masses is intensified in regions of horizontal narrowing of straits and vertical shallowing of sills, such as the Makassar Strait and the Dewakang Sill (800 m) south of the Makassar Strait. The shallow depth of the Dewakang Strait prohibits deeper water from entering by this route. Part of the upper layer exits directly to the Indian Ocean by the Lombok Strait (300 m) located south of the Makassar Strait between Bali and Lombok.[339] The rest continues across the Flores Sea, [**Figure 4-44**] where North Pacific water masses are now attenuated by mixing [**Figure 4-47**: Station 4] into the Banda Sea.

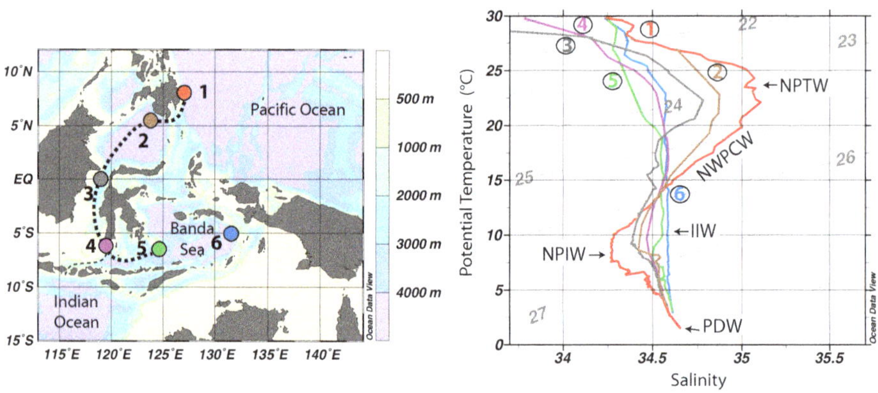

FIG. 4-47 – Evolution of θ-S diagrams along the western path, from the Mindanao Current (Station 1) as far as the eastern Banda Sea (Station 6). [North Pacific Tropical Water (NPTW), NW Pacific Central Water (NWPCW), North Pacific Intermediate Water (NPIW), Pacific Deep Water (PDW), Indonesian Intermediate Water (IIW).] (With ODV and World Ocean Database.)

The eastern path, which involves the deeper layers, has two possible itineraries. Both are fed by the New Guinea Coastal Undercurrent coming from the South Pacific [**Section V.4.3.3**]. The New Guinea Coastal Undercurrent is composed of South Pacific Tropical Water (SPTW), characterized by a salinity maximum, Southwest Pacific Central Water (SWPCW),

339 This is a zone of strong turbulence where many small-scale eddies can be observed.

and Antarctic Intermediate Water (AAIW), marked by a salinity minimum [**Figure 4-48**: Station 1]. Upper-layer thermocline waters practically do not enter the Banda Sea, but feed the North Equatorial Countercurrent eastward (see **Figure 5-31**). Lower-layer thermocline waters penetrate into the Banda Sea, either through the Maluku Sea and the Lifamatola Passage (sill at 1,950 m) and flow into the Seram Sea, or they follow a more direct but less deep path (sill at 510 m) through the Halmahera Sea. From there, they join the northern Banda Sea [**Figures 4-43, 4-44** and **4-48**]. After the passage through the Halmahera Sea, the Maluku Sea, and the Seram Sea, the salinity maximum of South Pacific Tropical Water and the salinity minimum of Antarctic Intermediate Water have eroded when they arrive in the Banda Sea [**Figure 4-48**: Station 5].

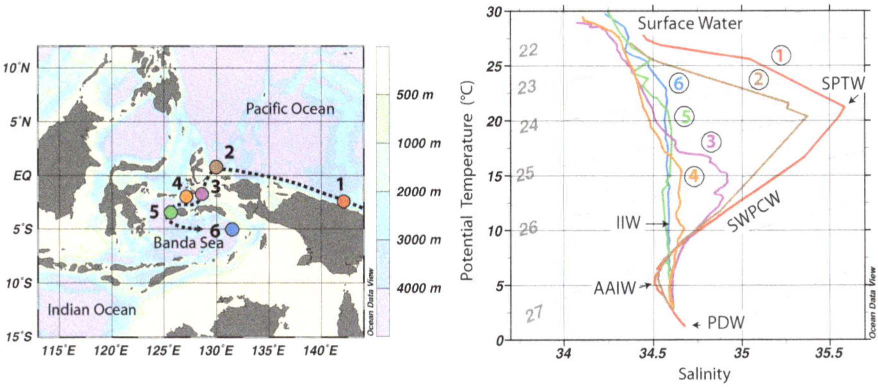

FIG. 4-48 – Evolution of θ-S diagrams along the eastern path, from the New Guinea Coastal Current (Station 1) as far as the eastern Banda Sea (Station 6). [South Pacific Tropical Water (SPTW), SW Pacific Central Water (SWPCW), Antarctic Intermediate Water (AAIW), Pacific Deep Water (PDW), Indonesian Surface Water (ISW), Indonesian Intermediate Water (IIW).] (With ODV and WOCE data.)

The Indonesian zone is a warm-water region [**Figure 4-27**]. Located under the strongest atmospheric convection on the planet, it receives abundant precipitation [**Figure 4-6**] which reduces sea-surface salinity [**Figure 4-29**]. Warm and low-salinity surface water has low density and the climate of the Indonesian region (heating and input of fresh water) can only diminish it and thereby intensify the stratification. However, in the Banda Sea, down to more than 5,000 m depth, oxygen values of the order of 100 µmol kg^{-1} occur, higher than in the North Pacific (or in the Arabian Sea, **Figure 4-42a**). These values indicate a renewal of Deep Water, which cannot take place through convection because surface water density is too low.

Two processes drive vertical mixing over great depths along the Indonesian Throughflow to the Indian Ocean. One is turbulence, generated during the passage over several sills. The other is the strong energy of internal tidal waves,[340] which are trapped in fragmented basins, where they reflect on basin walls. The internal tidal waves intensify vertical mixing through vertical diffusion[341]. As a result, mixing in the Banda Sea between the western branch and the eastern branch gives birth to a water mass that is relatively homogeneous in salinity in which source characteristics are no longer identifiable [**Figures 4-47** and **4-48**, Station 6). Salinity variations with depth are weak under the Surface Water, which undergoes seasonal climatic variations.

The outflow of Banda Sea Water to the Indian Ocean takes place through two paths. One, south of Timor, passes through the Leti Strait (1,250–1,300 m) or the less deep Babar Strait ($<$ 1,000 m), the Timor Sea and Timor Passage (1,890 m) [**Figures 4-43** and **4-49**]. The other, north of Timor, passes through the Wetar Basin, the Ombai Strait (3,250 m, between Alor and Timor) and the Savu Sea, which communicates with the Indian Ocean *via* the Sumba Strait (800 m, between Flores and Sumba), the Savu Strait (1,150 m, between Sumba and Savu) and the Dao Strait (about 1,150 m, between Savu and Roti).

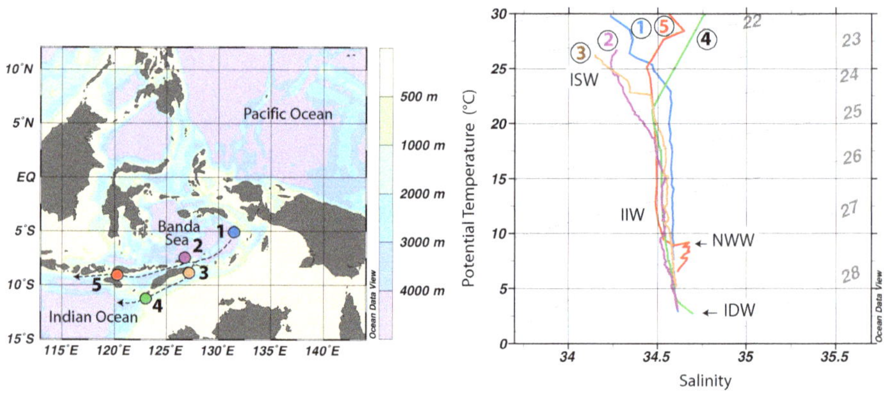

FIG. 4-49 – θ-S diagrams at the Throughflow exit, from the Banda Sea toward the Indian Ocean, Station 1 is in the eastern Banda Sea, Station 2 is in the southern Banda Sea, Station 3 is in the Timor Sea, Station 4 is in the Timor Passage, and Station 5 is in the Sumba Strait. [Indonesian Surface Water (ISW), Indonesian Intermediate Water (IIW), NW Indian Water (NWW), Indian Deep Water (IDW).] (With ODV and WOCE data.)

The Indonesian Throughflow at the outlets of the Indonesian straits is composed of Indonesian Surface Water (ISW), whose characteristics vary with the season, and Indonesian Intermediate Water (IIW), characterized

340 In this region topographically trapped internal tidal waves are most intense (ten times more than in the open ocean).

341 It is suggested that water-mass transformation in the Indonesian Seas could contribute to upwelling of cold water in the global conveyor belt.

by homogeneous salinity ($34.56 < S < 34.60$), from 600 m down to about 1,300 m [**Figure 4-46**, Station 3, **Figures 4-49** and **4-50**].

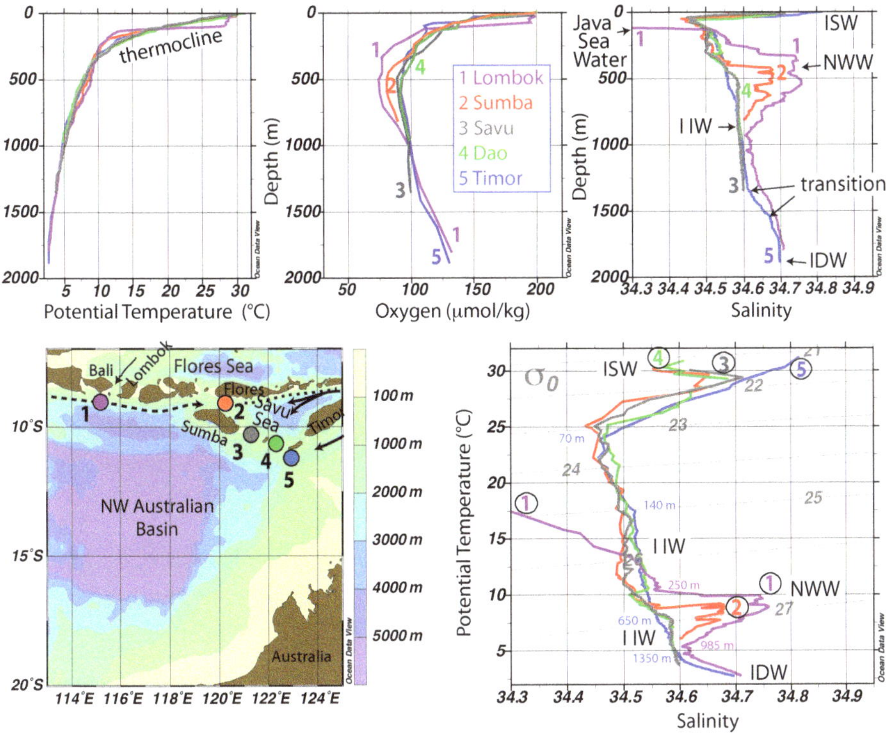

FIG. 4-50 – Profiles of θ, S, O_2, and θ-S diagrams in the straits leading into the Indian Ocean: the Lombok Strait (Station 1), the Sumba Strait (Station 2), the Savu Strait (Station 3), the Dao Strait (Station 4), and the Timor Passage (Station 5). [Indonesian Surface Water (ISW), Indonesian Intermediate Water (IIW), NW Indian Water (NWW), Indian Deep Water (IDW).] (With ODV and WOCE data.)

Indonesian Intermediate Water does not form at the surface as do most water masses. It forms in the interior of the Indonesian Seas through mixing of several water masses coming from the Pacific Ocean. Its formation region and characteristics are clearly distinctive.

The depth of the Timor Passage (1,890 m) is greater than that of the straits feeding it east of the Timor Sea. This allows more saline and more oxygenated Indian Ocean Deep Water to enter over the sill toward the Timor Sea [**Figure 4-50**, Station 5]. Blocked by shallower sills to the northeast, it cannot exit, and returns toward the Indian Ocean, mixing with the deep part of Indonesian Intermediate Water. The water thus formed is a transition between Indonesian Intermediate Water and Indian Ocean Deep Water, found between 1,350 m and 1,550 m in the Timor Passage [**Figure 4-50**].

Saline and oxygen-poor water from the northwest Indian Ocean (NWW) which flows along the coast of Java penetrates into the Savu Sea through the Sumba Strait [**Figure 4-49**, Station 5 and **Figure 4-50**, Stations 1 and 2]. It is characterized by a salinity maximum and a dissolved oxygen minimum between 425 m and the sill depth (850 m).

Direct current measurements provide simultaneous estimates of transport in the various passages. In the Strait of Makassar (Dewakang Sill, 680 m deep) the annual average transport southward is estimated at 11.6 Sv ± 3.3 Sv, to which should be added about 1 Sv coming from the South China Sea. In the Lifamatola Strait (1,940 m) transport is estimated at 2.5 Sv ± 1.5 Sv southward between 1,250 m and the bottom. The water involved must upwell in the Banda Sea in order to leave by the shallower sills. In the Lombok Strait (between Bali and Lombok, 300 m deep, 35 km wide) transport is estimated at 2.6 Sv ± 0.6 Sv. It varies between 1.8 Sv and 3.2 Sv depending on sea level on the Indian Ocean side. In the Ombai Strait (between Alor and Timor, 3,300 m deep, 37 km wide) [**Figure 4-43**], which opens onto the Savu Sea and the Sumba, Savu, and Dao straits, transport is estimated at 4.9 Sv, varying between 2.7 Sv and 5 Sv. In the Timor Passage (1,890 m deep, 160 km wide) transport reaches 7.5 Sv, between 6.2 Sv and 10.5 Sv.

In sum, westward transport of Indonesian Throughflow leaving through the Lombok Strait, the Ombai Strait, and the Timor Passage reaches an annual mean of 15 Sv, ranging between 10.7 Sv and 18.7 Sv, two thirds of which (11.2 Sv) flows in the 0-300 m layer.

Sea-level fluctuations between the entry straits on the Pacific Ocean side and the exit straits on the Indian Ocean side produce variations in transport between the two oceans. The variability is complex since it is subject to a number of influences: far-off, through waves, or nearby, through winds, whose effects cut back or increase the flow accordingly. The Throughflow undergoes interannual variations due to large-scale climatic anomalies of the ENSO type [**Section V.3.4**] and varies seasonally under the effect of monsoon winds which blow over the entire region. It also undergoes intra-seasonal variations linked to westerly winds gusts on the equator.

The interannual variability depends on the El Niño-Southern Ocean anomaly over the Pacific Ocean and on anomalies in the equatorial zone of the Indian Ocean called the Indian Ocean Dipole (IOD). These two climate anomalies are not always in phase.[342] In the El Niño[343] phase, there is less supply of water where the Throughflow enters and its transport diminishes. In the La Niña[344] phase, the supply of water increases and the Throughflow

342 In 1993–1994 and 1997–1998 IOD anomalies were in phase with El Niño anomalies over the Pacific.

343 The Pacific southeast trades weaken during El Niño.

344 The Pacific southeast trades intensify during La Niña.

intensifies. When the Indian Ocean Dipole is positive, Indian Ocean westerly winds at the equator during the inter-monsoonal period are replaced by easterly winds. Surface temperature, thermocline depth, and sea level decrease in the east and along the coast of Java. Throughflow transport then increases. In the inverse situation of reinforcement of Indian Ocean westerly winds on the equator, it diminishes.

Throughflow in the surface layer is dominated by the seasonal influence of local monsoon winds. During the northwest monsoon, winds have a tendency to reduce surface-layer transport toward the Indian Ocean in the Timor Passage, and to periodically reverse surface flow in the Lombok and Ombai straits. During surface reversals the Throughflow is found at the subsurface in the Ombai Strait. In the northern part of this strait, up against the island of Alor, a narrow surface filament flows toward the Banda Sea a large part of the year. The Throughflow occupies the rest of the strait south of this filament.

During the southeast monsoon, wind generates a divergence in the Banda Sea and along the exit straits, producing upwelling south of Java and lowering sea level. These two processes intensify the transport in the Throughflow surface layers. In the Timor Passage, which transports half the Throughflow, the majority of the transport is concentrated in the 0-200 m layer with a secondary maximum around 1,200 m.

Sea level at the exit of the Lombok Strait undergoes variations due to westerly wind events over the equatorial zone of the Indian Ocean during inter-monsoons, which propagate in the form of trapped waves[345] along the equator and the eastern coast as far as the exit straits. When sea level is high, it acts as a dam to the Throughflow. When it is low, the Throughflow flows easily into the Indian Ocean.

In the Timor Passage, current reversals appear between 1,400 m and the sill, also connected, probably, to intraseasonal equatorial waves, whose energy can propagate at depth, allowing them to reach the Timor sill.

The Throughflow feeds the South Equatorial Current and forms, at the thermocline level, a Hydrologic Front (SEC Front) with homogeneous salinity and oxygen around 13°S, forming a boundary between northern and southern Indian Ocean water masses [see **Section IV.5.3.3**].

---

[345] Equatorial Kelvin waves generated by wind events at the equator propagate eastward at the equator, giving rise to coastal Kelvin waves that propagate poleward along eastern coasts. Their energy penetrates to depth along their trajectory. Thus, above 300 m, energy escapes through the Lombok Strait. Above 800 m it enters the Savu Sea through the Sumba Strait. Above 1,150–1,250 m it enters into the Savu Sea through the Savu and Dao Straits south of Sumba. Below 1,150–1,250 m it can reach 1,890 m depth in the Timor Passage and can influence deep transport and trigger current reversals.

5.3 Indian Ocean Water Masses

5.3.1 Antarctic Intermediate Water and Subantarctic Mode Waters

Subantarctic Mode Waters, the densest of which is Antarctic Intermediate Water, represent the most important volume of water subducted into southern hemisphere anticyclonic gyres. Subantarctic Mode Waters correspond to thick mixed layers formed just to the north of the Subantarctic Front. They are characterized by a potential-vorticity minimum, that is, by a vertical density gradient minimum. They are transported eastward by the Circumpolar Current and are subducted in various places in the subtropical gyres.

In the Indian Ocean, Subantarctic Mode Water of density 26.6–26.7 subducts north of Kerguelen, 60°E–90°E, a region where the Subantarctic Front is near the Subtropical Front and feeds the western part of the subtropical gyre. Southwest of Australia, between 90°E and 120°E, this type of Subantarctic Mode Water is enriched in oxygen through convection and forms a thick mixed layer around the Subtropical Front. It feeds the eastern part of the subtropical gyre. Subantarctic Mode Water of density $\sigma \sim 26.8$, comes from southwest of Tasmania and enters through the southeastern Indian Ocean under less-dense Subantarctic Mode Water of the eastern Indian Ocean and above denser Antarctic Intermediate Water.

South of 15°S, on the meridional section between the Bay of Bengal and the Antarctic Ocean [**Figure 4-51**], the salinity minimum around 1,000 m depth between the Subantarctic Front and 15°S shows the influence of Antarctic Intermediate Water (AAIW) of density 27.1–27.3. Antarctic Intermediate Water forms mainly by convection and subduction in the southeast Pacific in the Subantarctic Front zone off the coast of Chile where surface salinity is low [**Section II**].

Antarctic Intermediate Water penetrates into the subtropical gyre of the Indian Ocean through two branches. One branch comes from the southeast Pacific Ocean through the southern Atlantic Ocean *via* the Malvinas Current confluence region. It is then entrained by the Circumpolar Current and enters the subtropical gyre around 80°E–90°E. The other, more important branch enters through the southeast Indian Ocean. It comes from the east in the current south of Tasmania, north of the Circumpolar Current. Its origin is also the southeast Pacific. These two components of Antarctic Intermediate Water are not locally ventilated. They receive colder and less saline water at depth from the polar zone by the intrusion of eddies across the Subantarctic Front.

In **Figure 4-52**, Station 1, the nearest to the Antarctic Ocean, the salinity minimum in Antarctic Intermediate Water is associated with an oxygen maximum since it is the station nearest to its source. To the north, the salinity minimum in Antarctic Intermediate Water is attenuated through mixing with Indian Ocean Deep Water (IDW) below and South Indian Central Water (SICW) above [**Figures 4-51** and **4-52**, Stations 2, 3, 4]. To the west, under the Agulhas Current, traces of more saline and less oxygenated Red Sea Water are found at the same density levels. They mix with Antarctic Intermediate Water to form interleaving layers that perturb the profiles. Indonesian Water entrained by the South Equatorial Current and forming the Hydrologic Front, limits the extension northward around 15°S of AAIW, as well as Subantarctic Mode Water [**Figure 4-52**, Station 5].

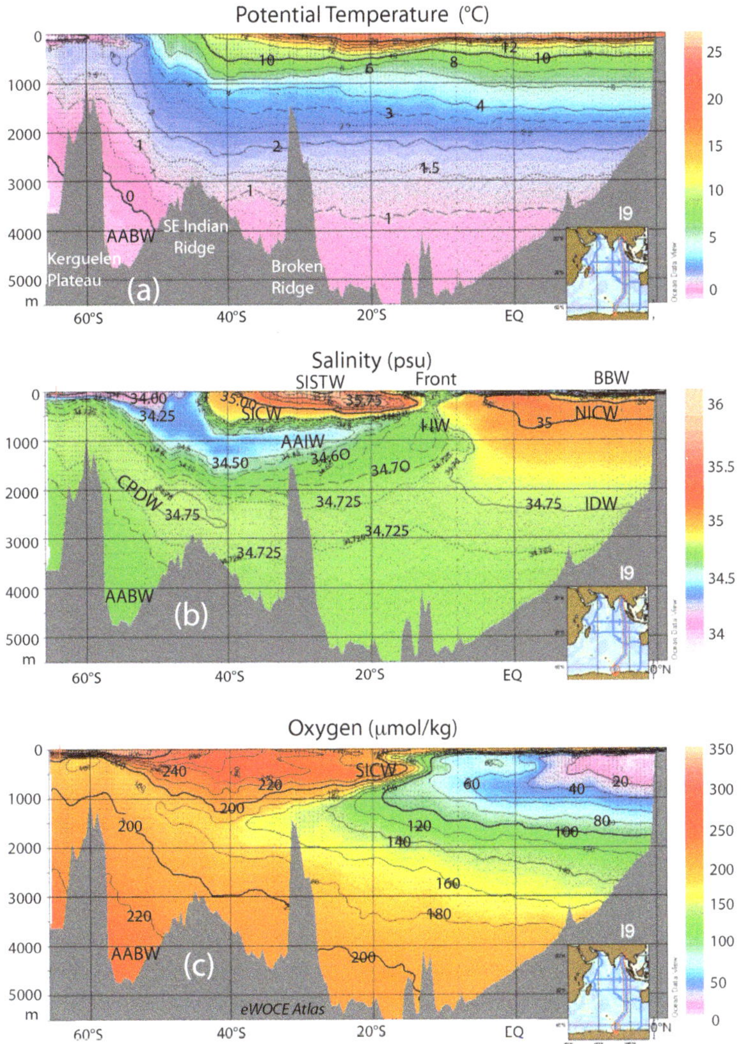

FIG. 4-51 – Meridional sections from the Bay of Bengal Bay to the Antarctic Ocean (**a**) potential temperature (°C), (**b**) salinity (psu), (**c**) dissolved oxygen (mmol/kg). [Bay of Bengal Water (BBW), South Indian Subtropical Water (SISTW), North Indian Central Water (NICW), South Indian Central Water (SICW), Antarctic Intermediate Water (AAIW), Indian Deep Water (IDW), Circumpolar Deep Water (CPDW), Antarctic Bottom Water (AABW).] (WOCE Section I09, *eWOCE Atlas*, Schlitzer, 2002.)

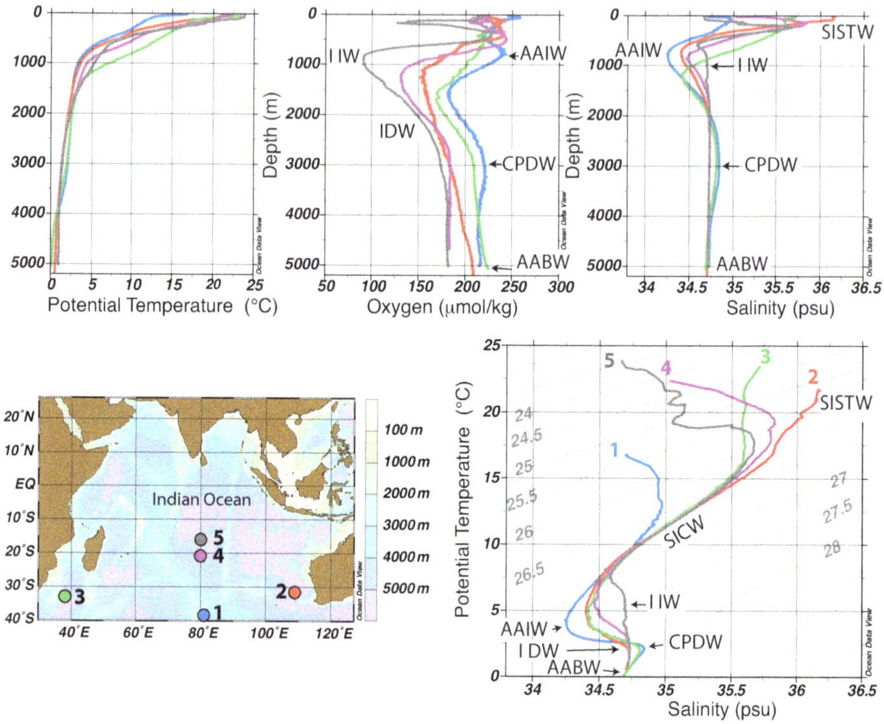

FIG. 4-52 – Profiles of potential temperature, salinity, dissolved oxygen, and θ-S diagrams in the southern Indian Ocean, showing the evolution of Antarctic Intermediate Water (AAIW). [Circumpolar Deep Water (CPDW), South Indian Central Water (SICW), Indian Deep Water (IDW), Antarctic Bottom Water (AABW), Indonesian Intermediate Water (IIW), South Indian Subtropical Water (SISTW).] (With ODV and WOCE data.)

5.3.2 South Indian Ocean Central Water, Subtropical Mode Waters, South Indian Ocean Subtropical Water

South Indian Ocean Central Water (SICW) is the water mass of the Indian Ocean thermocline south of about 13°S [the Hydrologic Front (SEC Front), **Section IV.5.3.3**, following]. It forms in the subtropical convergence zone where surface water tends to subduct (plunge) under lighter water, between 40°S and 30°S [**Section I.3.2**]. Under the effect of evaporation and low precipitation [**Section I.2.5.2**] surface salinity increases in this zone between 40°S and 30°S [**Figure 4-29**]. Seasonal temperature variations can reach 6 °C to 7 °C, producing density variations great enough to induce convective motion in winter leading to the formation of several Subtropical

Mode Waters of different densities depending on their salinity and temperature. The coldest of the Subtropical Mode Waters is that which forms southernmost, around the Subtropical Front (40°S). Mode waters are characterized by a vertical density gradient minimum [potential-vorticity minimum, **Section I.5.3.4**], traces of their formation through convection, and by an oxygen maximum [**Figure 4-53**, Station 5, **Figure 4-54**, and **Figure 4-55**, Station 6]. These mode waters are subducted into the interior of the anticyclonic gyre under wind influence. At 30°S, Subtropical Mode Waters that form South Indian Ocean Central Water are found under the subtropical salinity maximum and reach a depth of 500 to 600 m with temperatures between 10 °C and 17 °C and salinities between 34.7 and 35.8.

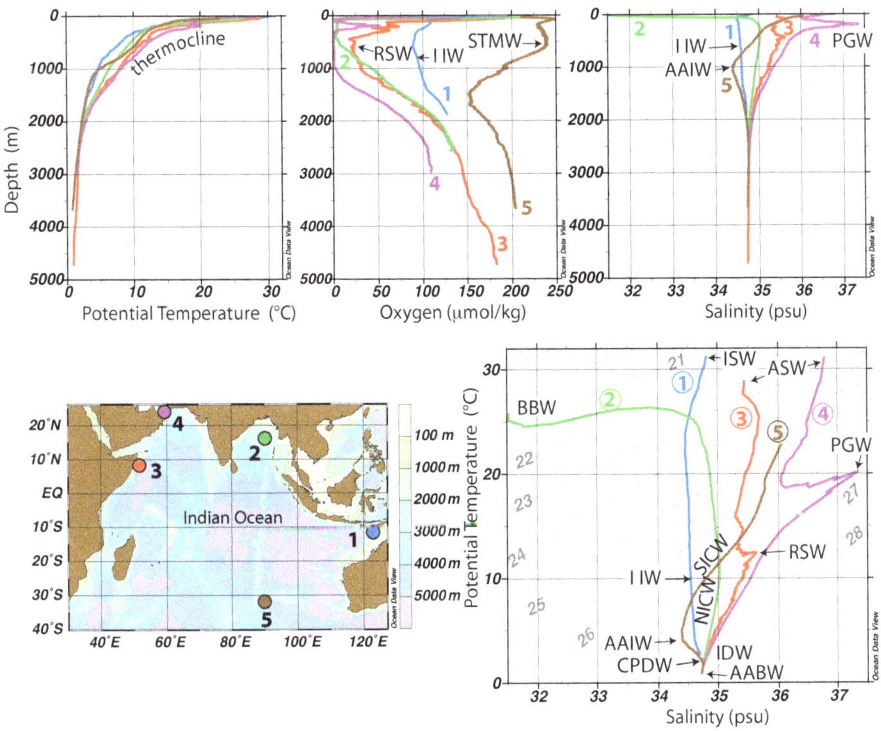

FIG. 4-53 – Profiles of potential temperature, salinity, dissolved oxygen, and θ-S diagrams from stations in the Indian Ocean. [Bay of Bengal Water (BBW), Arabian Sea Water (ASW), Persian Gulf Water (PGW), Red Sea Water (RSW), Indonesian Surface Water (ISW), Indonesian Intermediate Water (IIW). South Indian Central Water (SICW) including Subtropical Mode Water (STMW), North Indian Central Water (NICW), Antarctic Intermediate Water (AAIW), Indian Deep Water (IDW), Antarctic Bottom Water (AABW).] (With ODV and WOCE data.)

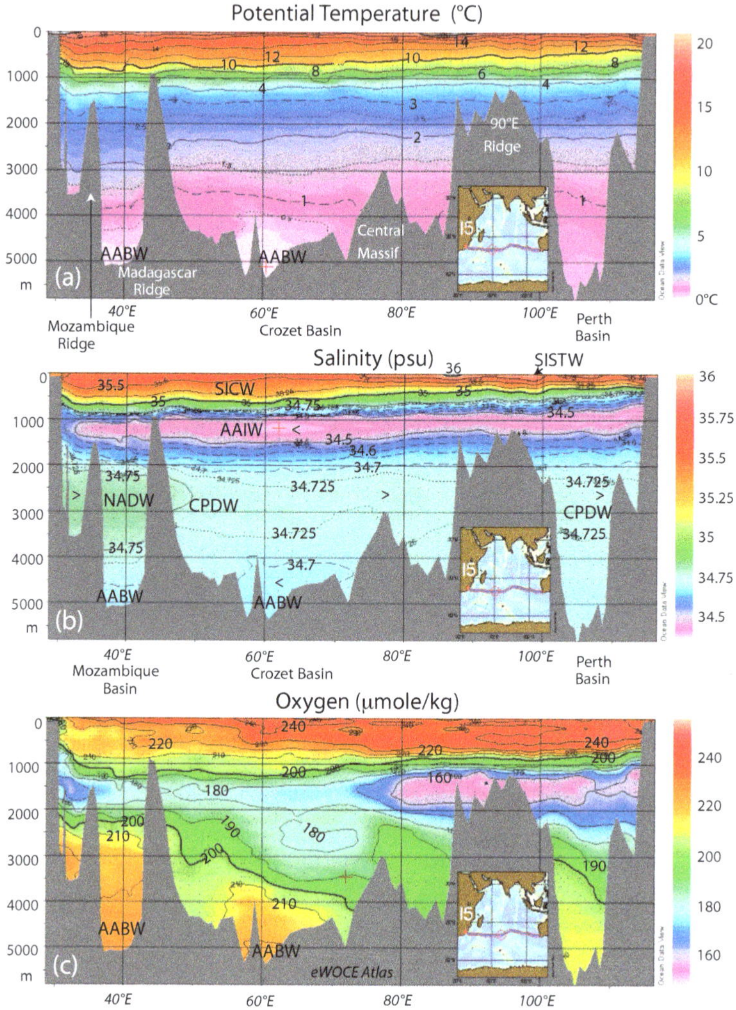

FIG. 4-54 – Zonal sections at 32°S in the Indian Ocean of **(a)** potential temperature, **(b)** salinity, and **(c)** dissolved oxygen. [South Indian Subtropical Water (SISTW), South Indian Central Water (SICW), Antarctic Intermediate Water (AAIW), North Atlantic Deep Water (NADW), Indian Deep Water (IDW), Circumpolar Deep Water (CPDW), Antarctic Bottom Water (AABW).] (WOCE Section I05, *eWOCE Atlas*, Schlitzer, 2002.)

The subtropical salinity maximum, located on the surface around 30°S [**Figure 4-55**, Station 5], marks the place of formation of South Indian Ocean Subtropical Water (SISTW). Indonesian Intermediate Water (IIW) and low-salinity surface water from the Hydrologic Front limit the northward extension of South Indian Ocean Subtropical Water [**Figure 4-55**, Stations 6 and 3). At Stations 3 and 6 the oxygen maximum between 300 m and 450 m shows traces of Subtropical Mode Water in South Indian Ocean Central Water, weakened by mixing with Indonesian Water having a lower dissolved oxygen concentration (≈ 95 μmol kg^{-1}).

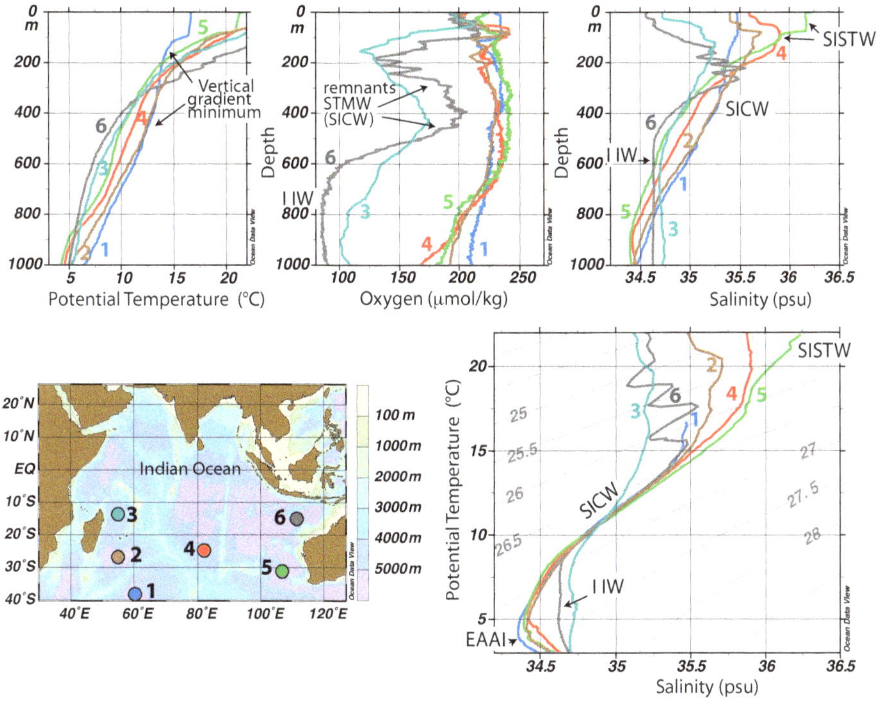

FIG. 4-55 – Profiles of potential temperature, salinity, dissolved oxygen, and θ-S diagrams of upper south Indian Ocean waters. [South Indian Central Water (SICW), Subtropical Mode Water (STMW), South Indian Subtropical Water (SISTW), Antarctic Intermediate Water (AAIW), Indonesian Intermediate Water (IIW).] (With ODV and WOCE data.)

High oxygen levels in South Indian Ocean Central Water correspond to the densest and coldest Subtropical Mode Water (O_2 > 225 μmol kg^{-1}, or 5 ml l^{-1}) and indicate a rapid circulation in the southern anticyclonic gyre. The characteristics of South Indian Ocean Central Water are similar to those of Central Water of the South Atlantic and Southwest Pacific.

5.3.3 The Hydrologic Front

The Indian Ocean Hydrologic Front (or South Equatorial Current Front), centered around 12°S–13°S, extends from the Indonesian straits as far as Madagascar [**Figures 4-51, 4-56** and **4-57**]. The Hydrologic Front is unique to the Indian Ocean. It exists due to the input of Indonesian Waters in the east and the presence of the South Equatorial Current at the same latitude.

Composed of Surface Water from the Indonesian seas (ISW) and Indonesian Intermediate Water (IIW) that form its thermocline, it is characterized by a salinity minimum. It represents the path of Indonesian Waters entrained by the South Equatorial Current. Low-salinity Surface Water is modified by regional seasonal climatic variations, especially in January and February when it is under the zone of strong ITCZ precipitation [**Figures 4-6** and **4-30**].

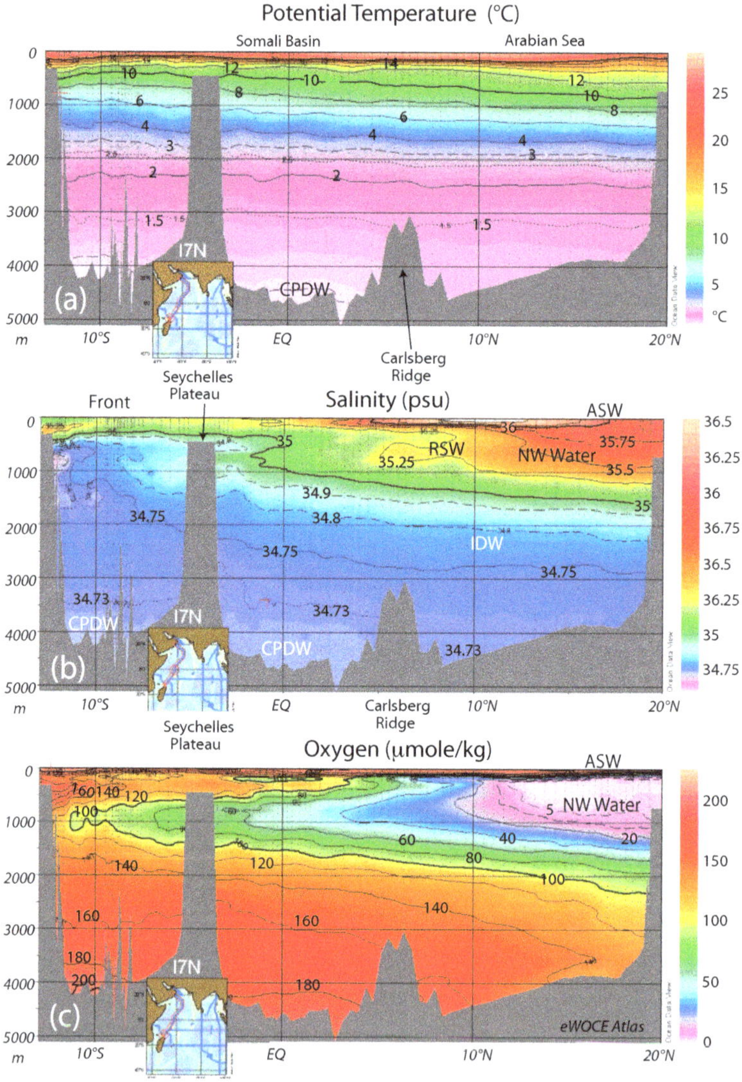

FIG. 4-56 – Meridional section from the Arabian Sea to Cape Amber (the northernmost point of Madagascar) of **(a)** potential temperature, **(b)** salinity, and **(c)** dissolved oxygen. [Arabian Sea Water (ASW), Red Sea Water (RSW), North West Water (NWW), Circumpolar Deep Water (CPDW), Indian Deep water (IDW)] (WOCE Section I07, *eWOCE Atlas*, Schlitzer, 2002.)

The Hydrologic Front separates saline waters formed in the northwest Indian Ocean from waters coming from the southern Indian Ocean, namely South Indian Ocean Central Water, South Indian Ocean Subtropical Water, and Antarctic Intermediate Water, which is less saline than Indonesian Intermediate Water. Under the Surface-Water layer, vertical salinity variations are weak down to the bottom [**Figures 4-51** and **4-57**]. The characteristics of the front evolve through mixing with adjacent waters westward and its components are entrained in different branches of the South Equatorial Current to the north and south.

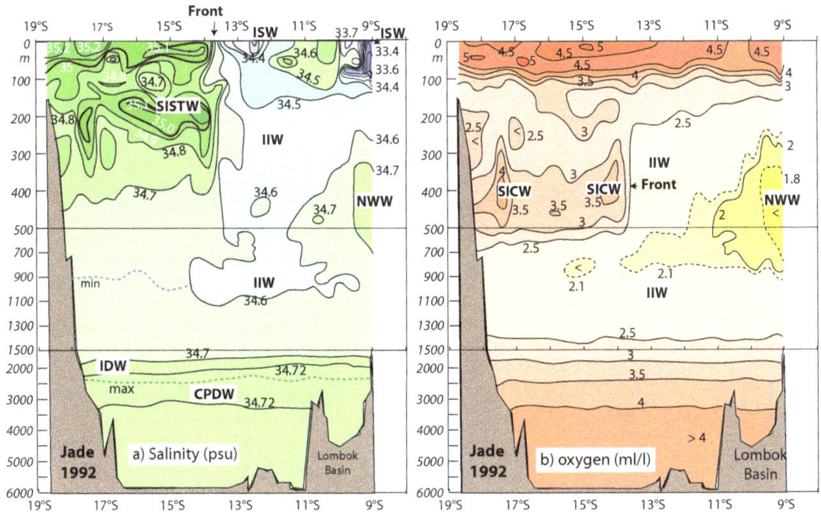

FIG. 4-57 – Section from the JADE[346]92 Expedition in the eastern Indian Ocean, between Bali and Australia around 116°E showing (**a**) salinity and (**b**) dissolved oxygen. The position of the section is shown in **Figure 4-58**. [South Indian Subtropical Water (SISTW), South Indian Central Water (SICW), Indonesian Surface Water (ISW), NW Indian Water (NWW), Indian Deep Water (IDW), Circumpolar Deep Water (CPDW).] There is a change of vertical scale at 500 m and at 1,500 m. (After Fieux *et al.*, 1996a.)

On the JADE section between Bali and the Australian coast, near the Indonesian straits, Indonesian Intermediate Water is found between 11°S and 14°S [**Figure 4-57**]. On the Australian side it is bordered on the south by South Indian Ocean Subtropical Water (SISTW), marked by a salinity maximum around 200 m under saline surface water[347] [**Figures 4-57b** and **4-58**], and by South Indian Ocean Central Water (SICW) of which the

[346] JADE: **J**ava **A**ustralia **D**ynamics **E**xperiment, 1989–1998.
[347] Strong evaporation in the region increases salinity.

coldest part is characterized by an oxygen maximum around 400 m due to traces of Subtropical Mode Water. On the north it is bordered by oxygen-poor Northwest Indian Ocean Water (NWW) between 250 m and 850 m. Oxygen levels in Indonesian Intermediate Water (IIW) fall between those of SICW and those of NWW. Indonesian Intermediate Water, with remarkably homogeneous salinity down to 1,100 m, represents a salinity minimum. It is less saline than the South Indian Ocean water masses, SISTW and SICW, and less saline than low-oxygen NWW. NWW is entrained by the Java Undercurrent and undergoes erosion from its place of formation. In a layer around 400 m the salinity of SICW is the same as that of NWW ($S \approx 34.7$) [**Figures 4-57** and **4-58**]. These water masses can be differentiated by their oxygen concentrations. The influence of a secondary current maximum in the Timor Passage around 1,000–1,100 m can be seen in the section by a spreading of the zone of salinity slightly less than 34.60 in the center of the section characterizing the influence of Indonesian Intermediate Water. The low flow of Indonesian Intermediate Water measured around 600–650 m through the Timor Passage results in a minimal influence of low-salinity Indonesian Water at this depth. At the longitude of the section, 116°E, in the northwest Australian Basin, Antarctic Intermediate Water can no longer be discerned.

The persistence of the Hydrologic Front as far as Madagascar is indicative of weak meridional exchanges in this zone. Exchanges between the southern hemisphere and the northern hemisphere take place through currents on the western boundary along the African coast. Thus a weak part of South Indian Ocean Central Water and Indonesian Intermediate Water penetrates into the northern hemisphere by the branch of the South Equatorial Current that passes north of Madagascar, then northward *via* the Zanzibar Current, and the Somali Current during the southwest monsoon. In the western region, their characteristics evolve rapidly through mixing with saline waters of the northwest [see below].

5.3.4 Arabian Sea Water, Northwest Indian Ocean Water

The oceanographic structure north of the Hydrologic Front differs from that to the south from the surface to about 2,000 m. In the Arabian sea, Red Sea Water (with a salinity maximum around 700–800 m), Persian Gulf Water (with a salinity maximum around 200–300 m), and Arabian Sea Surface Water (whose salinities reach 36.4 under the strong evaporation of the region) [**Figures 1-21, 4-29**, and **Section IV.3.2**] form a high-salinity layer of Northwest Indian Ocean Water (NWW) from 0 to 2,000 m [**Figures 4-39, 4-56** and **4-59**].

When the Somali Current flows north (during the southwest monsoon) NWW mixes in the southern Arabian Sea with South Equatorial Current Water, composed of a mixture of Indonesian Water and traces of South Indian

Ocean Central Water, *via* the Zanzibar Current and the Somali Current to form North Indian Ocean Central Water (NICW), which spreads toward the Bay of Bengal under the low-salinity surface layer. Away from the western region, the renewal of waters in the northern Indian Ocean occurs slowly. North of the equator, below the surface layer down to 1,000 m, oxygen is less than 50 μmol kg^{-1} (1.2 ml l^{-1}) [**Figure 4-59**].

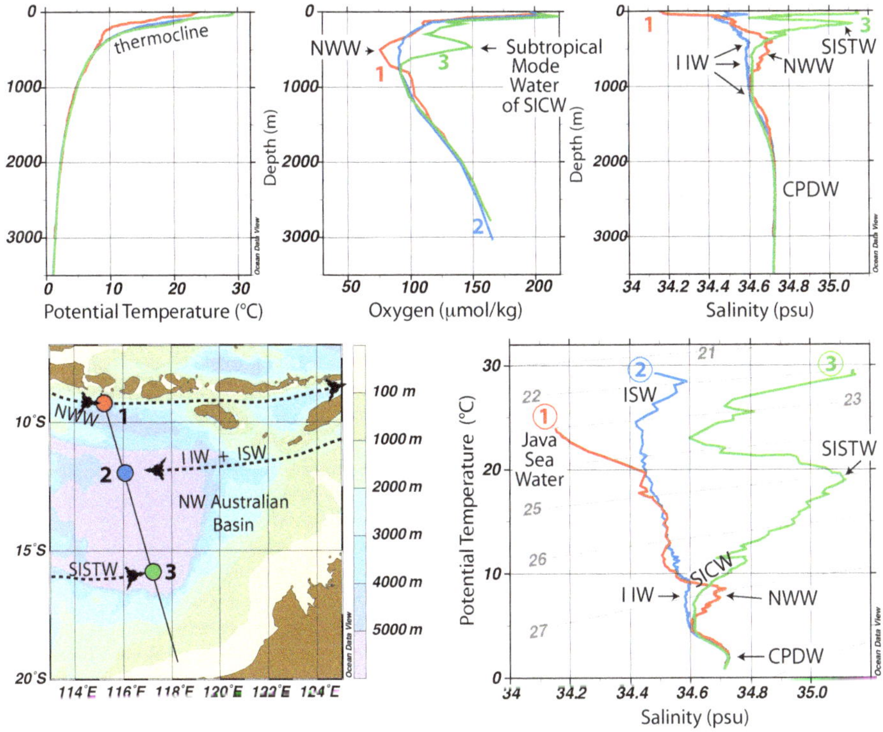

FIG. 4-58 – Profiles of potential temperature, dissolved oxygen, and salinity, and θ-S diagrams from three stations in the JADE section between Bali and Australia shown in **Figure 4-57**. [NW Indian Water (NWW), Indonesian Surface Water (ISW), Indonesian Intermediate Water (IIW), South Indian Subtropical Water (SISTW), South Indian Central Water (SICW), Circumpolar Deep Water (CPDW).] (With ODV and WOCE data.)

In the west, when the Somali Current flows south, during the northeast monsoon, Northwest Water mixes with Indonesian Intermediate Water and traces of South Indian Ocean Central Water (both less saline) in the equatorial zone, to form Equatorial Water (EW), entrained eastward by the zonal equatorial circulation. The salinity of Equatorial Water is relatively homogeneous, below a subsurface salinity maximum around 100 m depth, which characterizes the Equatorial Undercurrent, entraining Northwest Water eastward at the end of the northeast monsoon [**Figure 4-60**, Stations 3, 4, 5, 6]. The Equatorial Water characteristics are intermediate between that of water masses from the north (Station 2) and that of water masses from the south (Station 1).

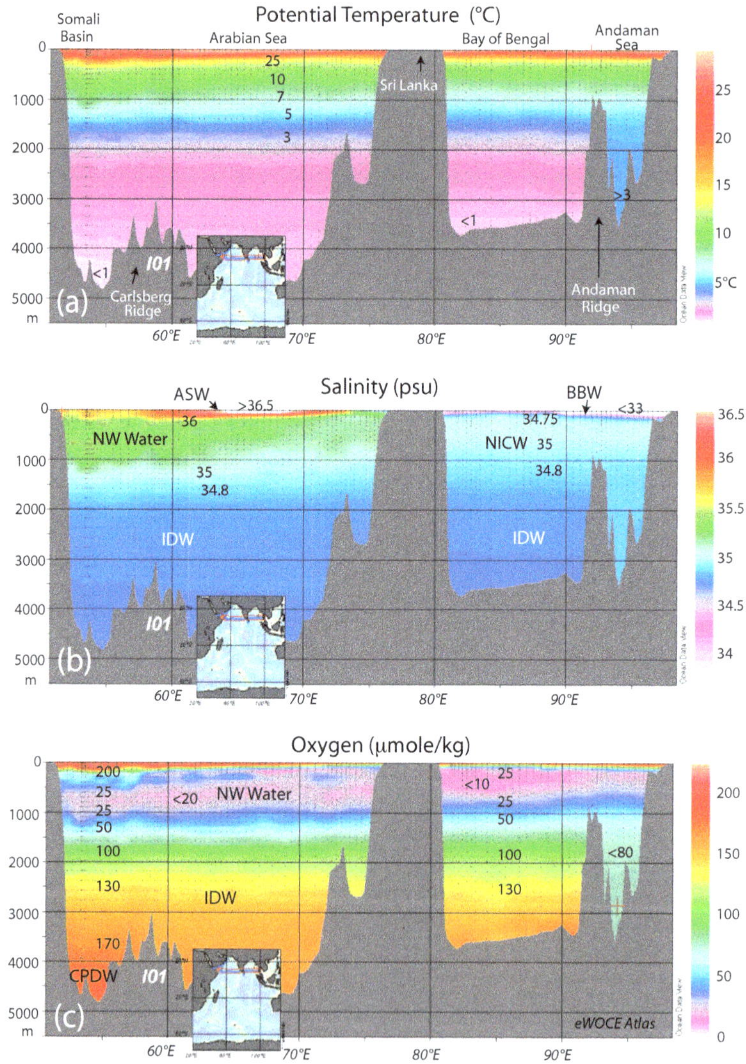

FIG. 4-59 – Zonal sections in the northern Indian Ocean at 10°N of **(a)** potential temperature (°C), **(b)** salinity (psu), and **(c)** dissolved oxygen (mmol/kg). [Arabian Sea Water (ASW), Bay of Bengal Water (BBW), NW Water (NWW), North Indian Central Water (NICW), Indian Deep Water (IDW).] (WOCE Section I01, *eWOCE Atlas*, Schlitzer, 2002.)

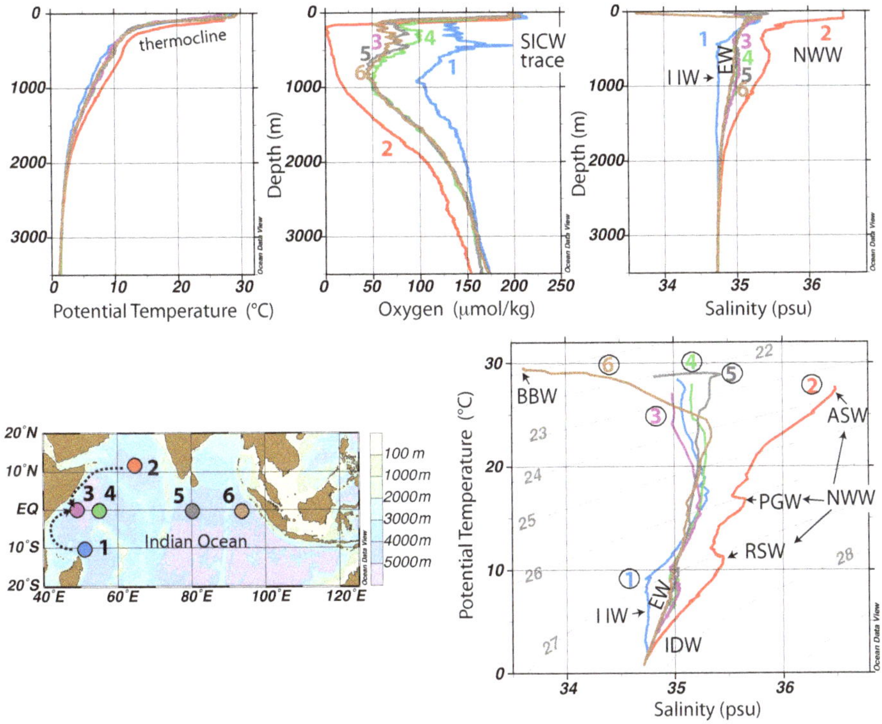

FIG. 4-60 – Profiles of potential temperature, dissolved oxygen, and salinity and θ-S diagrams from stations on the equator in the Indian Ocean (Stations 3, 4, 5, 6), in the Arabian Sea (Station 2), and in the Hydrologic Front (SEC) (Station 1). [Equatorial Water (EW), Bay of Bengal Water (BBW), South Indian Central Water (SICW), Indonesian Intermediate Water (IIW), Red Sea Water (RSW), Persian Gulf Water (PGW), Arabian Sea Water (ASW), NW Water (NWW), Indian Deep Water (IDW)] (With ODV and WOCE data.)

5.3.5 Bay of Bengal Water

In contrast to the Arabian Sea, the Bay of Bengal receives a number of river inputs which drain the Indian Peninsula, the Himalayan foothills and part of the Indochinese Peninsula [**Figure 4-2**]. Freshwater inputs create Bay of Bengal Water, a brackish layer of surface water ($S < 33$) about 50 to 100 m thick [**Figures 4-51, 4-52,** station 2, and **4-59**]. Oxygen levels are low in North Indian Ocean Central Water under this layer. The lowest values are found in the northern Bay of Bengal where they are less than 20 μmol kg^{-1} around 200 m [**Figure 4-51**] indicating a very slow renewal. Between the low-salinity surface layer and the underlying water a halocline (a layer with an accentuated vertical salinity gradient) forms in the interior of the thermally mixed layer above the thermocline, creating what is called a "barrier layer" [**Section V.4.2.5**]. The barrier layer limits heat exchanges between the surface layer and underlying water thus preventing surface water from exporting its heat downward. Bay of Bengal Surface Water extends its influence as far as the equator [**Figure 4-60**, Stations 5 and 6].

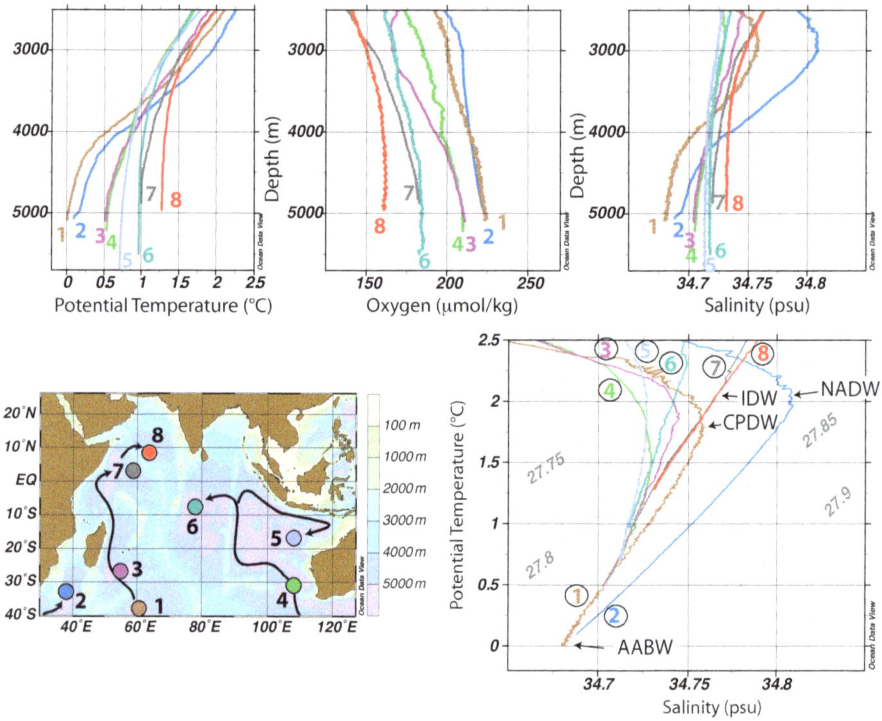

FIG. 4-61 – Profiles of potential temperature, dissolved oxygen, and salinity and θ-S diagrams of Bottom Water and Deep Water in each basin of the Indian Ocean. [Circumpolar Deep Water (CPDW), North Atlantic Deep Water (NADW), Indian Deep Water (IDW), Antarctic Bottom Water (AABW).] (With ODV and WOCE data.)

5.3.6 Indian Deep Water

Indian Deep Water has its source outside the Indian Ocean. It is located between 1,500–2,000 m and 3,800 m. In the south, the salinity maximum around 3,000 m depth [**Figure 4-61**, Stations 2 and 1] originates as North Atlantic Deep Water directly entrained by the South Indian Current and Circumpolar Deep Water of the Antarctic Circumpolar Current. It enters from the southwest and flows north in the form of a deep western boundary current following topography, as indicated by the slope of the 1.5 °C and 2 °C isotherms and of the 34.73 isohaline [**Figure 4-54**].

In the west, at 30°S, its salinity maximum is 34.85 (θ = 2 °C, σ_0 = 27.85, and O_2 = 4.7 ml l^{-1} or 204 µmol kg^{-1}). It attenuates through mixing eastward where the maximum is only 34.74 (θ = 1.45 °C, σ_0 = 27.81, and O_2 = 4.3 ml l^{-1} or 187 µmol kg^{-1}) [**Figure 4-54**]. In the course of its northward advance, it mixes with Bottom Water and Antarctic Intermediate Water in the south [**Figure 4-41a**, Stations 1, 2, 3, 4]. In the north, it mixes with Northwest Water (Stations 7 and 8) and forms Indian Ocean Deep Water. From the west of each deep basin, Indian Ocean Deep Water spreads eastward and upward to over-lying layers through diffuse upwelling. The evolution of its characteristics is indicated on the diagrams of **Figure 4-61**. Then it spreads southward and feeds the CPDW in its upper part forming Upper Circumpolar Deep Water characterized by an oxygen minimum.

5.3.7 Indian Ocean Bottom Water

Dense water masses, formed in the northern Red Sea and Persian Gulf cannot reach the bottom of the Indian Ocean because of the mixing they undergo when flowing out at shallow depths (< 110 m). Antarctic Bottom Water and Lower Circumpolar Deep Water form Indian Ocean Bottom Water, below about 3,800 m, penetrating northward with difficulty due to numerous ridges [**Figure 4-3**]. Antarctic Bottom Water reaches the Mozambique Basin and the Crozet Basin [**Figure 4-61**, Stations 2 and 1 and **Figure 4-62**] and disappears northward with passage over the ridges, replaced by Lower Circumpolar Deep Water.

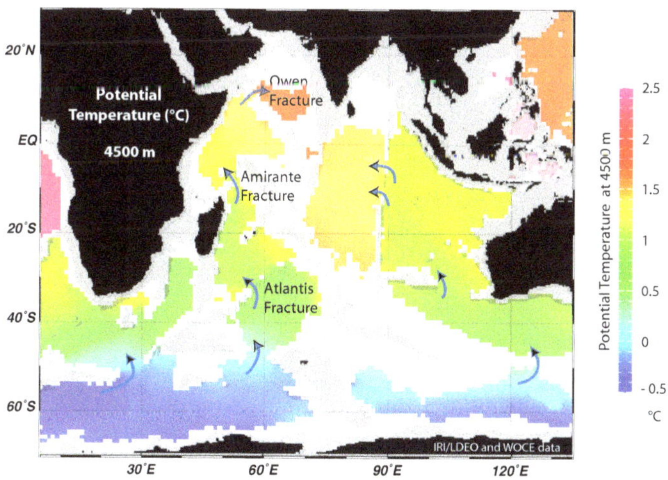

FIG. 4-62 – Evolution of potential temperature at 4,500 m and passages across the ridges in the Indian Ocean. (IRI/LDEO and WOCE data.)

Temperature distribution at 4,500 m shows several entry passages [**Figure 4-62**]. From the Weddell Sea, Antarctic Bottom Water first finds a passage across the Southwest Indian Ridge through a fracture zone around 32°E. It enters the Mozambique Basin at a temperature of 0.1 °C [**Figure 4-61**, Station 2][348] but is blocked to the north by the sill of the Mozambique Strait [**Figure 4-3**].

Bottom Water follows two paths northward in the Indian Ocean. The first passes through the Crozet Basin and the second through the South Australian Basin.

In the west, Antarctic Bottom Water penetrates into the Crozet Basin, between the Crozet Islands and the Kerguelen Plateau. The passage from the Crozet Basin to the Madagascar Basin is *via* the Atlantis Fracture Zone across the Southwest Indian Ridge around 30°S and 56°E–59°E [**Figure 4-61**]. The flow is intensified on the western boundary of the Madagascar and Mascarene Basins along the continental slope. The intensification of the current is indicated by the slope of the 1 °C isotherm along Madagascar on the 30°S temperature section [**Figure 4-54**]. Bottom Water continues its course to the Somali Basin through the Amirante Trench (trench depth > 5,200 m, sill around 4,200 m, transport of the order of 1 Sv), then arrives in the Arabian Sea through the Owen Fracture Zone across the Carlsberg Ridge. Its characteristics weaken with each ridge passage where the coldest water is blocked by topography. Past the Mascarene Basin the lower part of Circumpolar Deep Water is found on the bottom. It mixes with the overlying warmer and more saline part of Deep Water along its path, intensified in the fracture zones. Bottom Water has a potential temperature of 0 °C in the Crozet Basin [**Figure 4-61**, Station 1], 0.5 °C in the Madagascar and Mascarene Basins (Station 3), 1 °C in the Somali Basin (Station 7), and 1.3 °C in the Arabian Sea (Station 8).

In the east, Antarctic Bottom Water penetrates into the South Australian Basin by a fracture zone (the "Australian-Antarctic Discordance") across the Southeast Indian Ridge around 50°S and 124°E south of Australia. The lower part of Circumpolar Deep Water enters into the Perth Basin, the Wharton Basin, and the Northwest Australia Basin in the form of a western boundary current along the Ninety East Ridge [**Figure 4-62**]. It finds several passages into the Central Basin across the Ninety East Ridge, at 10°S and at 5°S. Transport across these passages is of the order of 0.5 Sv. On the zonal section at 30°S the oxygen level is lower in the Central Basin on the west of the Ninety East Ridge indicating that it is older than on the east [**Figure 4-54**].

Whether by the western route or the eastern route, temperature and salinity of Bottom Water increase north along the path and oxygen diminishes [**Figures 4-51, 4-54** and **4-62**].

[348] Flowing northward at 0.2 m s^{-1} under the Agulhas Current.

The Mozambique Basin, the Arabian Sea, and the Central Basin are dead ends for this much-eroded Antarctic Bottom Water, which rises through slow upwelling and mixes with Lower Circumpolar Deep Water.

The transition between Bottom Water and Deep Water is gradual. The evolution of their characteristics through mixing is shown on the θ-S diagrams of **Figure 4-61**. The transport of these water masses northward is estimated at ~ 5 Sv into the western basins, along Madagascar, and at ~ 6 Sv into the eastern basins, along the Ninety East Ridge. Topography allows the Central Basin to receive Circumpolar Deep Water directly from the south but Lower Circumpolar Deep Water must pass through the fractures across the Ninety East Ridge. In the Indian Ocean the high salinity core of the Lower Circumpolar Deep Water is still above the bottom layer, while in the Pacific Ocean it is on the bottom.

A return southward is needed to balance the inputs of Antarctic Bottom Water and Circumpolar Deep Water coming from the exterior of the Indian Ocean. This return takes place as Indian Ocean Deep Water.

In differentiating between Bottom Water of Antarctic origin and Deep Water from the North Atlantic, the Indian Ocean is in an intermediate position with respect to the Atlantic Ocean where the two water masses are quite distinct, and with respect to the North Pacific Ocean where the two water masses are mixed.

The research ship *Marion Dufresne*, named in honor of 18th century French explorer Marc-Joseph Marion du Fresne is also a supply vessel for the French Antarctic Territories (off Bali, JADE 89 cruise).

Chapter 6
Distinctive features of the Indian Ocean

Because it is closed to the north by the Asian continent around the tropic of Cancer, the Indian Ocean is the only ocean to undergo a seasonal reversal of winds north of 10°S. In response to this monsoon regime, the ocean circulation reverses, generating the strongest oceanic variability in the planetary ocean. In contrast to the Pacific and Atlantic Oceans, wind and current systems shift southward in the northern winter and the Intertropical Convergence Zone passes into the southern hemisphere.

Communication with the Pacific Ocean in the equatorial zone generates the southward-flowing Leeuwin Current along the eastern coasts, a feature that is unique to the Indian Ocean. Consequently, there is no upwelling on the eastern boundary as in other oceans (except along the Java coast during the SE monsoon and off the NW coast of Australia during the NW monsoon). On the contrary, in the northwest during the southwest monsoon, upwelling enriches and cools waters off the Somali and Arabian coasts. Rather than equatorial upwelling, there is an intense eastward equatorial jet during inter-monsoonal periods, another distinctive characteristic of the Indian Ocean.

The Indian Ocean has, along with the western Pacific, the highest ocean surface temperatures on the planet. It constitutes the most important "furnace" for the atmosphere. It undergoes the strongest climatic seasonal variability of the planet under the effect of the monsoon system, but also possesses interannual climatic variations linked to the variability of pressure systems over the region. Blocked to the north by the Asian Continent, northward heat export is prevented and only weak ventilation is allowed from the north. It supplies heat southward for the world ocean.

Part V
The Pacific Ocean

Fig. 5-1 – Pacific Ocean map of Abraham Ortelius (Maris Pacifici novissima descriptio, from Atlas thectrium orbis terrarum, 19th edition, Anvers, 1592). (From *Cartes géographiques anciennes*, Kupcik, 1981.)

Chapter 1
Introduction

Europeans knew of an ocean east of China from the time of Claude Ptolemy in the 2nd century AD [**Figure 3-1**] but the existence of the American continent was unknown.

Certainly, the Polynesians voyaged eastward on their slight pirogues, advancing from island to island as far as Easter Island and were able to make the islands habitable. They could return without map or instruments. However, their astonishing non-written knowledge remained unknown to Europeans.

For many centuries the Arabs and Chinese had navigated between the western Pacific and the Indian Ocean using the seasonally reversing winds blowing along their coasts. Travelling merchant-explorers such as Marco Polo during his trip to China in the 13th century provided descriptions allowing cartographers to include the Chinese and Japanese coasts as well as some Indonesian islands. In 1513 Vasco Nuñez de Balboa saw the Pacific Ocean for the first time. But there is no account of an ocean crossing before that of Fernando de Magellan in 1520. He died in the Philippines in 1521, but his ship completed a tour of the world for the first time.

On the map of Abraham Ortelius in 1592 [**Figure 5-1**] the Asian and American coasts, New Guinea, and the Indonesian islands may be identified. Australia is attached to the Antarctic continent (called "Terra Australis") and greatly overflows the southern Pacific Ocean. After Magellan's crossing, numerous voyages followed, such as those of James Cook[349] (1768–1771, 1772–1775, and 1776–1779) and Jean-François de Galaup de La Pérouse.[350] These and others discovered, through expedition after expedition, the true limits of this vast ocean.

[349] Cook was killed by Hawaiian natives in 1779.

[350] La Pérouse died following the sinking of his ship, *Astrolabe*, at Vanikoro (Vanuatu) in 1788.

Chapter 2
Geographic characteristics

The Pacific Ocean [**Figure 5-2**] is the most extensive and deepest of all the oceans (average depth 4,200 m). To the east, the mountainous coasts of the American continent bound it. To the west, eastern Australia, the Sunda Islands, and the coasts of SE and NE Asia bound it. To the north, the continental masses of Asia and America close in to the narrow (85 km) and shallow (55 m) Bering Strait located at 66°N. To the south it is widely open to the Antarctic Ocean. It extends over 135° of longitude (that is, 12,000 km)

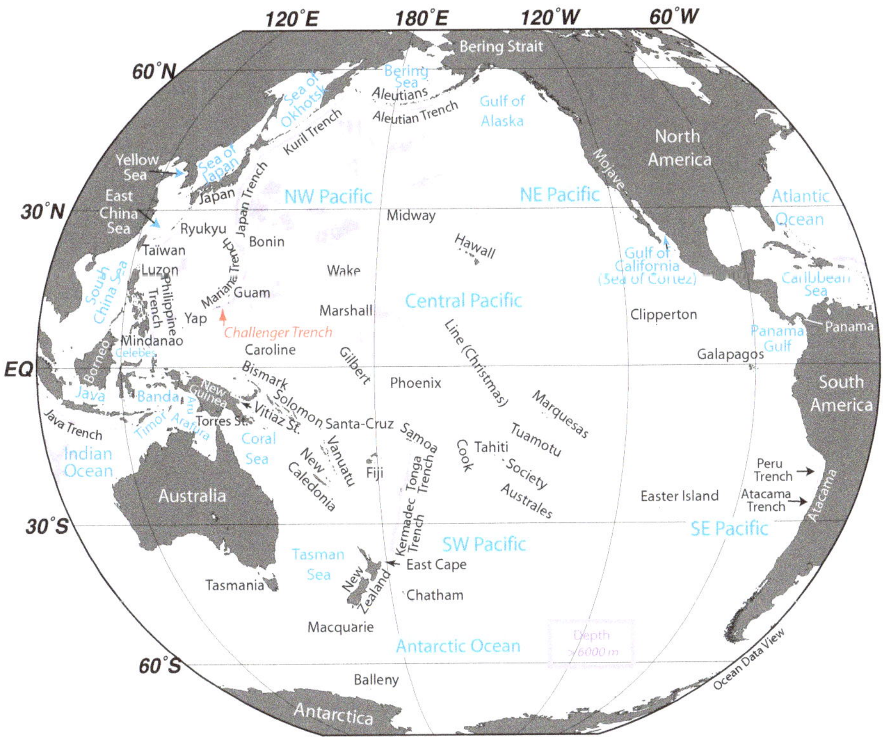

FIG. 5-2 – Map of the Pacific Ocean. Depths greater than 6,000 m are shown in violet.

at 40°S between Australia and South America. The limit between the two oceans is the Subtropical Front, or the 40°S parallel to the east where the Front disappears [**Section II**]. In the tropical zone it extends zonally over 180° of longitude, or half Earth's circumference (20,000 km between the Malacca Strait and Panama). The Bering Strait is separated by 106° of latitude from the Subtropical Front (11,800 km).

If all adjacent seas are included, the Pacific Ocean covers an area of 147 million square kilometers, or 40% of the planetary ocean's surface, equivalent to the total area of all land above sea level.

The long global mid-oceanic ridge system formed by tectonic motions in the terrestrial crust divides up the ocean floor and hinders bottom-water circulation [**Figure 5-3**]. In the Atlantic and Indian Oceans the ridge lies in a mid-ocean position while in the Pacific it is close to the eastern boundary. The East Pacific Ridge extends the Antarctic-Pacific Ridge. It underpins Easter Island (27°06S, 109°22W), passes between the Albatross (or Galapagos) Plateau to the east and Clipperton Island to the west, joins the Gulf of California (or Sea of Cortez) and is prolonged on the continent. Offshoots from this ridge, the Chile Ridge, the Nazca Ridge, and the Galapagos Plateau define the Chile Basin, the Peru Basin, and the Panama Basin. The Peru-Chile (Atacama) Trench of 8,065 m depth, 5,900 km long, borders the Andes Cordillera with peaks rising to 7,000 m. The East Pacific Ridge is cut with fracture zones that are oceanographically important as they provide passages for bottom water. The position of the ridge reduces deep communication between the Antarctic Ocean and the South Pacific. Linkages between the Antarctic-Australian Basin, the Tasman Sea and Southwest Pacific Basin are blocked below 3,400 m. A low point in the East Pacific Ridge around 130°W allows a passage down to 4,000 m between the Southeast Pacific Basin and the Southwest Pacific Basin. The Chile Basin communicates with the Southeast Pacific Basin down to about 3,800 m. The Peru Basin is isolated beyond 3,500 m.

The central region of the Pacific is bounded on the east by the North American coast, on the north by the Aleutians (with a trench at 7,700 m), and on the west by the line of great trenches [**Figure 5-2**]. These run from the Kamchatka Peninsula as far as New Zealand. They pass the Kuril Islands (10,500 m), Japan (9,800 m), the Bonin Islands (9,700 m), the Mariana Islands (10,850 m), the Caroline Islands, the Bismarck Archipelago (including New Britain), the Solomon Islands, Fiji, Tonga (10,800 m), and the Kermadec Islands (10,000 m). The greatest ocean depths are found along this line in the Challenger Deep (which reaches 10,994 m at the southwest extremity of the Mariana Trench between the islands of Guam and Yap). In the north of this central region a chain of underwater volcanic seamounts, the Hawaii-Emperor Seamount Chain, extends southward from the western end of the Aleutians, then southeastward along the Hawaiian Islands [**Figure 5-3**]. Similar chains exist between 20°N and 20°S such as the Christmas (Line), Marquesas, Tuamotu, Society (Tahiti), and Austral island chains.

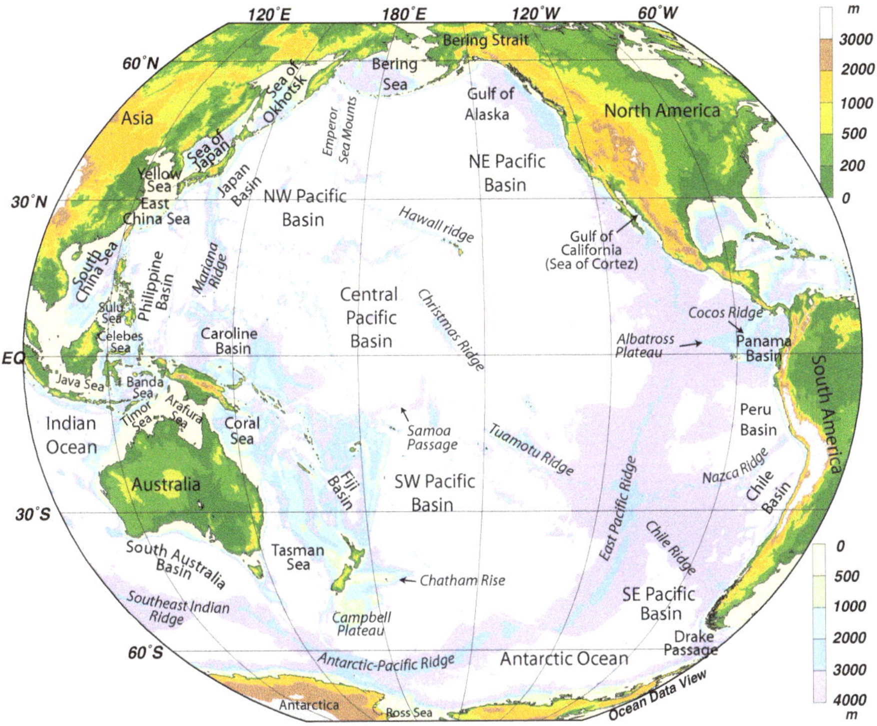

FIG. 5-3 – Pacific Ocean topography. Depths greater than 4,000 m are white.

West of the line of trenches, the region is broken up. It includes a great variety of basins, with varying depths and degrees of isolation. The Yellow Sea, the East China Sea, the Gulf of Thailand, the Java Sea, and the Arafura Sea are shallow continental shelves. The Sulu Sea, the Celebes Sea, the Banda Sea, the Philippine Basin, the Caroline Basin, the Coral Sea, the South Fiji Basin, and the Tasman Sea are deep seas (4,000 to 7,500 m). Finally, the Bering Sea, the Sea of Okhotsk, the Sea of Japan, and the South China Sea are mixed basins having only a portion that is deep. In contrast with the adjacent seas of the Atlantic or Indian Oceans, the adjacent seas of the Pacific are all on the western side, and have only a weak oceanographic influence. The Indonesian Seas play an important role in the transfer of heat and mass from the Pacific to the Indian Ocean, influencing the oceanographic character of the Indian Ocean[351] [**Section IV**].

The Pacific is strewn with numerous volcanic islands but also with a large number of guyots, volcanic seamounts rising to more than 2,000 or

[351] Pacific Ocean temperatures in oceanic general-circulation models become too high if the Indonesian passages are closed.

3,000 m above the ocean floor. Some have a nearly flat summit covered with a thick deposit of coral origin, the sign of slow sinking. They are most numerous in the northwest and central Pacific. They do not seem to have an important effect on the rather slow horizontal circulation in these regions, but they must have an effect in the dissipation of tidal energy, which induces turbulent mixing of deep waters leading to slow upwelling.

The Pacific is a region of strong volcanic and seismic activity. Undersea earthquakes generate tsunami waves, which propagate at velocities of more than 800 km per hour and become destructive when arriving on coasts. Because of the great number of islands in the tropical zone where the surface water temperature is never less than 20 °C, coral activity covers a vast region known as the Coral Belt, which includes innumerable atolls typical of the Pacific and the Australian Great Barrier Reef.

Chapter 3 Climatology

3.1 Pressure distribution

The theoretical pressure distribution [**Figure 1-53**], composed of subtropical high pressure centered on 30°N and 30°S, equatorial low pressure, and subpolar low pressure, is modified by the presence of continents (in particular the Asian continent), which generate strong seasonal variations [**Figures 5-4** and **5-5**]. Because most planetary land mass is in the northern hemisphere, there is an average displacement of the entire pressure system northward by several degrees of latitude.

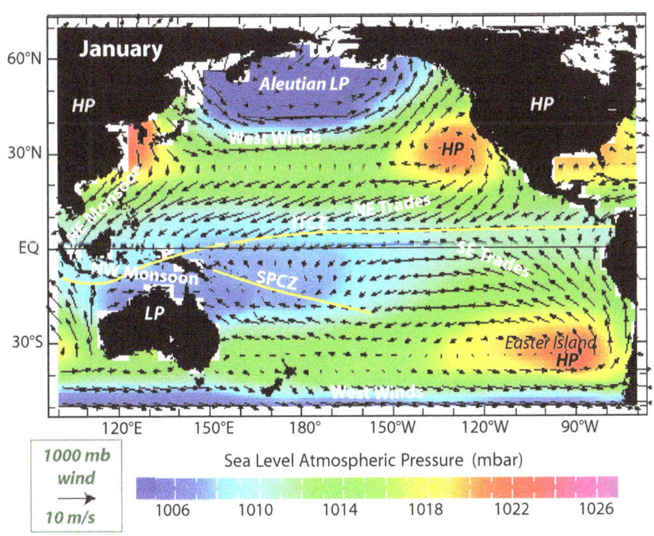

FIG. 5-4 – January atmospheric pressure and winds over the Pacific Ocean. (IRI/LDEO, and ECMWF climatology, Trenberth, 1992.)

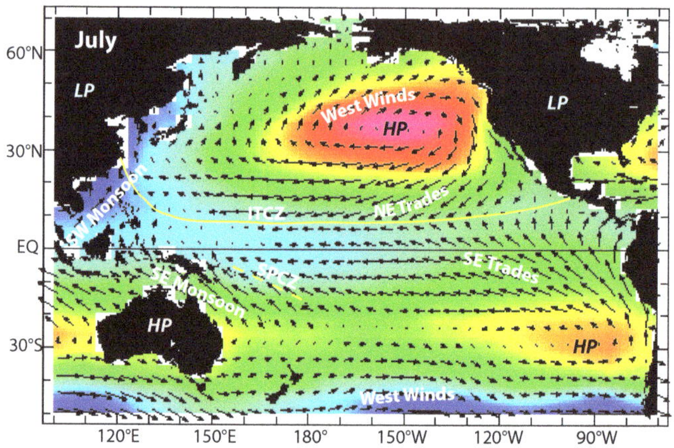

FIG. 5-5 – July atmospheric pressure and winds over the Pacific Ocean (same scale as **Figure 5-4**). (IRI/LDEO, and ECMWF climatology, Trenberth, 1992.)

In January [**Figure 5-4**], the Aleutian low pressure centered on 50°N, 180° dominates the northern Pacific from 30°N to 70°N. The subtropical high is then divided in two. The eastern anticyclonic region is reduced in size and shifted southeast (32°N, 135°W). In the west, the strong high-pressure center over Asia spills out over the ocean. The zone of equatorial low pressure, corresponding to the Intertropical Convergence Zone (ITCZ) is located north of the equator.

In the South Pacific, it is summer in January and the high-pressure region is displaced southward (35°S, 100°W) by the presence of low pressure over Australia and New Guinea, which extends out over the Coral Sea as far as New Caledonia and Fiji. Low pressure extends southeast and is associated with a secondary convergence zone, known as the South Pacific Convergence Zone (SPCZ) [**Figure 5-4**]. In its northwest part, the SPCZ merges with the ITCZ. In its western part, it merges with the South Indian Convergence Zone (SICZ) which corresponds to the ITCZ in January. The SPCZ is most active in the southern hemisphere summer. Wind convergence and high sea-surface temperature entrain an upward motion of air masses generating clouds [**Figure 5-6**] and heavy precipitation. In the east, the ITCZ is centered around 5°N and in the west the two convergence zones extend from about 5°N–10°N (ITCZ) around 150°E to 25°S–30°S (SPCZ) around 160°W.

In July,[352] the subtropical high-pressure zone centered on 40°N, 150°W covers the entire northeast Pacific between 20°N and 50°N [**Figure 5-5**], from the American coast as far as 160°E. In this season relatively low pressure covers North America and very low pressure reigns over the Asian continent.[353] The area of subpolar low pressure over the Bering Sea is reduced.

[352] Pressure systems are farthest north in the northern hemisphere summer.

[353] The Asian pressure system is linked to the more rapid warming of land masses.

In the South Pacific, it is winter in July. High pressure dominates over Australia and South America and subtropical high pressure is continuous between 15°S and 40°S with a maximum over the Easter Island region. The zone of equatorial low pressure (ITCZ) centered over 8°N–10°N covers the entire width of the Pacific Ocean and extends over the Indonesian Seas region, from New Guinea as far as Southeast Asia.

3.2 Winds

Anticyclonic atmospheric circulation [**Figures 5-4** and **5-5**] is found around high-pressure centers, altered by the influence of the monsoon in the west [**Section IV.3.1**] and the seasonal extension of the Aleutian low pressure in the north. In the entire western part of the Pacific Ocean, from the northern Coral Sea to the Sea of Japan, the ocean is subject to the monsoon regime, with a wind reversal twice a year along the western side of the ocean.

On annual average, the NE trades, the southern arm of the north anticyclonic system, dominate [**Figure 1-58**].

In January [**Figure 5-4**], the NE trade winds blow from the Tropic of Cancer (23°27′N) as far as 8°N, along the entire width of the Pacific Ocean, from North America to SE Asia, where they converge and join the NE monsoon circulation around the Asian high pressure which spills out over the ocean. They are the most intense during this season. West of 170°W the NE trades reach the equator. Along the coast of California, the wind is predominantly from the north, with high variability. The North Pacific Ocean is dominated, from 30°N to the Bering Strait, by a cyclonic system around the Aleutian subpolar low. Westerly winds, strongest at this season, blow between subtropical high pressure and subpolar low pressure. In the east the cyclonic system drives southerly winds along the coasts of Canada and Alaska. To the north, NE polar winds, most intense at this season, are found over the Bering Sea and the Sea of Okhotsk. Over the Sea of Japan the monsoon, deflected by Aleutian low pressure, is from the NW, while to the south, over the South China Sea, it is from the NE.

In the ITCZ zone of low pressure, centered on 5°N, winds are weak and variable.[354]

To the south, the SE trade winds, the northern arm of the anticyclonic system centered around 35°S, are weaker than the NE trade winds but stable[355] east of 170°W. They cross the equator in the central region, between 110°W and 170°W, as far as the ITCZ. West of 170°W the region is under the influence of Australian low pressure and of the South Pacific Convergence Zone (SPCZ). The trades then have a NNE component as far as about

[354] Sailors call the ITCZ region "the doldrums".

[355] This is why early western navigators baptized this region as "pacific".

20°S, east of the SPCZ. North of the Coral Sea, the monsoon has a NW component. South of 40°S, in the Antarctic zone, gales blow from the west in the "roaring forties", the southern arm of the anticyclonic system.

In July [**Figure 5-5**] in the northern hemisphere, the cyclonic region is greatly reduced over the Bering Sea and the anticyclonic region covers the NE Pacific. Winds are from the WSW between the two zones, weaker than in January. Along the California coast, winds are from the north and are more regular than in winter. They continue with the NE trade winds, weaker than in January, which extend as far as the Philippines where they converge with the SSW monsoon winds generated by Asian low pressure.

In the southern hemisphere north of the subtropical high pressure belt centered around 28°S[356], SE trade winds cover the entire Pacific from the Tropic of Capricorn (23°27′S) to the equator in the west and up to 8°N in the east. They converge with SE monsoon winds, blowing north of the Coral Sea in the region of the SPCZ [**Figure 5-5**]. Along the South American coast between 10°S and 30°S, winds are from the south, parallel to the coast (as in January). Further north, winds diverge between 80°W and 90°W, with SW winds toward the Gulf of Panama and SE trades further offshore. South of 30°S, winds mesh with the great system of westerlies of the Antarctic Ocean.

3.3 Precipitation, evaporation, and heat exchanges

The western Pacific undergoes strong seasonal variations under the influence of the monsoon regime. In the northern hemisphere winter (January–March) NW winds are cold and dry, particularly over the Sea of Okhotsk, the Sea of Japan, and the Yellow Sea [**Figure 5-4**]. This is a continental regime. In the northern hemisphere summer (July–September) winds from the S to SE [**Figure 5-5**] are warm and humid with extensive precipitation. Thus, this is an oceanic regime.

Northeastern and eastern regions of the Pacific, from southern Alaska to northern California receive year-round winds from offshore charged with humidity having oceanic characteristics, with little seasonal variation. This produces a strong climatic asymmetry between the eastern and western coasts of the North Pacific, which is even more pronounced than in the Atlantic at the same latitudes [**Figure 1-20a, Section I.2.5.1**].

Examples of this asymmetry are:

- At 56°N, in the west over the Sea of Okhotsk in winter, the average temperature is about −15 °C while in the east, at Sitka, in the Gulf of Alaska, it is always above freezing. Precipitation is also quite different: 400 to 1,000 mm/year over the Sea of Okhotsk, *versus* 2,000 mm/year at Sitka.

[356] Near Easter Island.

- At 43°N, at Vladivostok, north of the Sea of Japan, the sea is frozen in winter, while there is never ice in the Gulf of Alaska as far north as 60°N.
- At 35°N, at Yokohama, Japan, seasonal temperature variations are large, going from 5 °C in winter to 27 °C in summer, with 1,800 mm/year in precipitation. At the same latitude in San Francisco, temperatures vary little: 12 °C in winter and 15 °C in summer and precipitation is 500 mm/year.
- In the region of the westerlies, the passage of weather disturbances provides a precipitation maximum of 2,200 mm/year in the west to 1,400 mm/year in the east [**Figure 1-20**].

The two subtropical anticyclonic zones are shifted eastward (because of the predominant influence of the Asian continental mass). Here evaporation exceeds precipitation, which does not exceed 100 mm/year[357] [**Figures 1-20, 1-21** and **1-22**]. The largest part of precipitation falling on the American continent flows into the Atlantic because the continental divide is nearer the Pacific. In contrast, in the west, precipitation and inputs by Chinese and eastern Siberian rivers that empty into the adjacent seas dominate over evaporation and have a strong impact on the oceanography. In the NE Pacific numerous Canadian and American rivers[358] flow into the Pacific and influence the circulation.

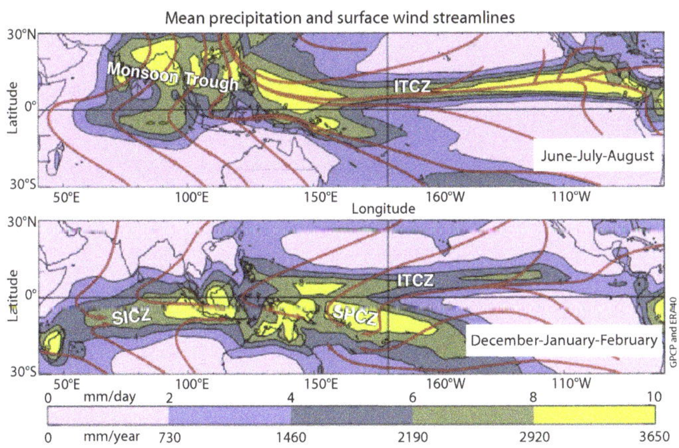

FIG. 5-6 – Mean precipitation in mm/day (from GPCP) and surface wind streamlines (from ERA40) in the equatorial Pacific and Indian Oceans for June–July–August and for December–January–February [Intertropical Convergence Zone: ITCZ; South Pacific Convergence Zone: SPCZ; South Indian Convergence Zone: SICZ, Monsoon Trough: low pressure zone with heavy precipitation]. (Adapted from E. Vincent, 2011.)

[357] The coasts bordering these regions are often deserts: Atacama, Mojave.

[358] The total flow of these rivers equals that of the Mississippi.

The ITCZ [**Figures 1-54, 5-4,** and **5-5**] is a zone of strong precipitation [**Figure 1-20a**], which can exceed 3,000 mm/year to which should be added the SPCZ [**Figure 5-6**] where precipitation varies between 1,500 mm/year and 3,000 mm/year. The SPCZ is most marked in the southern hemisphere summer during the NW monsoon. Over Borneo,[359] New Guinea, and the Solomon Islands, precipitation can even reach more than 4,000 mm/year. In the zone of periantarctic westerlies to the south, precipitation is of the order of 1,000 to 1,400 mm/year.

Because of its form and global position, 60% of the Pacific Ocean is in the tropical and equatorial zone (30°S–30°N). The heat exchange balance among absorbed solar radiation [**Figure 1-9**], infrared losses [**Figure 1-10**], evaporation [**Figure 1-11**], and conduction [**Figure 1-12**] shows a heat gain in upwelling zones (equatorial, Peru, California) which is particularly intense in the eastern equatorial zone [**Figure 1-13**].

3.4 Climatic anomalies over the Pacific: El Niño

The ocean has variability on all time and space scales. The purpose of this text is principally to describe the average state of the ocean and not its variability. However, here is a brief example of climatic anomaly on the scale of the largest ocean, the El Niño-Southern Oscillation.

Sir Gilbert Walker introduced the term "Southern Oscillation" (SOI) in 1928. It is related to the strong anti-correlation between atmospheric pressure at Tahiti, near the center of subtropical high pressure, and pressure at Darwin, near ITCZ low pressure. In the 1960's, Jacob Bjerknes connected variations in the Southern Oscillation with the arrival of warm water along the coast of Peru, known as El Niño, accompanied by anomalous precipitation. The climatic-anomaly ensemble is called "El Niño-Southern Oscillation" (ENSO).

The first Southern Oscillation Index (SOI) is based on the monthly mean sea level pressure difference between Tahiti and Darwin (chosen because there are long time series of measurements there). As sea level pressure difference is linked to trade-wind variations and associated with sea-surface temperature (SST) changes, several SOI indices are based on SST in different equatorial zones [**Figure 5-7a**]. A Multivariate ENSO Index is based on sea-level pressure, surface wind, sea-surface temperature, surface air temperature and cloudiness[360] [**Figure 5-7b**].

These ENSO anomalies occur irregularly, every 2 to 7 years.

[359] Borneo is also known as Kalimantan in Indonesia.

[360] All these variables can be observed.

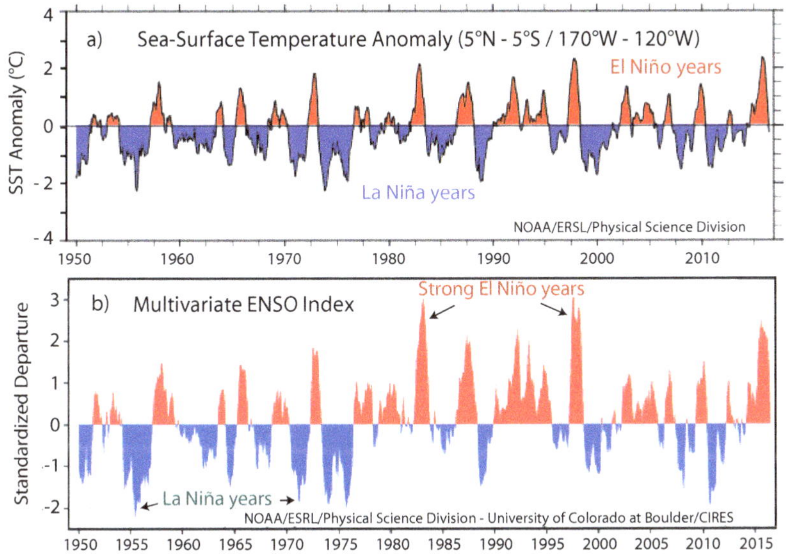

FIG. 5-7 – **(a)** Sea-surface temperature anomaly for the zone 5°N-5°S/170°W-120°W. (NOAA/ESRL/PSD.) **(b)** Standardized departure of the multivariate El Niño-Southern Oscillation Index. (NOAA/ESRL/PSD University of Colorado at Boulder/CIRES.)

In the average situation, the pressure gradient between subtropical South Pacific high pressure and equatorial low pressure drives the SE trade winds [**Figure 5-4**]. When the equatorial low pressure shifts north of the equator, the trades blow over the equatorial zone of the Pacific Ocean piling up lighter warm water westward, and generating an oceanic surface slope of about 50 cm between east and west [**Figure 5-8a**]. Thus the western Pacific (as well as the eastern Indian Ocean) is the greatest reservoir of warm water in the world (called the "warm pool"). Temperature here is always greater than 28 °C, over a depth that can exceed 100 m.

The presence of this warm water generates an atmospheric zonal circulation cell between America and Indonesia, called the Walker Cell [**Figure 5-8a**], with the active ascending branch (humid air) over the western Pacific, and the descending branch (dry air) over the American side. Trade winds form the surface branch of the cell.

Air warmed from below is unstable and rises, where it cools and contracts, producing a condensation of water vapor and strong precipitation. When the water vapor condenses, the heat extracted from the ocean during evaporation is transmitted to the atmosphere. This region acts as the most important "furnace" for the atmosphere.

In the ocean, the thermocline, which separates the warm surface layer from the deeper cold layer, develops an inverse zonal slope to compensate for the surface slope [**Figure 5-8a**]. It is found around 200 m in the west and 50 m in the east. Under the effect of the trade winds and the Coriolis force, which changes sign at the equator, equatorial divergence produces upwelling of underlying colder water in the east [**Section I.5.2.2**]. In the west, the warm layer is sufficiently thick so that, even with divergence, surface temperature does not drop.

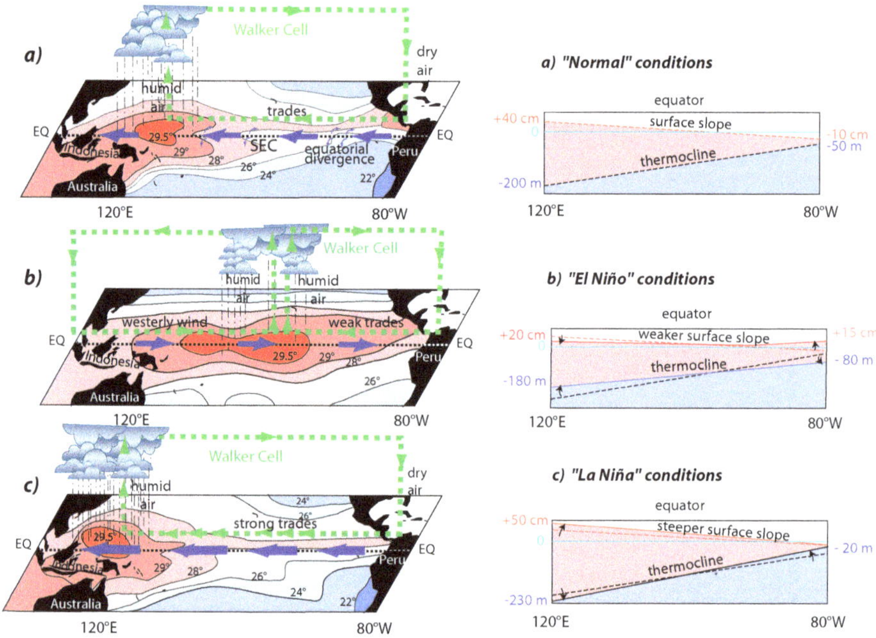

FIG. 5-8 – (**a**) Average state over the Pacific Ocean. (**b**) El Niño conditions. (**c**) La Niña conditions (From NOAA PMEL, 2009). The diagrams on the right show (exaggerated) surface and thermocline vertical oscillations along the equator.

In general, toward December, the trades as well as the persistent southerly winds along the Peruvian coast that generate fish-rich upwelling weaken somewhat [**Section V.4.4.1**]. The normal seasonal weakening of the winds produces a warm current from the north over the upwelling region. Peruvian fishermen called the current "El Niño" ("the boy" in Spanish, referring to baby Jesus) since the current arrives around Christmas. However, the name El Niño has since been attached to the interannual climatic anomaly on a Pacific-wide scale[361] and not just to the local seasonal variability.

[361] Infrared sensors on early sea-surface temperature measurements by satellite showed an anomaly of several degrees in the central Pacific. Thinking that it could be a calibration error, a ship sent to the area found that the anomaly was real, and that it involved not just the surface but a thick underlying layer.

In some years, the pressure difference between Tahiti and Darwin drops so that the SOI is negative [**Figure 5-7b**]. In that case the SE trades weaken significantly (even reversing in the west) and the zonal equatorial surface slope caused by the accumulation of warm water in the west is no longer in balance with the wind stress.

In the equatorial region, the ocean responds quickly[362] to changes in wind stress. Within a few months an eastward return of warm water takes place [**Figures 5-8b** and **5-9a**]. Sea level drops in the western Pacific and rises in the eastern Pacific. The surface slope decreases, as does the thermocline slope. Rain associated with the atmospheric convection zone located above warm surface water shifts to the center of the Pacific. Some islands in the central Pacific Ocean experience intense rainfall while droughts occur in northern Australia and Indonesia.

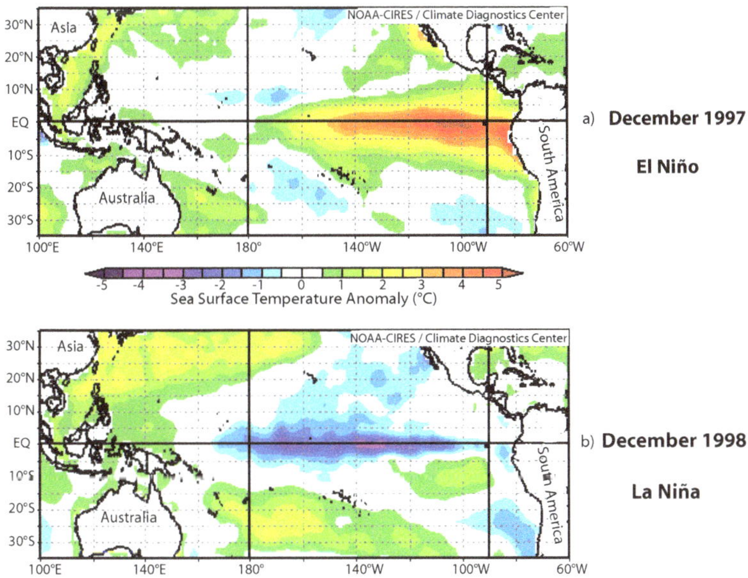

FIG. 5-9 – Surface temperature anomaly in the equatorial Pacific Ocean (**a**) during the December 1997 El Niño (warm phase), (**b**) during the inverse situation of the December 1998 La Niña (cold phase). (NCEP/NCAR reanalysis, NOAA-CIRES/ Climate Diagnostics Center.)

Along the coast of Peru, water warms so quickly that the anchovy fishery collapses. In this region of normally little rainfall downpours ravage the coasts, sometimes producing landslides that bury entire villages. These El

[362] The oceanic response is associated with the propagation of equatorial Kelvin waves.

Niño years, referred to as warm episodes, can be catastrophic. Peruvian fishermen catch only few fish and often have no roof for shelter, such as happened with the intense El Niño events of 1993 and 1997, which were devastating [**Figures 5-7** and **5-9a**].

When the difference between the pressure at Tahiti and Darwin increases (positive SOI, **Figure 5-7b**) the trades become stronger than average, the accumulation of warm surface water in the west increases so that sea level rises, and precipitation is more abundant over Indonesia. Along the coast of Peru, sea level drops and upwelling intensifies. On the equator, the strengthening of the trades increases upwelling in the central and eastern Pacific and sea-surface temperature is abnormally low. In contrast to El Niño, this is called a La Niña year, also referred to as a cold episode [**Figures 5-8c** and **5-9b**].

The El Niño-Southern Oscillation phenomenon is an example of ocean-atmosphere interaction that modifies climate. The intense variation in trade winds modifies ocean circulation, modifying thereby the distribution of sea-surface temperature in the equatorial basin. The atmosphere responds to these sea-surface temperature variations by a displacement of atmospheric convection regions which follow the warm water and which produce deficits or excesses of precipitation. The displacement of convection zones is linked to atmospheric pressure changes and thus to winds which in their turn modify ocean circulation and so on.

Atmospheric circulation modifications at such scales produce impacts of varying degree on the atmospheric circulation of the entire planet.

There are other large-scale climatic anomalies, such as the North Atlantic Oscillation[363] or the Indian Ocean Dipole[364]. They demonstrate the strong interactions and the great variability of the ocean-atmosphere system.

---

[363] NAO, North Atlantic Oscillation.

[364] IOD, Indian Ocean Dipole.

Chapter 4
Surface circulation

4.1 General circulation

Two atmospheric anticyclonic systems, whose principal elements are the westerlies and the trade winds, drive an ocean circulation constrained by continental boundaries in the form of two anticyclonic gyres [**Section I.5.1**], separated by the North Equatorial Countercurrent which extends across the width of the Pacific [**Figure 5-10**]. The cyclonic atmospheric system around the Aleutian low drives a cyclonic ocean circulation. The many islands and adjacent seas in the western Pacific lessen the intensity of western boundary currents.

The north anticyclonic gyre includes the North Pacific Current (or Drift), the California Current flowing southward, the North Equatorial Current (NEC) flowing westward, and the western boundary currents comprised of the Philippine Current and the Kuroshio, particularly intensified along the Japanese coast.

The north subpolar cyclonic gyre is constrained by the presence of the Aleutians. It includes the northern part of the North Pacific Current flowing eastward, the Alaska Current flowing north and west along the Alaska coast, and the Alaskan Stream flowing southwest along the Aleutians. The southern part of the Kamchatka Current and the Oyashio flowing southward (between 60°N and 45°N) are the western boundary currents of the cyclonic gyre, which terminate by merging with the Kuroshio.

The South Pacific Current represents the southern branch of the south anticyclonic gyre. To the east, the Chile-Peru current flows alongside the South American coast (offshore it is called the Humboldt Current).[365] At lower latitudes, the South Equatorial Current flows westward in several branches. The South Equatorial Current splits when arriving on the

[365] The Humboldt Current is named for the German explorer Alexander von Humboldt (1769–1859).

Australian coast feeding the North Queensland Current (NQC) northward and the East Australian Current (EAC) southward. On the eastern coast of New Zealand, the East Cape Current, prolonged by the East Auckland Current, converges with the current flowing north along the South Island transporting cold water coming from the Antarctic Circumpolar Current.

The North Equatorial Countercurrent forms the boundary between the two anticyclonic gyres and is fed by the two systems.

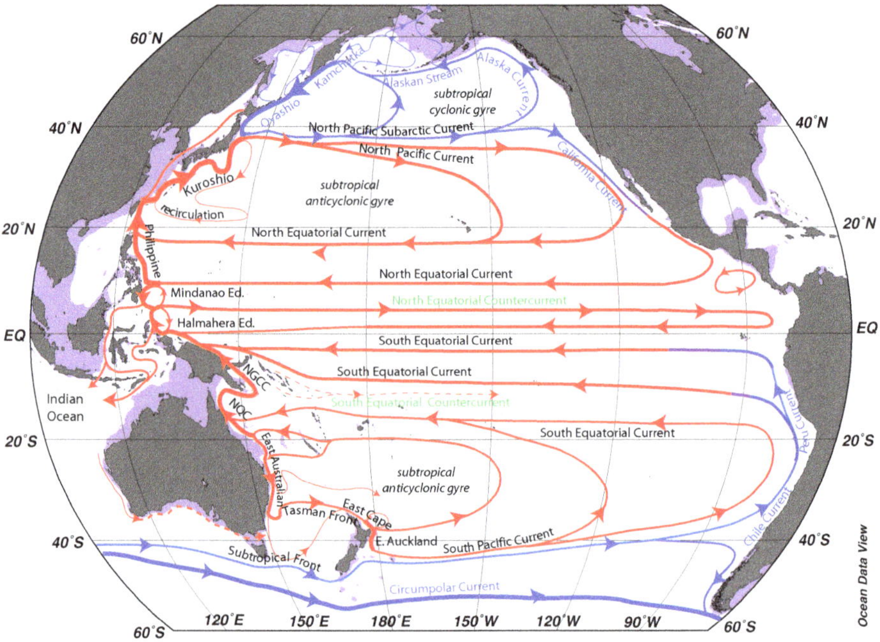

FIG. 5-10 – Pacific Ocean surface circulation from May to December. [New Guinea Coastal Current (NGCC), North Queensland Current (NQC), East Auckland Current (EAuckC), Mindanao Eddy, Halmahera Eddy and East Cape Anticyclonic Eddy.] Cold currents are blue, warm currents are red, depths less than 250 m are violet. Fronts are found at the convergence between cold and warm currents. From January to April, the upper part of the NGCC changes direction. (After Tomczak and Godfrey, 1994, see the legends of **Figures 5-26** and **5-31**.)

4.2 Equatorial current systems

Because of the extent of the Pacific Ocean, equatorial currents, which are principally zonal, can here develop freely and can qualitatively approach their theoretical form [**Figure 1-70**] though they are shifted north of the equator [**Figure 5-11**].

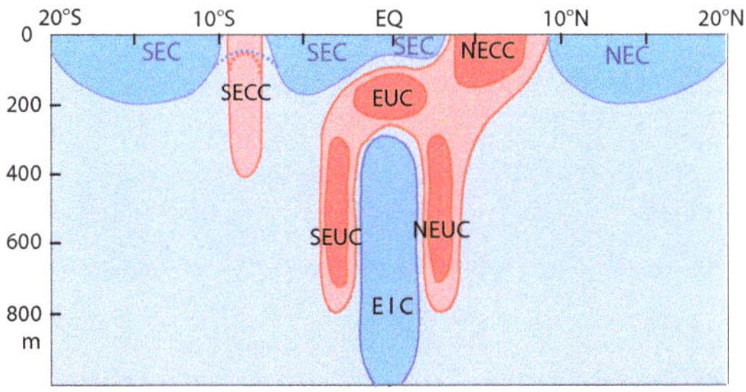

FIG. 5-11 – Schematic equatorial current systems between 20°N and 20°S. Eastward currents are in red, westward in blue. The South Equatorial Countercurrent (SECC) does not always reach the surface. [South Equatorial Current (SEC), North Equatorial Current (NEC), North Equatorial Countercurrent (NECC), Equatorial Undercurrent (EUC), Equatorial Intermediate Current (EIC), North and South Equatorial Undercurrents (NEUC, SEUC).]

4.2.1 The North Equatorial Current

Driven by the NE trade winds,[366] the North Equatorial Current (NEC) flows westward [**Figures 5-10** and **5-11**] over about 150–200 m depth. It arises near the coast of Central America at around 7°N–8°N in January, the northern winter, the season when the trades are most intense, and 10°N in July, the northern summer. It is fed during its path by the California Current coming from the north, and extends as far as the Philippines with a current maximum around 15°N. Arriving on the west coast of the Philippines, around 12°N, it splits into two western boundary currents. One feeds the Philippine Current northward, representing the beginning of the Kuroshio. The other, to the south, is the source of the Mindanao Current. The NEC reaches its full extent in February, in mid-Pacific, between 9°N–10°N and 20°N–25°N with speeds of the order of 0.3 m s^{-1}. Its transport varies seasonally depending on the strength of the trades from 20 Sv to 45–55 Sv, with a maximum flow in winter and a minimum in summer.

4.2.2 The South Equatorial Current

Driven by the southern hemisphere SE trade winds, which overflow into the northern hemisphere, the South Equatorial Current (SEC) flows westward from 4°N to about 20°S, depending on the season [**Figures 5-10** and **5-11**]. Several branches can be seen in the westward flow.

The north branch of the SEC is centered on 2°N at 155°W [**Figure 5-12**]. Its transport increases from the east (10 Sv ± 2 Sv) to the central Pacific. It reaches a maximum in July (southern hemisphere winter) around 140°W,

[366] The process is identical to that in the Atlantic Ocean [**Section III.4.1**].

with speeds greater than 1 m s^{-1} and a transport of the order of 24 Sv ± 4 Sv, then decreases westward from 165°E.

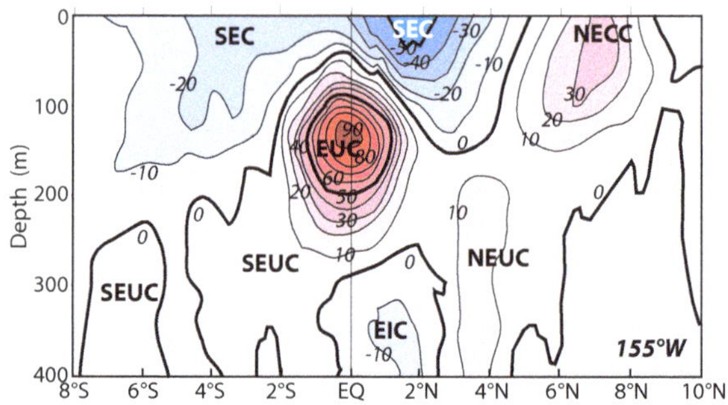

FIG. 5-12 – Meridional section of zonal equatorial currents at 155°W down to 400 m, between 8°S and 10°N, based on the 1985–2000 average (cm s^{-1}). Eastward currents are red, westward are blue. [South Equatorial Current (SEC), Equatorial Undercurrent (EUC), North Equatorial Countercurrent (NECC), Equatorial Intermediate Current (EIC), North Equatorial Undercurrent (NEUC), South Equatorial Undercurrent (SEUC).] (After Johnson *et al.*, 2002.)

The south branch of the SEC is centered on 3°S and ranges from 8°S to the equator. Its transport increases from 6 Sv ± 2 Sv in the east to a maximum of 28 Sv ± 4 Sv between 140°W and 180°W and diminishes westward from 156°E, where the current merges with the New Guinea Coastal Current (19 Sv ± 4 Sv).

The meridional slope of the thermocline, which changes sign at the equator (though the current does not), is an indicator of the SEC [**Figure 5-13**]. It can be detected up to the boundary with the eastward North Equatorial Countercurrent where the slope reverses.

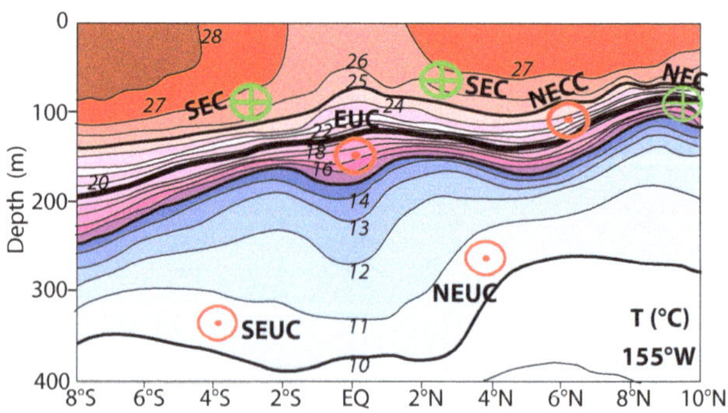

FIG. 5-13 – Meridional temperature section at 155°W. The thermocline slope indicates the direction of equatorial currents. Eastward currents are shown with green crosses, eastward currents with red points (see abbreviations, **Figure 5-12**). (After Johnson *et al.*, 2002.)

The southernmost branch of the SEC splits in two around 18°S on the Australian coast, where it forms the western boundary currents of the South Pacific [**Figure 5-10**]. It feeds the East Australian Current to the south and the North Queensland Current (NQC) to the NW during the SE monsoon, in July. The North Queensland Current flows alongside the Great Barrier Reef, crosses the Coral Sea following the topography, and is prolonged as the New Guinea Coastal Current (NGCC). The NGCC flows northwestward passing through the Vitiaz Strait between New Guinea and New Britain and feeds the North Equatorial Countercurrent forming the Halmahera Eddy. During the NE monsoon, in January, the upper layer of the NGCC reverses to the SE. The SEC also feeds the Great Barrier Reef Undercurrent, which is prolonged as the New Guinea Undercurrent (NGUC). Both are western boundary undercurrents, which flow year-round to the NW and feed the Equatorial Undercurrent [**Section V.4.2.5**].

4.2.3 The North Equatorial Countercurrent

Under the ITCZ, enclosed between the westward NEC and SEC, there develops[367] across the whole width of the Pacific, the North Equatorial Countercurrent (NECC) flowing eastward, centered between 5°N–7°N [**Figures 5-10, 5-11** and **5-12**]. It is fed in the west, depending on the season, by western boundary currents: the Mindanao Current transporting southward NEC water, and the New Guinea Coastal Current transporting northward SEC water. The NECC extends down to 150 m–200 m depth.

From January to April, the NECC is fed mostly by the NEC, *via* the south-flowing Mindanao Current. It extends from 2°30′N to 8°N in the west with speeds of the order of 0.2 m s^{-1}, and from 4°30′N to 8°N in the east. Its transport varies between 20 Sv and 25 Sv ± 4 Sv in the west, and then decreases eastward from 155°W. At 110°W, the core of the NECC deepens to 150 m and it nearly disappears from the surface (its transport drops to 8 Sv ± 2 Sv) then it rises again to the surface around 95°W.

From May to December, the NECC extends from 5°N to 10°N with speeds of 0.4 m s^{-1} to 0.6 m s^{-1}. Currents from the two hemispheres feed it: the Mindanao Current from the north, and the New Guinea Coastal Current from the SE, an extension of the SEC.

The NECC mean annual transport decreases from 45 Sv at 135°E in the west to 10 Sv at 90°E in the east, through entrainment by the NEC and especially by the SEC. At 165°E its transport has been estimated at 27 Sv ± 14 Sv.

Thermocline slope differentiates the NECC from the SEC and the NEC [**Figure 5-13**]. When it arrives on the coast of Central America, the NECC turns northward and forms a cyclonic circulation called the Costa Rica

[367] See the similar process in the Atlantic [**Section II.4.1.3**].

Dome.[368] The dome structure can be seen by a rise in the isotherms at its center around 8°N, 88°W [**Figure 5-14**].

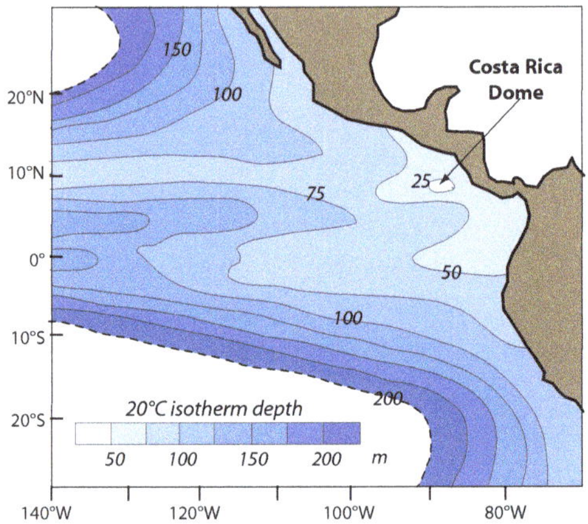

FIG. 5-14 – Average 20 °C isotherm depth revealing the Costa Rica Dome. (After Voituriez, 1981.)

4.2.4 The South Equatorial Countercurrent

The South Equatorial Countercurrent (SECC) is found in the interior of the SEC, around 8°S–10°S [**Figure 5-10**]. It is poorly represented on current charts derived from ship drift. As with the NECC, it is located under a region of weak winds. It is strongest during the season from February to April, at the end of the NW monsoon, a period during which the South Pacific Convergence Zone (SPCZ) is well established. To the east, where the SPCZ disappears, the current rapidly diminishes and disappears in the eastern Pacific. Its transport may reach 10 to 15 Sv.

4.2.5 The Equatorial Undercurrent

In 1952 Townsend Cromwell and Raymond Montgomery discovered the most powerful of the equatorial currents, the Equatorial Undercurrent (EUC), or Cromwell Current.[369] The EUC flows eastward on the equator under the westward-flowing SEC and is located in the core of the thermocline [**Figure 5-15**]. It is a thin ribbon of current centered on the equator, about 200 m thick [**Figure 5-12**], with a maximum width of 400 km, extending in length more than 14,000 km.

[368] The Costa Rica Dome is similar to the Guinea Dome in the Atlantic Ocean.
[369] As they paid out the line carrying their measuring instruments, they were surprised to see it suddenly strongly slope to the east.

The NE and SE trade winds pile up warm surface water in the west producing a difference in elevation of about 50 cm between the east and west [**Figure 5-16**] and lowering the thermocline in the west. The mechanisms are the same as those in the Atlantic Ocean [**Section III.4.1.5**].

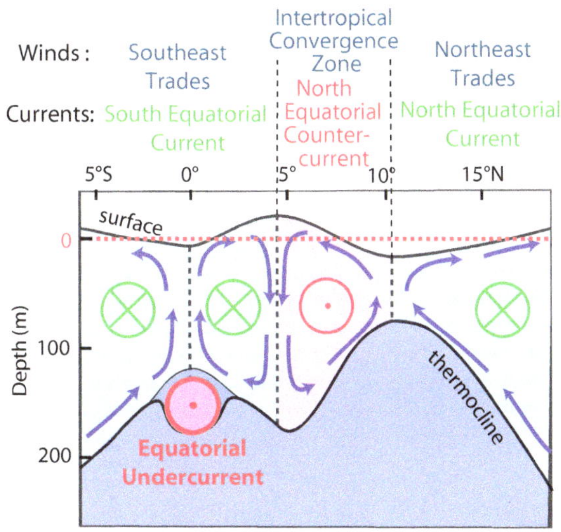

FIG. 5-15 – Schematic meridional section of (exaggerated) surface and thermocline slopes associated with equatorial currents and winds. Meridional cells are induced by convergence (downwelling) and divergence (upwelling). Westward currents are green-blue, eastward are red-pink.

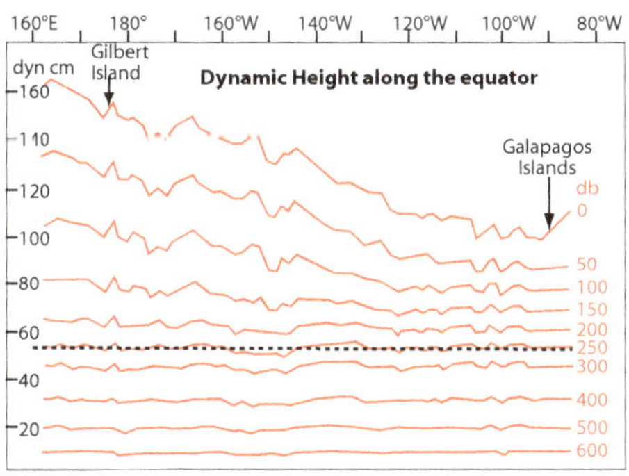

FIG. 5-16 – Dynamic height relative to 700 db at various levels along the equator. (After Lemasson and Piton, 1968.)

Variations in dynamic height [**Figure 5-16**] at different points along the equator from which the gradient of zonal pressure can be calculated show that the gradient disappears around 250 m, a dynamic height that is the same across the width of the Pacific. This is the depth at which the Equatorial Undercurrent disappears. Away from the equator, north and south, the zonal pressure gradient acting under the Coriolis Effect, generates a current equatorward, which feeds the Equatorial Undercurrent.

The core of the Equatorial Undercurrent is located in the thermocline [**Figure 5-17**], deeper in the west than in the east, where it sometimes reaches the surface when the trades are strong [**Figure 5-18**]. The Equatorial Undercurrent follows the slope of the thermocline [**Figure 5-19**]. It is found around 200 m in the west and rises to 50 m in the east, with speeds that can reach 1.5 m s^{-1}.

Under the effect of current shear and the adjustment of the density field to the current, the thermocline deepens in the lower part of the undercurrent and rises in the upper part. This creates a characteristic isotherm spreading [**Figure 5-17**], corresponding to a weakening of the vertical temperature gradient.

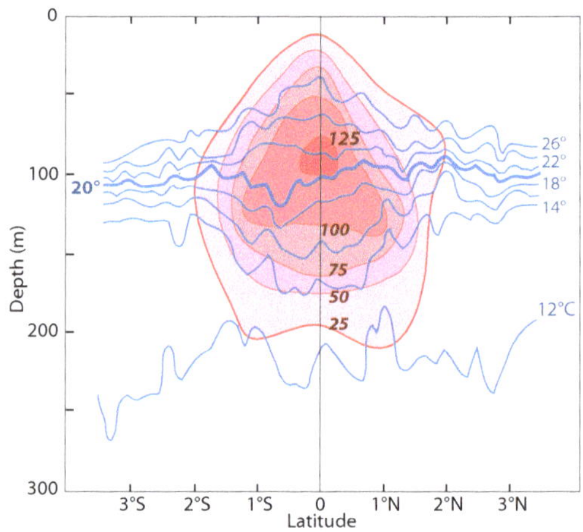

FIG. 5-17 – Eastward velocity and temperature section across the core of the Equatorial Undercurrent at 140°W. Isotachs in cm s^{-1} are red, isotherms are blue. (After Knauss, 1960.)

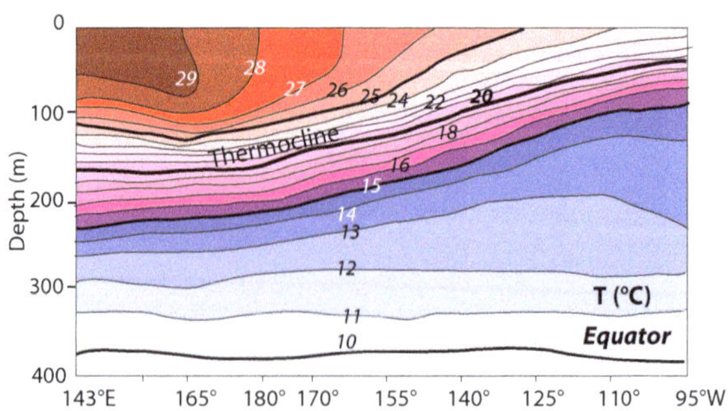

FIG. 5-18 – Zonal temperature section along the equator (1985–2002 average). (After Johnson *et al.*, 2002.)

The study of the oceanographic characteristics of the Equatorial Undercurrent and the distribution of tracers such as salinity or tritium show that it is principally fed by the SEC. The SEC (more saline and less rich in tritium than the NEC) feeds the Equatorial Undercurrent *via* the Great Barrier Reef Undercurrent and the New Guinea Coastal Undercurrent, passing through the Strait of Vitiaz. The Equatorial Undercurrent is marked by a subsurface salinity maximum whose depth diminishes eastward [Stations 1 and 5, **Figure 5-20**]. It represents the boundary between thermocline waters of the South Pacific (for 70%) and of the North Pacific (for 30%).

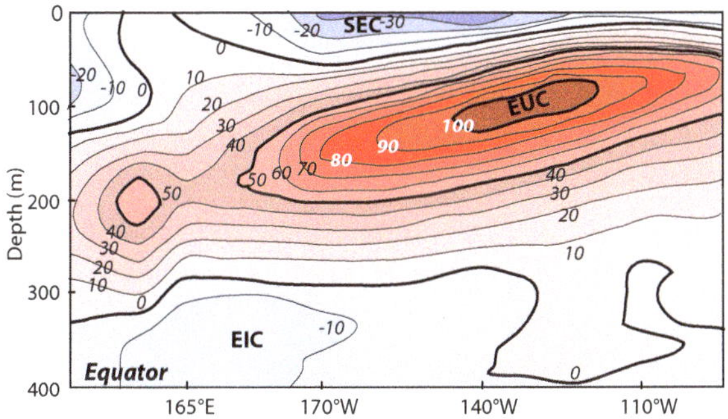

FIG. 5-19 – Zonal component of current (in cm s^{-1}) along the equator from direct current measurements (1985–2000 average). Eastward velocity is in red [Equatorial Undercurrent, EUC], westward velocity is in blue [South Equatorial Current, SEC, Equatorial Intermediate Current, EIC]. (After Johnson *et al.*, 2002.)

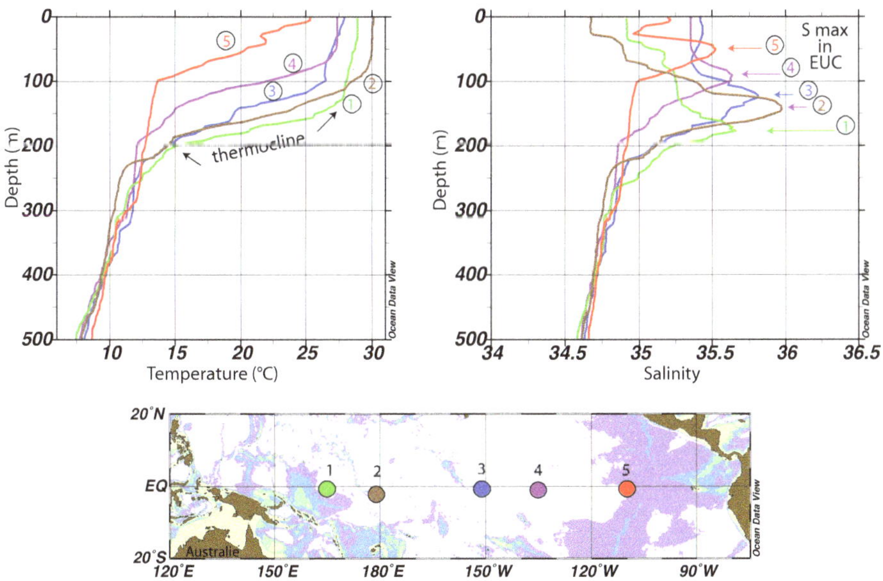

FIG. 5-20 – Temperature and salinity profiles on the equator. (With ODV and WOCE data.)

Equatorial Undercurrent transport, of the order of 12 Sv ± 3 Sv around 143°E, increases to a maximum of 35 Sv ± 3 Sv between 155°W–125°W, the region of strongest trade winds, which produce the strongest zonal pressure gradient [**Section I.5.2.2**]. The intensity of the Equatorial Undercurrent undergoes strong seasonal variations [**Figure 5-21**], with a maximum in May in the east, and in June in the west, with significant interannual variability. During an El Niño event, with weaker trades that can even reverse in the west [**Section V.3.4**], it can completely disappear for several months.

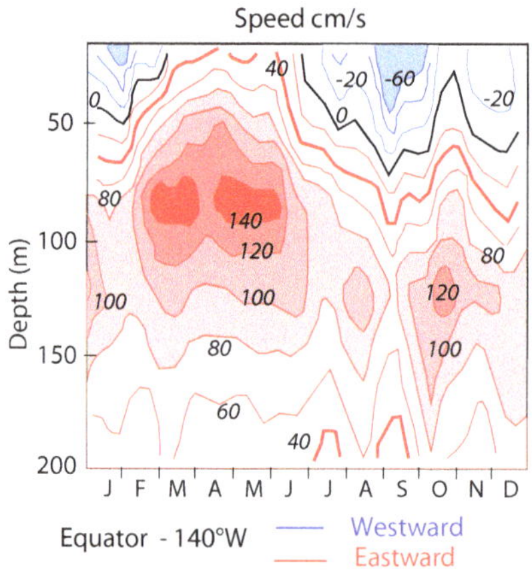

FIG. 5-21 – 15-day averaged zonal current component recorded at 140°W showing the variability of the Equatorial Undercurrent during 1984. (After Halpern and Weisberg, 1989.)

Temperature and salinity profiles at stations 1 and 2 located in the west [**Figure 5-20**] show a layer that is homogeneous in temperature but not in salinity. In the example shown in **Figure 5-22**, the homogeneous temperature layer has a thickness of 100 m while the mixed layer, which is homogeneous in temperature, salinity, and density, is only 35 m thick. Salinity increases at 35 m depth forming a halocline, associated with a pycnocline. This kind of stratification is called a "barrier layer" (between 35 and 100 m in this example) because it hinders heat exchange between the homogeneous temperature layer and the colder subsurface layer (below 100 m). Generally, the heat stored in the mixed layer, mixed due to wind and thus homogeneous in temperature and salinity, is dissipated through entrainment of cold water from the underlying layer. In barrier-layer stratification, the loss of heat from the surface layer receiving solar radiation downward cannot take place since the temperature is the same in the layer of higher density just beneath (between 35 and 100 m). In this case, only horizontal currents can play a role in the removal of heat stored in the surface layer.

Stratification that includes a barrier layer is found in regions where low salinity of surface water does not reach the full thickness of the homogeneous temperature layer. Such stratification is found in the western equatorial Pacific, in the Bay of Bengal, and in the equatorial Atlantic.

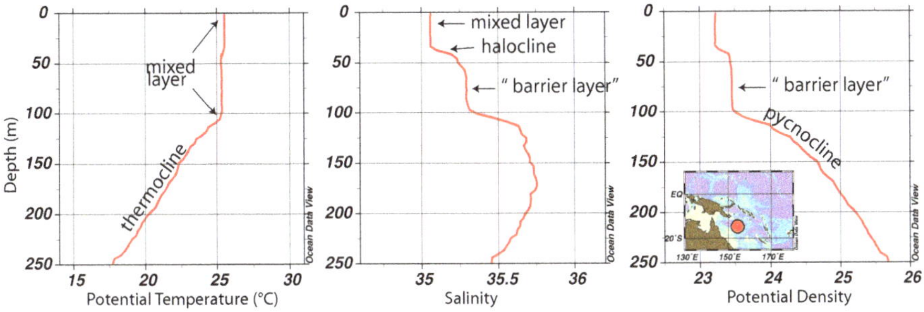

FIG. 5-22 – Temperature, salinity, and density profiles characteristic of a barrier layer in the western Pacific Ocean. (With ODV and WOCE data.)

4.2.6 The Equatorial Intermediate Current and equatorial jets

An Equatorial Intermediate Current (EIC) is found under the EUC, flowing westward centered on the equator. At 165°E, it extends from 200–280 m to 450–500 m with speeds of 0.2 m s^{-1} around 350–400 m depending on the season [**Figures 5-12, 5-23a** and **5-23b**] and has a transport of about 7 Sv ± 4.8 Sv. The upper part is called the Equatorial Intermediate Current (EIC) and the lower part the Lower Equatorial Intermediate Current (LEIC).

Sections crossing the equator at nine longitudes from 146°E to 86°W show zonal currents below 400 m similar to those found in a 16-month mean section on 159°W [**Figures 5-23a** and **5-23b**].

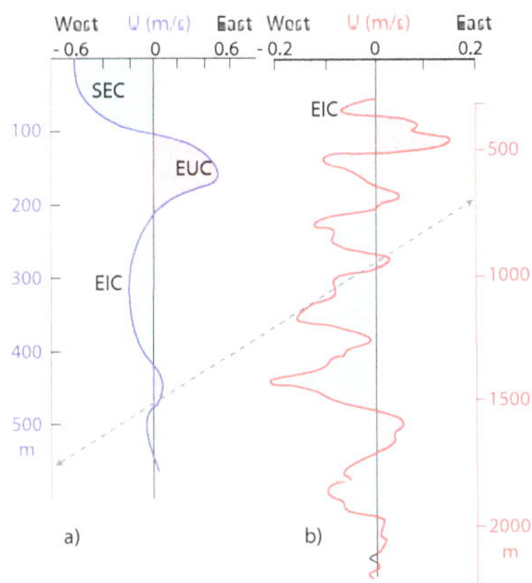

FIG. 5-23A – Zonal current component profiles at the equator around 165°E. **(a)** From 0 to 700 m. (After Delcroix and Hénin, 1988.) **(b)** From 250 m to 2,200 m. (After Firing, 1987.)

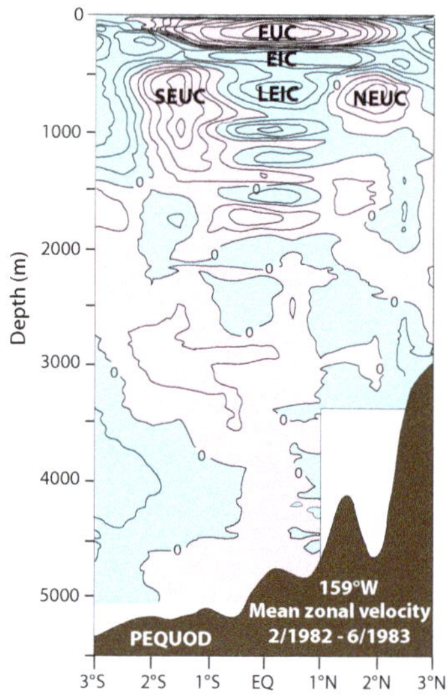

FIG. 5-23B – Mean zonal current from 41 sections along 159°W, February 1982 to June 1983, from the Pacific Equatorial Ocean Dynamics (PEQUOD) program. Contour interval is 2.5 cm s^{-1}. Regions of westward flow are in blue, and eastward flow in pink. Equatorial Undercurrent (EUC), Equatorial Intermediate Current (EIC), Lower Equatorial Intermediate Current (LEIC), North and South Equatorial Undercurrents (NEUC, SEUC). (After Firing *et al.*, 1998.)

At depth on the equator, the current structure alternates as in the Atlantic and Indian Oceans with a stack of alternating zonal currents, called equatorial jets, having speeds that reach up to 0.2 m s^{-1} [**Figures 5-23a** and **5-23b**].

4.2.7 The North Equatorial Undercurrent and the South Equatorial Undercurrent

As in the Atlantic, the equatorial current system includes a North Equatorial Undercurrent (NEUC) and a South Equatorial Undercurrent (SEUC). They flow to the east,[370] flanking the EIC between 2°N and 4°N and 2°S and 4°S in the west, under the thermocline in the depth range 500–1,500 m [**Figures 5-11** and **5-23b**]. At 165°E, their transport reaches 15 ± 7 Sv for the NEUC and 9 ± 3 Sv for the SEUC. Farther east they decrease: at 155°W their transports are estimated at 8 Sv for the NEUC and 4 Sv for the SEUC. They continue farther east moving away from the equator as their depths diminish, as in the Atlantic.

[370] These are also called *Tsuchiya jets*. Mizuki Tsuchiya first described them in the 1970's.

The equatorial current system includes numerous reversals, both at the surface and below [**Figures 5-11, 5-15** and **5-23**]. Below the surface, away from wind influence, the system is symmetric about the equator (EUC, EIC. NEUC, SEUC) in contrast to the surface currents (NEC, SEC, NECC, SECC) that are tied to trade winds and convergence zones (ITCZ, SPCZ) which do not have symmetric positions about the equator.

4.2.8 Equatorial upwelling

Even though the zonal components of equatorial currents are the most intense, the meridional and vertical components cannot be neglected [**Figure 5-15**]. At the equator, the SE trades and the change of sign of the Coriolis force produce divergence [**Section I.5.2.2**], which generates equatorial upwelling in the upper 100–200 m. The upwelling brings colder underlying water to the surface, particularly in the eastern part of the Pacific basin where the thermocline is nearer the surface [**Figures 5-18** and **5-24**]. Below the surface, equatorial upwelling is fed by horizontal inflow that converges toward the equator.

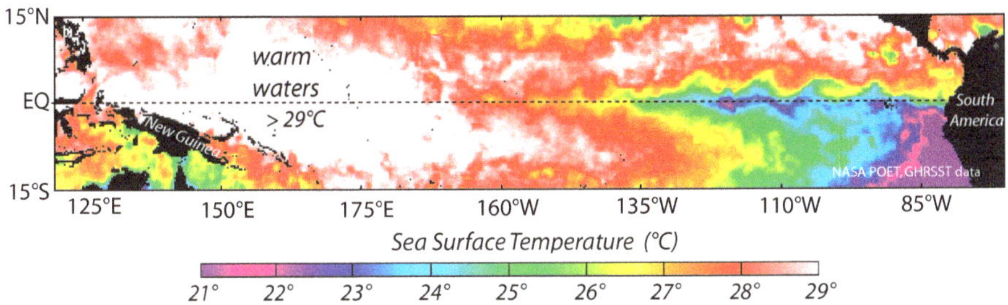

FIG. 5-24 – Equatorial Pacific Ocean surface temperature on 30 June 2004 as seen from satellite. **(a)** In the east, equatorial upwelling and instability waves can be seen. **(b)** In the west, the region of temperatures greater than 29 °C, called the "warm pool" is in white. (NASA POET, and GHRSST data.)

In the west, equatorial upwelling is not visible. The warm mixed layer is thicker and the thermocline is deeper than in the east. Furthermore, winds are often weaker than in the east.

Between the South Equatorial Current (flowing west) and the North Equatorial Countercurrent (flowing east) Coriolis force drives convergence that produces downward vertical motion, or downwelling [**Figure 5-15**]. The resulting meridional circulation cells must be taken into account in the balance. For example, heat stored in the ocean surface at the equator is balanced by upwelling cold water and by horizontal motion. Indirect estimates of Pacific Ocean equatorial upwelling, using current-meter networks and hydro stations, give values of about 47 Sv.

4.2.9 Instabilities

Westward-propagating meanders appear in the eastern equatorial Pacific from June to December at the border between upwelled cold water and warmer water to the north. To a lesser extent they are also seen on the southern border of the cold water. They result from the propagation of eddies associated with unstable waves (Legeckis waves) seen on satellite images of surface temperature [**Figure 5-24**]. They are the consequence of current shear between the NECC (eastward), the SEC (westward), and the EUC (eastward). They have a wavelength of about 1,000 km, a period of 15 to 30 days, and a propagation speed westward of 0.3–0.5 m s^{-1}. They are especially evident in the eastern Pacific between 110°W and 140°W, where currents are strongest and where there are surface temperature gradients generated by equatorial upwelling. The meander structure results from strong horizontal and vertical motions. The current convergence on the attack side of an eddy produces a downflow of water coming from the equator and intensifies the front. These instabilities disappear from March to May when current speeds lessen. They are also not observed during El Niño events because of the weakening of equatorial currents and upwelling.

4.2.10 Variability

The entire system of equatorial currents is highly variable. When the trades weaken so much that they are replaced by westerlies over the western Pacific (the El Niño situation) the EUC can completely disappear for several months and the surface current flows to the east. Similarly, NECC transport can exceed 70 Sv one year and fall to 2 Sv a year later. Seasonally, the ITCZ and the SPCZ have a great influence on surface currents, notably on the SECC, which can disappear in the southern winter (May–July). The SEC and the NECC strengthen late in the year. Below the surface, the EUC is at a maximum in the southern winter when it is nearest the surface in the east. Seasonal variations can differ between east and west. Intraseasonal variability dominates in the western equatorial Pacific where periods of westerly winds, more frequent between November and April, produce a surface current going eastward, similar to the Wyrtki Jet in the Indian Ocean. Thus, because of the great intraseasonal, seasonal, and interannual variability of equatorial currents, the transport values given are only averages.

4.3 Western boundary currents

4.3.1 The Kuroshio

The Kuroshio ("black current" in Japanese, since its waters are dark blue) is the warm western boundary current of the North Pacific anticyclonic gyre[371] [**Figure 5-25**]. It is the most intense western boundary current in the Pacific Ocean. It extends from 20°N to 38°N [**Figures 5-10** and **5-26**]. It is fed by the Philippine Current, which borders the island of Luzon, a continuation of the north branch of the North Equatorial Current. During the NE monsoon (January) a meander forms south of Taiwan, then rejoins the main current, which follows the exterior of the island. It passes the Ryukyu Ridge between Taiwan and Kyushu (of less than 1,000 m depth, which underlies the Ryukyu Islands) to rejoin the continental slope in the East China Sea. Crossing the ridge generates a meander to the southeast.

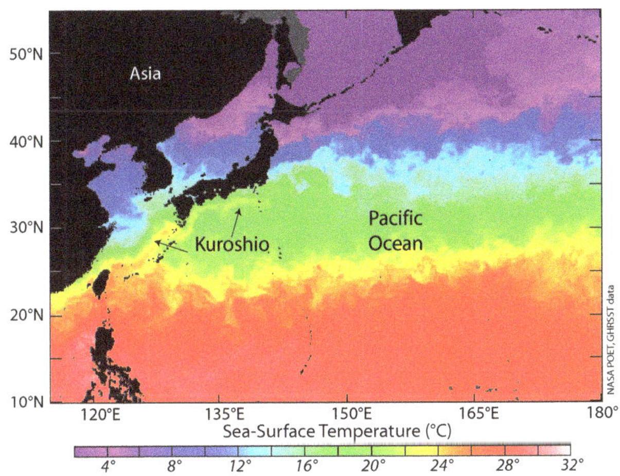

FIG. 5-25 – Sea-surface temperature in the NW Pacific Ocean. (Satellite image of January 19, 2008, NASA POET, and GHRSST data.)

The Tsushima Current detaches from the Kuroshio around 30°N, to enter the Sea of Japan, which it crosses to the north, with speeds of the order of 1 m s^{-1}, and a width of 50 to 100 km. The principal branch of the Tsushima Current (warm at this latitude) exits the Sea of Japan by the Tsugaru Strait [**Figure 5-26**] and turns southward, meeting the cold Oyashio and the main branch of the Kuroshio past Japan. A smaller part of the Tsushima Current joins the Sea of Okhotsk through the La Pérouse (Soya) Strait [**Section V.5.6.3**].

[371] The Kuroshio, like the Gulf Stream, is on the edge of the subtropical anticyclonic gyre where positive wind curl generates downwelling. Waters there are thus nutrient-poor and have low productivity, giving them their dark blue color.

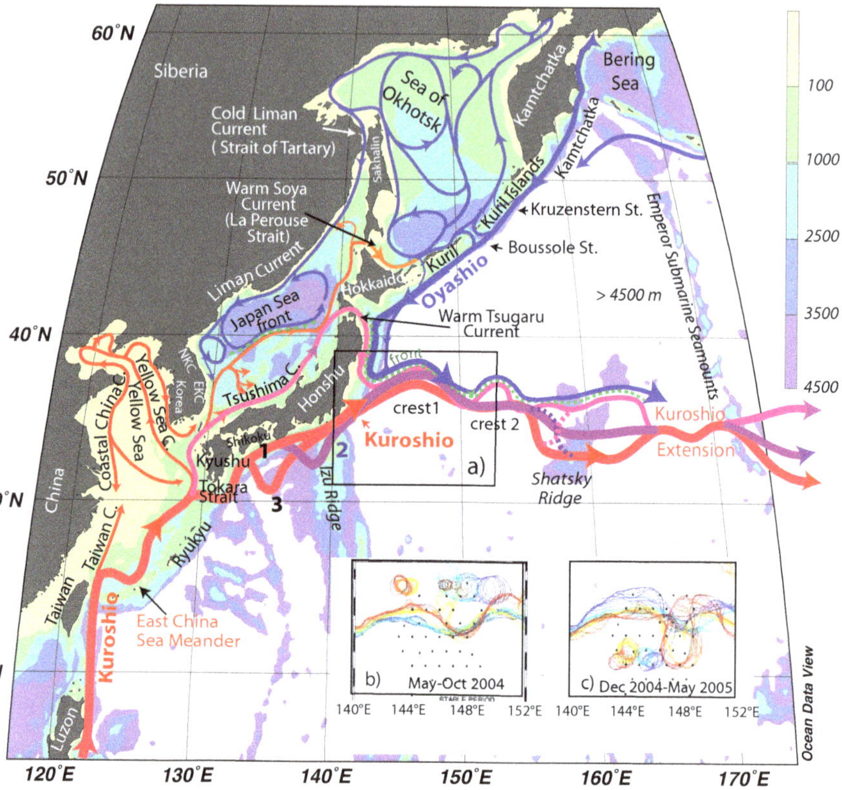

FIG. 5-26 – Circulation in the NW Pacific Ocean. The front formed by the convergence of the Kuroshio, the Oyashio, and the warm current of Tsugaru is shown dotted green. The different paths of the Kuroshio south of Japan (1, 2, 3) are shown as well as crests 1 and 2. [EKC = East Korean Warm Current, NKC = North Korean Cold Current.] (After Kawai, 1972; Mizuno and White, 1983; Senjyu, 1999.) The variability of crest 1 is illustrated in the insets **(a)** and **(b)** for summer, 2004, and in **(c)** for winter, 2004–2005. (After Kess, 2008.)

The main part of the Kuroshio branches eastward, south of the island of Kyushu, through the Tokara Strait (about 1,000 m deep) and follows the eastern continental shelf of Japan transporting warm water of tropical origin northward. Northeast of the Tokara Strait the Kuroshio extends more than 600 m deep with speeds greater than 1 m s^{-1} at the surface and 0.5 m s^{-1} at 400 m. Below, an undercurrent flows in the opposite direction (southwestward).

The Izu Ridge south of Honshu is an obstacle that varies the Kuroshio trajectory. The Kuroshio preferentially takes one of the three trajectories shown in **Figure 5-26**. Trajectory 1 follows the continental slope and crosses the ridge with little deformation. Trajectory 2 forms a meander over the Izu Ridge. Trajectory 3 forms a large southward meander before the Izu Ridge then crosses it to the northeast. On a perpendicular temperature section across the current at 165°E [**Figure 5-27a**], the current position is indicated by a sharp rise of the isotherms and isopycnals around 35°N.

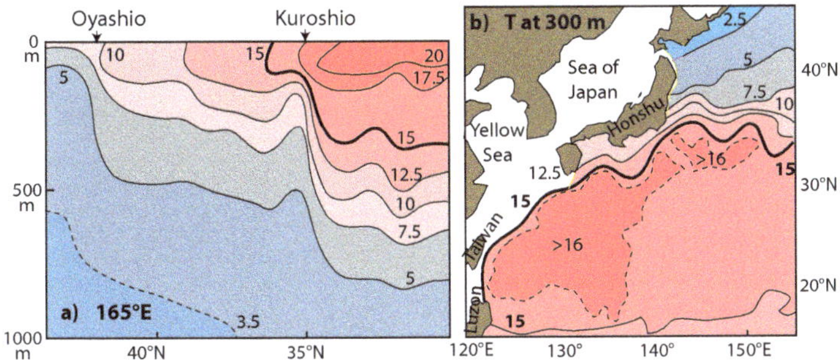

FIG. 5-27 – **(a)** Temperature sections at 165°E. Around 34°N, the strong slope of the isotherms (5 °C, 10 °C, 15 °C) signals the presence of the Kuroshio extension eastward. Around 41°N, the slope of the 5 °C isotherm indicates the Oyashio extension. (After Joyce, 1987.) **(b)** Temperature at 300 m depth in the Kuroshio region. (After Stommel and Yoshida, 1972.)

On a 300 m temperature chart, the convergence of isotherms between 15 °C and 7.5 °C shows the current [**Figure 5-27b**]. Northeast of Taiwan, the 15 °C isotherm at 300 m is often used as an indicator of Kuroshio position.

On a Kuroshio cross-section, 1,000 km long, going south from the Island of Shikoku [**Figure 5-28**] the current core is between 32°N and 32°20′N with speeds greater than 1.1 m s^{-1} over the upper 100 m. The current still reaches 20 cm s^{-1} at 1,000 m depth. The surface level difference (close to the dynamic height, in dynamic meters), between the two sides of the Kuroshio is of the order of a meter, close to that for the Gulf Stream.

When the Kuroshio leaves the continental slope around 35°N, it is called the "Kuroshio Extension". It becomes unstable and forms meanders (called crests, **Figure 5-26**), rings, and eddies [**Figure 5-29**] as in the Gulf Stream [**Section III.4.2.1**]. The variability of the crests is illustrated in **Figure 5-26,** insets a, b, c.

The thermal front between the Kuroshio and Oyashio is visible on satellite images and can be used to follow trajectories, meanders, and the formation of warm and cold rings.

North of Luzon the Kuroshio transports about 20 to 30 Sv and speeds are of the order of 0.5 m s^{-1} to 1 m s^{-1}. During its path, it entrains adjacent water and reaches 1.5 m s^{-1} to 2 m s^{-1} off Japan with a transport that varies between 33 Sv and 88 Sv, with a 7-year average of 60 Sv. It undergoes seasonal variations as does the subtropical anticyclonic wind system, but the Izu Ridge attenuates the signal coming from the east. Its transport is maximal in December and minimum (10 to 15% less) in September.

The North Pacific Current, the northern arm of the anticyclonic gyre, prolongs the Kuroshio Extension [**Figure 5-10**].

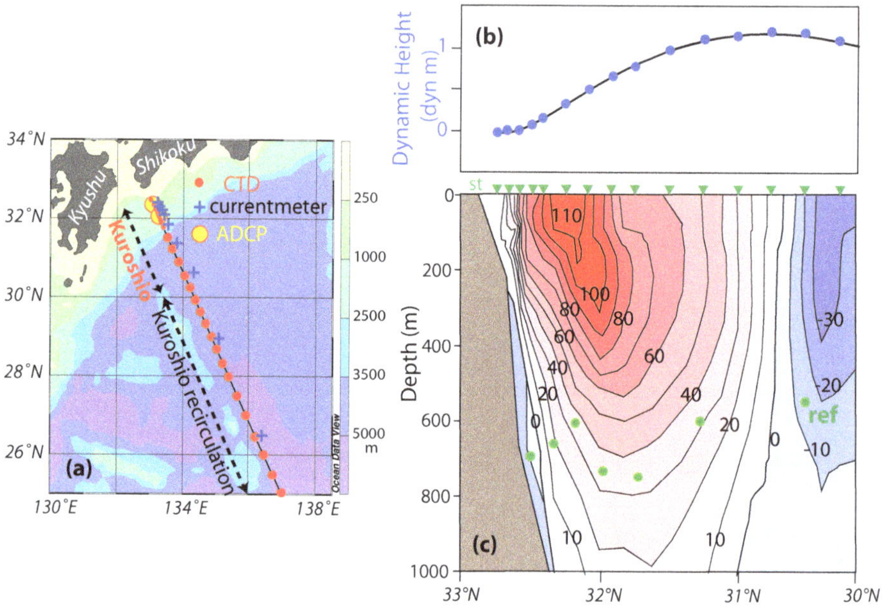

FIG. 5-28 – (**a**) Position of the cross-Kuroshio section south of Shikoku, from October, 1993 to November, 1995 using moored current meters (blue crosses), Doppler-effect acoustic profilers (yellow circles), and hydro stations (red dots). (**b**) Surface dynamic height variations calculated using hydro station data. (**c**) Geostrophic velocity section, in cm s^{-1} in November 1993. Velocities were calculated using hydro station data at the stations marked with triangles and calibrated using direct current measurements, indicated by the green points. (After Imawaki *et al.*, 1997.)

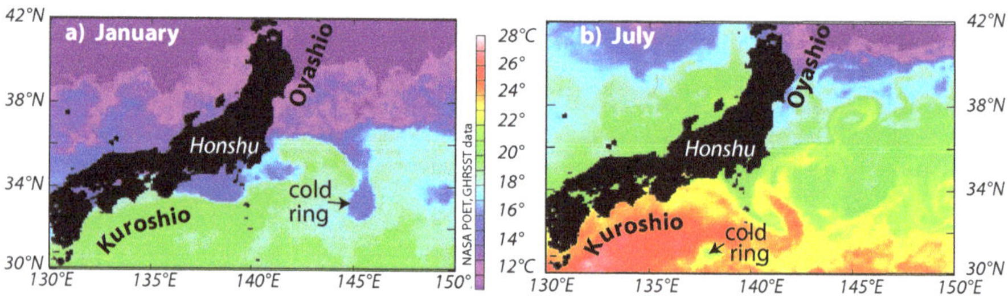

FIG. 5-29 – The (warm) Kuroshio and the (cold) Oyashio front, as seen from satellite. Surface temperatures (**a**) on 7 January 2008, and (**b**) on 1 July 2008, show meanders over the Izu Ridge and cold rings of the Oyashio. (NASA POET, and GHRSST data.)

4.3.2 The subpolar cyclonic circuit and the Oyashio

The subpolar cyclonic gyre is north of the Kuroshio and its extension [**Figure 5-10**], under the Aleutian Low.

The eastern arm of the subpolar cyclonic gyre is the Alaska Current, also called the Alaska Coastal Current, which flows northward. Fresh water inputs from Alaskan rivers diminish the density of coastal water and increase the pressure gradient perpendicular to the current. The current is thus intensified on the continental shelf near the coast. It is strongest in winter, with speeds of about 0.3 m s^{-1}. It is less deep than the Alaskan Stream which prolongs it, gradually turning to the west, with speeds of 0.2 to 0.3 m s^{-1}, extending to the sea floor. The Alaskan Stream, with a width of about 150 km, flows alongside the Aleutians. Its transport from 155°W to 175°E between 0 and 1,000 m has been estimated between 5 and 12 Sv.

The Kamchatka Current, which drains cold water from the Bering Sea, and the Alaskan Stream merge to form the Oyashio[372] [**Figure 5-10**], the western boundary current of the North Pacific subpolar cyclonic gyre. Its nutrient-rich green water contrasts with the blue water of the Kuroshio.

The Oyashio splits south of Hokkaido. One branch mixes with the warm Tsugaru Current, a branch of the Kuroshio that has turned off to the Sea of Japan [**Figure 5-26**]. The other branch joins the Kuroshio near the extremity of the first crest to form the unstable Subpolar Front. It is a region of mixing where the oceanography is complex. Meanders that sometimes transform into eddies form here [**Figure 5-29**]. The front is located around 39°N–40°N, exceptionally reaching 36°N. It is the boundary where wind-stress curl is zero [**Section I.5.3.2**], between the subpolar cyclonic circuit with positive curl and the anticyclonic subtropical circuit with negative curl. On a section perpendicular to the current at 165°E [**Figure 5-27a**], the slope of the 5 °C isotherm shows the presence of the Oyashio around 41°N with a transport of the order of 22 Sv.

An eastward continuation of the Oyashio, the North Pacific Subarctic Current, closes the cyclonic circuit on the south. The North Pacific Current is the prolongation of the Kuroshio Extension to the east [**Figure 5-10**]. The subpolar front is located between the two.

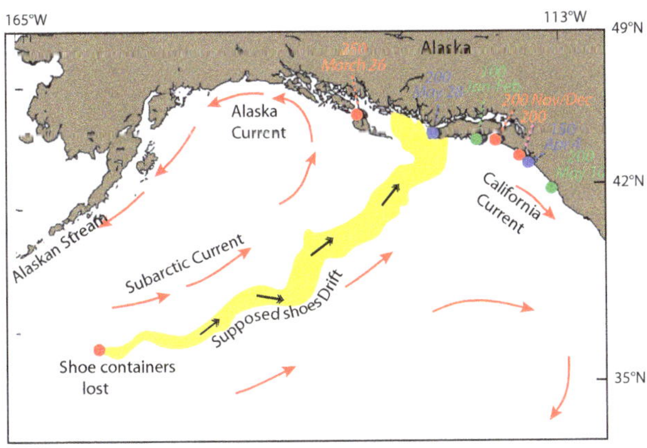

FIG. 5-30 – Drift of tennis shoes lost in May 1990 in the North Pacific Current. (After Ebbesmeyer and Ingraham, 1992.)

[372] Oyashio in Japanese means *parent current* because it is rich in nutrients and nourishing.

In 1990 a ship was caught in a storm in the North Pacific Current and lost its containers filled with tennis shoes. About a year later, shoes were found all along the Canadian and American coasts [**Figure 5-30**]. The Alaska Current took some northward and the California Current others southward although they all started from the same point.[373] This gives an example of the turbulent state of the ocean circulation as followed by unexpected tracers!

4.3.3 The Mindanao Current and the New Guinea Coastal Current

Off the Philippines, the principal branch of the NEC turns northward to form the Philippine Current, which becomes the source of the Kuroshio. The other branch turns southward as the Mindanao Current, which feeds the NECC and the Throughflow into the Indian Ocean *via* the Celebes Sea. It has a width of about 100 km and narrows to the south. As Mindanao Current turns to the east to form the NECC, part of the current recirculates to the north to create the cyclonic Mindanao Eddy [**Figures 5-10, 5-31** and **5-32**].

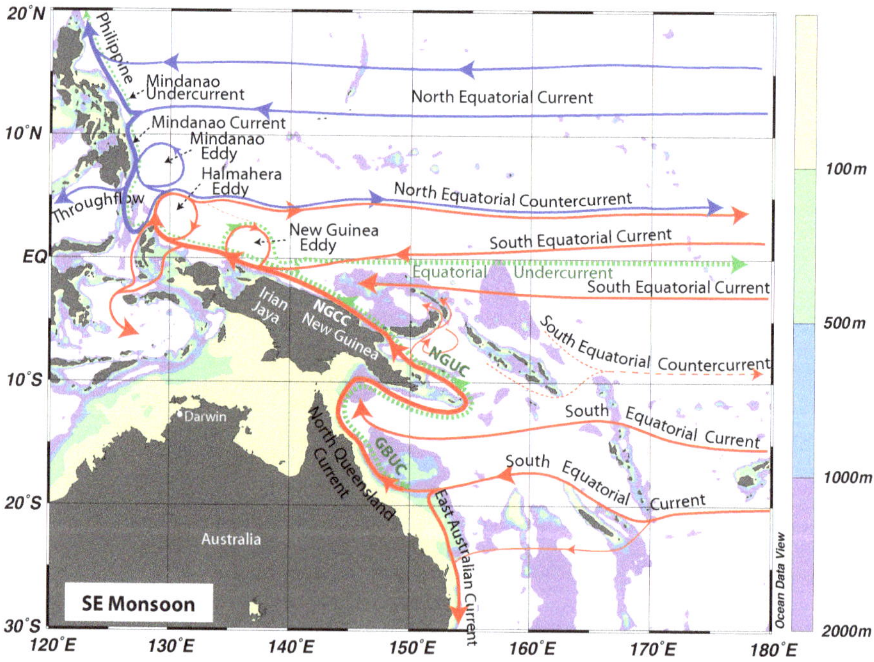

FIG. 5-31 – Western Pacific equatorial circulation during the SE monsoon. Water masses of the northern hemisphere are in blue, southern hemisphere currents are in red and undercurrents are in green [New Guinea Coastal Current (NGCC), New Guinea Undercurrent (NGUC), Great Barrier Undercurrent (GBUC)]. (After Delcroix *et al.*, 1992; Fine *et al.*, 1994; Lagerloef *et al.*, 1999; Wijffels *et al.*, 2001; Johnson *et al.*, 2002; Kashino *et al.*, 2007.)

[373] Much later, shoes were even found in Hawaii and the Philippines!

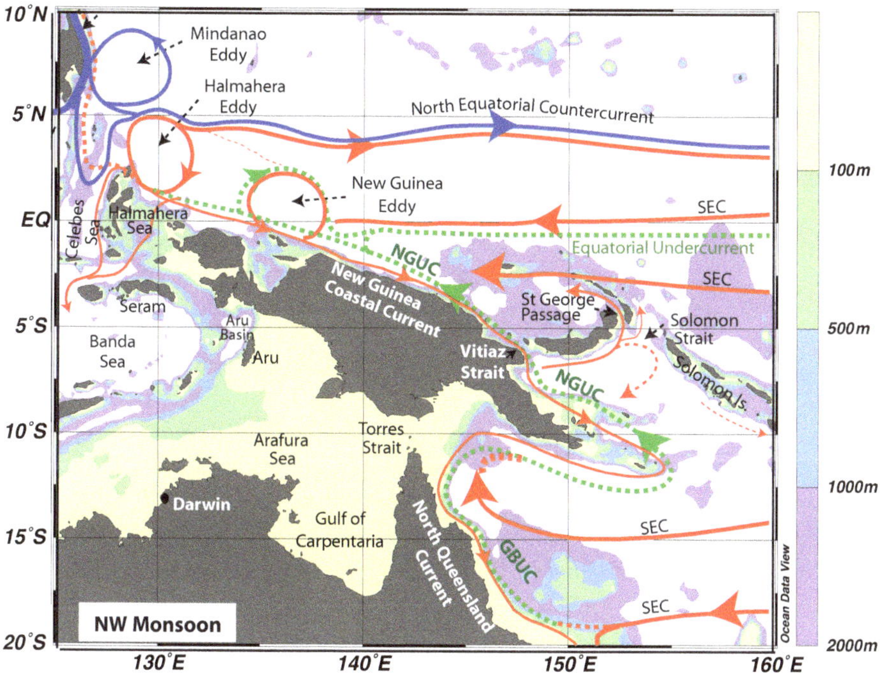

FIG. 5-32 – Western Pacific equatorial circulation during the NW monsoon. The New Guinea Coastal Current reverses in the surface layer. (See the references to **Figure 5-31**.)

The Mindanao Current undergoes seasonal and interannual variations depending on the strength of the winds. Its transport varies between 25 Sv and 35 Sv. It is a western boundary current 250 m deep. Below it, an undercurrent flows northward, between 250 m and 500 m, transporting about 18 Sv.

To the south, NE of the Island of Halmahera, the anticyclonic Halmahera Eddy is fed by the New Guinea Coastal Current (NGCC) and the New Guinea Coastal Undercurrent (NGCU) whose transport is from 8 to 14 Sv through the Vitiaz Strait, to which is added a weaker branch which flows through the Saint George Passage north of New Britain. These two currents are fed by the North Queensland Current and the Great Barrier Reef Undercurrent themselves fed by the SEC. The Halmahera Eddy is well developed during the SE (northern summer) monsoon when the SEC partly feeds the NECC, *via* the New Guinea Coastal Current and the Great Barrier Reef Undercurrent (GBUC) [**Figure 5-31**]. During the NW monsoon, the upper part of the New Guinea Coastal Current reverses to the southeast [**Figure 5-32**]. The powerful New Guinea Coastal Undercurrent continually flows to the northwest, feeding the EUC.

4.3.4 The East Australian Current

In the South Pacific Ocean the southern branch of the SEC feeds the East Australian Current, a western boundary current of the South Pacific subtropical gyre [**Figures 5-31** and **5-33**]. It is the weakest of the western boundary currents because of water lost to the Indian Ocean across the Indonesian Seas and a topography broken up by numerous islands, in particular New Zealand [**Figures 5-2** and **5-3**]. On average it only transports 15 Sv at 30°S. It follows the Australian coast as far as 34°S where one part splits off as the Tasman Current, heading north of New Zealand [**Figure 5-10**]. It produces a thermal front between warm Coral Sea water and cold Tasman Sea water, whose position fluctuates between 31°S and 37°S. The bifurcation region is subject to strong instability. The East Australian Current throws off a number of warm anticyclonic rings to the south (about 3 per year) and only a few cold cyclonic rings northward [**Figure 5-33**]. Rings can drift in the Tasman Sea for several months. There are regularly 4 to 6 at one time. Winter cooling erases their surface signature. Since East Australian Current transport is weak, the energy in these rings can exceed that of the current itself.

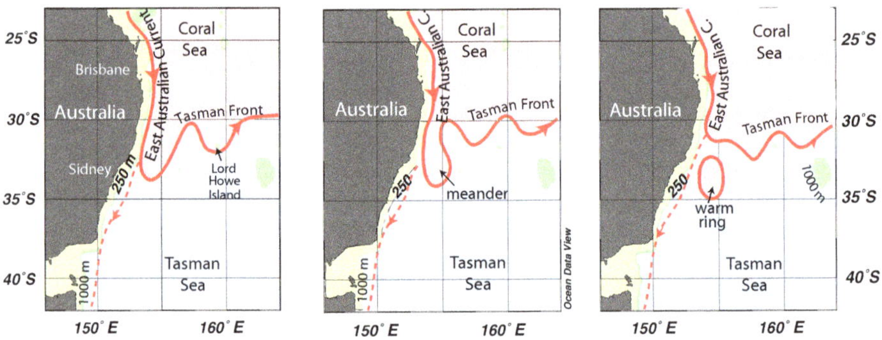

FIG. 5-33 – Formation of a warm eddy in the East Australian Current in the Tasman Sea. The Tasman Front separates Coral Sea warm water and Tasman Sea cold water. 250 m and 1,000 m isobaths are shown. (After Nilsson and Cresswell, 1981.)

During the southern hemisphere summer (December to March) the current is stronger (1.5 m s^{-1}) than in winter and is found nearer the coast, with a core 30 km offshore at 30°S. In winter the speed decreases (0.7 m s^{-1}) and there is sometimes a northward coastal countercurrent. On charts of the variability of satellite-derived sea-surface topography, this region stands out with a maximum due to strong eddy activity [**Figure 3-32**].

The Tasman Current extends north then east of New Zealand from December to March, *via* the East Cape Current and the East Auckland Current flowing along the coast southward, entraining warm water along the north island as far as the Chatham Plateau. This, too, is a western boundary current. In winter, it forms a stationary anticyclonic eddy north of East Cape and leaves the coast eastward, forming a thermal front with cold water coming from the south. The tropical front, located around 25°S, marks the northern limit of eastward flow in the south subtropical anticyclonic gyre [**Figure 5-10**].

4.4 Eastern boundary currents and coastal upwelling

Generally, currents in eastern regions of subtropical and subpolar circuits are more diffuse and slower than currents in western regions [**Section I.5.3.3**]. Eastern subtropical regions are the sites of coastal upwelling generated by winds that blow equatorward parallel to the coast [**Section I.5.2.1**].

4.4.1 Peruvian upwelling

In the southern hemisphere, under the action of southerly winds [**Figures 5-4** and **5-5**], the equatorward Peru Current comprises the Peru Coastal Current[374] (PCC) associated with a system of coastal upwelling that extends from 40°S as far north as 4°S [**Figures 1-59** and **5-34**]. The PCC is marked by surface temperatures that are several degrees less than the zonal mean. Upwelling intensity, which involves about 50 meters, varies according to the topography of the continental slope and seasonal and interannual wind variations. The inversion of the slope of subsurface isotherms[375] indicates the existence of an undercurrent, the Peru Coastal Undercurrent (PCUC) [**Figure 5-34**], in the opposite sense from 10°S to 43°S with speeds of 0.1 m s^{-1} around 10°S that diminish southward. The upwelling brings subsurface water rich in nutrients to the surface. The high productivity of this region is the source of the great fisheries of Peru and Chile and, through the intermediary of birds, which feed on fish, of the intensive exploitation of guano deposits, a popular fertilizer. When the trade winds weaken, during El Niño events [**Section V.3.4**], the northern portion of the cold water is invaded by warm waters, which temporarily destroy the region's productivity.

Offshore, the oceanic Chile-Peru Current (or Humboldt Current) closes the eastern part of the anticyclonic circuit in a more diffuse way.

---

[374] Transport estimated between 0.4 and 1.3 Sv.

[375] The isotherms adjust geostrophically to the density field.

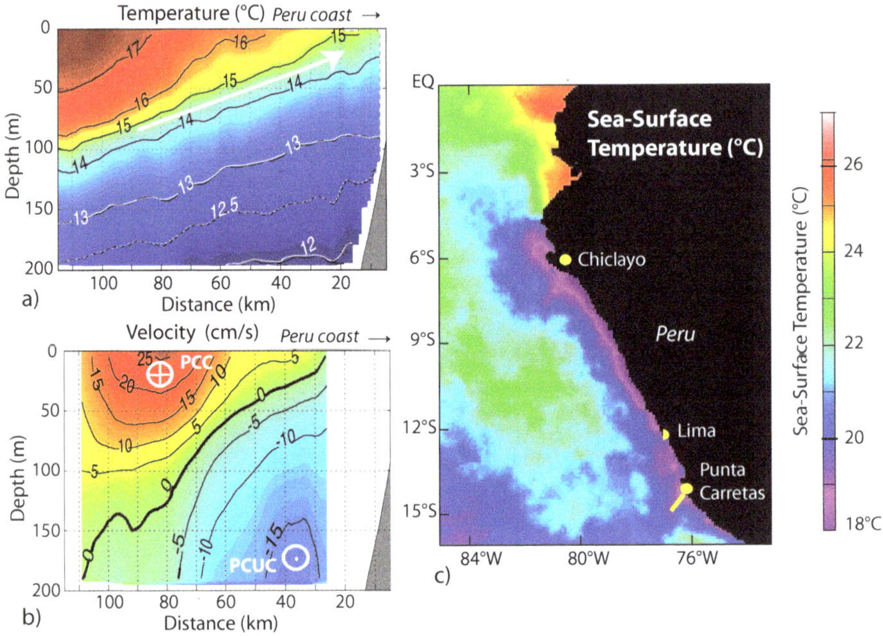

FIG. 5-34 – (**a**) Temperature section across the Peru upwelling off Punta Carretas. (After Pietri, 2016.) (**b**) Meridional component of the current for the same section, Peru Coastal Current (PCC), Peru Coastal Undercurrent (PCUC). (After Pietri, 2016.) (**c**) Satellite sea-surface temperature in the Chile-Peru upwelling. (1 January 2008 satellite image, NASA POET, and GHRSST data.)

4.4.2 California upwelling

In the North Pacific, the California upwelling system [**Figure 5-35**] is not as regular or as intense as that of Peru because of the strong variability of winds in the region. Entrainment by north winds competes with the meridional pressure gradient. When the winds diminish, the northward undercurrent increases. When the winds strengthen, the southward surface current increases, upwelling is stronger and the undercurrent decreases.

From April to September north winds blow and upwelling dominates. Surface temperatures then decrease to 15–14 °C along the coast. On the central and south coasts of California a cyclonic eddy arises between the California Current and the coast, leading to a northward coastal current.

Winds are variable and sometimes from the SE during the rest of the year. Off northern California a northward coastal current (the Davidson Current) appears when winds weaken or come from the SE as they do between October and March, with a maximum in January–February. The current disappears when north winds strengthen in April.

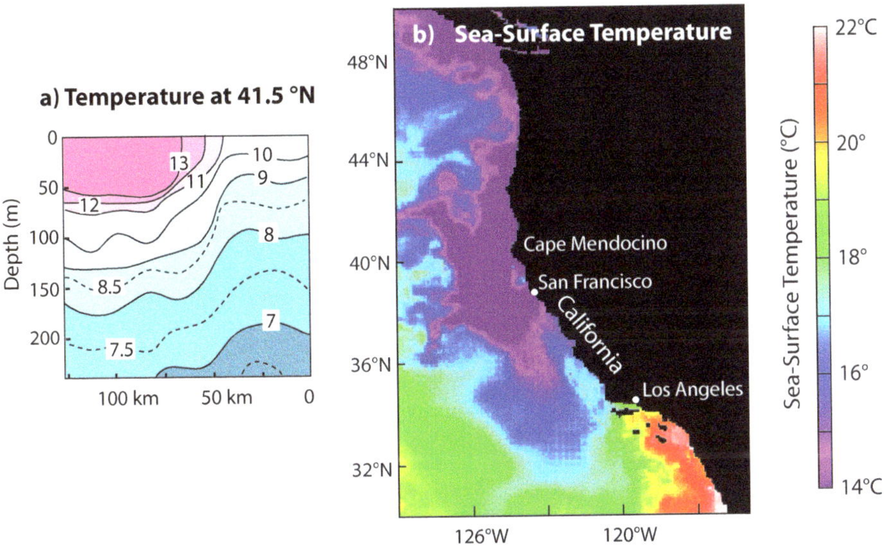

FIG. 5-35 – **(a)** Temperature section at 41.5°N in the California upwelling in June 1987. (After Kosro *et al.*, 1991.) **(b)** Sea-surface temperature showing cold water along the coast of California. (14 September 2008 satellite image, NASA POET, and GHRSST data.)

Before overfishing in the 1930's the California upwelling region had an important sardine fishery.

The cooling of air in contact with cold water produces frequent fogs on the coastal strip. The low humidity in the air condenses in contact with the ocean resulting in very dry air over the continent.

4.5 The South Pacific Current

As in the South Atlantic and Indian Oceans, the South Pacific Current [**Figure 5-10**], located north of the Subtropical Front, is distinct from the Circumpolar Current. It extends the East Cape and East Auckland Currents, passing northeast of New Zealand. It represents the southern branch of the southern anticyclonic gyre. Approaching the barrier of the South American continent, which extends as far as 56°S, the current turns northward, entraining cold water along the Chilean coast and feeding the oceanic Chile-Peru (Humboldt) Current. The Subtropical Front disappears in this region. South of 45°S, the Circumpolar Current flows around Antarctica through the Drake Passage.

Chapter 5
Water properties

5.1 Surface water

5.1.1 Surface temperature

Surface temperature distribution is the result of heat balance between ocean and atmosphere [**Section I.2.2.5, Figure 1-13**] modified by ocean circulation [**Figure 5-10**]. Surface isotherms [**Section I.2.4, Figure 1-17**] do not follow parallels but open fan-like westward in the equatorial zone, entrained by circulation. The thermal equator follows the ITCZ from 5°N in the west to 15°N in the east, with annual average temperatures greater than 27 °C. The North and South Equatorial Currents accumulate warm water to the west. The highest temperatures are found in the western equatorial Pacific[376] [**Figure 5-24**], with an annual average that exceeds 29 °C. The Kuroshio entrains this water northward and the East Australian current southward.

The eastern equatorial region of the Pacific Ocean is colder since it receives cold water from the Peru Current and equatorial upwelling cools surface water.

At mid-latitudes, it is the inverse. The fan of isotherms is opened to the east by the Alaska Current northward and by the California Current southward. To the south, the South Pacific Current flows northward along the coasts of Chile and Peru, entraining subpolar cold water northward. California and Peru upwellings amplify the effect. However, California upwelling is not continuous, and its influence on annual average temperature is less marked than off the coast of Peru.

The strongest horizontal temperature gradients (fronts) are found off Japan, at the convergence of the two western boundary currents. The Kuroshio transports warm tropical water and the Oyashio transports cold

[376] Equally warm water is found in the Indian Ocean during inter-monsoon periods and even warmer in the NW adjacent seas.

subpolar water, forming the Subpolar Front. In the south, the effect of the East Australian Current is less marked, since its transport is weaker.[377]

Seasonal variations of surface temperature are weak in the equatorial and tropical zones [**Figure 5-36**]. They are greater in temperate latitudes and decrease in high latitudes. The monsoon regime affects the entire northwest region and reinforces these seasonal variations enough to produce extreme values between 10 °C and 25 °C. The greatest variations are found in the northern Sea of Japan and in the Yellow Sea.

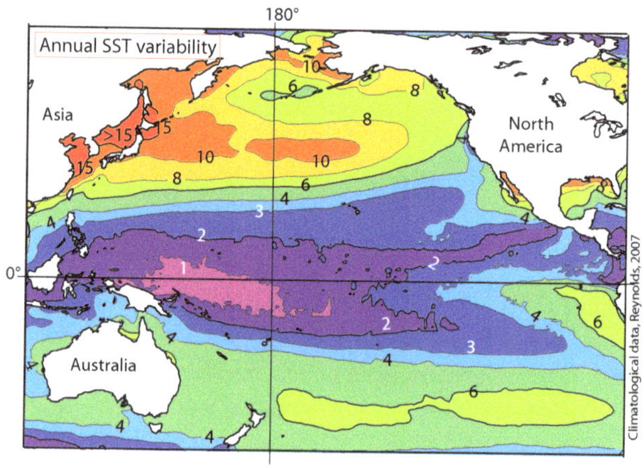

FIG. 5-36 – Amplitude of annual sea-surface temperature variability (°C) in the Pacific Ocean. Maximum amplitude is 24.5 °C. (Courtesy of C. de Boyer Montégut, climatological data, Reynolds, 2007.)

5.1.2 Surface salinity

Surface salinity distribution [**Section I.2.6, Figure 1-23**] is the result of interaction between the ocean and regional climate, which is a balance between evaporation, precipitation, and stream inflow [**Section I.2.5.2, Figure 1-22**], modified by ocean circulation [**Figure 5-10**].

A low-salinity band is found under the ITCZ, between 5°N and 10°N, with a minimum in the Gulf of Panama ($S < 33$). This minimum is produced by strong rainfall from air masses charged with humidity over the Atlantic Ocean. Low-salinity water is entrained by the NEC and influences the entire North Pacific. This band of low salinity extends southward under the South Pacific Convergence Zone between New Guinea and Fiji.

[377] The energy of the East Australian Current is mainly in the turbulent rings that it generates.

Salinity maxima are found in anticyclonic centers where evaporation dominates over precipitation [**Figure 1-22**]. Northward, in the Gulf of Alaska, salinity is low because of rainfall and inputs of fresh water from numerous rivers along the coasts of Canada and Alaska. Salinity is also low in the Bering Sea and in adjacent seas of the western Pacific for the same reasons. Even in regions of maximum subtropical salinity, salinity is lower in the north (35.2 maximum) than in the south Pacific (36.2 maximum) [**Figure 1-23**]

High precipitation in the equatorial western Pacific creates a special type of stratification in the surface layer characteristic of a barrier layer [**Figure 5-22**]. It blocks heat exchange between the surface, which receives solar radiation, and colder subsurface water.

5.2 Subtropical salinity maximum, North and South Tropical Waters

Two saline water masses form in subtropical regions where the surface salinity is maximal [**Figure 1-23**]. They are North Pacific Tropical Water (NPTW) and South Pacific Tropical Water (SPTW). Under the influence of wind curl, they subduct equatorward under lighter water of the equatorial region [**Section I.3.2**]. In the two regions where the subtropical salinity maximum is at the surface [**Figures 5-37** and **5-39**], θ-S diagrams underline the differences in water properties between the North Pacific and the South Pacific. North Pacific Tropical Water is less saline (S = 35.3–35.4) than South Pacific Tropical Water (S = 36.4–36.7).

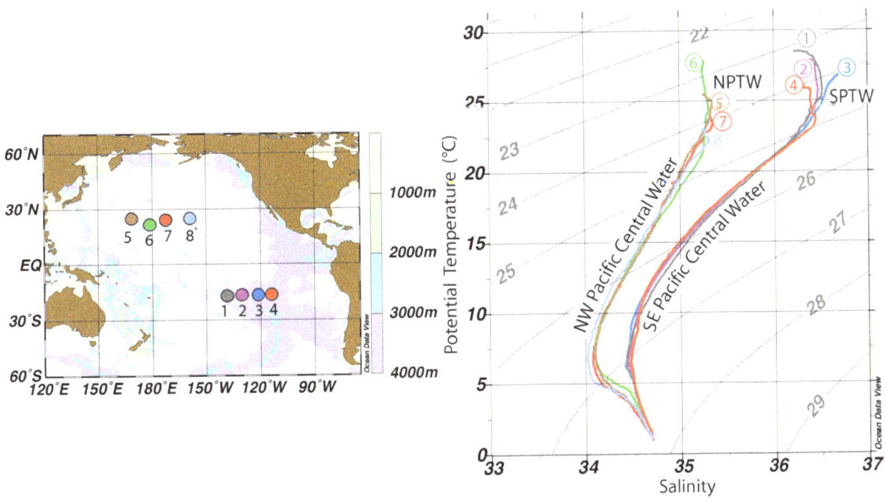

FIG. 5-37 – θ-S diagrams from stations in the two regions where the subtropical salinity maximum is at the surface. Stations 1 to 4 are in the south anticyclonic gyre; stations 5 to 8 are in the north anticyclonic gyre. (With ODV and WOCE data.)

5.3 Central and Equatorial Waters

Central Water, or Thermocline Water, is found under Tropical Water [**Section I.3.2**]. Central Water forms in regions of subtropical convergence then is subducted under lighter water. It is represented as a straight segment on a θ-S diagram since its characteristics cover a large range of temperature and salinity [**Figures 5-37** and **5-38**]. It exists only in the region between the two zones of north and south subtropical convergence. Because of the great extent of the Pacific Ocean, and of the variability of surface atmospheric conditions, several kinds of Central Water with different θ-S characteristics are formed. Six kinds of thermocline water can be distinguished: NW Pacific Central Water (NWPCW), NE Pacific Central Water (NEPCW), North Pacific Equatorial Water (NPEW), South Pacific Equatorial Water (SPEW), SW Pacific Central Water (SWPCW), and SE Pacific Central Water (SEPCW) [**Figure 5-38**]. The average spatial distribution of each of these water masses is sketched in **Figure 5-39**, as well as the subduction region of South Pacific Tropical Water (SPTW) where surface and subsurface salinity maxima exceed 36 in the southern anticyclonic gyre.

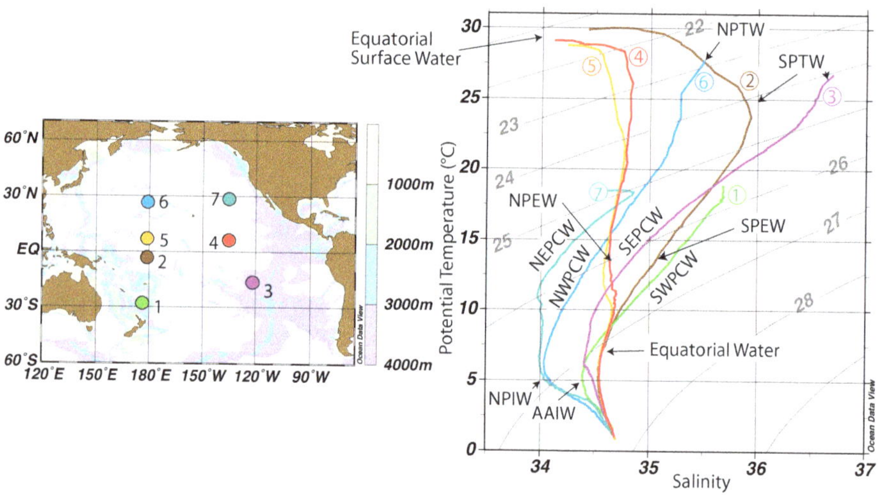

FIG. 5-38 – θ-S diagrams from 7 stations between the two subtropical fronts corresponding to different thermocline water masses in the Pacific Ocean. [NE Pacific Central Water (NEPCW), NW Pacific Central Water (NWPCW), SE Pacific Central Water (SEPCW), SW Pacific Central Water (SWPCW), South Pacific Equatorial Water (SPEW), North Pacific Equatorial Water (NPEW), North Pacific Tropical Water (NPTW), South Pacific Tropical Water (SPTW), Antarctic Intermediate Water (AAIW), North Pacific Intermediate Water (NPIW).] (With ODV and WOCE data.)

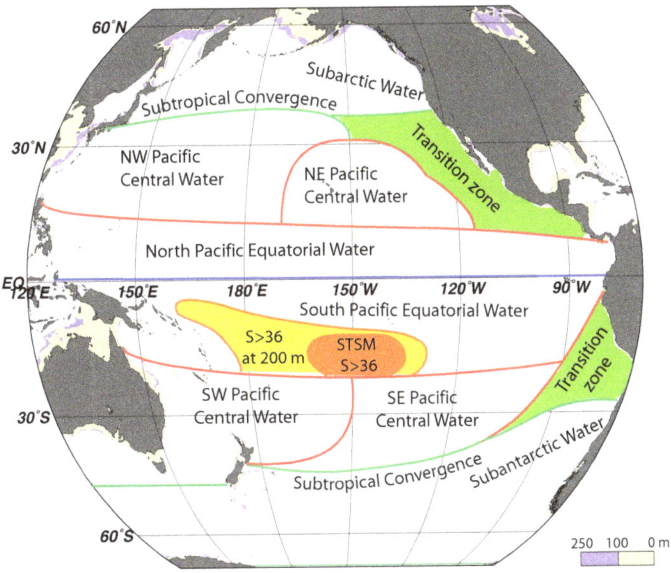

FIG. 5-39 – Distribution of thermocline waters in the Pacific Ocean. Green areas are transition regions with Subarctic Water and Subantarctic Water. Average salinity greater than 36, the region of subtropical salinity maximum (STSM), is orange when at the surface and yellow when at 200 m. (After Tomczak and Godfrey, 1994.)

5.3.1 Southwest and Southeast Pacific Central Waters

Southwest Pacific Central Water (SWPCW), formed in the subtropical convergence zone between Tasmania and New Zealand, is the most saline [**Figure 5-38**, Station 1]. It is the variety that most approaches Central Water of the southern Atlantic Ocean and the southern Indian Ocean. Southeast Pacific Central Water (SEPCW) is less saline than SWPCW. Surface θ-S characteristics corresponding to SE Central Water show that it forms between 180° and 150°W. East of 110°W, thermocline salinity decreases through the influx of Subantarctic Water by the Chile-Peru Current [**Figure 5-10**], where the subtropical convergence zone is diffuse. It is a region of mixing between SE Central Water and Subantarctic Water in a so-called transition zone [**Figure 5-39**].

5.3.2 Northwest and Northeast Pacific Central Waters

The most important Central Water in the North Pacific is Northwest Pacific Central Water (NWPCW). It forms in the north subtropical convergence zone (southeast of the Kuroshio), where surface salinity is less than in the south [**Figure 1-23**], then is entrained into the anticyclonic circulation.

NW Pacific Central Water is less saline than the two Central Waters in the south. Northeast Pacific Central Water can be differentiated as the least saline of all Central Waters in the Pacific, with temperature less than 17 °C (Station 7, **Figure 5-38**). Its low salinity results from mixing with less saline Subarctic Water. The mixing region between the two North Pacific Central Water masses is found around 170°W.

5.3.3 North and South Pacific Equatorial Waters

Equatorial Water occupies the greatest volume of the Pacific Ocean thermocline. It has no equivalent in the other oceans. It is one of the rare water masses that does not form entirely at the surface under the effect of exchanges with the atmosphere.

In the deep layer (2 °C $< \theta <$ 8 °C) from 2,000 m to 500 m, there is only a single Equatorial Water [**Figure 5-38,** Stations 2, 4, and 5]. Equatorial Water is formed by mixing of Central and Intermediate Waters and by upwelling of Deep Water that greatly attenuates the Intermediate Water salinity minimum.

Above 500 m (about 8 °C) two kinds of Equatorial Water can be seen. They are South Pacific Equatorial Water (SPEW) [**Figure 5-38**, Station 2] and North Pacific Equatorial Water (NPEW) [**Figure 5-38**, Stations 4 and 5]. Their characteristics are intermediate between those of North Pacific Central Water and South Pacific Central Water.

The upper layer of South Pacific Equatorial Water ($\theta >$ 17 °C) is formed by subduction of surface water from the subtropical high-salinity region ($S >$ 36, $T >$ 26 °C) (South Pacific Tropical Water) [**Figures 1-23, 5-38**, Station 2, and **Figure 5-29**] under the lighter surface water, low salinity and warm, below the ITCZ and SPCZ [**Figure 5-6**]. The lower layer (8 °C $< \theta <$ 17 °C) is formed through mixing between the two South Pacific Central Waters: Southeast Pacific Central Water [**Figure 5-38,** Station 3] and Southwest Pacific Central Water [**Figure 5-38,** Station 1].

The surface layers of North Pacific Equatorial Water form through mixing of NW Central Water and South Pacific Equatorial Water (between 10 °C and 15 °C) in the western equatorial Pacific in regions of strong shear between the North Equatorial Countercurrent and the South Equatorial Current. Above 18–20 °C, it represents the lowest-salinity thermocline water in the Pacific Ocean [**Figure 5-38**, Stations 4 and 5] because it receives the strong precipitation of the ITCZ [**Figure 5-6**]. The boundary between the two Equatorial Waters, of about 250 km width, is along the equator, showing the lack of communication that exists between North and South Pacific circulations. The transition between North Pacific Equatorial Water and NW Pacific Central Water is more gradual. Along the front between the two water masses, NW Pacific Central Water is above the North Pacific Equatorial Water, with the front at 250 m around 12°N and at 500 m around

18°N. North of 20°N, only NW Pacific Central Water is found. The diagrams of three stations from 4°S to 26°N show a gradual transition from South Pacific Equatorial Water [**Figure 5-40**, Station 1] to North Pacific Equatorial Water (Station 2) and to NW Pacific Central Water (Station 3).

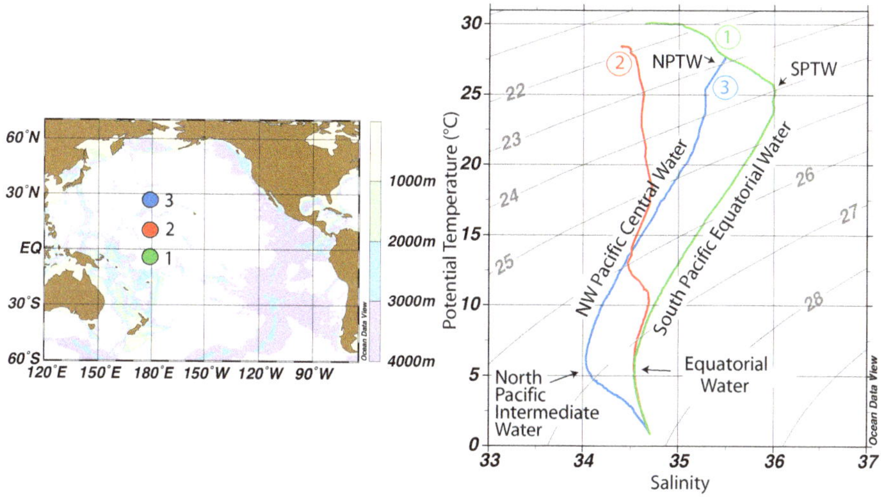

FIG. 5-40 – θ-S diagrams from 3 stations at 179°E: Station 1 at 4°S; Station 2 at 10°N; Station 3 at 26°N. (With ODV and WOCE data.)

The hydro-oceanographic vessel *Beautemps-Beaupré*, put in service in 2003, was co-financed by the French ministries of Defense and Research. (Photo by G. Bessero, © SHOM 2006.)

5.4 Intermediate Waters

The salinity minima between 60 m and 1,200 m on the meridional section at 165°W [**Figure 5-41b**] show the presence of clearly separated intermediate water masses.

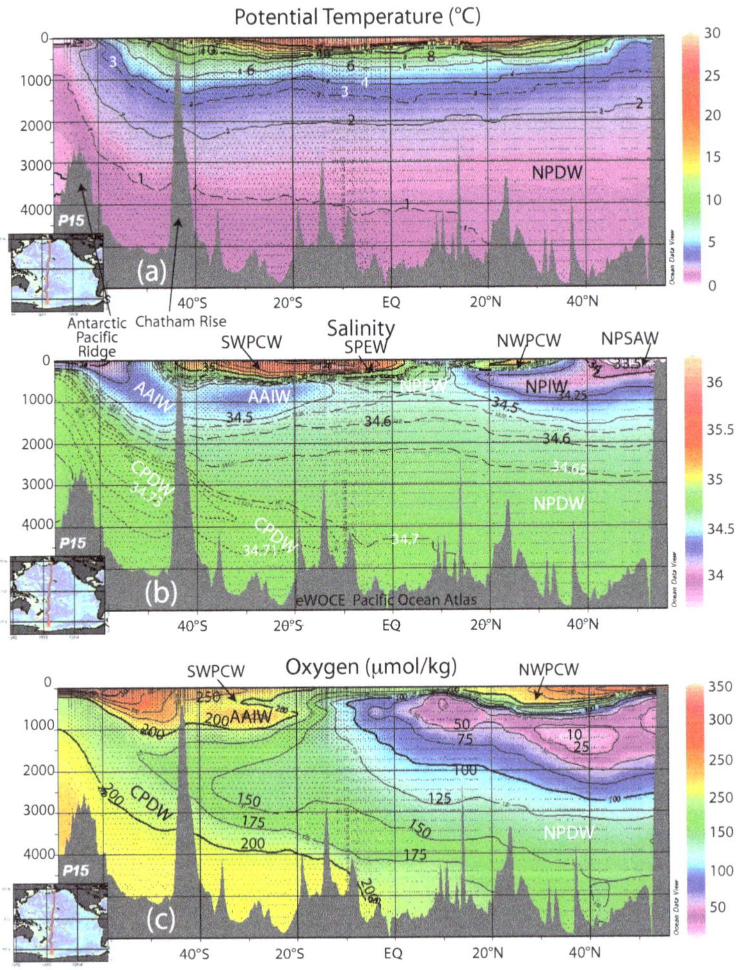

FIG. 5-41 – Meridional sections at 165°W in the western Pacific of (**a**) temperature, (**b**) salinity, and (**c**) dissolved oxygen. [Circumpolar Deep Water (CPDW), Antarctic Intermediate Water (AAIW), North Pacific Intermediate Water (NPIW), NW Pacific Central Water (NWPCW), SW Pacific Central Water (SWPCW), South Pacific Equatorial Water (SPEW), North Pacific Equatorial Water (NPEW), North Pacific Subarctic Water (NPSAW).] (Section P15, *eWOCE Atlas of the Pacific Ocean*, Talley L.D., 2007.)

5.4.1 Antarctic Intermediate Water

In the south, Antarctic Intermediate Water (AAIW) forms in the Antarctic Ocean at the Subantarctic Front, in particular in the southeast Pacific, through winter convection, in a region where the surface salinity is low [**Section II.6.3.2**]. It is characterized by a salinity minimum with the lowest salinity in the SE Pacific Basin [**Figure 5-42**, Station 3]. Antarctic Intermediate Water is entrained into the southern anticyclonic circulation in a large westward movement [**Figure 5-43**] during which it is eroded through mixing with water above and below [**Figure 5-42**, Stations 2, 3, 4]. Toward the equator (Stations 5 and 6) it is completely eroded through mixing with Deep Water, below, and Central Water, above. Equatorial Water replaces it [**Figure 5-42,** Station 6].

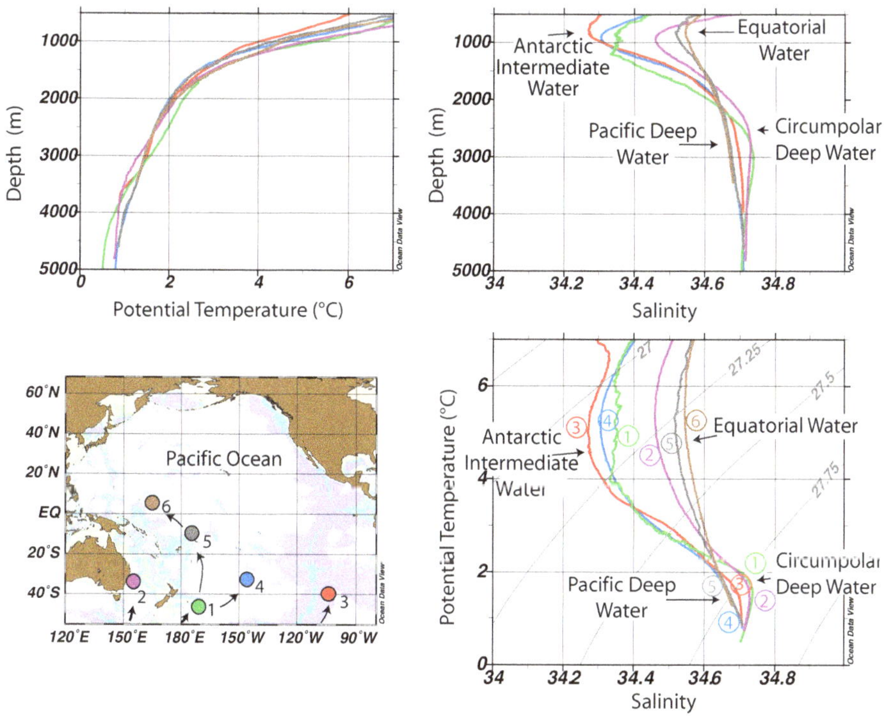

FIG. 5-42 – The evolution of θ-S characteristics in Circumpolar Deep Water and Antarctic Intermediate Water in the South Pacific. (With ODV and WOCE data.)

Travelling westward, a branch returns southward *via* the western boundary, where it converges with Antarctic Intermediate Water of the southern Tasman Sea [**Figure 5-44**, Station 1]. The extension of the East Australian Current produces a front where the two water masses mix. Evidence of mixing can be seen on a θ-S diagram by the rapid salinity variations in the envelope represented by the two Intermediate Water masses (Station 1 and the group of Stations 5, 6, 7, 8, 9).

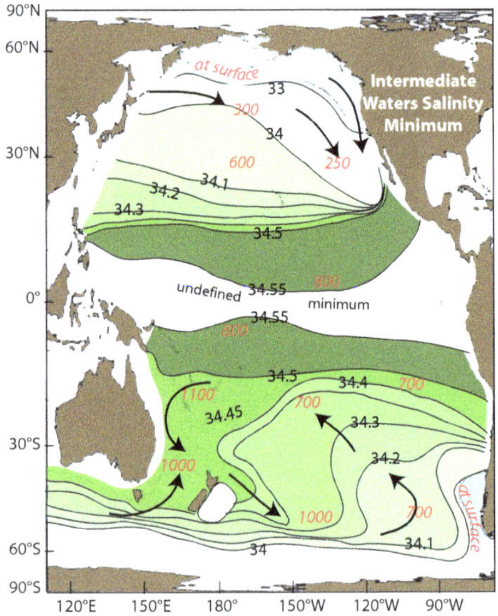

FIG. 5-43 – Salinity at the depth of the salinity minimum, showing the expansion of Intermediate Waters in the Pacific Ocean. The depth of the minimum is given in meters in red. Regions where the surface salinity is lower than that of the minimum of Intermediate Waters are shown in sky blue. (After Tomczak and Godfrey, 1994.)

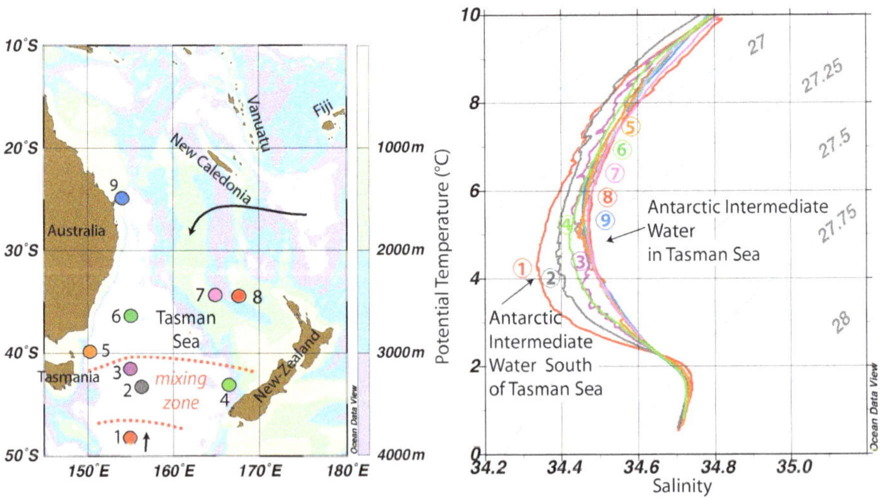

FIG. 5-44 – θ-S diagrams in the Tasman Sea showing interleaving south of 40°S (Stations 2, 3, 4) indicating that Antarctic Intermediate Water from the SW (Station 1) is mixing with Antarctic Intermediate Water coming from the north and east (Stations 5, 6, 7, 8, 9). (With ODV and WOCE data.)

5.4.2 Subarctic Water and North Pacific Intermediate Water

A low-salinity [**Figure 1-23**] and cold [**Figure 1-17**] water mass called Subarctic Water is found north of the Subtropical Convergence in the North Pacific, comparable to Subantarctic Water in the Antarctic Ocean. The low salinity of the Subarctic Water is due to precipitation and coastal river runoff in the Gulf of Alaska. In the subarctic region, the gradual transition from Subarctic Water to Deep Water is marked by an increase in salinity and a decrease in temperature [**Figure 5-45**, Station 2, end of summer] except in winter where surface water can become colder than the water below [**Figure 5-45**, Station 1, end of winter]. Station 3 is typical of the transition between cold and low-salinity Subarctic Water and more saline Central Water.

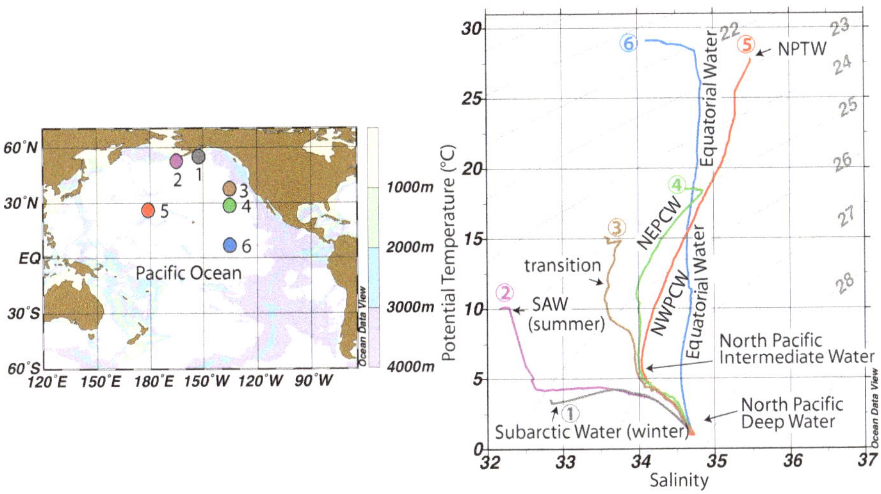

FIG. 5-45 – θ-S diagrams from stations located in the North Pacific. [Subarctic Water (SAW), NW Pacific Central Water (NWPCW), North Pacific Intermediate Water (NPIW), NE Pacific Central Water (NEPCW), salinity maximum of North Pacific Tropical Water (NPTW).] (With ODV and WOCE data.)

North Pacific Intermediate Water (NPIW) forms in the energetic front between the Kuroshio and Oyashio with water formed at the surface in the northwestern Sea of Okhotsk (see **5.6.3**) during sea ice formation, which undergoes strong mixing when exiting through the Kuril Island straits. On its way eastward, it entrains some subpolar low-salinity water. It diffuses southward into the north anticyclonic gyre through subduction below North Pacific Central Waters. It is characterized by a salinity minimum [**Figure 5-45**, Stations 4 and 5] at about 150 m depth at 30°N in the east [**Figures 5-46b** and **5-43**] (S = 33.5) and around 800 m in the west (S = 34.1), eroded by mixing with NE Pacific Central Water and NW Pacific Central Water. Its minimum salinity core lies around σ_0 ~ 26.8. The NPIW is the densest water formed in the Pacific Ocean.

The core of North Pacific Intermediate Water is less deep and less saline than that of Antarctic Intermediate Water [**Figure 5-41**]. In the equatorial region, it is no longer possible to discern the salinity minima of either North Pacific Intermediate Water or of Antarctic Intermediate Water. Deep Water diffuses slowly through upwelling in the overlying layers and vertical salinity variations in North Pacific Equatorial Water become weak [**Figure 5-45**, Station 6].

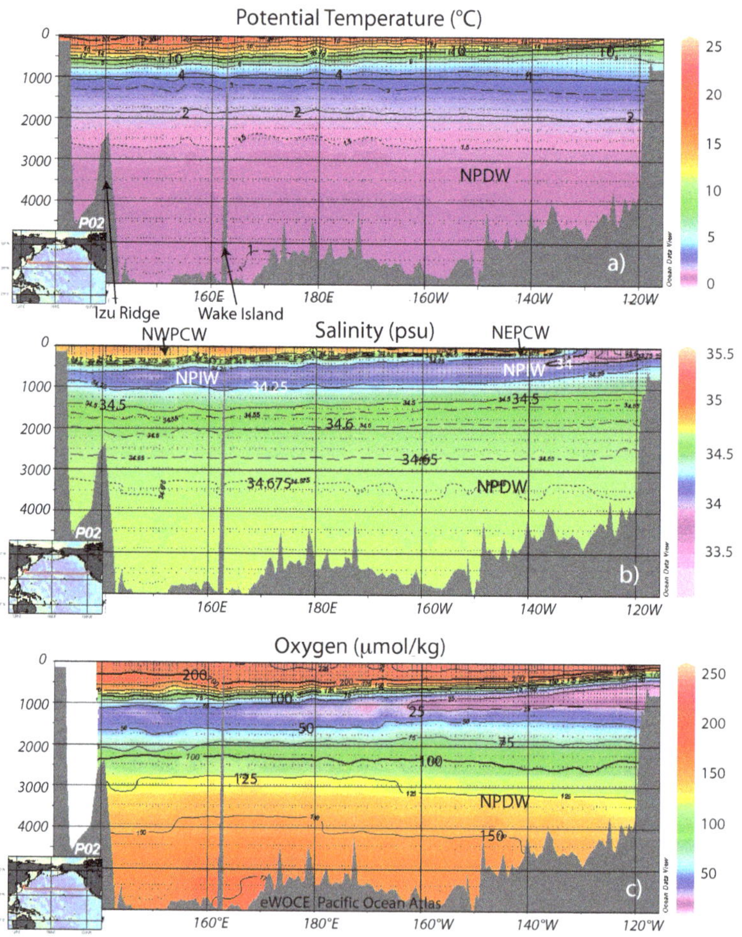

FIG. 5-46 – Sections at 30°N in the Pacific Ocean of **(a)** potential temperature, **(b)** salinity, and **(c)** dissolved oxygen. [North West Pacific Central Water (NWPCW), North East Pacific Central Water (NEPCW), North Pacific Intermediate Water (NPIW), North Pacific Deep Water (NPDW).] (Section P2, from the *eWOCE Pacific Ocean Atlas*, Talley L.D., 2007.)

5.5 Deep Water and Bottom Water

On the meridional section at 165°W the isotherms [**Figure 5-41a**] and isohalines [**Figure 5-41b**] show that there is no formation of Bottom Water or of Deep Water in the Pacific Ocean[378] since deep isopleths do not reach the surface anywhere except south of 40°S in the Antarctic Ocean.

Even though the Ross Sea in the Antarctic is a center of dense deep-water formation near the Pacific Ocean, the undersea topography of the Pacific-Antarctic Ridge [**Figure 5-3**] prevents this dense water from extending into the Pacific Ocean. To the north, the Bering Strait, a few tens of meters deep, does not allow Arctic Bottom Water to penetrate into the Bering Sea. Bottom Water and Deep Water in the Pacific Ocean are formed elsewhere. They arrive from the western Antarctic Ocean and enter much altered in the southwest Pacific Ocean. In the course of the passage eastward around the Antarctic continent the characteristics of Antarctic Bottom Water are attenuated, by mixing with Lower Circumpolar Deep Water [**Section II.6.3.4**]. In the southwest Pacific, Antarctic Bottom Water first finds a passage toward the Tasman Sea across the Pacific-Antarctic Ridge south of Australia but is blocked by topography around 20°S [**Figure 5-47**].

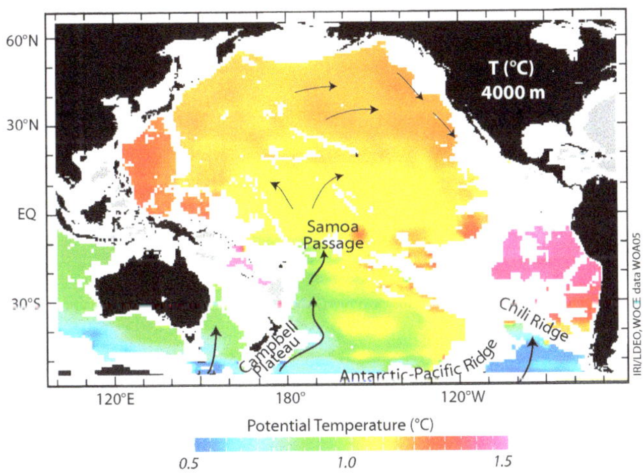

FIG. 5-47 – Potential temperature at 4,000 m. Depths less than 4,000 m are white. (IRI/LDEO, WOCE data WOA05.)

The principal passage, along the eastern edge of the Campbell Plateau, east of New Zealand, feeds the entire Pacific Ocean [**Figure 5-48**, Station 1]. The most characteristic branch is up against the western side of the basin between 3,500 m and the bottom [**Figure 5-49**], as a deep western boundary current. The third northward passage is east of the East Pacific Ridge, allowing the

378 North of the Subtropical Front, around 40°S.

entry of much altered Antarctic Bottom Water whose water characteristics are close to those of Lower Circumpolar Deep Water [**Figures 5-47** and **5-51**, Station 1]. It remains confined to the Southeast Basin since the Chile Ridge blocks passage northward. The current is intensified on the western boundary of each basin. On the zonal section at 32°S [**Figure 5-49**] the general slope of the isopleths indicates a northward current below 2,500 m, in particular against the Kermadec Ridge, which extends from New Zealand (around 180°). On the bottom, along this ridge, salinities less than 34.72, temperatures less than 0.6 °C, and oxygen levels greater than 200 μmol kg^{-1} show the track of altered Antarctic Bottom Water (close to Lower Circumpolar Deep Water).

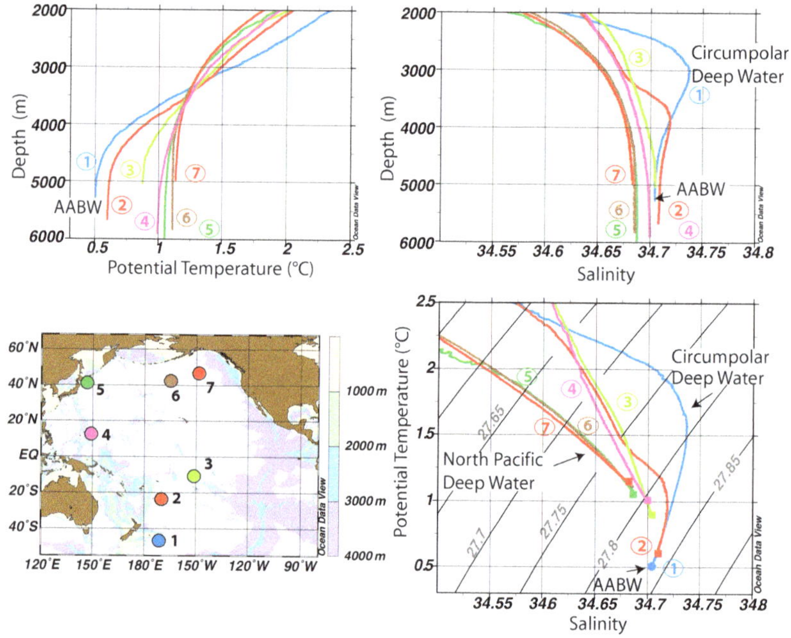

FIG. 5-48 – Evolution of θ-S characteristics of Deep Water, going from its entry in the southern Pacific Ocean (Station 1) up to the northern Pacific. (With ODV and WOCE data.)

From east of New Zealand, Antarctic Bottom Water, already much mixed with Lower Circumpolar Water, extends northward into the deep basins through the Samoa Passage (10°S, 169°W, sill depth 4,800 m) where northward transport of water colder than θ = 1.2 °C is around 8.4 Sv. Along its trajectory to the north and NE, under the effect of topographic and mixing constraints, the temperature of bottom water increases, and its salinity and oxygen levels decrease [**Figures 5-47** and **5-48**, Stations 2 to 7]. In the east, the salinity maximum of Lower Circumpolar Deep Water, though attenuated, does not reach the topographically isolated basins along South America [**Figure 5-50**]. Bottom temperatures are from 0.3 °C to 0.5 °C more elevated [**Figures 5-47** and **5-51**,

Stations 3, 4, 5, 6] than in the central basin (Station 2). The bottom temperature of each basin depends on the depth of the sill that feeds it. Below sill level, water characteristics are homogeneous. Bottom water of the Panama Basin is the warmest (Station 6, **Figure 5-51**). In the southeastern Pacific Ocean [**Figure 5-50**] the 34.7 isohaline and the 1 °C isotherm, indicators of Lower Circumpolar Deep Water, do not pass the East Pacific Ridge, or the Chile Ridge while, in the west, they reach 15°N [**Figure 5-41**]. Antarctic Bottom Water is limited to south of the Chile Ridge [**Figure 5-47**].

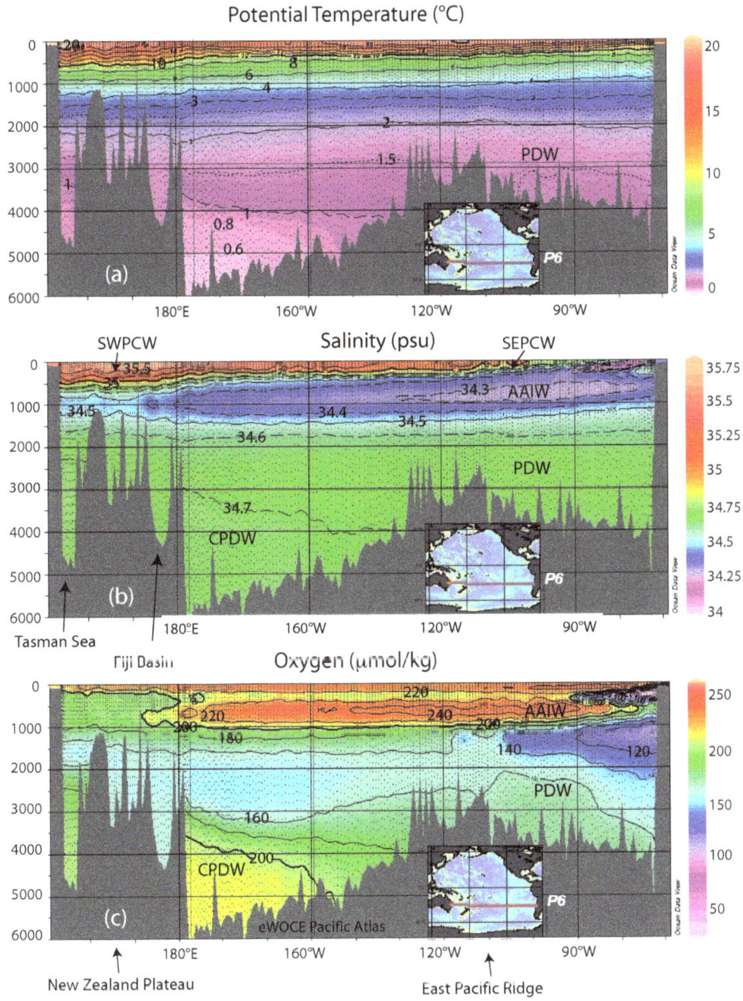

FIG. 5-49 – Zonal sections at 32°S of (**a**) potential temperature, (**b**) salinity, and (**c**) dissolved oxygen. [Circumpolar Deep Water (CPDW), Pacific Deep Water (PDW), Antarctic Intermediate Water (AAIW), SW Pacific Central Water (SWPCW), SE Pacific Central Water (SEPCW).] (Section P6, from the *eWOCE Pacific Ocean Atlas*, Talley L.D., 2007.)

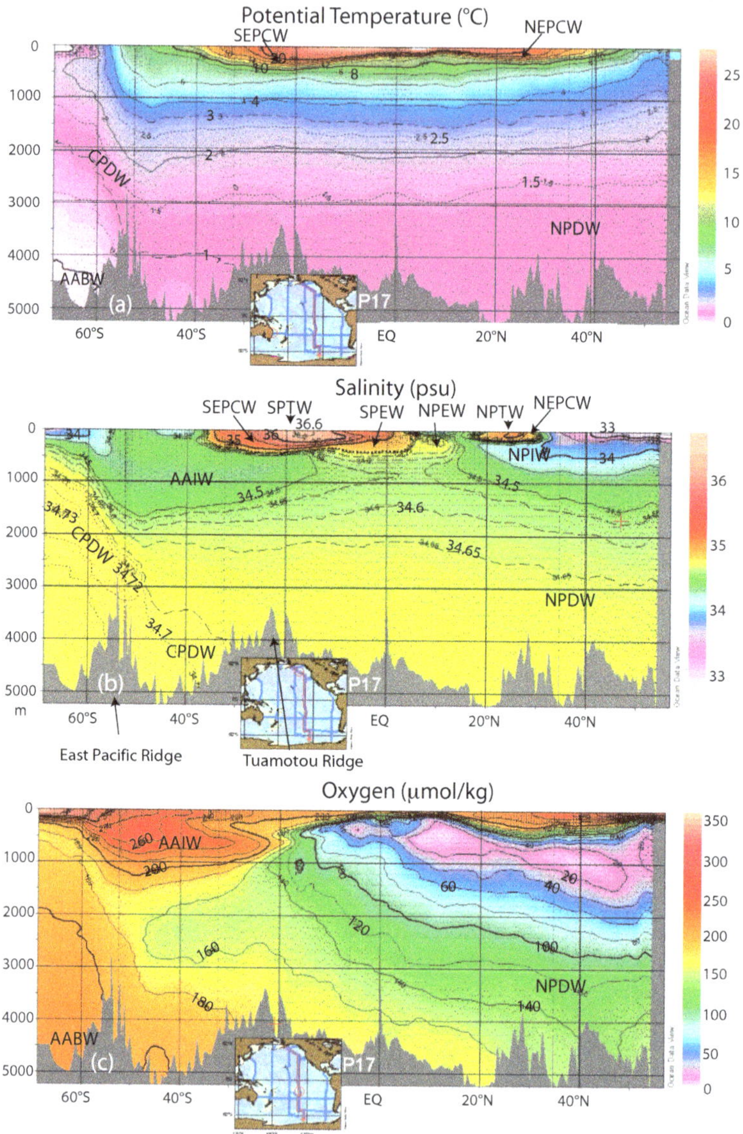

FIG. 5-50 – Meridional section at 135°W in the eastern Pacific of **(a)** potential temperature, **(b)** salinity, and **(c)** dissolved oxygen. [Circumpolar Deep Water (CPDW), Antarctic Bottom Water (AABW), salinity minimum of North Pacific Intermediate Water (NPIW) and of Antarctic Intermediate Water (AAIW), salinity maximum of North Pacific Tropical Water (NPTW) and of South Pacific Tropical Water (SPTW), NE Pacific Central Water (NEPCW), SE Pacific Central Water (SEPCW)] (Section P17, from the *eWOCE Pacific Ocean Atlas*, Talley L.D., 2007.)

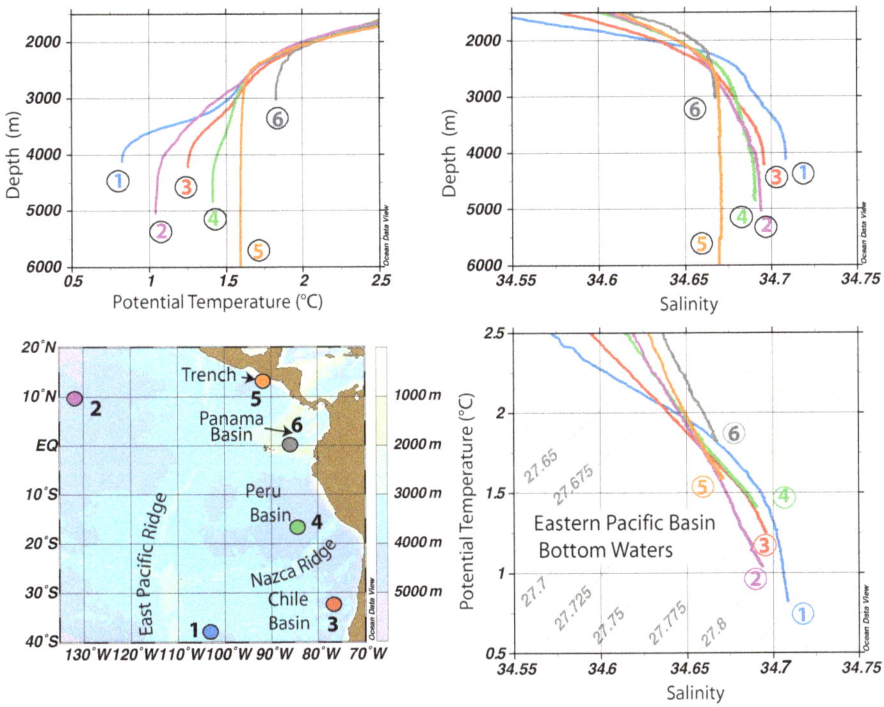

FIG. 5-51 – Deep profiles and θ-S diagrams in the eastern basins of the Pacific Ocean, confined by topography. (With ODV and WOCE data.)

Deep Water presents a unique uniformity below 2,000–2,500 m [**Figure 5-50**], especially in the northern ocean [**Figure 5-48**, Stations 5, 6, 7]. During their passage northward, Antarctic Bottom Water and Lower Circumpolar Deep Water mix and are no longer discernible. The salinity maximum of Lower Circumpolar Deep Water disappears north of the equator. The upper part of Circumpolar Deep Water mixes with the lower level of Antarctic Intermediate Water, lowering its salinity. North Pacific Deep Water is a homogeneous water mass also called "common water",[379] since it represents an important volume with respect to other water masses. Below 2,500 m, temperature varies between 1 °C and 2 °C and salinity between 34.63 and 34.69 [**Figures 5-48** and **5-52**]. In the volumetric T-S diagram Pacific Deep Water dominates with a peak at 1.1–1.2 °C, 34.68–34.69 [**Figure 1-30**]. Along its path, oxygen concentration decreases to less than 1.1 ml l^{-1} (<50 µmol kg^{-1}) in the layer under North Pacific Intermediate Water where renewal is slow. Natural carbon-14 measurements show that North Pacific Deep Water left the surface about a thousand years ago! It is the oldest water of the global ocean.

[379] It is called "common" since it is so often encountered.

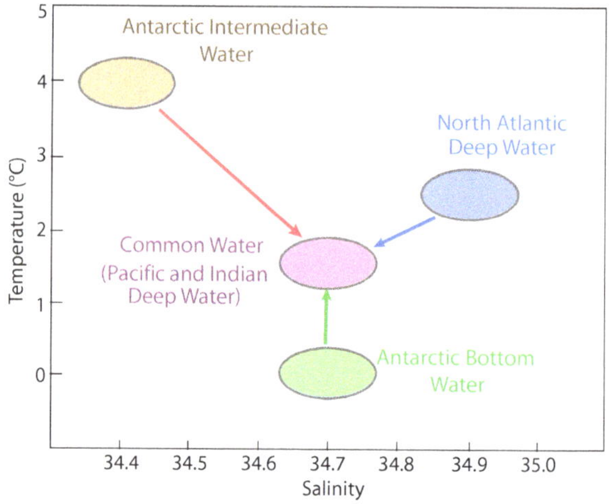

FIG. 5-52 – θ-S diagram of North Pacific Deep Water (or Common Water) components. (After *Ocean Circulation*, 1989.)

The Pacific Ocean Deep Water (PDW) is formed at depth through slow upwelling and diffusion and flow slowly southward towards the Antarctic Ocean. PDW is marked by low oxygen, no CFCs and high nutrients. In the Southern Ocean, the PDW merges with the Indian Ocean Deep Water and mixes with CPDW to form the Upper Circumpolar Deep Water marked by low oxygen.

In cold regions where dense water could form, surface density is so low (σ = 25.5 in the Bering Sea) because of low salinity that water dense enough to reach the bottom cannot form [**Figure 5-53**]. No deep or bottom water is formed in the Pacific. It is an ocean where renewal of water in the basin itself does not exceed 1,000–1,200 m depth.

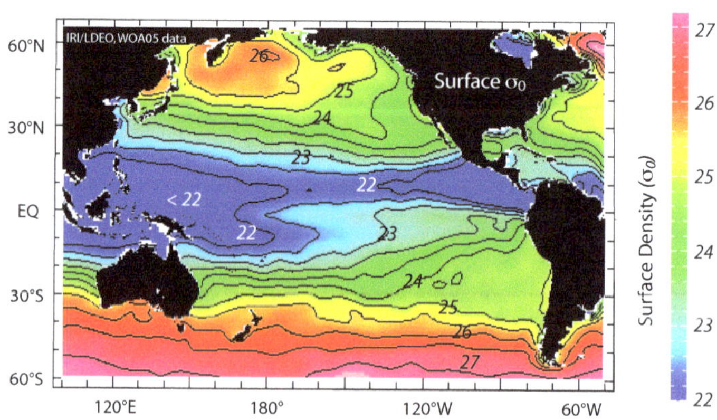

FIG. 5-53 – Pacific Ocean surface density. (IRI/LDEO, and WOA05 data.)

5.6 Some Pacific Ocean Adjacent Seas

5.6.1 The Sea of Japan

The Sea of Japan[380] is between the Korean Peninsula, the coast of China and Siberia, and an island arc composed of Sakhalin, Hokkaido, Honshu, and Kyushu [**Figure 5-54**]. In the south, the Yamato Ridge separates the Yamato Basin on the SE and the Ulleung Basin on the SW. The Japan Basin, of more than 3,500 m depth, lies to the north of them. The Sea of Japan is connected to the Pacific by a number of shallow straits. The Korea Strait between the Korean Peninsula and the island of Kyushu, the Tsugaru Strait between Honshu and Hokkaido, and the La Pérouse Strait between Hokkaido and Sakhalin are all less than 100 m depth. The Strait of Tartary, around 20 m deep, separates Sakhalin and the Siberian coast. The shallow depths of these many openings to the Pacific Ocean limit exchanges mainly to the surface layers.

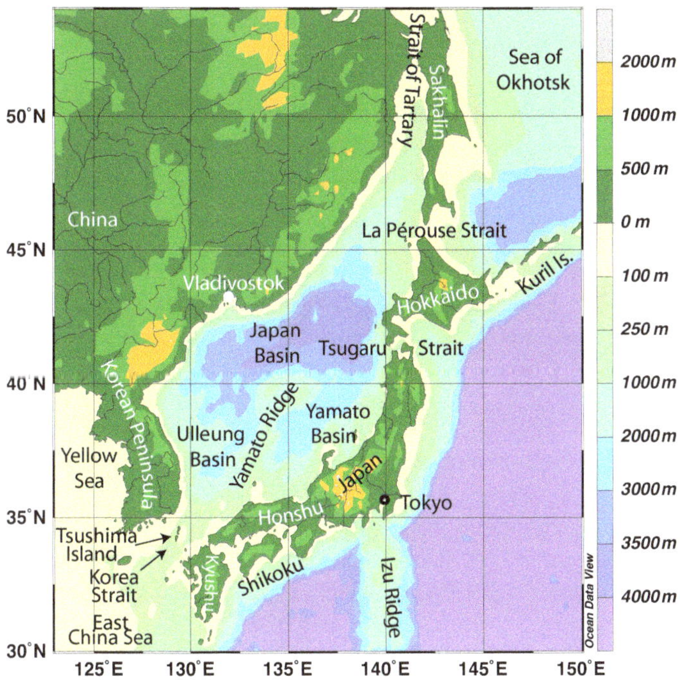

FIG. 5-54 – Sea of Japan topography.

[380] The Sea of Japan is called *East Sea* in Korean nomenclature.

The Sea of Japan is fed by warm water from the East China Sea to the south and by cold waters from the Okhotsk Sea to the north [**Figure 5-55**].

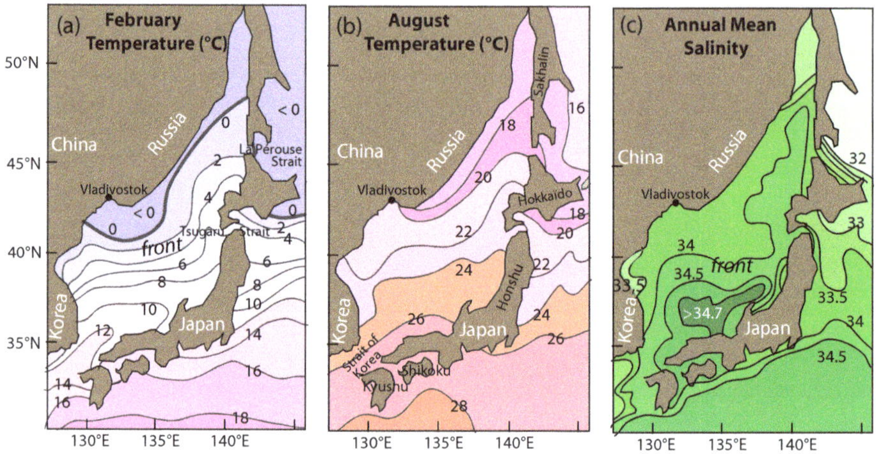

FIG. 5-55 – Sea of Japan surface temperature, (**a**) in February, (**b**) in August, (**c**) annual average surface salinity. (After Gamo *et al.*, 1986.)

To the south, the Sea of Japan receives North Pacific Tropical Water *via* the warm Tsushima Current, a branch of the Kuroshio [**Figure 5-26**], which penetrates through the Korea Strait connecting the East China Sea to the Yamato and Ulleung Basins over a sill of about 110 m [**Figure 5-54**]. It has a maximum transport of about 1.3 Sv in August. It splits into two branches around the Tsushima Islands. The western branch is a western boundary current called the East Korean Warm Coastal Current (EKC). The eastern branch of the Tsushima Current follows the coast of Japan.

To the north, the Sea of Japan receives cold water from the Sea of Okhotsk through the Strait of Tartary. It is entrained southward by the cold Liman Current along the Siberian coast to form a cyclonic circuit [**Figure 5-26**] from which a branch detaches to the south called the North Korean Current (NKC). The convergence between cold low-salinity water from the north and warm saline water from the south creates an unstable front that generates eddies around 40°N, similar to the front between the Kuroshio and the Oyashio.

The warm Tsugaru Current, a prolongation of the Tsushima Current, exits through the Tsugaru Strait, of about 100 m depth. This is the principal current leaving the Sea of Japan [**Figure 5-26**]. It converges with the Oyashio, which comes from the north along the Kuril Islands, and forms a second front that joins the Kuroshio front to the south. Between Hokkaido and Sakhalin the warm Soya Current, an extension of the northern branch of the Tsushima Current, exits through the La Pérouse Strait and joins the cold water south of the Okhotsk Sea, and the Pacific Ocean across the Kuril Islands.

Large inputs of fresh water from the Yellow Sea [**Figures 5-26** and **5-55c**] lower the salinity in the surface layer (0–30 m) of the entering Tsushima

Current from 34.6 in winter to less than 32.5 in summer [**Figure 5-56**]. The Sea of Japan is one of the rare regions where the surface temperature variation between summer and winter can reach more than 18 °C [**Figures 5-55a** and **b**]. This is due to the summer increase of incoming warm tropical water by the Tsushima Current and winter cooling of surface water by Siberian air flowing out over the Sea of Japan [**Figure 5-4a**] where the sea is often frozen as far south as Vladivostok (43°N).

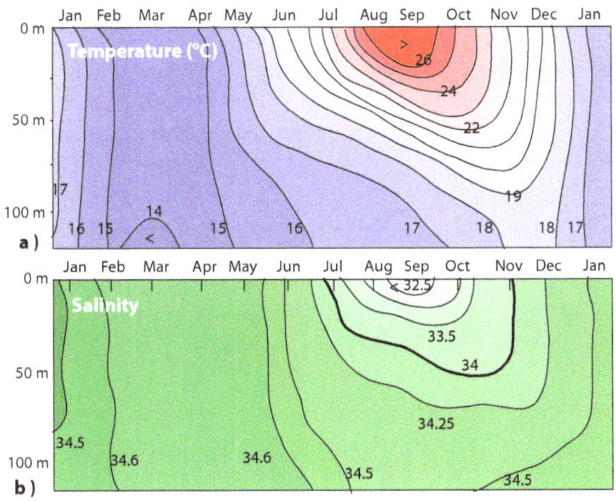

FIG. 5-56 – Seasonal variations in the Strait of Korea of **(a)** temperature, and **(b)** salinity. (After Inue *et al.*, 1985.)

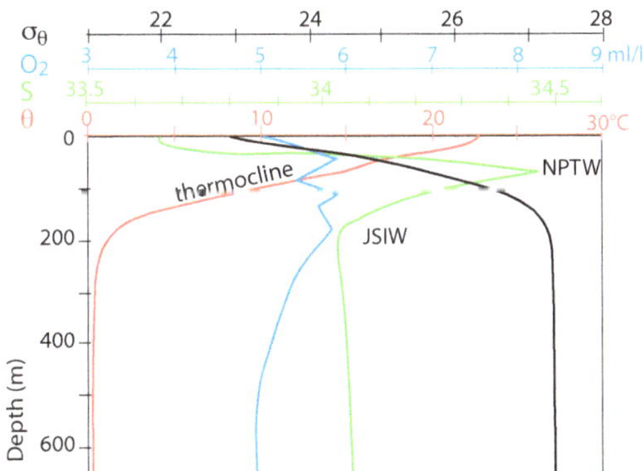

FIG. 5-57 – Typical profiles of potential temperature, salinity, dissolved oxygen, and potential density in the southern Sea of Japan, Ulleung Basin (see the map, **Figure 5-54**). [North Pacific Tropical Water (NPTW), also called Tsushima Current Warm Water (TWW), Japan Sea Intermediate Water (JSIW), Japan Sea Deep Water or Japan Sea Proper Water (JSPW).] (After Senjyu, 1999.)

Typical profiles [**Figure 5-57**] of a station in the south of the Sea of Japan [**Figure 5-58a**] show that below about 200 m, oceanographic characteristics vary little. 90% of the volume of the Sea of Japan has temperatures in the range 0 °C < θ < 1 °C.

Warm and saline North Pacific Tropical Water (S > 34.10, 34.50 maximum) transported by the Tsushima Current, called Tsushima Current Warm Water lies below Surface Water freshened by the influence of water from the Yellow Sea in summer [**Figures 5-57. 5-58** and **5-59a**].

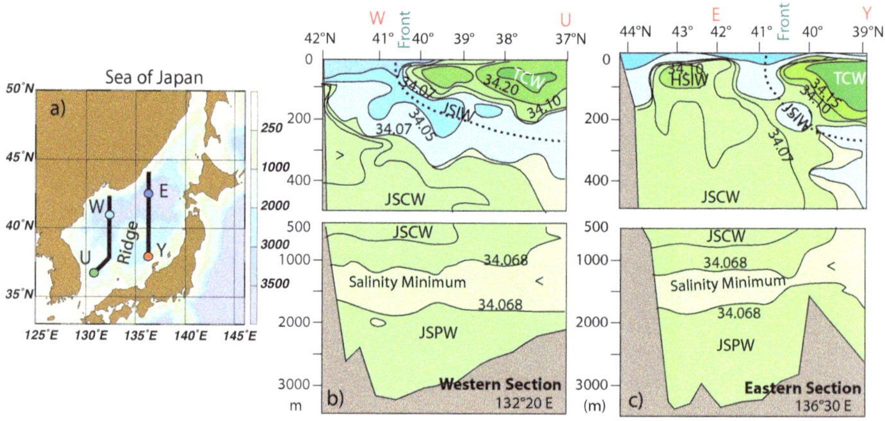

FIG. 5-58 – (**a**) Position of the two salinity sections in the Sea of Japan, (**b**) Stations W and U at 132°20E and (**c**) Stations E and Y at 136°30E. [Japan Sea Central Water (JSCW), High Salinity Intermediate Water (HSIW).] (See legend, **Figure 5-57.**) (After Kim *et al.*, 2004.)

Japan Sea Intermediate Water, characterized by a salinity minimum, forms in the north and is subducted along the front around 40°N [**Figures 5-58b** and **c**].

In the eastern Sea of Japan, around 200 m depth [**Figure 5-58c**] north of the front, a salinity maximum (S > 34.07) characterizes High Salinity Intermediate Water (HSIW), which is high in salinity, rich in oxygen, and whose potential temperature range is 1 °C < θ < 5 °C. Its source is Tsushima Warm Water transported by the eastern branch of the Tsushima Current, cooled and eroded by mixing with less saline and colder Japan Sea Intermediate Water.

Japan Sea Proper Water (JSPW or Japan Sea Deep Water) extends from 250 m to the bottom. Its exceptionally uniform characteristics (0 °C < θ < 1 °C and 34.06 < S < 34.09) are the result of the shallow depth of the sills connecting it to other basins [**Figures 5-54** and **5-57**]. However, a slight salinity minimum (S < 34.068) is apparent (except in the Yamato Basin) in the interior of Deep Water around 1,500 m, separating Japan Sea

Central Water and true Japan Sea Proper Water [**Figure 5-59c**]. Japan Sea Proper Water forms by winter convection north of 43°N and in the central region (40°N–43°N, 132°E–134°E) but for the most part remains confined to the Sea of Japan. Local renewal of Japan Sea Proper Water is evidenced by a high dissolved oxygen (and tritium) concentration down to the bottom [**Figure 5-57**]. Japan Sea Proper Water is represented by a tiny segment on θ-S diagrams [**Figure 5-59a**].

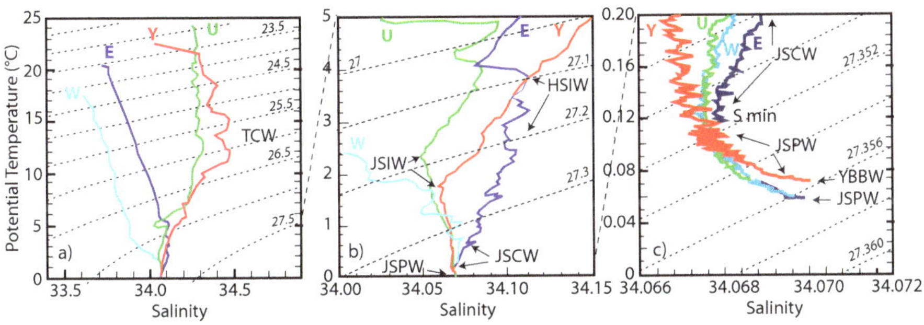

FIG. 5-59 – **(a)** θ-S diagrams in the Sea of Japan basins: Station E in the eastern Japan Basin, Station W in the western Japan Basin, Station U in the Ulleung Basin, and Station Y in the Yamato Basin. Station positions are shown on the **Figure 5-58a** map. **(b)** θ-S diagrams between 0 °C and 5 °C. **(c)** θ-S diagrams between 0 °C and 0.2 °C. See **Figure 5-57** caption. (After Kim *et al.*, 2004.)

Below 2,000 m [**Figure 5-59c**] a slightly higher potential temperature is evident in the Yamato Basin (Station Y) linked to the presence of the Yamato Ridge that prevents the coldest Japan Sea Proper Water from penetrating into the basin.

In the north, despite low winter temperatures which create sea-ice formation in the NW Sea of Japan, bottom water density does not exceed $\sigma_0 = 27.356$ [**Figure 5-59c**] because salinity is too low.

Warm water in the Sea of Japan transfers its heat to cold Siberian air masses in winter. It is a region of strong oceanic winter heat loss [**Figure 1-14**].

5.6.2 The Bering Sea

The Bering Sea, in the northern Pacific Ocean between the Alaskan coast, the Aleutians, Siberia, and the Bering Strait [**Figure 5-3**] extends over 2.3×10^6 km^2. The continental shelf occupies half the area. The deep part, to the SW, is cut by the Shirshov Ridge at 171°E and Bowers Ridge, which forms an arc around 179°E [**Figure 5-60**]. East of 171°E, the sill depths between the Aleutian Islands are generally less than 1000 m.

In the east, Alaskan Stream water enters into the shallow part of the Bering Sea, in particular through the Unimak Pass at 165°W (sill depth ≈ 100 m) with a mean transport of 0.2 Sv. It forms filaments of current on the shelf, which vary according to the wind, and join the Anadyr Current (speed ≈ 0.3 m s^{-1}), which exits through the Bering Strait into the Chukchi Sea and the Arctic Ocean and whose transport varies between 0.6 Sv and 1 Sv. A branch of the Alaskan Stream transporting North Pacific Deep Water penetrates first by a passage (≈1,600 m) east of Bowers Ridge. It turns sharply eastward to form a cyclonic circuit in the deep basin of the Bering Sea following the continental slope [**Figure 5-60**]. The current along the slope is associated with a countercurrent. The Shirshov Ridge induces a second cyclonic circuit in the southwest part of the deep basin, fed by a vein of the eastern circuit and by the entry of water through the western Aleutians. The principal outflow takes place southward by the Kamchatka Current, along the Kamchatka Peninsula.

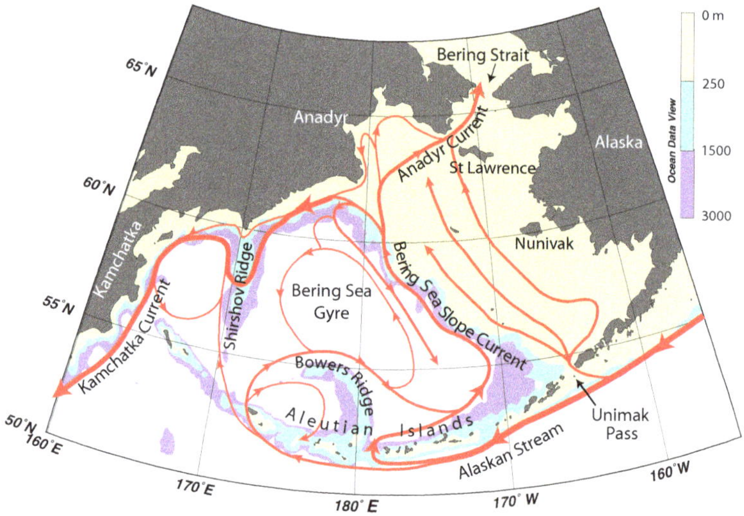

FIG. 5-60 – Bering Sea circulation. (After Tomczak and Godfrey, 1994.)

The Bering Sea has a subpolar climatic regime. In January, pack ice extends over the continental shelf and along the Kamchatka coast [**Figure 3-67**]. In July, the sea is entirely ice-free.

Water masses of the Pacific Ocean (Station 1, **Figure 5-61**) enter by numerous passages across the Aleutians. They are modified by the climatic conditions of the region, in particular on the continental shelf where low-salinity water (S = 32.6, Station 2) is near freezing temperature in winter. The oceanographic structure of the Bering Sea cyclonic circuit is formed of three layers. The Deep Water layer comes from the Pacific through the western

Aleutian deep passages. Bottom temperature in the deep basin (Station 3) corresponds to the temperature of Deep Water at the entry sill depths. It is thus slightly higher than on the bottom of the North Pacific (Station 1). The intermediate layer, characterized by a temperature minimum (Stations 4 and 5) forms through winter convection on the continental shelf (Station 2) from where it flows into the deep basin and mixes with North Pacific Deep Water (NPDW). Surface-layer salinity is low ($32.5 < S < 33$) and increases sharply in the first 250 m. A temperature maximum (about 3.8 °C) found around 250–300 m depth corresponds to the upper part of NPDW (Stations 3, 4, 5) whose temperature decreases through mixing with shelf water. Surface density is higher in the center of the cyclonic circuit than on the periphery due to geostrophic equilibrium but, because of low salinity, it cannot form water that is dense enough in winter to sink to the bottom. Renewal of Bering Sea Deep Water takes place over the sills.

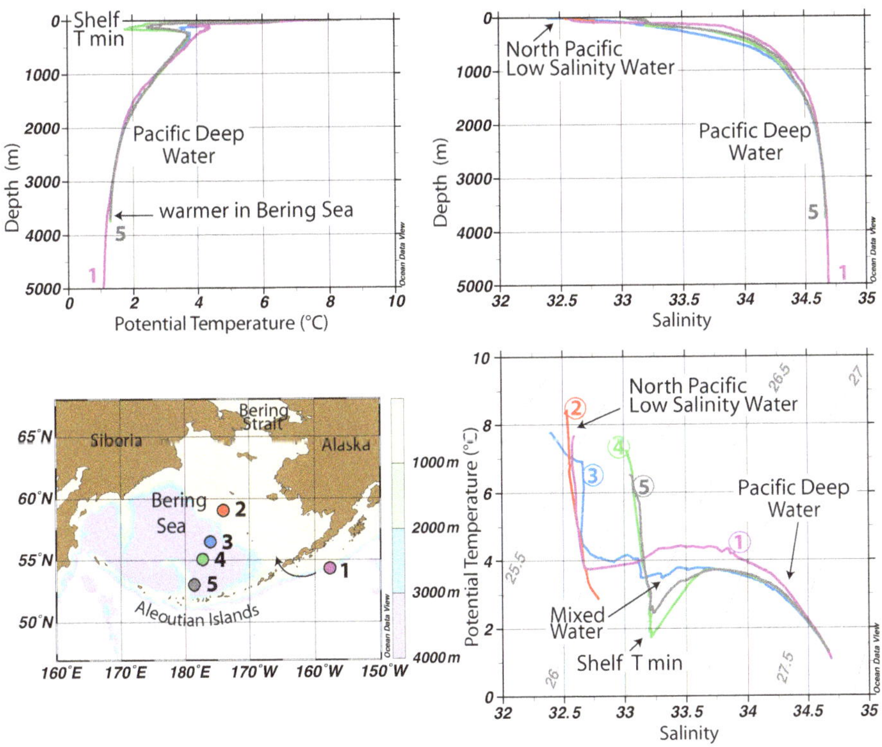

FIG. 5-61 – July profiles of salinity, potential temperature, and θ-S diagrams inside the Bering Sea (Stations 2, 3, 4, 5) and outside in the Alaskan Stream (Station 1). (With ODV and WOCE data.)

5.6.3 The Sea of Okhotsk

The Sea of Okhotsk lies between Sakhalin Island on the west, the coast of Siberia on the north, and the Kamchatka Peninsula on the east, prolonged by the Kuril Islands and Hokkaido to the south [**Figure 5-26**]. Like the Bering Sea, it has a shallow region to the north and a deep region to the south. It connects to the Pacific Ocean through numerous passages across the Kuril Islands, the most important being the Boussole Strait[381] (sill at 2,320 m) and the Krusenstern Strait (sill at 1,920 m) through which Kamchatka Current water enters [**Figures 5-62** and **5-63**]. Two passages connect the Sea of Okhotsk and the Sea of Japan. Between the coast of Siberia and Sakhalin Island the Tartary Strait (~ 20 m depth) allows cold and low-salinity water (σ_0 = 26.8) formed on the Sea of Okhotsk continental shelf [**Figure 5-63,** Station 1] to enter the northern Sea of Japan. Between Hokkaido and Sakhalin, the La Pérouse Strait (< 200 m) allows warm and saline water from the Sea of Japan to enter *via* the Soya Current, a northern branch of the Tsushima Current. It mixes with water formed on the shelf of the Sea of Okhotsk.

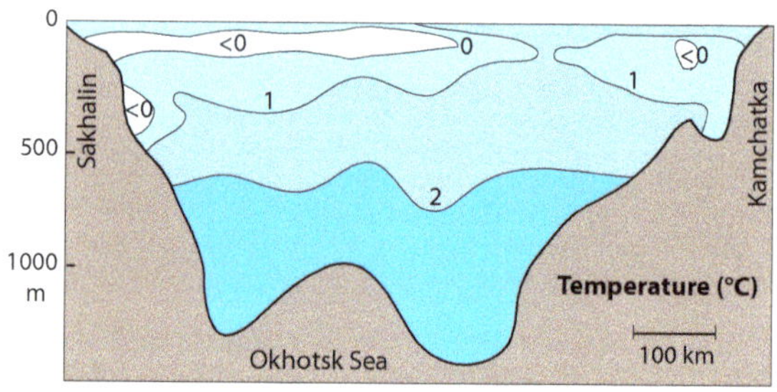

FIG. 5-62 – Temperature section south of the Sea of Okhotsk. (Position is shown in **Figure 5-63**.) (After Tomczak and Godfrey.)

The northern Sea of Okhotsk is ice-covered for much of the year, as is the Bering Sea [**Figure 3-67**]. Circulation is cyclonic [**Figure 5-26**] with extensions over the continental shelf. The subsurface layer of negative temperature formed on the continental shelf is entrained westward then southward by the cyclonic circulation [**Figure 5-62**]. The outgoing current is found mainly on the south side of the Boussole Strait. The three-layer oceanographic structure [**Figure 5-63**] is similar to that of the Bering Sea [**Figure 5-61**]. It includes a thick layer of North Pacific Deep Water, a temperature minimum layer ($T < 0$ °C), trace of the water formed on the continental shelf in winter

[381] The *Boussole*, the French word for "compass", was one of the vessels of the expedition (1785–1787) of Jean-François de Galaup, Count de La Pérouse (the other being the *Astrolabe*).

during ice formation, and a surface layer of Pacific origin, whose temperature varies seasonally [**Figures 5-62** and **5-63**]. In the Kuril Basin to the south the intermediate layer is influenced by the warm and saline Soya Current, which enters through the La Pérouse Strait (Station 4, **Figure 5-63**) and which represents a source for North Pacific Intermediate Water (σ_0 ~ 26.8).

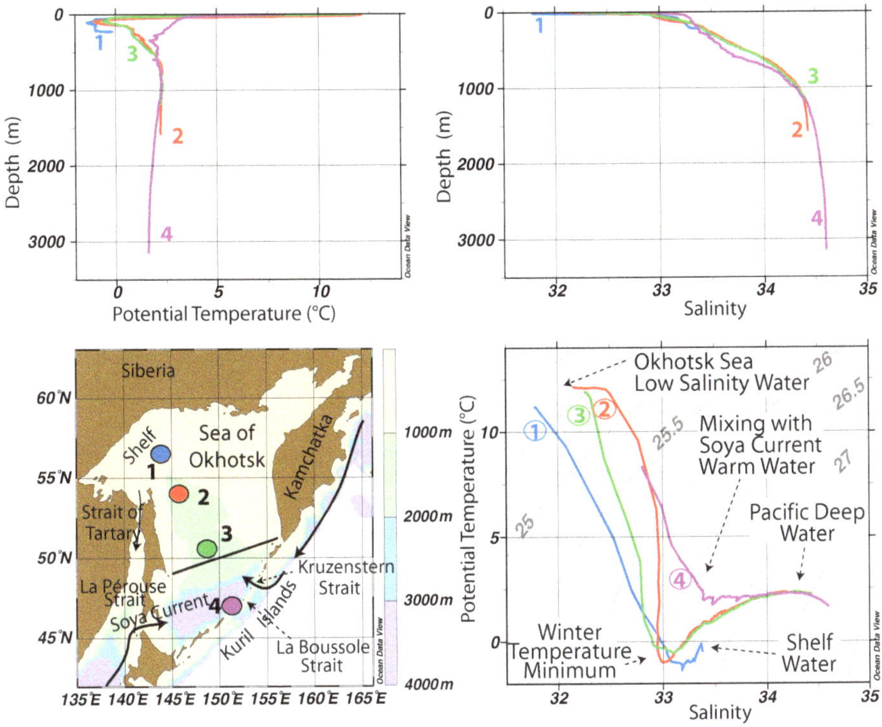

FIG. 5-63 – Profiles of potential temperature, salinity, and θ-S diagrams in the Sea of Okhotsk. (With ODV and WOCE data.)

Chapter 6
Distinctive features of the Pacific Ocean

The Pacific is the vastest ocean, and the closest to theoretical schemes of the ocean. Trade winds develop freely here, as does the equatorial current system that results. The distribution of water masses in the 0–1,000 m layer is relatively symmetric about the equator.

However, a number of alterations to the theoretical schemes are due to the proximity of continental masses at the Bering Strait, to the communication with the Indian Ocean in the equatorial zone, to the presence of boundary seas in the west, and to the influence of Asian and Australian monsoons, which drive north and south subtropical high-pressure systems eastward. Because of high surface temperatures in the Intertropical Convergence Zone, which extend over the western equatorial Pacific, precipitation is abundant. In the north, precipitation generally dominates over evaporation due to the transport of humidity from the equatorial zone by the Asian summer monsoon. Furthermore, the narrowness of Central America allows trade winds from the Atlantic to transfer fresh water, drawn from the Atlantic Ocean, into the Pacific. In the north, exchanges between the subtropical gyre and the subpolar gyre are relatively weak. The ensemble of factors results in salinity that is clearly lower in the North Pacific than in the North Atlantic.

Winter continental effects that produce strong exchanges with the atmosphere in the northwest Pacific do not have the same consequences as in the North Atlantic. Deep Water formed in the Sea of Japan reaches a density in the range of that of Antarctic Intermediate Water, which is much less than that of Lower Circumpolar Deep Water and is confined to the Sea of Japan. The deepest water formed in the Pacific Ocean is North Pacific Intermediate Water. Below, Pacific water masses are formed elsewhere. Deep Water, created in the North Atlantic follows a long circuit *via* the Antarctic Ocean and Circumpolar Deep Water. Deep Water in the northern Pacific Ocean left the surface many centuries ago. This is the dead end of the world ocean deep circulation.

The Pacific is an immense ocean where continental influences are limited, resulting in no formation of deep water. Nevertheless, it is the site of intense exchanges with the atmosphere in the equatorial zone. The western Pacific region represents the strongest zone of atmospheric convection on the globe, where the ocean transfers a large part of its accumulated heat to the atmosphere. An anomaly in this region results, by chain reaction, in anomalies over the whole of the planet. Predicting these interannual variations is fundamental in order to understand climatic variability, not only of this region, but also for the entire globe.

Part VI
Conclusions

We have tried in this text to describe the water properties and circulation of each ocean that reflect the planetary ocean response to exchanges with the atmosphere. The overview that follows highlights the major characteristics described in the preceding parts.

Chapter 1
Planetary ocean water properties

In spite of the oneness of the planetary ocean and the persistence of mixing and exchange from one region to another, each ocean has unique features resulting from its morphology and global situation.

To summarize, these features can be represented with the horizontal global distribution of temperature and salinity at various levels [**Figures 6-1a** to **f**], with global θ-S diagrams [**Figure 6-2**], and with comparisons among meridional sections [**Figure 6-3**].

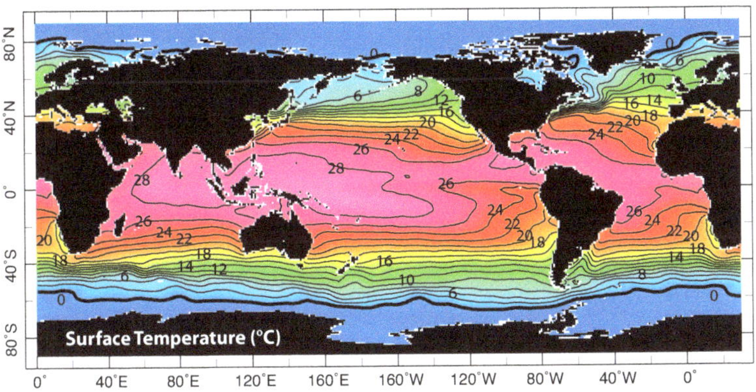

FIG. 6-1A – Average surface temperature. (IRI/LDEO, WOCE WOA05 data.)

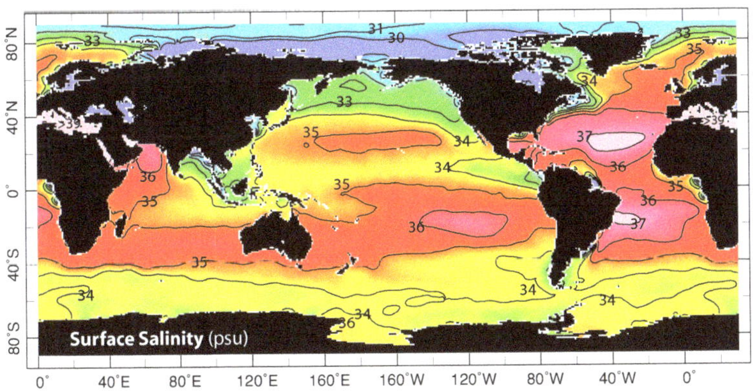

FIG. 6-1B – Average surface salinity. (IRI/LDEO, WOCE WOA05 data.)

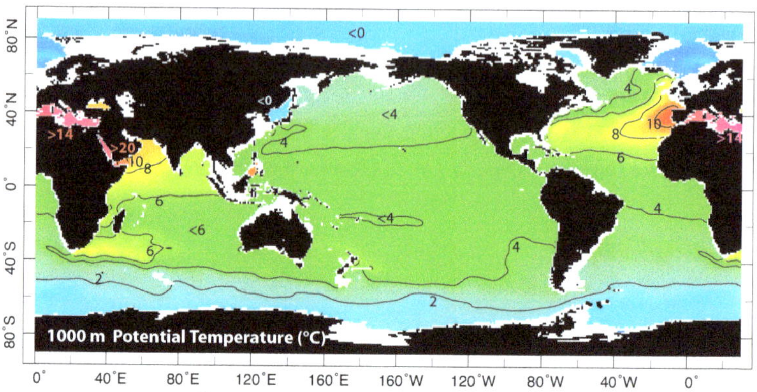

FIG. 6-1C – Average potential temperature at 1,000 m. Depths less than 1,000 m are in white. (IRI/LDEO, WOCE WOA05 data.)

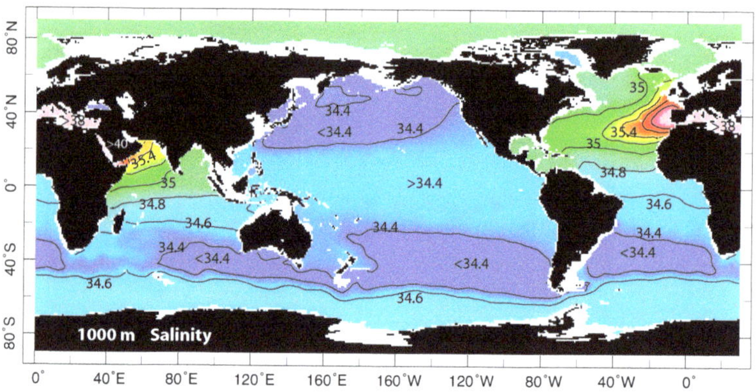

FIG. 6-1D – Average salinity at 1,000 m. Depths less than 1,000 m are in white. (IRI/LDEO, WOCE WOA05 data.)

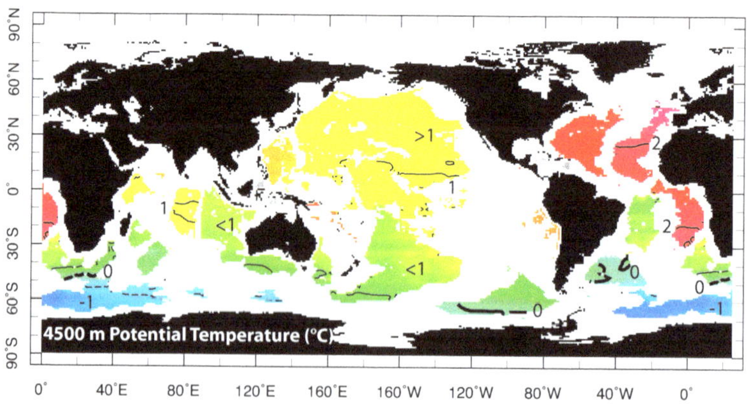

FIG. 6-1E – Average temperature at 4,500 m. Depths less than 4,500 m are in white. (IRI/LDEO, WOCE WOA05 data.)

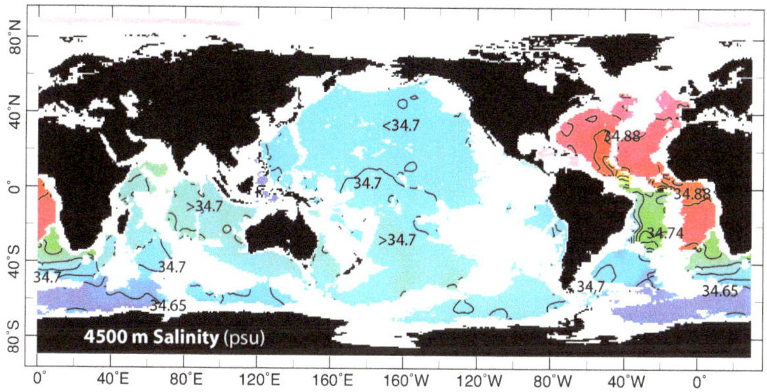

FIG. 6-1F – Average salinity at 4,500 m. Depths less than 4,500 m are in white. (IRI/LDEO, WOCE WOA05 data.)

1.1 Temperature and salinity distribution

Mean sea-surface temperature [**Figure 6-1a**] has a maximum in the tropical region of the western Pacific and eastern Indian Ocean (the "warm pool"). The eastern Pacific and Atlantic show coastal and equatorial upwelling as a temperature minimum, which is not the case in the Indian Ocean. Isotherms are particularly tight on the western sides, where there is confluence between currents of subpolar origin and currents of tropical origin. Tight isotherms also delineate the Subantarctic Front and the North Subpolar Fronts.

Mean surface salinity [**Figure 6-1b**] shows maxima in the center of anticyclonic gyres and significantly low values in the Pacific Ocean, particularly in the North Pacific. High salinity water entrained by the North Atlantic Drift and the Norwegian Current is responsible for the formation of Deep Water, which does not exist in the North Pacific.

The characteristics of the ocean surface have a direct impact on climatic variability in modulating ocean-atmosphere exchanges.

Horizontal gradients at 1,000 m [**Figures 6-1c** and **6-1d**] are much weaker than at the surface. The two maxima show the influence of the Mediterranean and the Red Sea, with no equivalent maximum in the Pacific Ocean. A temperature minimum is confined to the Sea of Japan. The salinity distribution at 1,000 m [**Figure 6-1d**] shows minima in subtropical regions due to intermediate waters (Antarctic Intermediate Water and North Pacific Intermediate Water). No such minima are found in the North Atlantic and in the northern Indian Ocean because of the influence of the salty Mediterranean and Red Sea Waters.

The continuity of property distribution at all depths shows the role of the Antarctic Ocean, which provides communication on the south between the three other oceans.

Potential temperature at 4,500 m [**Figure 6-1e**] shows the propagation of Antarctic Bottom Water, the coldest and densest in the open ocean, in the western South Atlantic Basin, channeled by topography. Outside those basins with sills where the coldest water cannot penetrate (the Caribbean, the eastern and western basins of the Pacific, and the Arabian Sea) Bottom Water of the eastern and northeastern Atlantic is markedly warmer. This warming is due to mixing between water formed in the Nordic Seas, influenced by Mediterranean Water as it overflows the sills, and the upper part of Antarctic Bottom Water.

The distribution of salinity at 4,500 m [**Figure 6-1f**] shows the influence in the Atlantic of saline water from the Nordic Seas, contrasting with Antarctic Bottom Water, blocked in great part by the mid-Atlantic Ridge and the Walvis Ridge. For temperature as well as salinity, contrasts disappear in the Pacific Ocean, the sign of the absence of local deep-water formation.

1.2 The ensemble of θ-S diagrams

The envelopes of θ-S diagrams for each ocean, extended to the Antarctic sector, integrate the variations in the pairing of potential temperature and salinity. Their comparison below surface water [**Figure 6-2**] shows that the Atlantic Ocean has the highest salinities, due to the influence of Mediterranean Water. The Pacific Ocean has the lowest salinities and the Indian Ocean, where an intermediate salinity maximum marks the influence of Red Sea Water, lies between the other two. Characteristics converge near the bottom and correspond to Antarctic Bottom Water. We note that the formation region of Antarctic Bottom Water, essentially the Weddell Sea, is relatively restricted with respect to the volume of ocean that it occupies.

1.3 Specific features of each ocean

The Atlantic Ocean, with its great meridional extent, is the principal connection between the two polar basins. Its relative narrowness reinforces continental effects and amplifies exchanges with the atmosphere in regions where air masses with continental properties outflow. The thin band of the Isthmus of Panama allows Atlantic trade winds to reach the Pacific charged with fresh water picked up over the Atlantic. All these processes result in an increase in surface salinity [**Figure 6-1b**] and density, which becomes sufficiently great in winter to produce convective motion over thousands of meters in the north. The Mediterranean, a region of strong evaporation, provides saline water to the entire North Atlantic at depths of about 1,000–1,500 m [**Figure 6-1d**]. Surface and deep circulation provide strong meridional exchanges. The salinity distribution at depth [**Figures 6-3a** and **b**, Section A16] shows the formation and southward expansion of North

Atlantic Deep Water, characterized by a salinity maximum, as far as the Antarctic Ocean. As the preferred location for transforming surface water into deep water, the Atlantic Ocean is a key element in the planetary thermohaline circulation.

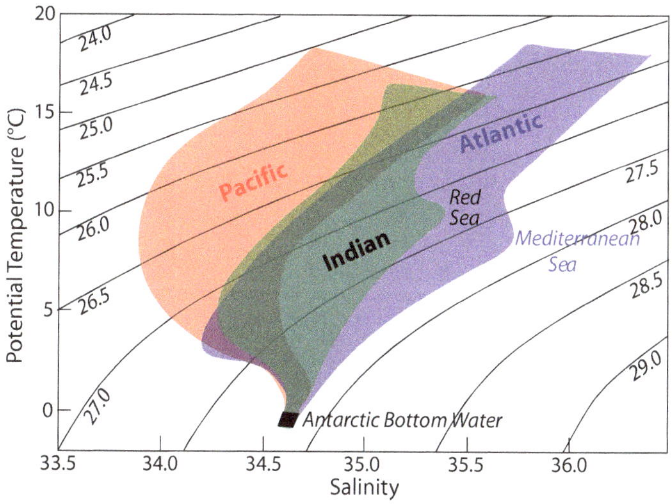

FIG. 6-2 – Characteristic θ-S envelopes below the warm variable surface layer for the Atlantic (blue), Indian (green), and Pacific (red) oceans. (After *Ocean Circulation*, 1989.)

The Indian Ocean, closed to the north by the Asian continent, differs noticeably from the Atlantic and Pacific Oceans [**Figure 6-3b,** Section I9]. Seasonal atmospheric pressure variations over the Asian continent create the strongest monsoon regime on earth. This system produces strong variations in oceanic circulation north of 10°S, most notably the reversing Somali Current. Topography and climate of the surrounding lands produce an asymmetry between surface salinity of the Arabian Sea and that of the Bay of Bengal [**Figure 6-1b**], which receives the greatest part of runoff water from India and the Himalayas. Lower thermocline water in the northern Indian Ocean is as little ventilated as that of the North Pacific, but is significantly more saline [**Figure 6-1d**] because of the input of water formed in the Red Sea and Persian Gulf. The Indian Ocean receives two types of water at depth: North Atlantic Deep Water, passing south of Africa, already somewhat eroded, and Circumpolar Deep Water, having the same origin but altered through mixing, which it undergoes in the course of its circuit around Antarctica in the Circumpolar Current [**Figure 6-3b**]. During inter-monsoonal periods, an equatorial jet accumulates warm water eastward, unlike what occurs in the Atlantic and Pacific.

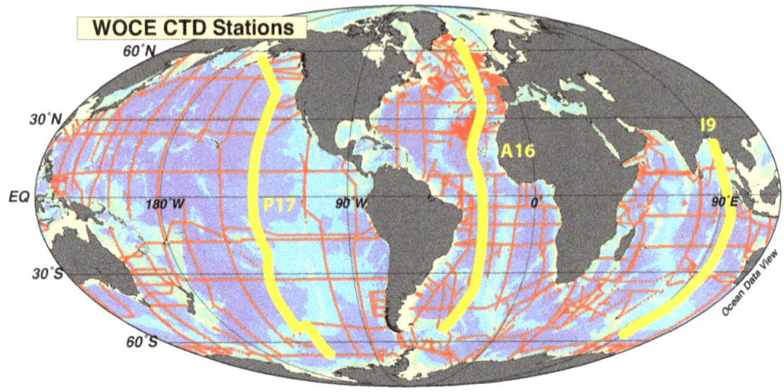

FIG. 6-3A – Location of the WOCE core hydrographic stations, along Pacific P17, Atlantic A16, and Indian I9 sections. (NOAA WOA05.)

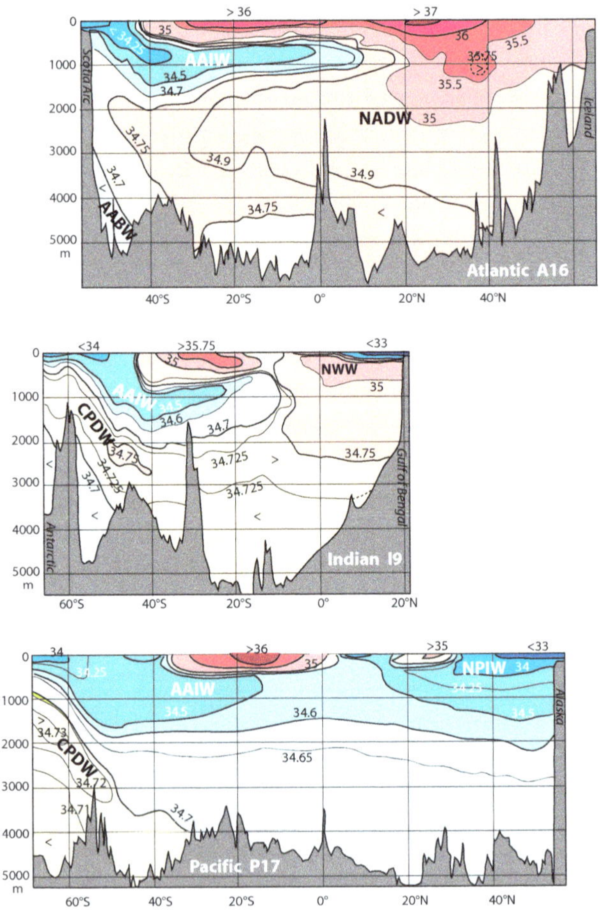

FIG. 6-3B – Simplified salinity sections along WOCE sections Atlantic A16, Indian I9, and Pacific P17. (NOAA WOA05.)

The Pacific Ocean is the vastest. It is largely open to the Antarctic Ocean, while exchanges with the Arctic Ocean are minimal. In the equatorial zone it communicates with the Indian Ocean to which it sends warm and low-saline water. It receives a lot of fresh water, particularly that carried from the Atlantic by trade winds in the equatorial zone. From the equatorial zone, the southwest monsoon drives an extensive amount of humidity northward. This results in lowered salinity and thus the surface density is much lower in the North Pacific than in the North Atlantic [**Figure 6-1b** and **Figure 6-3b**, Section P17]. Therefore wintertime continental effects do not have the same consequences as in the North Atlantic. Ventilation reaches down to only a few hundred meters. At greater depths, Pacific water masses are not formed in this ocean: Antarctic Intermediate Water is formed in the Antarctic Ocean; Deep Water comes from Antarctic Circumpolar Deep Water, whose origin is North Atlantic Deep Water; Bottom Water, formed in the Antarctic cannot be distinguished from Lower Circumpolar Deep Water except in the extreme southern part of the ocean. In the North Pacific, Deep Water left the surface many centuries ago. This is the dead end of deep circulation in the world ocean.

The equatorial Pacific Ocean is the site of intense exchanges with the atmosphere. The western Pacific region and the eastern Indian Ocean, where surface temperatures exceed 29 °C in the "warm pool", is the region of the strongest atmospheric convection on the planet, where the ocean transfers to the atmosphere a large part of its accumulated heat.

In the Antarctic Ocean, the circulation of zonal winds and currents is intense. The extreme climate drives a vertical circulation that is particularly active under the effect of sea freezing, responsible for a weak stratification, without a thermocline, quite different from that of the tropical oceans. The dominant zonal circulation of the Antarctic Circumpolar Current extends to the sea floor and is subject to topographic influences. Meridional circulation can be schematically summarized as an input of warm and saline North Atlantic Deep Water, which rises to the surface layer in the Antarctic divergence zone in the form of Circumpolar Deep Water. One part is transformed to the south near the Antarctic continent into the coldest and densest Antarctic Bottom Water. Another part, to the north in the Polar Frontal Zone, becomes less dense and less saline and forms Antarctic Intermediate Water, which spreads out northward around 800 to 1,000 m depth. These two water masses influence all the other oceans [**Figures 6-1d, 6-1e, 6-1f** and **6-3b**]. Thus, an anomaly in the formation of Bottom Water or of Antarctic Intermediate Water is propagated throughout the ocean.

The Austral Ocean is the regulator of water and heat exchanges between the Atlantic, Indian, and Pacific Oceans. It is an active ocean, which plays a major role in the equilibrium of the planetary ocean and its variability. As a case in point, **Figure 6-4** shows a census of North Atlantic Deep Water. The method is based on identifying water properties and their place of formation, using major water sources such as North Atlantic Deep Water, Antarctic Intermediate Water, or Antarctic Bottom Water. Determining the percentage of the presence of these water masses throughout the ocean is calculated by a multivariate analysis of the tracer (θ, S) concentrations. **Figure 6-4** shows the trace of North Atlantic Deep Water in the southern Indian and Pacific Oceans, altered by mixing in the Antarctic Circumpolar Current. The volume occupied by Antarctic Bottom Water in the global ocean is estimated at about twice that of North Atlantic Deep Water.

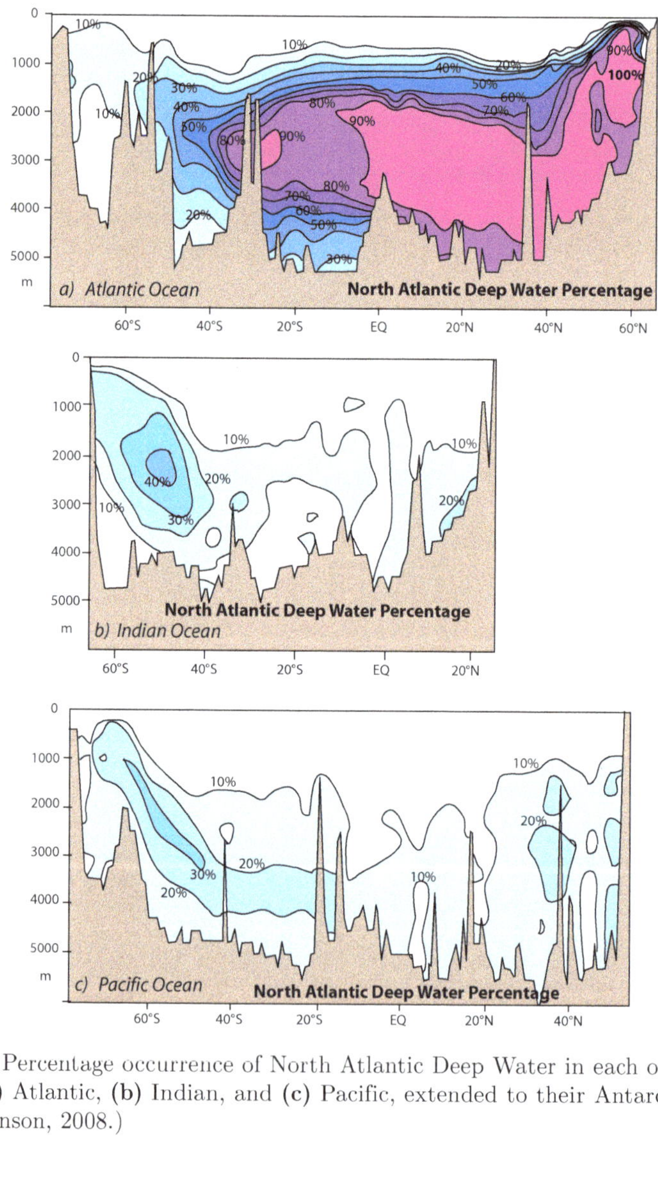

FIG. 6-4 – Percentage occurrence of North Atlantic Deep Water in each of the three oceans: **(a)** Atlantic, **(b)** Indian, and **(c)** Pacific, extended to their Antarctic sector. (After Johnson, 2008.)

Chapter 2
Circulation of the planetary ocean

With its enormous thermal capacity, the ocean stores energy from the sun, transports it, and transmits it to the atmosphere which, in return, provides the energy that puts the ocean in motion.

Through the various processes of heat and water exchange with the atmosphere, the ocean plays an important role in the climate system and in its long-term memory. The renewal times of water in different ocean layers are: of the order of a season for the surface mixed layer; of the order of 10 to 20 years for the main thermocline (the "oceanic troposphere"); of the order of 100 to 1,000 years for the intermediate and deep layers (the "oceanic stratosphere").

Knowledge of global oceanic circulation has evolved greatly thanks to advances in observational techniques (satellite sensors giving a global vision, long-term moorings, profiling drifters...), to large-scale international experiments (such as WOCE, **Figure 6-3a**), and to the ocean models required to understand such a complex system. For example, observations have shown that turbulent energy may be as great as the energy of the mean currents and that heat and salt transfers by mesoscale eddies are of the same order as those carried out by the mean circulation.

Models allow the exploration of the various processes involved in oceanic circulation. Progress in parameterizing these processes in global models and increases in computer power have allowed model grids to become fine-scale, to assimilate large amounts of data and to obtain predictions over long periods. Though such models are not discussed, some examples of model output are presented here.

Figure 6-5 is an example of ocean model output. It used a $\frac{1}{4}$-degree grid, precise bathymetry, and initial oceanographic conditions to give sea-level topography (dynamic height calculated by integrating oceanic properties from the sea floor to the surface) in response to seasonal atmospheric forcing (by winds, heat flux, fresh-water flux...). This type of model is of great value in a predictive approach.

Models permit an exploration of interannual oceanic variability that would be impossible using only observations.

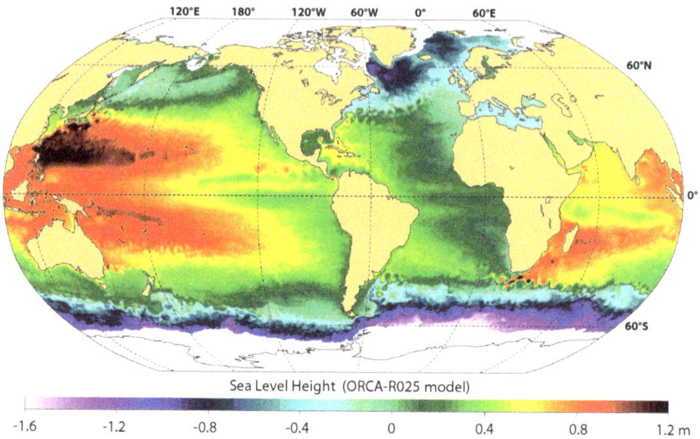

FIG. 6-5 – Ocean surface topography and ice cover (in white) for 30 May, 2007, simulated by a global-circulation numeric model (ORCA) with about 20 km resolution. The topography can indicate direction, intensity, current fluctuations, and surface eddies. (Courtesy of J.M. Molines, M. Juza, and B. Barnier, LEGI-CNRS, 2009.)

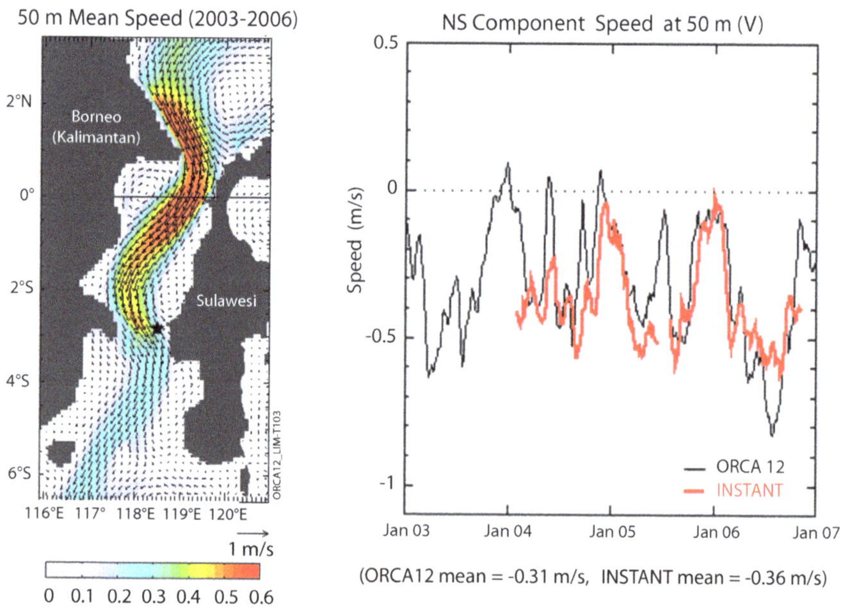

FIG. 6-6 – **(a)** Average current at 50 m in the Makassar Strait of the Indonesian Archipelago, obtained with the eddy-resolving global model ORCA 12, having a resolution grid of 1/12 degree at the equator, developed by Mercator-Ocean in Toulouse, France. The black star indicates the position of the mooring. **(b)** Meridional component of velocity at 50 m as a function of time from the INSTANT mooring (red line) and as simulated (black line) by Mercator ORCA12. Measured and simulated velocities at the mooring point were averaged over 3 days and filtered over 30 days to indicate a semi-annual cycle. (Courtesy of E. Durand, © MERCATOR OCEAN. Work-up and interpretation were done using MERCATOR products, all rights reserved, and INSTANT 2009 data.)

Another example [**Figure 6-6a**] shows the distribution of mean current in the Strait of Makassar and its variability between 2003 and 2007, obtained with the aid of the predictive Mercator model. The comparison of meridional velocity variations between model output and current meter measurements [**Figure 6-6b**] confirms the validity of the model that allows a global image of the circulation at several depths in the strait.

An example shows the simulation of North Atlantic salinity [**Figure 6-7a**] using the ocean model NEMO, forced by wind stress, and heat and fresh-water fluxes during the period 1958–2001, compared to observations [**Figure 6-7b**].

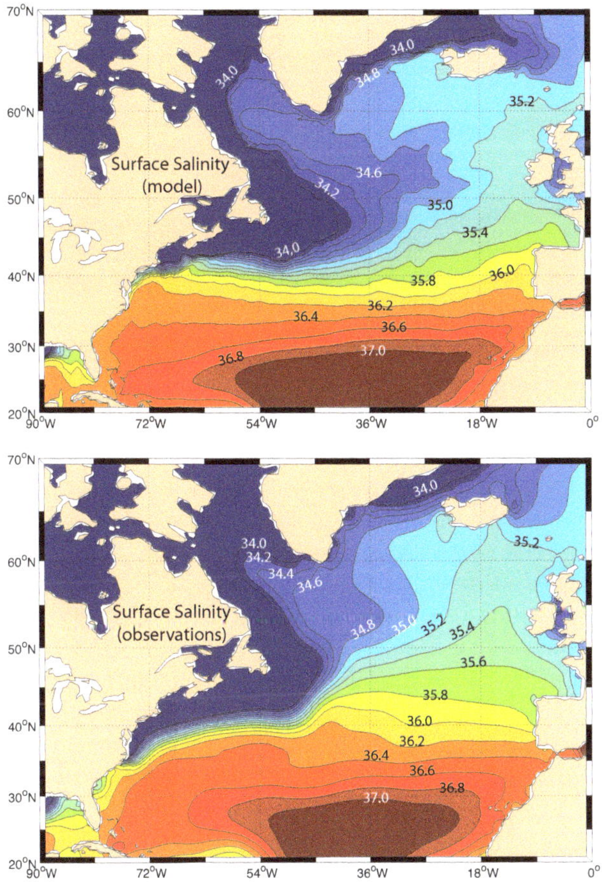

FIG. 6-7 – Average surface salinity 1958–2001 obtained (**a**) with the ocean model NEMO driven by the ERA40-ECMWF reanalysis and (**b**) using Levitus data. (Courtesy of C. Herbaut and M.-N. Houssais, LOCEAN-CNRS Paris VI University, 2009.)

Models and observations show that the circulation on the scale of the planetary ocean is much more complex than the simplified diagram [**Figure 1-51**] of large-scale circulation cells of the Thermohaline Circulation

(THC), also called Meridional Overturning Circulation (MOC). The diagram of [**Figure 6-8**] is an example.

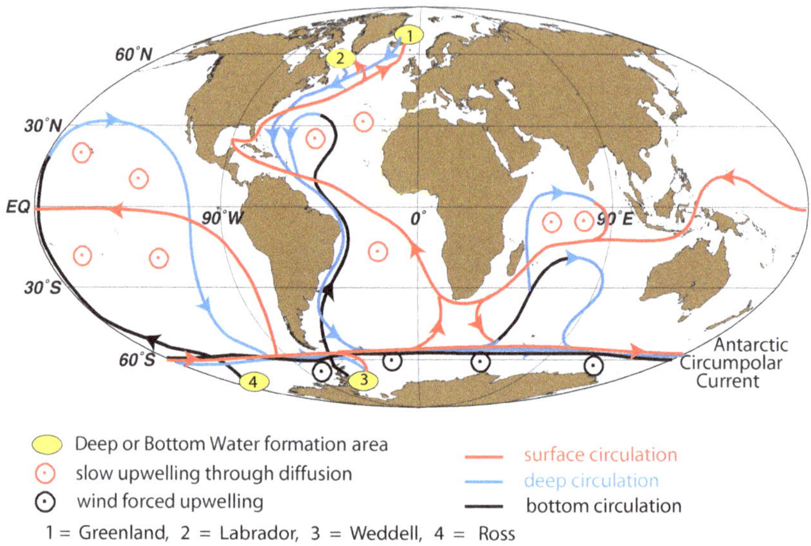

FIG. 6-8 – Diagram of large-scale circulation cells. Zones of equatorial and coastal upwelling are not shown since they have little or no involvement in Deep Water conversion processes. (After Kuhlbrodt *et al.*, 2007.)

A number of circulation cells, on a range of scales, are connected to the principal THC, which involve connections among different water masses through vertical convection, downwelling, upwelling, diffusion, and horizontal motions. **Figure 6-9** shows the trajectories of hundreds of thousands of water particles simulated in a general circulation model illustrating THC branches.

The North Atlantic deep-convection regions make up the descending branch of the thermohaline cell, forming North Atlantic Deep Water, which feeds Circumpolar Deep Water. The latter is redistributed to the Indian and Pacific Oceans in which it slowly diffuses to the surface layers or is reinjected into Upper Circumpolar Deep Water.

An equatorial route north of Australia across the Indonesian seas and the Indian Ocean (the warm route) and an austral route through the Drake Passage, south of South America (the cold route) provide the return of the surface and intermediate layers from the Pacific Ocean toward the North Atlantic. A third route (the intermediate route) joins in south of Australia, allowing a return current westward, north of the Antarctic Circumpolar Current. It provides a connection between the subtropical gyres of the South Atlantic, the South Indian, and the South Pacific Oceans [**Figures 6-9** and **6-10**] and is called the "super gyre". Antarctic Intermediate Water and

Subantarctic Mode Water represent the deep and cold elements of the return routes. All these routes cross the South Atlantic Ocean northward and converge in the Gulf Stream to close the immense belt of the THC.

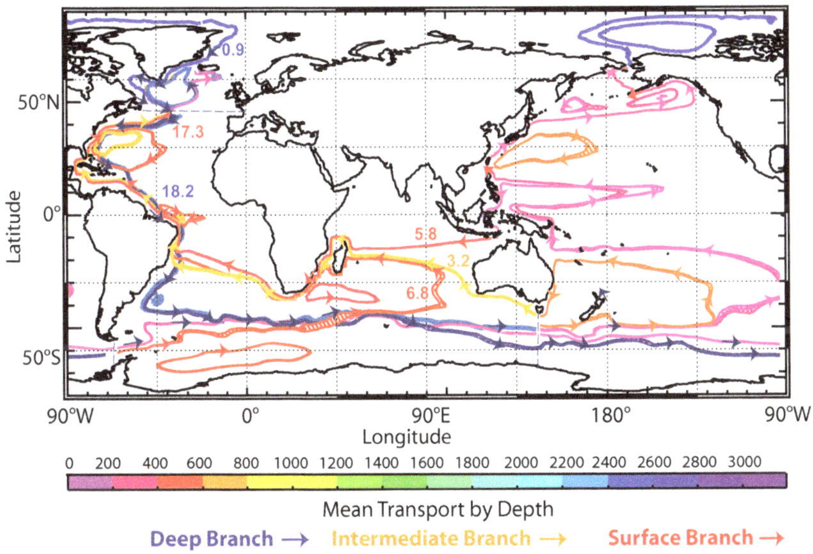

FIG. 6-9 – Thermohaline Circulation (THC) reconstruction (MOC or AMOC, Atlantic Meridional Overturning Circulation) with the aid of millions of particle trajectories simulated in the oceanic general circulation numerical model, ORCA2. The graphic represents the mean trajectory of all the particles injected into the model, which makes isolated circulations that are not in the model appear. The Internet site with animation is indicated in the references. Red, orange, and yellow show surface water and thermocline water that are transformed in the Nordic Seas into North Atlantic Deep Water, in blue. The figures give estimated transport of the various branches in Sv. (After Speich *et al.*, 2007.)

Grafted onto the principal THC, part of Circumpolar Deep Water enters two cells of meridional circulation to form Antarctic Intermediate Water and Antarctic Bottom Water [**Figures 2-44** and **6-10**].

Other shallower cells are also connected to the THC, connecting subtropical convection zones to equatorial or coastal upwelling regions.

Ocean heat transport carried out through the action of these various cells is of the same order as transport by the atmosphere, even though it is much slower.

On average, the Pacific Ocean transports heat stored in the equatorial zone to the north and south where it redistributes it to the atmosphere. The Indian Ocean stores heat in the north and transports it to the south. On average, the Atlantic Ocean transports heat at all latitudes from south to north where it transmits it to the atmosphere. Even though it has extremely low temperatures, the Antarctic Ocean, *via* Circumpolar Deep Water, provides heat to the atmosphere.

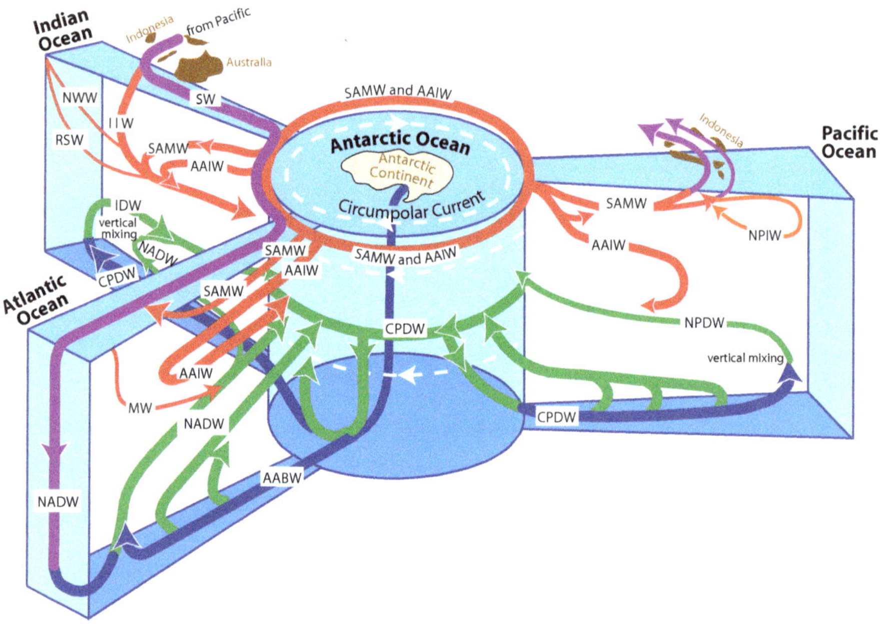

FIG. 6-10 – Schematic of the circulation around the Antarctic and the circulation branches attached to each ocean. [Surface Water (SW), Subantarctic Mode Water (SAMW), Antarctic Intermediate Water (AAIW), Mediterranean Water (MW), North Atlantic Deep Water (NADW), Antarctic Bottom Water (AABW), Circumpolar Deep Water (CPDW), NW Indian Water (NWW), Indonesian Intermediate Water (IIW), Red Sea Water (RSW), Indian Deep Water (IDW), North Pacific Intermediate Water (NPIW), North Pacific Deep Water (NPDW).] (After Schmitz, 1996.)

A way to explore the oceans is now available thanks to Mercator Ocean and the European Copernicus services. A 3D analysis and forecasting of the global ocean and regional basins is routinely produced, using the most advanced numerical models assimilating a wide range of satellite and in situ observations, and serving thousands of users world wide. Visit mercator-ocean.eu and marine.copernicus.eu and dive with the Ocean Forecasters into our oceanic planet.

Chapter 3
Thoughts on ocean variability, climatic implications

Improvements in the precision of observational techniques, the advent of satellite sensors, and the accumulation of repeated long-term measurements allow us to approach the immense complexity of the ocean milieu and its variability.

This descriptive oceanography text is limited to the present knowledge of the present ocean. Since the ocean and atmosphere are in perpetual readjustment, it is difficult to define an average state.

Below, we briefly present examples of ocean variability other than the ENSO anomalies already discussed in **Section V**.

3.1 Oceanic response to global atmospheric warming

The ocean plays a fundamental role in regulating climate. The oceanic response to human perturbations is one of the crucial questions of our age. Sea-level rise due to increasing heat content (the steric effect) [**Figure 6-11**] and melting of glaciers linked to greenhouse warming are globally established even if they pose sampling problems. Ocean acidification also occurs as a result of increasing concentration of carbon dioxide.

Oceanic warming seems real in spite of large error-bars [**Figure 6-11b**] but the evolution of oceanic temperature distribution is still a subject of controversy regarding evaluation methods and the regions and ranges of depths involved. The role of circulation is central to the redistribution of heat exchanges with the atmosphere and can only be determined by using a model approach associated with appropriate essential long-term measurements.

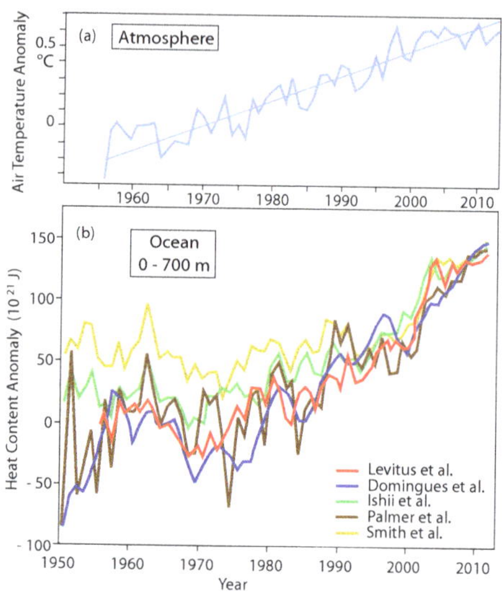

FIG. 6-11 – **(a)** Global mean surface air temperature evolution from NOAA, as anomalies relative to 1900-1999 plotted with linear trends for 1970-2013 (After Trenberth *et al.*, 2014.) **(b)** Observation-based estimates of global mean upper ocean heat content (0 – 700 m) updated from Levitus *et al.*, 2012 (red), Ishii and Kimoto, 2009 (green), Domingues *et al.*, 2008 (blue), Palmer *et al.*, 2007 (brown), Smith and Murphy, 2007 (yellow brown). The uncertainty of the estimates is of the order of the dispersion amongst the five. Estimates are shifted to align for 2006-2010, five years well measured by Argo floats, and then plotted relative to the resulting mean of all curves for 1971, the starting year for trend calculations. (After IPCC, Climate Change 2013.)

3.2 Examples of interannual variability

The North Atlantic is the site of strong interannual atmospheric variations having major repercussions on ocean convection processes. For example, the intensity of convection in the Labrador Sea is correlated with the North Atlantic Oscillation (NAO) [**Figure 6-12**]. The NAO index is the anomaly of the pressure difference between the Icelandic low and the Azores high [**Figure 6-12**]. A period of cooling and freshening of Labrador Sea Water during 1988–1995 was associated with a strong NAO index. In 1994, convection reached a depth never before observed, giving rise to a new component of Labrador Sea Water that has been tracked as far as the tropical Atlantic.

Interannual anomalies have led to changes in the fresh-water flux into the Nordic Seas. For example, the decrease in surface salinity during the 1970's: the "great salinity anomaly". The Gulf Stream includes water coming from the South Atlantic and feeding the Nordic Seas where it "densifies" through atmospheric exchanges, sinks to great depth, and then begins the passage toward the Antarctic. The return branch toward the Antarctic represents the descending branch of the large-scale thermohaline circulation (THC or MOC). Anomalies that reduce the "densification" of surface water from the Gulf Stream branch into the Nordic Seas could also reduce the level of production of Deep Water and in turn could diminish Gulf Stream transport.

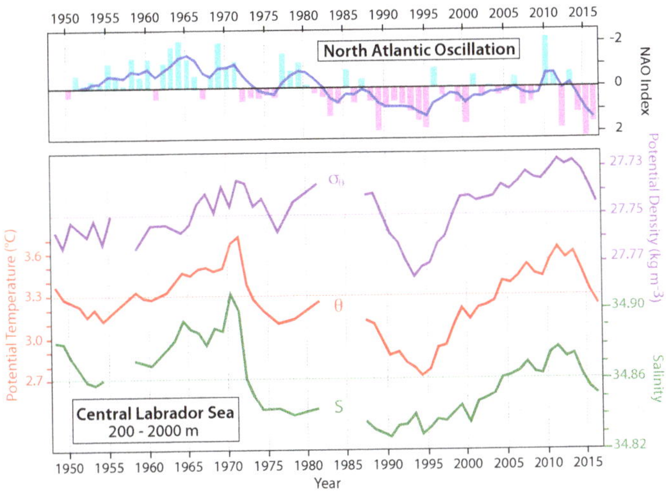

FIG. 6-12 – The North Atlantic Oscillation from 1987 to 2016 (scale inverted). Comparable evolution of annual-means for the 200 – 2000 m vertical interval in the central Labrador Sea: annual-mean potential temperature (θ, °C) (red), salinity (S) (green) and corresponding potential density anomaly (σ_{1000}, kg m^{-3}) (blue). (From Yashayaev and Loder, 2016.)

Indian Ocean climatic variability is mainly seasonal, linked to the monsoons. Coupled with the seasonal signal, an interannual component is connected with the Indian Ocean Dipole (IOD), an atmospheric anomaly. The dipole is a coupled ocean-atmosphere phenomenon characterized by abnormal surface temperature cooling in the SE equatorial Indian Ocean and abnormal warming in the western equatorial Indian Ocean. Associated with these changes, atmospheric convection in the eastern Indian Ocean shifts westward, resulting in high precipitation over east Africa and severe drought in Indonesia.

Seasonal and interannual variations in ice-pack extent, modulated by Antarctic circulation, have considerable influence on global climatology. Their time scale may vary, through the intermediary of atmospheric cells (rapid) and oceanic cells (slow). Variability in the northward transport of Antarctic Bottom Water is observed as far as the northwest Atlantic, no doubt linked to climatic conditions along the Antarctic continent.

In the Arctic Ocean, ice extent and thickness have considerably diminished in a few decades. To such an extent that the northwest passage, north of Canada, and the northeast passage, north of Siberia, to the Pacific Ocean are often open to shipping in summer.

Understanding these interannual variations is essential in order to predict climate variability, not only for a region, but for the entire planet.

3.3 In the form of an epilog

These kinds of oceanic anomalies, observations of which are beginning to allow us to explore, are only examples.

Climatic perturbations will continue to affect ocean circulation and water-mass characteristics in the decades to come. The resulting changes in the planetary ocean will in turn be the catalysts or regulators of initiating climatic effects.

This text does not attempt to address the predictability of the impact of climatic variability on how the ocean functions but we recall here some of the questions with which scientists are confronted and to which they will have to respond in a precise manner.

Great uncertainty persists at the moment on the response time scale of the ocean to human perturbations, on the amplitude of their effects on the atmosphere and climate. It is urgent to better evaluate the variations of the thermal content of the ocean, of sea level (as a result of glaciers melting and thermal expansion), and of the amplifying effect of El Niño events.

Oceanographers and climatologists address these questions through coupled ocean-atmosphere models, trying to put the relevant feedback processes involved in the climate system in the form of equations.

The enormous mass of the ocean, its thermal capacity, and its chemical composition give it a central role as a potential regulator of human perturbations (and a possible absorber of excess carbon dioxide). Can the planet modulate, or even eliminate human impacts or, on the contrary, will human disturbances upset the mechanism with dramatic consequences on the continents and their population?

Throughout this text, we have sought to establish a basic understanding of the ocean. However, it is a medium that is not easy to understand. The more one observes, the more one is confronted with its immense complexity. The ocean is still mysterious. It is one of reasons why it attracts us so passionately. All sailors will say so.

References, acronyms, web sites

References

Aagaard, K., Swift, J.H., Carmack, E.C., 1985. Thermohaline circulation in the arctic mediterranean seas. *J. Geophys. Res.*, **90**, 4833-4846.

Aagaard, K., Fahrbach, E., Meincke, J., Swift, J.H., 1991. Saline outflow from the Arctic Ocean: its contribution to deep waters of the Greenland, Norwegian, and Iceland Seas. *J. Geophys. Res.*, **96** (C11) 20433-20441.

Aleem, A. A., 1980. On the history of Arab navigation. In: "Oceanography, The Past", Proceedings of the third International Congress on the History of Oceanography held at Woods Hole Oceanographic Institution. M. Sears and D. Merriman, eds., New York: Springer-Verlag, 582-595.

Andrié, C., 1996. Chlorofluoromethanes in the deep equatorial Atlantic revisited. In: "The South Atlantic: Present and Past circulations", G. Wefer, W. Berger, G. Siedler, D. Webb, eds., Berlin: Springer-Verlag, 273-288.

Andrié, C., Ternon, J.-F., Messias, M.-J., *et al.*, 1998. Chlorofluoromethane distributions in the deep equatorial Atlantic during January-March 1993. *Deep-Sea Res. I*, **45**, 903-930.

Andrié, C., Gouriou, Y., Bourles, B., Ternon, J.-F., Braga, E.S., Morin, P., Oudot, C., 2003. Variability of AABW properties in the equatorial channel at 35°W. *Geophys. Res. Lett.*, **30** (5) 8007, doi: 10.1029/2002GL015766.

Apel, J.R., 1987. "Principles of Ocean Physics", Academic Press, New York, 634 pp.

Arhan, M., Mercier, H., Bourlès, B., Gouriou, Y, 1998. Hydrographic sections across the Atlantic at 7°30N and 4°30S. *Deep-Sea Res. I*, **45**, 829-872.

Arhan, M., Heywood, K., King, B., 1999. The deep waters from the Southern Ocean at the entry to the Argentine Basin. *Deep-Sea Res. II*, **46**, 475-499.

Arhan M., Mercier H., Park Y.H., 2003. On the deep water circulation of the eastern South Atlantic Ocean. *Deep-Sea Res. I*, **50**, 889-916.

Armi, L., Stommel, H., 1983. Four views of a portion of the north Atlantic subtropical gyre. *J. Phys. Oceanogr.*, **13**, 858-857.

Arnault, S., 1987. Tropical Atlantic geostrophic currents at the equator. *J. Geophys. Res.*, **6**, 287-297.

Arnault, S., Menard, Y., Merle, J., 1990. Observing the tropical Atlantic Ocean in 1986-1987 from altimetry. *J. Geophys. Res.*, **95**, 17921-17945.

Association Internationale des Sciences Physiques de l'Océan (IAPSO), 1974. "Océanographie Physique de la mer Rouge", Symposium de l'.... Paris, 312 pp. Publications du Centre National pour l'Exploitation des Océans. *Actes de Colloques Series*, N° 2.

Atmadipoera, A., Molcard, R., Madec, G., Wijffels, S., Sprintall, J., Koch-Larrouy, A., Jaya, I., Supangat, A., 2009. Characteristics and variability of the Indonesian throughflow water at the outflow straits. *Deep-Sea Res. Part I*, **56** (11), 1942-1954.

Bacon, S., Gould, W.J., Jia, Y., 2003. Open-ocean convection in the Irminger Sea. *Geophys. Res. Lett.*, **30** (5) 1246, doi: 10.1029/222GL016271.

Barnier, B., Madec, G., Penduff, T., *et al.*, 2006. Impact of partial steps and momentum advection schemes in a global ocean circulation model at eddy-permitting resolution. *Ocean Dynamics*, **56** (5-6) 543-567. (Special Issue in honour and in memory of Christian Le Provost: Modelling, observing, and forecasting sea level, ocean tides and ocean circulation: reviews and recent progress.)

Baumgartner, A., Reichel, E., 1975. "The World Water Balance", Elsevier, Amsterdam, 179 pp.

Beranger, K., Drillet, Y., Houssais, M.N., Testor, P., Bourdalle-Badie, R., Bozec, A. Mortier, L. Bouruet-Aubertot, P., Crepon, M., 2010. Impact of the spatial distribution of the atmospheric forcing on water mass formation in the Mediterranean Sea. *J. Geophys. Res.*, **115** (C12041) 22, doi:10.1029/2009JC005648.

Bigg, G.R., 1986. "The Oceans and Climate", Cambridge University Press, Cambridge, UK, 266 pp.

Bingham, F.M., Lukas, R., 1994. The southward intrusion of North Pacific Intermediate Water along the Mindanao coast. *J. Phys. Oceanogr.*, **24**, 141-154.

Birol, F., Morrow, R., 2001. Source of baroclinic waves in the southeast Indian Ocean. *J. Geophys. Res.*, 106 (C9) 145-160.

Bjerknes, J., 1969. Atmospheric teleconnections from the equatorial Pacific. *Mon. Weather Rev.*, **97**, 163-172.

Boden, B.P., Duncombe Rae, C.M., Lutjeharms. J.R.E. 1988. The distribution of the diatoms of the south-west Indian Ocean surface waters between Cape Town and the Prince Edwards Islands Archipelago. *South African Journal of Science*, **84**, 811-818.

Boorstin, D., 1983. "The Discoverers: A History of Man's Search to Know His World and Himself", Random House, New York, 747 pp.

Borenas, K.M., Lake, I.L., Lundberg, P.A, 2001. On the intermediate water masses of the Faeroe-Bank channel overflow. *J. Phys. Oceanogr.*, **31**, 1904-1914.

Bougeault, P., Sadourny, R., 2001. "Dynamique de l'atmosphère et de l'océan". Palaiseau: Les Editions de l'Ecole Polytechnique, Palaiseau, France, 309 pp.

Bourles, B., Molinari, R.L., Johns, E., Wilson, W.D., Leaman, K.D., 1999. Upper layer currents in the western tropical North Atlantic (1989-1991). *J. Geophys. Res.*, **104** (C1) 1361-1375.

Bourles, B., D'Orgeville, M., Eldin, G., Chuckla, R., Gouriou, Y., Du Penhoat, Y., Arnault, S., 2002. On the thermocline and subthermocline eastward currents evolution in the Eastern Equatorial Atlantic. *Geophys. Res. Lett.*, **29** (16). doi:10.1029/2002GL015098.

Bourles, B., Andrié, C., Gouriou, Y. *et al.*, 2003. The deep equatorial currents in the eastern Equatorial Atlantic Ocean. *Geophys. Res. Lett.*, **30** (5) 8002. doi:10.1029/2002GL015095.

Bovis, A., 2009. "Hydrodynamique navale: théorie et modèles". Les Presses de l'ENSTA, Paris, *Collection Les Cours*, 431 pp.

Boyer, T.P., Antonov, J.I., Baranova, O.K., Coleman, C., Garcia, H.E., Grodsky, A., Johnson, D.R., Locarnini, R.A., Mishonov, A.V., O'Brien, T.D., Paver, C.R., Reagan, J.R., Seidov, D., Smolyar, I. V., and Zweng, M.M., 2013: World Ocean Database 2013, NOAA Atlas NESDIS 72, S. Levitus, Ed., A. Mishonov, Technical Ed.; Silver Spring, MD, 209 pp., http://doi.org/10.7289/V5NZ85MT

Brambilla, E., Talley, L.D., 2008. Subpolar Mode Water in the northeastern Atlantic. Part 1: Properties and circulation. *J. Geophys. Res.*, **113**, C04025. doi:10.1029/2006JC004062.

Brambilla, E., Talley, L.D., Robbins, P.E., 2008. Subpolar Mode Water in the northeastern Atlantic. Part 2: Origin and transformation. *J. Geophys. Res.*, **113**, C04026. doi:10.1029/2006JC004063.

Brandt, P., Schott, F.A., Provost, C., Kartavtseff, A., Hormann, V., Bourles, B., Fischer, J., 2006. Circulation in the central equatorial Atlantic: Mean and intraseasonal to seasonal variability. *Geophys. Res. Lett.*, **33**, L07609. doi:10.1029/2005GL025498.

Bray, N., Ochoa, J., Kinder, T., 1995. The role of the interface in exchange through the Strait of Gibraltar. *J. Geophys. Res.*, **100**, 10755-10776.

Bray, N., Wijffels, S.E., Chong, J.C., Fieux, M., Hautala, S., Meyers, G., Morawitz, W.M.L., 1997. Characteristics of the Indo-Pacific throughflow in the eastern Indian Ocean. *Geophys. Res. Lett.*, **24** (21) 2569-2572.

Brewer, P., 1983. "Oceanography: the Present and Future", Springer-Verlag, New York, 392 pp.

Brewer P.G. and Dyrssen D., 1985. Chemical oceanography of the Persian Gulf. *Prog. Oceanogr.*, **14**, 41-55.

Broecker, W.S., 1979. A revised estimate for the radiocarbon age of North Atlantic Deep Water. *J. Geophys. Res.*, **84**, 3218-3266.

Broecker, W.S., 1991. The great ocean conveyor. *Oceanography*, **4**, 79-89.

Bruce, J., Fieux, M., Gonella, J., 1981. A note on the continuance of the Somali eddy after the cessation of the SW monsoon. *Oceanologica Acta,* **4** (1) 7-9.

Bruun, A.F., Greve, S.V., Mielche, H., Sparck, R., 1956. "The Galathea deep sea expedition 1950-1952", Allen and Unwin, London, 196 pp.

Budyko, M.I., 1986. "The evolution of the Biosphere", Reidel, Dordrecht, 423 pp.

Campbell, G.G., Vonder Haar, T.H., 1980. Climatology of radiation budget measurements from satellites. *Atmospheric Science Paper.* n° 323, 74 pp. Colorado State University, Department. of Atmospheric Sciences, Fort Collins.

Capet, X., Estrade P., Machu E., Ndoye S., Grelet J., Lazar A., Marié L., Dausse D., and Brehmer P., 2017. On the dynamics of the southern Senegal upwelling center: observed variability from synoptic to super-inertial scales, *J. Phys. Oceanog.*, **47**, 155-180.

Carissimo, B.C., Oort, A.H., Vonder Haar, T.H., 1985. Estimating the meridional energy transports in the atmosphere and oceans. *J Phys. Oceanogr.*, **15**, 82-91.

Carmack, E., Chapman, D.C., 2003. Wind-driven shelf/basin exchange on an Arctic shelf: The joint roles of ice cover extent and shelf-break bathymetry. *Geophys. Res. Lett.*, **30** (14) 1778, doi: 10.1029/2003GL017526.

Coachman, L.K., Aagaard, K., 1988. Transports through Bering Strait: annual and interannual variability. *J. Geophys. Res.*, **93**, 15535-15539.

Colin De Verdière, A., 1983. Lagrangian eddy statistics from surface drifters in the eastern North Atlantic. *J. Mar. Res.*,**41** (3) 375-398.

Collin, A., 1996. Canadian Arctic Archipelago and Baffin Bay. In: "*The Encyclopedia of Oceanography*", R. Fairbridge, ed., New York: Reinhold, pp. 157-160.

Copin-Montégut, G., 1996. "Chimie de l'eau de mer", Institut Océanographique, Paris, *Collection Synthèses*, 319 pp.

Coste, B., Fiuza, A. F. G., Minas, H. J., 1986. Conditions hydrologiques et chimiques associées à l'upwelling côtier du Portugal en fin d'été. *Oceanol. Acta,* **9**, (2), 149-158.

Crease, J., 1967. The flow of Norwegian Sea water through the Faeroe Bank Channel. *Deep-Sea Res.*, **12**, 143-150.

Crépon, M., 1996. "Initiation à la Dynamique de l'Océan", Institut Océanographique, *Oceanis* Paris, **22**, (2) 271 pp.

Crépon, M., Boukthir, M., 1987. Effect of deep water formation on the circulation of the Ligurian Sea. *Annales Geophysicae*, **58**, 43-48.

Cresswell, G. R., and T. J. Golding, 1980: Observations of a south-flowing current in the southeastern Indian Ocean. *Deep-Sea Res.,* **27**, 449–466.

Daniault, N., 1984. Apport des techniques spatiales à la connaissance des courants de surface: Application à l'océan Antarctique, *Thesis*, University of Bretagne Occidentale, Brest, 69 pp. + [46] pp. of plates.

Da Silva, A., Young, A.C., Levitus, S., 1994. Atlas of Surface Marine Data, Vol. 1, Algorithms and Procedures, *NOAA/NESDIS Technical Report*, n° 6., US Dept. of Commerce, Washington, DC.

Deacon, G., 1984. "The Antarctic circumpolar ocean", Studies in Polar Research, Cambridge University Press, Cambridge, UK, 180 pp.

Deacon, G., 1937. The hydrology of the Southern Ocean, *Discovery Reports*, vol. 15, Cambridge University Press, Cambridge, UK, 124 pp.

De Boyer Montégut, C., Madec, G., Fischer, A.S., Lazar, A., Ludicone, D., 2004. Mixed layer depth over the global ocean: An examination of profile data and a profile base climatology. *J. Geophys. Res.*, **109**, C12003, doi: 10.1029/2004JC002378.

Delcroix, T., Hénin, C., 1988. Observations of the Equatorial Intermediate Current in the western Pacific Ocean (165°E). *J. Phys. Oceanogr.*, **18**, 363-366.

Delcroix, T., Eldin, G., Radenac, M.-H., Toole, J., Firing, E., 1992. Variation of the western equatorial Pacific ocean 1986-1988. *J. Geophys. Res.*, **97**, 5423-5445.

Delcroix, T., Picaut, J., Eldin, G., 1991. Equatorial Kelvin and Rossby waves evidenced in the Pacific Ocean through Geosat sea level and surface currents anomalies. *J. Geophys. Res.*, **95**, 13151-13158.

Deniaud Lherminier, P., 1998. Convection Profonde en Mer du Groenland: Etude Expérimentale des Phases de Préconditionnement et de Mélange, *Thesis*, University of Paris 6, 169 pp.

Dickson, R., Brown, J., 1994. The production of North Atlantic Deep Waters: sources, rates and pathways. *J. Geophys. Res.*, **99** (6) 12319-12341.

Dickson, R., Meincke, J., Malmberg, S.A., Lee, A.J., 1988. The "Great Salinity Anomaly" in the northern North Atlantic 1968-1982. *Progr. Oceanogr.*, **20**, 103-151.

Dietrich, G., 1969. "Atlas of the hydrography of the northern Atlantic Ocean". International Council for the Exploration of the Sea, Copenhagen, 140 pp.

Dietrich, G., Kalle, K., Krauss, W., Siedler, G., 1980. "General Oceanography", 2nd ed., Wiley Interscience, New York, 626 pp.

Domingues, C.M., Church, J.A., White, N.J., *et al.*, 2008. Improved estimates of upper-ocean warming and multi-decadal sea-level rise, *Nature*, **453**, 1090-1095.

Dong, S., Sprintall, J., Gille, S.T., Talley, L, 2008. Southern Ocean mixed-layer depth from Argo float profiles. *J. Geophys. Res.*, **113**, C06013, doi:10.1029/2006JC004051.

Donguy, J. R., Hénin, C., 1975. Evidence of the South Tropical Countercurrent in the Coral Sea. *Austr. J. Mar. Freshw. Res.*, **96**, 405-409.

Döös, K., Coward, A., 1997. The southern ocean as the major upwelling zone of the North Atlantic Deep Water, *WOCE Newsletter*, n° 27, 3-17.

D'Ortenzio, F., Ludicone, D., De Boyer Montégut, C., Testor. P., Antoine D., Marullo, S., Santoleri, R., Madec G., 2005. Seasonal variability of the mixed layer depth in the Mediterranean Sea as derived from in situ profiles. *Geophys. Res. Lett.*, **32**, L12605, doi:10.1029/2005GL022463.

Duing, W., Schott, F., 1978. Measurements in the source region of the Somali Current during the monsoon reversal. *J. Phys. Oceanogr.*, **8**, 278-289.

Duplessy, J. C., Morel, P., 1992. "Gros temps sur la planète", Odile Jacob, Paris, *Points*, 337 pp.

Duplessy, J.C., 1996. "Quand l'océan se fâche". Odile Jacob, Paris, 277 pp.

Ebbesmeyer, C.C., Ingraham, W.J., Jr., 1992. Shoe spill in the north Pacific. *EOS*, 1992, **73**, 361.

Ekman, V.W., 1905. On the influence of the earth's rotation on ocean currents, *Arch. Math, Astron. Phys.*, **2** (11) 1-53.

Emile-Geay, J., Cane, M., Naik, N. *et al.*, 2003. Warren revisited: Atmospheric freshwater fluxes and "why is no deep water formed in the North Pacific". *J. Geophys. Res.*, **108** (C6), 3178, doi:10.1029/2001JC001058.

England, M.H., 2001. Review of "Ocean Circulation and Climate: Observing and Modelling the Global Ocean", Siedler, G, Church, J., Gould, J., Eds, San Diego: Academic Press, *Austr. Met. Mag.*, **50**, 317-318.

Escoffier, C., 1998. Sur le forçage atmosphérique et la formation des eaux modales subtropicales dans l'Atlantique Sud Ouest. *Thesis*, University of Paris 6, 302 pp.

Fellous, J.-L., Gauthier, C. (eds.), 2007. "Comprendre le changement climatique". Odile Jacob, Paris, *Sciences*, 297 pp.

Feng, M., Meyer, G., Pearce, A., Wijffels, S., 2003. Annual and interannual variations of the Leeuwin Current at 32°S. *J. Geophys. Res.*, **108** (C11) 3355, doi:10.1029/2002JC001763.

Ffield, A., Gordon, A., 1996. Tidal mixing signatures in the Indonesian seas. *J Phys Oceanogr.*, **26**, 1,924-1,937.

Fieux, M., Stommel, H., 1977. Onset of SW monsoon over the Arabian Sea from marine reports of surface winds. Structure and variability. *Mon. Weather Rev.*, **105** (2) 231-236.

Fieux, M., Levy, C., 1983. Seasonal observations in the Western Indian Ocean. In: Hydrodynamics of the Equatorial Ocean, J.C.J. Nihoul, ed., *Elsevier Oceanography Series*, **36**, 17-29.

Fieux, M., Kartavtseff, A., 1985. Campagne SINODE à bord du *Marion Dufresne*, Volume I, Routines 1980-1983. *Les Rapports de Campagnes à la Mer*, Mission de Recherche des TAAF, N°85-02, 126 p.

Fieux, M., 1987. "Océan Indien et Mousson, Conference in memory of Anton Bruun", UNESCO, Paris, 15 pp.

Fieux, M., Swallow, J.C., 1988. Flow of deep water into the Somali Basin. *Deep-Sea Res.*, **35** (2) 303-309.

Fieux, M., Andrié C., Delecluse P., Ilahude A.G., Kartavtseff A., Mantisi F., Molcard R., Swallow J. C., 1994. Measurements within the Pacific-Indian Oceans Throughflow region, *Deep-Sea Res. I*, **41**, 1091-1130.

Fieux, M., Andrié, C., Charriaud, E., Ilahude, A., Metzl, N., Molcard, R., Swallow, J., 1996a. Hydrological and chlorofluoromethane measurements of the Indonesian throughflow entering the Indian Ocean. *J. Geophys. Res.*, **101** (C5) 12433-12454.

Fieux, M., Molcard, R., Ilahude, A.,1996b. Geostrophic transport of the Pacific-Indian Oceans throughflow. *J. Geophys. Res.*, **101** (C5) 12421-12432.

Fieux, M., 1997. La circulation océanique. In: "Océans et Atmosphère", Hachette Education, *Collection Synapses*, pp. 28-52.

Fieux, M., 2001a. Indian Ocean equatorial currents, in: "Encyclopedia of Ocean Sciences", (J. Steele, S. Thorpe, and K. Turekian, eds.) Academic Press, doi:10.1006/rwos.2001.0367.

Fieux, M., 2001b. Indian Ocean western boundary – Somali Current, in: "Encyclopedia of Ocean Sciences", J. Steele, S. Thorpe, and K. Turekian, eds., Academic Press, doi:10.1006/rwos.2001.0364.

Fieux, M., Reverdin, G., 2001. Current Systems in the Indian Ocean, in: "Encyclopedia of Ocean Sciences", J. Steele, S. Thorpe, and K. Turekian, eds., Academic Press, doi:10.1006/rwos.20 01.0363.

Fieux, M., Molcard, R., Morrow, R., 2005. Water properties and transport of the Leeuwin Current and Eddies off Western Australia. *Deep-Sea Res.* I, **52**, 1617-1635.

Fieux, M., Molcard, R., Morrow, R., Kartavtseff, A., and Ilahude, A. G., 2005. Variability of the throughflow at its exit in the Indian Ocean. *Geophys. Res. Lett.* 32, doi:10.1029/2005GL022836

Fine, R.A., 1993. Circulation of Antarctic Intermediate Water in the south Indian Ocean, *Deep-Sea Res.* I, **40**, 2021-2042.

Fine, R.A., Peterson, W.H., Ostlund, H.G., 1987. The penetration of tritium into the tropical Pacific. *J. Phys. Oceanogr.*, **17**, 553-564.

Fine, R., Lukas, R., Bingham, F.M., Warner, M., Gammon, R., 1994. The western equatorial Pacific: a water mass crossroads. *J. Geophys. Res.*, **99** (C12) 25063-25080.

Firing, E., 1987. Deep zonal currents in the central equatorial Pacific. *J. Mar. Res.*, 45, 791-812.

Firing, E., Wijjfels, S.E., Hacker, P., 1998. Equatorial subthermocline currents across the Pacific. *J. Geophys. Res.*, 103, C10, 21413-21423.

Fiuza, A. F. G., 1983. Upwelling patterns off Portugal, in: *Coastal upwelling,* E. Suess and J. Thiede, eds., Plenum Publ. Corp, New York, 85-98.

Freudenthal, S., Andrié, C., 2002. The arrival of a "new" Labrador Sea Water signal in the tropical Atlantic in 1996, *Geophys. Res. Lett.*, **29**, n°15, doi: 10.1029/2002GL015062.

Gamberoni, L., Geronimi, J., Jeannin, P.F., Murail, J.F., 1982. Study of frontal zones in the Crozet- Kerguelen region. *Oceanologica Acta*, **5**, 289-299.

Gamo, T., Nozaki, Y., Sakai, H., Nakai, T., Tsubota, H., 1986. Spatial and temporal variations of water characteristics in the Japan Sea bottom layer. *J. Mar. Res.*, **44**, 781-793.

Ganachaud, A., Wunsch, C., 2000. Improved estimates of global ocean circulation, heat transport and mixing from hydrographic data. *Nature*, **408**, 453-457.

Garcia, H.E., Locarnini, R.A., Boyer, T.P., Antonov, J.I., Baranova, O.K., Zweng, M.M., Reagan, J.R., Johnson, D.R., 2014. World Ocean Atlas 2013, Volume 3: Dissolved Oxygen, Apparent Oxygen Utilization, and Oxygen Saturation. S. Levitus, Ed., A. Mishonov Technical Ed.; NOAA Atlas NESDIS 75, 27 pp.

Garzoli, S. L., 1993. Geostrophic velocity and transport variability in the Brazil-Malvinas confluence, *Deep Sea Res.*, Part I, 40(7), 1379-1403.

Garzoli, S., Gordon, A.L., 1996. Origins and variability of the Benguela current. *J. Geophys. Res.*, **101**, 897-906.

Garzoli, S., Katz, E.L., 1983. The forced annual reversal of the Atlantic North Equatorial Current. *J. Phys. Oceanogr.*, **13**, 2082-2090.

Garzoli S. L., Ffield A., Johns W.E., Yao, Q., 2004. North Brazil Current retroflection and transports. *J. Geophys. Res.*, **109**, C01013, doi:10.1029/2003JC001775.

Gascard, J. C., 1978. Mediterranean deep water formation, baroclinic instability and oceanic eddies. *Oceanologica Acta*, **1** (3) 315-330.

Geistdoerfer, P., 2002. "Océanographie Générale". Infomer, Rennes, France, *Bibliothèque de l'Institut Français d'aide à la formation professionnelle maritime*, 269 pp.

Geistdoerfer, P., 2005. "La mer et les océans". Editions Ellipses, Paris, *l'esprit des Sciences*, 128 pp.

Gill, A.E., 1982. "Atmosphere-Ocean Dynamics". Academic Press, New York, 1982, 662 pp.

Godfrey, J.S., 1996. The effect of the Indonesian Throughflow on ocean circulation and heat exchange with the atmosphere: A review. *J. Geophys. Res.*,**101**, 12217-12237.

Godfrey, J. S., Golding, T. J., 1981. The Sverdrup relation in the Indian Ocean and the effect of Pacific- Indian through flow on Indian Ocean circulation and on the East Australian Current. *J. Phys. Oceanogr.*, **11**, 771-779.

Goedecke, E., Smed, J., Tomczak, G., 1967. Monatskarten des Salzgehaltes der Nordsee dargestellt für verschiedene Tiefenhorizonten. *Ergänzungsh. Deutsche Hydrographische. Zeitschrift*, **B 9**, 1-110.

Goldner, D. R., 1999. On the uncertainty of the mass, heat, and salt budgets of the Arctic Ocean. *J. Geophys. Res.*, **104** (C12) 29757–29770.

Gonella, J., 1969. Analyse des mesures de courants et de vents à la Bouée Laboratoire (position B), juillet 1968. *Cahiers Océanogr.*, **21** (9) 855-862.

Gonella, J., Fieux, M., Philander, G., Mise en évidence d'ondes de Rossby équatoriales dans l'océan Indien au moyen de bouées dérivantes. *Comptes Rendus Académie des Sciences, Paris*, T. 292.

Gordon, A.L., Tchernia, P., 1972. Waters of the continental margin off Adelie Coast. In: Antarctic Oceanology II: The Australian-New Zealand sector. D.E. Hayes, ed., American Geophysical Union, *Antarctic Research Series*, **19**, 59-69.

Gordon, A.L., Molinelli, E., Baker, T., 1978. Large-scale relative dynamic topography of the southern ocean. *J. Geophys. Res.*, **83** (C6) 3023-3032.

Gordon, A. L., 1985. Indian-Atlantic transfer of thermocline water at the Agulhas retroflection. *Science*, **227**, 1030-1033.

Gordon. A. L., 1986. Interocean exchanges of thermocline water. *J. Geophys. Res.*, **91**, 5037-5046.

Gordon, A.L., Lutjeharms, J. R. E., Grundlingh, M. L., 1987. Stratification and circulation at the Agulhas Retroflection. *Deep-Sea Res., A*, **34** (4) 565-599.Gordon, A.L., Bosley, K.T., 1991. Cyclonic gyre in the tropical South Atlantic. *Deep-Sea Res.*, **38** suppl., 323-343.

Gordon, A.L., Weiss, R. F., Smethie, W. M., Warner, M. J., 1992. Thermocline and Intermediate Water communication between the South Atlantic and Indian Oceans. *J. Geophys. Res.*, **97** (C5) 7223-7240.

Gordon, A.L., Ma, S., Hacker, P., Talley, L.D., Olson, D.B., Ffield, A., and Wilson, D., 1997. Advection and diffusion of Indonesian throughflow water within the Indian Ocean South Equatorial Current. *Geophys. Res. Let.*, **24**, 2573-2576.

Gordon, A.L., Barnier, B., Speer, K., Stramma, L, 1999. World Ocean Circulation Experiment: South Atlantic results. *J. Geophys. Res.*, **104** (C9) 20859-20861.

Gordon, A.L., Susanto, R.D., Ffield, A., 1999. Throughflow within Makassar Strait. *Geophys. Res. Lett.*, **26**, 3325-3328.

Gordon, A., Visbeck, M., Comiso, J., 2007. A possible link between the Weddell polynya and the Southern Annular Mode. *J. Climate*, **20** (11) 2558-2571.

Goureski, V., Koltermann, K., 2007. How much is the ocean really warming? *Geophys. Res. Lett.*, **34**, L01610, doi:10.1029/2006GL027834.

Gourdeau, L., Kessler, W.S., Davis, R.E., Sheman, J., Maes, C., Kestenare, E., 2008. Zonal jets entering the Coral Sea. *J. Phys. Oceanogr.*, **38**, 715-725, doi:10.1175/2007JPO3780.1.

Gouriou, Y., Andrié, C., Bourles, B., Freudenthal, S., Arnault, S., Aman, A., Eldin, G., Dupenhoat, Y., Baurand, F., Gallois, F., Chuchla, R., 2001. Deep circulation in the Equatorial Atlantic Ocean. *Geophys. Res. Lett.*, **28** (5) 819-822.

Grundlingh, M.L., 1984. Occurrence of Red Sea water in the southwestern Indian Ocean. *J. Phys. Oceanogr.*, **15**, 207-212.

Grundlingh, M.L., 1987. Cyclogenesis in the Mozambique Ridge. *Deep-Sea Res.*, **34**, 89-103.

Guay, C. K. H., Falkner, K.K., Muench, R.D., *et al.*, 2001. Wind-driven transport pathways for Eurasian Arctic river discharge. *J. Geophys. Res.*, **106** (C6) 11469–11480.

Halpern, D., Weisberg, R.H., 1989. Upper ocean thermal and flow fields at 0°, 28°W (Atlantic) and 0°, 140°W (Pacific) during 1983-1985. *Deep-Sea Res.*, **36**, 407-418.

Hanawa, K., Talley, L.D., 2001. Mode waters. In: "Ocean circulation and Climate", G. Siedler and J. Church, eds., *International Geophysics Series*, Academic Press, pp. 379-386.

Hastenrath, S., 1982. On meridional heat transports in the World Ocean. *J. Phys. Oceanogr.*, **12**, 922-927.

Hautala, S. L., Sprintall, J., Potemra, J., Ilahude, A. G., Chong, J.C., Pandoe, W., Bray, N., 2001. Velocity structure and transport of the Indonesian Throughflow in the major straits restricting flow into the Indian Ocean. *J. Geophys. Res.*, **106**, 19527-19546.

Hisard, P., 1975. La circulation superficielle dans la partie occidentale du golfe de Guinée. *Doc. Scient. Centre Recherche Océanogr. Abidjan*, **6**, 41-57.

Hisard, P., and Rual, P., 1970. Courant Equatorial Intermédiaire de l'Océan Pacifique et contre-courants adjacents, *Cah. O.R.S.T.O.M. Ser. Oceanogr.*, 8, 21-45.

Holland, D. M., Jenkins, A., 1999: Modeling Thermodynamic Ice–Ocean Interactions at the Base of an Ice Shelf. *J. Phys. Oceanogr.*, **29**, 1787–1800.

Hua, B.L., Moore, D.W., Le Gentil, S., 1997. Inertial nonlinear equilibration of equatorial flows. *J. Fluid Mech.*, **331**, 345–371.

Imawaki, S., Uchida, H., Ichikawa and ASUKA Group, 1997. Time series of the Kuroshio transport derived from field observations and altimetry data. *International WOCE Newsletter* n° 25, Feb.1997, pp. 15-18.

Imbert, B., Lorius, C., 2006. "Le grand défi des pôles". Gallimard, Paris, *Collection Découvertes,* 15, 160 pp. Inue, N., Miita, T., Tawara, S., 1985. Tsuchima Strait II: physics. In: "*Coastal Oceanography of Japonese Islands*", H. Kunishi *et al.*, eds. Tokai University Press, pp. 914-933.

IPCC, *Climate Change 2013, The Physical Science Basis, Chapter 3, Observations: Ocean* (see https://www.ipcc.ch/pdf/assessment-report/ar5/wg1/WG1AR5_Chapter03_FINAL.pdf).

Iselin, C. O'D., 1936. A study of the circulation of the western North Atlantic, *Papers in Physical Oceanography and Meteorology* (MIT and WHOI), **4**, (4) 101 pp. Ishii, M., Kimoto, M., 2009. Reevaluation of historical ocean heat content variations with time-varying XBT and MBT depth bias corrections. *J. Oceanogr.*, **65**, 287-299.

Ishii, M. and Kimoto, M., 2009. Reevaluation of historical ocean heat content variations with time-varying XBT and MBT depth bias corrections. *J. Oceanog.*, **65**, 3, 287-299.

Jackett, D. R., Macdougall, T.J., 1997. A neutral density variable for the world's oceans. *J. Phys. Oceanogr.*, **27**, 237-263.

Ivanoff, A., 1972. "Introduction à l'Océanographie, propriétés physiques et chimiques des eaux de mer". Vuibert, Paris, vol. 1, 208 pp, vol. II, 338 pp.

Jeandel, C., Thouron, D., Fieux, M., 1999. Concentrations and isotopic compositions of Neodyme in the eastern Indian Ocean and Indonesian Straits. *Geoch. Cosmochim. Acta.* **62** (15) 2597-2607.

John, H.C., Mohrholz, V., Lutjeharms, J.R.E., *et al.*, 2004. Oceanographic and faunistic structures across an Angola Current intrusion into northern Namibian waters. *J. Mar. Sys.*, **46**, 1-22.

Johns, W.E., Yao, F., Olsen, D.B., Josey, S.A., Grist, J.P., Smeed, D.A., 2003. Observations of seasonal exchange through the Straits of Hormuz and the inferred heat and freshwater budgets of the Persian Gulf. *J. Geophys. Res.* 108 (C12), 3391. Doi:10.1029/2003JC001881.

Johnson, G., 2008. Quantifying Antarctic Bottom Water and North Atlantic Deep Water volumes. *J. Geophys. Res.*, **113**, C05027, doi:10.1029/2007JC004477.

Johnson, G., 2008. Reduced Antarctic meridional overturning circulation reaches the North Atlantic Ocean. *Geophys. Res. Lett.*, **35**, L22601, doi:10.1029/2008GL035619.

Johnson, G., Sloyan, B., Kessler, W.S, Mctaggart, K.E., 2002. Direct measurements of upper ocean currents and water properties across the tropical Pacific during the 1990s. *Progr. Oceanogr.*, **52**, 31-61.

Jones, E.P., Rudels, B., Anderson, L.G., 1995. Deep waters of the Arctic Ocean: origins and circulation. *Deep-Sea Res. I*, **42** (5) 737-760.

Jones E.P., 2001. Circulation in the Arctic Ocean. *Polar Res.*, **20**, 139-146.

Jones, E.P., Anderson, L.G., Juttertström, S., Mintrop, L., Swift, J., 2008. Pacific freshwater, river water and sea ice meltwater across Arctic Ocean basins: Results from the 2005 Beringia Expedition. *J. Geophys. Res.*, **113**, CO8012, doi:10.1029/2007JC004124.

Joussaume, S., 1993. "Climat d'hier à demain". CNRS Editions/CEA, Paris, *Science au Présent*, 143 pp.

Joyce, T., 1987. Hydrographic sections across the Kuroshio Extension at 165E and 175W. *Deep-Sea Res.*, **34**, 1331-1352.

Kashino, Y., Ueki, I., Kuroda, Y., Purwandani, A., 2007. Ocean variability north of New Guinea derived from TRITON buoy data. *J. Oceanogr.*, **63**, 545-559.

Kawai, H., 1972. Hydrography of the Kuroshio Extension. In: "*Kuroshio, Physical Aspects of the Japan Current*", H. Stommel and K. Yoshida, eds., Seattle, University of Washington Press, 235-352.

Kim, K., Kim, K.R., Kim, Y.G., *et al.*, 2004. Water masses and decadal variability in the East Sea (Sea of Japan). *Progr. Oceanogr.*, **61**, 157-174.

Knauss, J.A., 1960. Measurement of the Cromwell current. *Deep-Sea Res.*, **6**, 265-286.

Knauss, J.A., 1997. "Introduction to Physical Oceanography". Prentice-Hall, New York, 309 pp.

Knox, R.A., 1976. On a long series of measurements of Indian Ocean equatorial currents near Addu Atoll. *Deep-Sea Res.*, **23**, 211-221.

Koch-Larrouy, A., Madec, G., Bouruet-Aubertot, P., Gerkema, T., Bessieres, L., Molcard, R. 2007. On the transformation of Pacific Water into Indonesian Throughflow Water by internal tidal mixing. *Geophys. Res. Lett.*, **34**, L04604, doi:10.1029/2006GL028405.

Koch-Larrouy, A., Madec, G., Blanke, B., Molcard, R., 2009. Water mass transformation along the Indonesian throughflow in an OGCM. *Ocean Dynamics*, **58** (3) 289-309.

Koch-Larrouy, A., Madec, G., Iudicone, D., Atmadipoera, A., Molcard, R., 2009. Physical processes contributing to the water mass transformation of the Indonesian Throughflow, *Ocean Dynamics*, **58** (3) 275-288.

Kolodziejczyk, N., Bourles, B., Marin, F., Grelet, J., Chuchla, R., 2009. The seasonal variability of the Equatorial Undercurrent at 10°W as inferred from recent in situ observations. *J. Geophys. Res.*, **114**, C06014, doi:10.1029/2008JC004976.

Koltermann, K.P., Gouretski V.V. and Jancke K., 2011. Hydrographic Atlas of the World Ocean Circulation Experiment (WOCE). Volume 3: Atlantic Ocean (eds. M. Sparrow, P. Chapman and J. Gould). International WOCE Project Office, Southampton, UK, ISBN 090417557X (available on-line at: https://doi.org/10.21976/C6RP4Z).

Kosro, P.M., Huyer, A., Ramp, S.R., Smith, R.L., Chavez, F.P., Cowles, T.J., Abbott, M.R., Strub, P.T., Barber, R.T., Jessen, P., Small, L.F., 1991. The structure of the transition zone between coastal waters and the open ocean off northern California, winter and spring 1987. *J. Geophys. Res.*, Oceans, 96, C8, 14707-14730.

Kuhlbrodt, T., Griesel, A., Montoya, M., *et al.*, 2007. On the driving processes of the Atlantic meridional overturning circulation. *Reviews of Geophysics*, **45**, RG2001, doi:10.1029/2004RG000166.

Kupcik, I., 1981. "Cartes Géographiques Anciennes. Evolution de la représentation cartographique du monde : de l'Antiquité à la fin du XIXe siècle". Gründ, Paris, 240 pp.

Lacan, F., 2002. Masses d'eau des Mers Nordiques et de l'Atlantique Subarctique tracées par les isotopes du néodyme, *Thesis*, University of Toulouse III, 293 pp.

Lacan, F., Jeandel, C., 2004. Denmark Strait water circulation traced by heterogeneity in neodymium isotopic compositions, *Deep-Sea Res. I*, **51**, 71-82 pp.

Lacombe, H., 1965. "Cours d'Océanographie Physique". Gauthier-Villars, Paris, 302 pp.

Lacombe, H., Gascard, J.C., Gonella, J., Béthoux, J.P., 1980. Response of the Mediterranean to the water and energy fluxes across its surface, on seasonal and interannual scales, *Oceanologica Acta*, **4**, n 2, 247-255.

Lacombe, H., Tchernia, P., Gamberoni, L., 1985. Variable bottom water in the western Mediterranean basin. *Progr. Oceanogr.*, **14**, 319-338.
Lagerloef, G., Mitchum, G., Lukas, R., Niiler, P., 1999. Tropical Pacific near surface currents estimated from altimeter, wind and drifter data. *J. Geophys. Res.*, **104**, 23313-23326.
Large, W., Bourassa, M., 2001. "To improve the basic description of the surface boundary conditions and the exchanges of physical properties with the atmosphere, and to establish their uncertainties". US WOCE Office, Texas A&M University. College Station, Texas, *US WOCE Implementation Report* n°13, 64pp.
Leaman, K.D., Johns, E., Rossby, T., 1989. The average distribution of volume transport and potential vorticity with temperature at three sections across the Gulf Stream. *J. Phys. Oceanogr.*, **19**, 36-51.
Legeckis., R., 1977. Long waves in the eastern equatorial Pacific ocean: a view from geostationary satellite. *Science*, **197**, 1179-1181.
Lemasson, L., Piton, B., 1968. Anomalie dynamique de la surface de la mer le long de l'équateur dans l'océan Pacifique. *Cah. ORSTOM, Sér. Océanogr.*, **6**, 39-45.
Levitus, S., 1982. "Climatological Atlas of the World Ocean". US Department of Commerce, Washington, DC, *NOAA Professional Paper* 13, 173 pp.
Levitus, S., Burgett, R., Boyer, T.P., 1994, "World Ocean Atlas 1994. Volume 3, Salinity". U.S. Department of Commerce, Washinton, DC, *NOAA NESDIS Atlas,* 3. 99 pp.
Levitus, S., Boyer, T.P., 1994. "World Ocean Atlas 1994. Volume 4, Temperature". U.S. Department of Commerce, Washinton, DC, *NOAA NESDIS Atlas,* 4, 117 pp.
Levitus, S., Antonov, J., Boyer, T., 2005. Warming of the world ocean, 1955-2003, *Geophys. Res. Lett.*, **32**, L02604, doi:10.1029/2004GL021592.
Levitus, S., Antonov, J., Boyer, T., et al, 2009. Global ocean heat content 1955-2008 in light of recently revealed instrumentation problems, *Geophys. Res. Lett.*, **36**, L07608, doi:10.1029/2008GL037155.
Levitus, S., Antonov, J.I., Boyer, T.P., Baranova, O.K., Garcia, H.E., Locarnini, R.A., Mishonov, A.V., Reagan, J.R., Seidov, D., Yarosh, E.S., Zweng, M.M., 2012. World ocean heat content and thermosteric sea level change, *Geophys. Res. Lett.*, **39**, 10, 5p, L10603, doi:10.1029/2012GL051106
Lévy M., Shankar D., André, J-M., Shenoi, S.S.C., Durand, F., and Boyer Montégut, C., 2007. Basin-wide seasonal evolution of the Indian Ocean's phytoplankton blooms. *J. Geophys. Res.*, **112**, doi:10.1029/2007JC004090
LIWEX Group, Malanotte-Rizzoli, P., *et al.,* 2003. The Levantine Intermediate Water Experiment (LIWEX) Group: Levantine basin—A laboratory for multiple water mass formation processes, *J. Geophys. Res.,* **108** (C9), 8101, doi:10.1029/2002JC001643.
Locarnini, R.A., Mishonov, A.V., Antonov, J.I., *et al.*, 2006. "World Ocean Atlas 2005. Volume 1, Temperature". US Department of Commerce, Washington, DC, *NOAA NESDIS Atlas*, 1,182 pp.

Locarnini, R.A., Mishonov, A.V., Antonov, J.I., Boyer, T.P., Garcia, H.E., Baranova, O.K., Zweng, M.M., Paver, C.R., Reagan, J.R., Johnson, D.R., Hamilton, M., and Seidov, D., 2013. World Ocean Atlas 2013, Volume 1: Temperature. S. Levitus, Ed., A. Mishonov Technical Ed.; *NOAA Atlas NESDIS 73*, 40 pp.

Lukas, R., Firing, E., Hacker, P., Richardson, P.L., Collins, C.A., Fine, R., Gammon, R., 1991. Observations of the Mindanao current during the Western Equatorial Pacific Ocean Circulation Study. *J. Geophys. Res.*, **96**, 7089-7104.

Lukas, R., Lindstrom, E., 1987. The mixed layer of the Western Equatorial Pacific Ocean. In: "*Proceedings 'Aha Huliko'a Hawaiian Winter Workshop on the Dynamics of the Oceanic Surface Mixed Layer*", Hawaii Institute of Geophysics, Honolulu.

Lutjeharms, J.R.E., Van Ballegooyen, R.C., 1988. The retroflection of the Agulhas Current. *J. Phys. Oceanogr.*, **18** (11) 1570-1583.

Luyten, J.R., Roemich, D.H., 1982. Equatorial currents at semi-annual period in the Indian Ocean. *J. Phys. Oceanogr.*, **12**, 406-413.

Luyten, J., Swallow, J., 1976. Equatorial Undercurrents. *Deep-Sea Res.*, **23**, 1005-1007.

Lythe, M. B., Vaughan, D.G., the BEDMAP Consortium, 2001. BEDMAP: A new ice thickness and subglacial topographic model of Antarctica. *J. Geophys. Res.*, **106** (B6) 11335–11351.

Madec, G., Lott, F., Delecluse, P., Crépon, M., 1996. Large scale preconditioning of Deep-Water Formation in the Northwestern Mediterranean Sea. *J. Phys. Oceanogr.*, **26**, 1393-1408.

Maier-Reimer, E., Mikolajewicz, U., Winguth, A., 1996. Future ocean uptake of CO_2: Interaction between ocean circulation and biology. *Climate Dynamics*, **12**, 711-721.

Maillard, C., 1974. Eaux intermédiaires et formation d'eau profonde en mer Rouge, In: "Symposium IAPSO, L'océanographie Physique en mer Rouge", Paris, *Actes de colloques CNEXO*, n°2, 105-133.

Malanotte-Rizzoli, P., 1991. The Northern Adriatic Sea as a Prototype of Convection and Water Mass Formation on the Continental Shelf, *In: Deep Convection and Deep Water Formation in the Oceans, Elsevier Oceanogr. Ser.*, **57**, 229-239.

Matthaus, W., Franck, H., 1990. The water volume penetrating into the Baltic Sea in connection with major Baltic inflows. *Gerlands Beiträge zur Geophysik*, **99**, 377-386.

McCartney, M.S., 1977. Subantarctic Mode Water. In: "A Voyage of Discovery, G. Deacon 70th Anniversary", M. Angel, ed., *Deep-Sea Res.*, suppl., 103-119.

McCartney, M.S., 1982. The subtropical recirculation of mode waters. *J. Mar. Res.*, **40** suppl, 427-464.

McCartney, M.S., Talley, L.D., 1984. Warm-to-cold water conversion in the northern Atlantic Ocean. *J. Phys. Oceanogr.*, **14**, 922-935.

McCartney, M.S., 1993. Crossing of the equator by the Deep Western Boundary Current in the Western Atlantic Ocean. *J. Phys. Oceanogr.*, **23**, 1953-1974.

McCartney, M.S., 1994. Atlantic Ocean Circulation, *Oceanus*, **37** (1) 1-29.

McDonald, A.M., Wunsch, C., 1996. An estimate of global ocean circulation and heat fluxes. *Nature*, **382**, 436-439.

MEDOC Group, 1970. Observation of deep water in the Mediterranean Sea. *Nature*, **227**, 1037-1040.

Memery, L., Arhan, M., Alvarez-Salgado, X.A., Messias, M.-J., Mercier, H., Castro, C., Rios, A., 2001. The water masses along the western boundary of the south and equatorial Atlantic. *Progr. Oceanogr.*, **47**, 69-98.

Mercier, H., Speer, K.G., 1998. The transport of Bottom Water through the Romanche Fracture Zone and the Chain Fracture Zone. *J. Phys. Oceanogr.*, **28**, 779-790.

Mercier, H., Arhan, M., Lutjeharms, J.R.E., 2003. Upper-layer circulation in the eastern Equatorial and South Atlantic Ocean in January-March 1995. *Deep-Sea Res. I*, **50**, 863-887.

Merle, J., 1980. Seasonal heat budget in the equatorial Atlantic Ocean. *J. Phys. Oceanogr.*, **10**, 464-469.

Merle, J., 2006. "Océan et Climat". Editions de l'Institut de Recherche pour le Développement, Paris, 222 pp.

Merle, J., Arnault, S., 1985. Seasonal variability of the surface dynamic topography in the tropical Atlantic Ocean. *J. Mar. Res.*, **43**, 267-288.

METMAR: N° 182 March 1999, N°190 March 2001, N°195 June 2002, N° 199 June 2003, N°203 June 2004, and the collection of the trimonthly from Meteo France, Paris, M. Hontarrède, Editor-in-chief.

Meyers, G., 1996. Variation of Indonesian Throughflow and the El Niño - Southern Oscillation, *J. Geophys. Res.*, **101**, 12255-12263.

Millot, C., 1992. Are there major differences between the largest mediterranean seas? A preliminary investigation. *Bull. Inst. Océanogr., Monaco*, n° 11 special, pp. 3-25.

Millot, C., 1985. Some features of the Algerian Current. *J. Geophys. Res.*, **90**, 7169-7176.

Millot, C., Taupier-Letage, I., 2005. Circulation in the Mediterranean Sea. In: "*The Handbook of Environmental Chemistry*", A. Saliot, ed., Berlin Heidelberg: Springer-Verlag, vol. 5, part K, 29-66.

Minster, J.-F., 1994. "Les océans". Flammarion, Paris, *Dominos Collection*, 126 pp.

Minster, J.-F., 1997. "La machine océan", Flammarion, Paris, *Nouvelle Bibliothèque Scientifique Collection*, 298 pp.

Mizuno, K., White, W.B., 1983. Annual and interannual variability in the Kuroshio Current system. *J. Phys. Oceanogr.*, **13**, 1847-1867.

Molcard, R., Fieux, M., Swallow, J.C., Banjarnahor, J., Ilahude, A.G., 1994. Low frequency variability of the currents in Indonesian Channels (Savu-Roti and Roti-Ashmore Reef). *Deep-Sea Res. I*, **41** (11/12) 1643-1661.

Molcard, R., Fieux, M., Ilahude, A.G., 1996. The Indo-Pacific Throughflow in the Timor passage. *J. Geophys. Res.*,**101** (C5) 12411-12420.

Molcard, R., Fieux, M., Syamsudin, F., 2001. The Throughflow within Ombai Strait. *Deep-Sea Res. I*, **48**, 1237-1253.

Molinari, R. L., 1982. Observations of Eastward Currents in the Tropical South Atlantic Ocean: 1978–1980. *J. Geophys. Res.*, **87** (C12) 9707–9714.

Molinari, R. L., Olson, D., Reverdin, G., 1990, Surface current distributions in the tropical Indian Ocean derived from compilations of surface buoy trajectories: The Indian Ocean drifters. *J. Geophys. Res.*, **95**, 7217-7238.

Molinari, R. L., Wilson, W.D., Leaman, K., 1985. Volume heat transports of the Florida current: April 1982 through August 1983. *Science*, **227**, 295-297.

Monin, A.S., 1975. The role of the oceans in climate models. In: "The Physical Basis of Climate and Climate Modelling", World Meteorol. Organ. Int. Counc. Sci. Unions, Geneva: *GARP Publ. Ser.*, n°16, Appendix 6, 201-205.

Morel, P., 1992. "Introduction à la Dynamique de l'Atmosphère, des Océans et du Climat". Paris: Institut Océanographique, Oceanis, Paris, 369 pp. *Cours d'Océanographie*, **18** (3).

Morrow, R., Birol, F., 1998. Variability in the southeast Indian ocean from altimetry: Forcing mechanisms for the Leeuwin Current, *J. Geophys. Res.*, 103, 18529-18544.

Morrow, R., Fang, F., Fieux, M., Molcard, R., 2003. Anatomy of Three Warm-core Leeuwin Current Eddies. *Deep-Sea Res. II*, **50**, 2229-2243.

Morrow, R., Donguy, J.R., Chaigneau, A., Rintoul, S.R., 2004. Cold-core anomalies at the subantarctic front, south of Tasmania. *Deep-Sea Res.* I, 1417-1440.

Munk, W.H., 1950. On the wind-driven ocean circulation. *J. Meteorol.*, **7**, 79-93.

Nebenzahl, K., 1991. "Atlas de Christophe Colomb et des Grandes Découvertes". Bordas, Paris, 175 pp.

Neumann, G., Pierson, W., 1996. "Principles of Physical Oceanography". Prentice-Hall, New York, 545 pp.

Nilsson, C., Cresswell, G., 1981. The formation and evolution of East Australian Current warm-core eddies. *Progr. Oceanogr.*, **9**, 133-183.

NOAA PMEL TAO Project Office, 2009. El Niño theme page. http://www.pmel.noaa.gov/tao/elnino/nino-home.html

Nowlin, W.D., Clifford, M., 1982. The kinematic and thermohaline zonation of the Antarctic Circumpolar Current at Drake Passage. *J. Mar. Res.*, **40** suppl., 481-507.

Nowlin, W.D., Klinck, J.M., 1986. Physics of the Antarctic Circumpolar Current. *Review of Geophysics*, **24**, 469-491.

Ollitrault, M., Lankhorst, M., Fratantoni, D., Richardson, P., Zenk, W., 2006. Zonal intermediate currents in the equatorial Atlantic Ocean. *Geophys. Res. Lett.*, **33**, L05605, doi:10.1029/2005GL025368.

Open University Course Team, 1989. "Ocean circulation". Pergamon, Oxford, 238 pp.

Open University Course Team, 1989. "Seawater: its composition, properties and behaviour", Pergamon, Oxford, 165 pp.

Orsi, A.H., Nowlin. W.D., Whitworth, T. III, 1993. On the circulation and stratification of the Weddell Gyre. *Deep-Sea Res. I*, **40** (1) 169-203.

Orsi, A.H., Whitworth, T. III, Nowlin, W.D, 1995. On the meridional extent and fronts of the Antarctic Circumpolar Current. *Deep-Sea Res. I*, **42**, 641-673.

Orsi, A.H., Johnson, G.C., Bullister, J.L., 1999. Circulation, mixing, and production of Antarctic Bottom Water. *Progr. Oceanogr.*, **43**, 55-109.

Orsi, A.H., Smethie, W.M., Bullister, J.L., 2002. On the total input of Antarctic waters to the deep ocean: A preliminary estimate from chlorofluorocarbon measurements. *J. Geophys. Res.*, **107**, C8, doi:10.1029/2001JC000976.

Orsi, A., Whitworth, T., 2005. Hydrographic Atlas of the World Ocean Circulation Experiment (WOCE), Volume 1, Southern Ocean., M. Sparrow, P. Chapman, and J. Gould, eds., WCRP, International WOCE Project Office, Southampton, U.K., ISBN 0-904175-49-9, 223 pp.

Ortelius, A., 1981. Atlas Theatrum orbis terrarum 19°édition Anvers, 1592. In: "Cartes Géographiques Anciennes", Paris: Kupcik I., Gründ ed, 240 pp.

Osterhus, S., Gammelsrod, T., 1999. The abyss of the Nordic Seas is warming. *J. Climate*, **12**, 3297–3304.

Ostlund, H.G., Grall, C., 1987. "North and Tropical Atlantic tritium and radiocarbon data". University of Miami, *Tritium Laboratory Data Report*, n°16, 277 pp.

Ostlund, H.G., Possner,t G., Swift, J.H, 1987. Ventilation rate of the deep Arctic ocean from carbon 14 data. *J. Geophys. Res.*, **92**, 3769-3771.

Ozaki, H., Obata, H., Naganobu, M., Gamo, T., 2009. Long-Term Bottom Water Warming in the North Ross Sea. *J. Oceanogr.*, **65**, 235-244.

Palmer, M., Haines, K., Tett, S., and Ansell, T., 2007. Isolating the signal of ocean global warming. *Geophys. Res. Lett.*, **34**, L23610.

Park, Y. H., 1989. Synthèse sur les caractères physiques du courant circumpolaire antarctique, *Ann. Hydro.*, n° 763, pp. 7-32.

Park, Y.H., Gamberoni, L., Charriaud, E., 1991. Frontal Structure and transport of the Antarctic Circumpolar Current in the south Indian Ocean sector, 40-80°E. *Mar. Chemistry*, **35**, 45-62.

Park, Y.H., Gamberoni, L., 1995. Large scale circulation and its variability in the south Indian Ocean from TOPEX/POSEIDON altimetry. *J. Geophys. Res.*, **100**, 24911-24929.

Park, Y.-H., Charriaud, E., Fieux, M., 1999. Thermohaline structure of the Antarctic Surface Water/Winter Water in the Indian sector of the Southern Ocean. *J. Mar. Sys.*, **17**, 5-23.

Park, Y.-H, Charriaud, E., Craneguy, P., 2001. Fronts, transports and Weddell Gyre at 30°E between Africa and Antarctica. *J. Geophys. Res.*, **106** (C2) 2857-2879.

Park, Y.-H., Roquet, H., Durand, I., Fuda, J.-L., 2008. Large scale circulation over and around the northern Kerguelen Plateau. *Deep-Sea Res. II*, 55, 566-581.

Parias, L-H, Stavrides, G., 2004. "Les Explorateurs: des Pharaons à Paul-Emile Victor". Robert Laffont, Paris, *Collection Bouquins*, 1216 pp.

Patzert, W.C., 1972. Seasonal reversal in Red Sea circulation. In: "L'Oceanographie Physique de la mer Rouge". Paris: UNESCO, 55–89.

Patzert, W.C., 1974. Wind-induced reversal in Red Sea circulation. *Deep-Sea Res.*, **21**, 109-121.

Pedlosky, J., 1987. "Geophysical Fluid Dynamics", 2nd ed. Springer-Verlag, New York 710 pp.

Peixoto, J.P., Oort, A., 1992. "Physics of Climate". American Inst. of Physics, New York, 520 pp.

Périgaud, C., Delecluse, P., 1993. Interannual Sea Level variations in the Tropical Indian Ocean from Geosat and Shallow Water Simulations. *J. Phys. Oceanogr.*, **23**, 1916-1934.

Peterson, R.G., Stramma, L., 1991. Upper-level circulation in the South Atlantic Ocean. *Progr. Oceanogr.*, **26**, 1-73.

Peterson, R.G., Whitworth, T., 1989. The Subantarctic and Polar Fronts in relation to deep water masses through the southwestern Atlantic. *J. Geophys. Res.*, **94**, 10817-10838.

Philander, G., 1990. "El Niño, La Niña and the Southern Oscillation". Academic Press, New York, 289 pp.

Pickard, G., Emery, W., 1982. "Descriptive Physical Oceanography". 4th ed., Pergamon, Oxford, 249 pp.

Piétri, A., Testor, P., Echevin, V., Chaigneau, A., Mortier, L., Eldin, G., Grados, C., 2013. Fine scale vertical structure of the upwelling system off Southern Peru as observed from glider data. *J. Phys. Oceanogr.*, **43**, 631–646. doi:10.1175/JPO-D-12-035.1

Pollard, R.T., Lucas, M.I., Read, J.F., 2002. Physical controls on biogeochemical zonation in the Southern Ocean. *Deep-Sea Res. II*, **49**, 3289-3305.

Pond, S., Pickard, G., 1993. "Introductory Dynamic Oceanography". Pergamon, Oxford, 329 pp.

Prinsenberg S.J., 1986. The circulation pattern and current structure of Hudson Bay. In: Canadian Inland Seas, Martini E.P., ed., *Oceanogr. Series*, 44, Elsevier, New York, 187–203.

Provost, C., Gana, S., Garcon, V., Maamaatuaiahutapu, K., England, M., 1995. Hydrographic conditions in the Brazil-Malvinas Confluence during Austral summer 1990. *J. Geophys. Res.*, **100** (C6), 10655-10678.

Provost, C., Escoffier C., Maamaatuaiahutapu, K., Kartavtseff, A., Garcon, V., 1999. Subtropical mode waters in the South Atlantic Ocean. *J. Geophys. Res.*, **104** (C9), 21033-21049.

Qiu, B., Lukas, R., 1996. Seasonnal and interannual variability of the North Equatorial Current, the Mindanao Current and the Kuroshio along the Pacific western boundary. *J. Geophys. Res.*, **101** (C5), 12315-12330.

Quadfasel, R.D., Schott, F., 1982. Water mass distributions at intermediate layers off the Somali coats during the onset of the southwest monsoon, 1979. *J. Phys. Oceanogr.*, **12**, 1358-1372.

Quadfasel, R.D., Schott, F., 1983. Southward subsurface flow below the Somali Current, *J. Geophys. Res.*, **88**, 5973-5979.

Reid J., 1994. On the total geostrophic circulation of the North Atlantic Ocean: Flow patterns, tracers and transports. *Progr. Oceanogr.*, **33** (1) 1-92.

Reid J., 2003. On the total geostrophic circulation of the Indian Ocean: Flow patterns, tracers and transports. *Progr. Oceanogr.*, **56**, 137-186.

Reppin, J., Schott, F., Fischer, J., Quadfasel, D., 1999. Equatorial currents and transports in the upper central Indian Ocean: Annual cycle and interannual variability. *J. Geophys. Res.*, **104** (67) 15495-15514.

Revault D'Allonnes, M., 1992. "L'océanographie Physique". Paris: Presses Universitaires de France, *Que sais-je?*, n°438, 125 pp.

Reverdin, G., 1987. The upper equatorial Indian Ocean: the climatological seasonal cycle. *J. Phys. Oceanogr.*, **17**, (7) 903-927.

Reverdin, G., Fieux, M., 1987. Sections in the western Indian Ocean - variability in the temperature structure. *Deep-Sea Res.*, **34**, (4A), 601-626.

Reverdin, G., McPhaden, M.J., 1986. Near-surface current and temperature variability observed in the equatorial Atlantic from drifting buoys. *J. Geophys. Res.*, **91** (5) 6569-6581.

Reverdin, G., Delecluse, P., Lévy, C., Andrich, P., Morlière, A., Verstraete, J.M., 1991. The near surface tropical Atlantic in 1982-1984: Results from a numerical simulation and data analysis. *Progr. Oceanogr.*, **23**, 273-340.

Reverdin, G., Fieux, M., Gonella, J., Luyten, J., 1983. Free drifting buoys measurements in the Indian Ocean equatorial jet. In: "Hydrodynamics of the Equatorial Ocean", J.C.J. Nihoul, ed., *Elsevier Oceanography Series*, **36**, 99-120.

Reverdin, G., Frankignoul, C., Kestenare, E., McPhaden, M.J., 1994. Seasonal variability in the surface currents of the equatorial Pacific. *J. Geophys. Res.*, **99** (C10) 20323-20344.

Reverdin, G., Kestenare, E., Frankignoul, C., Delcroix, T., 2007. Surface salinity in the Atlantic ocean (30°S-50°N). *Progr. Oceanogr.*, **73**, 311-340.

Reynolds, R., Smith, T., Liu, C., *et al.*, 2007. Daily high-resolution blended analyses for sea surface temperature. *J. Climate*, **20**, 5473-5496.

Reynolds, R. W., Smith, T.M., 1994. Improved global sea surface temperature analyses using optimum interpolation. *J. Climate*, **7**, 929-948.

Reynolds, R. W., Rayner, N.A., Smith, T.M., *et al.*, 2002. An improved in situ and satellite SST analysis for climate. *J. Climate*, **15**, 1609-1625.

Rhein, M., Stramma, L., Send, U., 1995. The Atlantic deep western boundary current: water masses and transport near the equator. *J. Geophys. Res.*, **100**, 2441-2457.

Richardson, P.L., 1980. The Benjamin Franklin and Timothy Folger Charts of the Gulf Stream. In: "*Oceanography, The Past*", M. Sears and D. Merriman, eds., New York: Springer-Verlag, 703-717.

Richardson, P.L., 1981. Gulf Stream trajectories measured with free-drifting buoys. *J. Phys. Oceanogr.*, **11**, 999-1010.

Richardson, P.L., 1983. Gulf Sream rings In: "*Eddies in Marine Science*", A.R.Robinson, ed., Berlin: Springer, 19-45.

Richardson, P.L., Maillard, C., Sanford, T., 1979. The life history of cyclonic Gulf Stream ring Allen. *J. Geophys. Res.*, **84** (C12), 7727-7741.

Richardson, P.L., Reverdin, G., 1987. Seasonal Cycle of Velocity in the Atlantic North Equatorial Countercurrent as Measured by Surface Drifters, Current Meters, and Ship Drifts. *J. Geophys. Res.*, **92**, (C4), 3691-3708.

Richardson, P.L., Walsh, D., 1986. Mapping climatological seasonal variations of surface currents in the tropical Atlantic using ship drifts. *J. Geophys. Res.*, **91**, 10537-10550.

Ridgway, K. R., Dunn, J. R., 2007. Observational evidence for a Southern Hemisphere oceanic supergyre. *Geophys. Res. Lett.*, **34**, L13612, doi: 10.1029/2007GL030392.

Rigor, I. G., Wallace, J. M., 2004. Variations in the age of Arctic sea-ice and summer sea-ice extent. *Geophys. Res. Lett.*, **31**, L09401, doi:10.1029/2004GL019492.

Rintoul, S. R., 1991. South Atlantic interbasin exchange. *J. Geophys. Res.*, **96** (C2) 2675-2692.

Rintoul, S., Donguy, J., Roemmich, D., 1997. Seasonal evolution of upper ocean thermal structure between Tasmania and Antarctica. *Deep-Sea Res.* Part I, **44**, 1185–1202.

Rintoul, S. R., 1998. On the origin and influence of Adélie Land Bottom Water. In: Ocean, Ice, and Atmosphere: Interactions at the Antarctic Continental Margin, Jacobs, S., Weiss, R., (Eds.). *Antarctic Res. Series* 75, American Geophysical Union, Washington, 151-171.

Rintoul, S.R., Sokolov, S., 2001. Baroclinic transport variability of the Antarctic Circumpolar Current south of Australia (WOCE repeat section SR3). *J. Geophys. Res.*, **106**, 2815–2832.

Rio, M.-H., Hernandez, F., 2004. A mean dynamic topography computed over the world ocean from altimetry, in situ measurements, and a geoid model, *J. Geophys. Res.*, **109**, C12032, doi:10.1029/2003JC002226.

Risien, C.M., and D.B. Chelton, 2008. A Global Climatology of Surface Wind and Wind Stress Fields from Eight Years of QuikSCAT Scatterometer Data. *J. Phys. Oceanogr.*, 38, 2379-2413 (see SCOW website).

Rixen, M., Beckers, J.-M., Levitus, S., Antonov, J., Boyer, T., Maillard, C., Fichaut, M., Balopoulos, E., Iona, S., Dooley, H., Garcia, M-J., Manca, B., Giorgetti, A., Manzella, G., Mikhailov, N., Pinardi, N., Zavatarelli, M., 2005. The Western Mediterranean Deep Water: A proxy for climate change. *Geophys. Res. Lett.*, **32**, L12608, doi:10.1029/2005GL022702.

Roach, A.T., Aagaard, K., Pease, C.H., *et al.*, 1995. Direct measurements of transport and water properties through the Bering Strait, *J. Geophys. Res.*, **100** (C9) 18443-18457.

Rooth, C., 1982. Hydrology and ocean circulation. *Progr. Oceanogr.*, **11**, 131-149.

Rochford, D., 1966. Distribution of Banda Intermediate Water in the Indian Ocean. *Aust. J. Mar. Fresh. Res.*, **17**, 61-76.

Roquet, F., Park, Y.H., Guinet, C., Bailleul, F., Charrassin, J.-B., 2009. Observations of the Fawn Trough Current over the Kerguelen Plateau from instrumented elephant seals. *J. Mar. Sys.*, doi:10.1016/j.jmarsys.2008.11.017.

Royer, T.C., Emery, W.J., 1987. Circulation in the Gulf of Alaska, 1981. *Deep-Sea Res.*, **34**, 1361-1377.

Rudels, B., Quadfasel, D., Friedrich, H., Houssais, M.-N., 1989. Greenland Sea convection in the winter of 1987-1988. *J. Geophys. Res.*, **94**, 3223-3227.

Rudels, B., Friedrich, H.J., Quadfasel, D.R., 1999. The Arctic circumpolar boundary current. *Deep-Sea Res. II*, **46**, 1023-1062.

Saenko, O.A., 2008. Influence of the enhanced mixing within the Southern Ocean fronts on the overturning circulation. *Geophys. Res. Lett.*, **35**, L09602, doi:10.1029/2008CL033565.

Saji, N.H., Goswami, B.H., Vinayachandran, P.N., Yamagata, T., 1999. A dipole mode in the tropical Indian Ocean, *Nature*, **401**, 360-363.

Sankey, T., 1973. The formation of deep water in the northwestern Mediterranean. *Progr. Oceanogr.*, **6**, 159-188.

Saraceno, M., Provost, C., Piola, A.R., Bava, J., Gagliardini, A., 2004. Brazil Malvinas Frontal System as seen from 9 years of Advanced Very High Resolution Radiometer data. *J. Geophys. Res.*, **109**, C05027.

Saunders, P. M., The flux of overflow water through the Charlie-Gibbs Fracture Zone. *J. Geophys. Res.*, **99** (C6) 12343-12355.

Schlichtholz, P., Houssais, M.N., 1999. An inverse modeling study in Fram Strait. Part II: water mass distribution and transports. *Deep-Sea Res. II*, **46**, 1137-1168.

Schlitzer, R., 2000. eWOCE - Electronic Atlas of WOCE Hydrographic and Tracer Data Now Available, *Eos Trans. AGU*, **8** (5) 45.

Schlitzer, R., 2002. Electronic atlas of WOCE Data: eWOCE Atlas. Available on-line at: www.ewoce.org

Schmid, C., Garraffo, Z., Johns, E., Garzoli, S.L., 2003. Pathways and variability at intermediate depths in the tropical Atlantic. In: "Interhemispheric Water Exchange in the Atlantic Ocean", Goni and P. Malanotte-Rizzoli, Eds., *Elsevier Oceanography Series*, **68**, 233-268.

Schmitz, W.J., 1995. On the interbasin-scale thermohaline circulation. *Rev. Geophys.*, **33** (2) 151-174.

Schmitz, W., 1996. "On the World Ocean Circulation: Volume I, Some global features". Woods Hole Oceanographic Institution, Woods Hole, Massachusetts, *WHOI-96-03*, 141 pp.

Schmitz, W.J., 1996. "On the World Ocean Circulation: Volume II, The Pacific and Indian oceans". Woods Hole Oceanographic Institution, Woods Hole, Massachusetts, *WHOI-96-08*, 237 pp.

Schmitz, W.J., McCartney, M., 1993. On the North Atlantic circulation. *Reviews of Geophysics*, **3** (1) 29-49.

Schott, F., Fieux, M., 1985. The Somali current in autumn 1984 before the onset of the northeast monsoon. *Nature*, **315**, 50-52.

Schott, F., 1987. Recent studies of western Indian Ocean circulation. In: "*Further Progress in Equatorial Oceanography*", E. Katz and J. Witte, eds., Nova University Press, 287-302.

Schott, F., Fieux, M., Kindle, J., Swallow, J., Zantopp, R., 1988. The boundary currents east and north of Madagascar: 2. Direct measurements and model comparison. *J. Geophys. Res.*, **93**, 4963–4974.

Schott, F., Swallow, J.C., Fieux, M., 1989. Deep currents underneath the equatorial Somali current. *Deep-Sea Res.*, **26** (8), 1191-1199.

Schott, F., Swallow, J.C., Fieux, M., 1990. The Somali current at the equator: annual cycle of currents and transports in the upper 1000 m and connection to neighbouring latitudes. *Deep-Sea Res.*, **37**, 1825-1848.

Schott, F., Fischer, J., 2000. The winter monsoon circulation of the northern Arabian Sea and Somali Current. *J. Geophys. Res.*, **105**, 6359-6376.

Schott, F.A., McCreary, J., 2001. The monsoon circulation of the Indian Ocean. *Progr. Oceanogr.*, **51**, 1-123.

Schott, F., Brandt P., Hamann, M., Fischer, J., Stramma, L., 2002. On the boundary flow off Brazil at 5°-10°S and its Connection to the Interior Atlantic. *Geophys. Res. Lett.*, **29** (17) 1840, doi:10.1029/2002GL014786.

Schott, F., Dengler, M., Brandt, P., Affler, K., Fischer, J., Bourles, B., Gouriou, Y., Molinari, R., Rhein, M., 2003. The zonal currents and transports at 35°W in the tropical Atlantic, *Geophys. Res. Lett.*, **30** (7) 1349, doi:10.1029/2002GL016849.

Schott, F., Zantopp, R., Stramma, L., *et al.*, 2004. Circulation and Deep Water export at the western exit of the subpolar North Atlantic. *J. Phys. Oceanogr.*, **34**, 817-843.

Schott, F.A., McCreary, J.P., Johnson, G.C., 2004. Shallow overturning circulations of the tropical - subtropical oceans. *Geophys. Monogr. Series* **147**, 261-285, doi:10.1029/147GM15.

Schott, F.A., Brandt, P., 2007. Circulation and deep water export of the subpolar North Atlantic during the 1990s. *Geophys. Monogr. Ser.*, **173**, 91-118.

Schott, F., Xie, S.-P., McCreary, J.M.P., 2009. Indian Ocean circulation and climate variability. *Rev. Geophys.*, **47**, RG1002, doi:10.1029/2007RG000245.

Sears, M., Merriman, D., eds., 1980. "*Oceanography, the Past*", Proceedings of the Third International Congress on the history of oceanography, Sept. 1980. Woods Hole, Springer-Verlag, New York, 812 pp.

Senjyu, T., 1999. The Japan Sea Intermediate Water; its characteristics and Circulation. *J. Oceanogr.*, **55**, 111-122.

Siedler, G., 1969. General circulation of the water masses in the Red Sea. In: "*Hot Brines and Recent Heavy Metal Deposits in the Red Sea*", E. Degens and D. Ross, eds., New York: Springer-Verlag, 131–137.

Siedler G., Church J., Gould J., eds., 2001. "Ocean circulation and climate, observing and modelling the Global Ocean". Academic Press, New York, *Internat. Geophys. Series*, vol. 77, 715 pp.

Siedler, G., Rouault, M., Lutjeharms, J.R.E., 2006. Structure and origin of the subtropical South Indian Ocean Countercurrent. *Geophys. Res. Lett.*, **33**, L24609, doi:10.1029/2006GL027399.

Shankar D., Vinayachandran P.N., Unnikrishnan A.S., 2002, The monsoon currents in the north Indian Ocean, *Progr. Oceanogr.*, **52**, 63-120.

Sloyan, B.M., Rintoul, S.R., 2001. Circulation, renewal and modification of Antarctic mode and intermediate water. *J. Phys. Oceanogr.*, **31**, 1005-1030.

Smethie, W.M., Chipman, D.W., Smith, J.H., Koltermann, K.P., 1988. Chlorofluoromethanes in the Arctic Mediterranean Seas: evidence for formation of bottom water in the Eurasian Basin and deep-water exchange through Fram Strait. *Deep-Sea Res.*, **35**, 347-369.

Smith, D.M., and Murphy, J.M., 2007. An objective ocean temperature and salinity analysis using covariances from a global climate model. *J. Geophys. Res. Oceans*, **112**, C02022.

Song, Q., Gordon, A. L., Visbeck, M., 2004. Spreading of the Indonesian Throughflow in the Indian Ocean. *J. Phys. Oceanogr.*, **34**, 772-779.

Sparrow, M., Chapman, P., Gould, J., eds., 2005-2009. "The World Ocean Circulation Experiment (WOCE) Hydrographic Atlas Series, (4 volumes). International WOCE Project Office, Southampton, UK.

Speer, K., McCartney, M., 1991. Tracing Lower North Atlantic deep water across the equator. *J. Geophys. Res.*, **96**, 20443-20448.

Speich, S., Blanke, B., Cai, W., 2007. Atlantic meridional overturning circulation and the Southern Hemisphere supergyre. *Geophys. Res. Lett.*, **34**, L23614, doi:10.1029/2007GL031583.

Speich, S., Blanke, B., De Vries, P., Drijfhout, S., Döös, K., Ganachaud, A., Marsh, R., 2002. Tasman leakage: A new route in the global ocean conveyor belt. *Geophys. Res. Lett.*, **29**, 1416, doi:10.1029/2001GL014586.

Sprintall, J., Gordon, A.L., Murtugudde, R., Susanto, R., 2000. A semi-annual equatorial western Indian Ocean forced Kelvin wave observed in the Indonesian Seas, May 1997. *J. Geophys. Res.*, **105** (C7) 17217-17230.

Sprintall, J., Tomczak, M. 1993. On the formation of Central Water and thermocline ventilation in the southern hemisphere. *Deep-Sea Res.*, **4**, 827-848.

Sprintall, J., Wijffels, S., Gordon, A.L., Ffield, A., Molcard, R., Dwi Susanto, R., Soesilo, I., Sopaheluwakan, Y., Surachman, Y., Van Aken, H., 2004. INSTANT: A new international array to measure the Indonesian Throughflow, *EOS*, **85** (39) 369.

Sprintall, J., Wijffels S., Molcard, R., Jaya, I., 2009. Direct estimates of the Indonesian Throughflow Entering the Indian Ocean: 2004-2006. *J. Geophys. Res.*, **114**, C07001, doi:10.1029/2008JC005257.

Sprintall J., Wijffels S., Gordon A.L, *et al.*, 2004. A new international array to measure the Indonesian Throughflow, *EOS*, **85** (39) 369, 376.

Stabeno, P.J., Reed R.K., Napp J.M, 2002. Transport through the Unimak Pass, Alaska. *Deep Sea Res. II*, **49** (26), 5919-5930.

Steele, M., Morison, J., Ermold, W., Rigor, I., Ortmeyer, M., Shimada, K., 2004. Circulation of summer Pacific halocline water in the Arctic Ocean. *J. Geophys. Res.*, 109.

Stommel, H. 1948. The westward intensification of wind-driven ocean currents. *Trans. Amer. Geophys. Union*, **29**, 202-206.

Stommel, H., 1957. A survey of ocean current theory. *Deep-Sea Res.*, **4**, 149-184.

Stommel, H., 1972. "The Gulf Stream: A Physical and Dynamical Description". 2nd edition. University of California Press, Berkeley, 248 pp.

Stommel, H., 1986. "A view of the sea". Princeton University Press, 165 pp.

Stommel, H., 1989. The Slocum mission. *Oceanography*, **2**, 22-25.

Stommel, H., Aarons, A.B., 1960. On the abyssal circulation of the world ocean - I. Stationary planetary flow patterns on a sphere. *Deep-Sea Res.*, **6**, 140-154.

Stommel, H., Yoshida, K., 1972. "Kuroshio, Physical Aspects of the Japan Current". University of Washington Press, pp 235-352.

Stramma, L., 1992. The South Indian Ocean current. *J. Phys. Oceanogr.*, **22**, 421-430.

Stramma, L., England, M., 1999. On the water masses and mean circulation of the south Atlantic Ocean. *J. Geophys. Res.*, **104**, 20863-20883.

Stramma, L., Lutjeharms, J.R.E., 1997. The flow field of the subtropical gyre of the South Indian Ocean. *J. Geophys. Res.*, **102** (C3) 5513–5530.

Stramma, L., Peterson, R.G., 1990. The South Atlantic Ocean Current. *J. Phys. Oceanogr.*, **20**, 846-859.

Stramma, L., Schott, F., 1996. Western equatorial circulation and interhemispheric exchange. In: "The warm water sphere of the North Atlantic Ocean", W. Krauss, ed., Berlin: Gebrüder Borntraeger, 195-227.

Stramma L., Schott F., 1999. The mean flow field of the tropical Atlantic Ocean. *Deep-Sea Res. II*, **46**, 279-303.

Stramma, L., Fischer, J., Brand,t P., Schott, F., 2003. Circulation, variability and near-equatorial meridional flow in the central tropical Atlantic. In: "Interhemispheric Water Exchange in the Atlantic Ocean" G. Goni and P. Rizzoli, eds., Amsterdam: *Elsevier Oceanography Series*, n° 68, 1-22.

Stramma, L., Peterson, R.G., Tomczak, M., 1995. The South Pacific Current. *J. Phys. Oceanogr.*, **25**, 77-91.

Suga, T., Talley, L., 1995. Antarctic Intermediate Water circulation in the tropical and subtropical south Atlantic. *J. Geophys. Res.*, **100** (C7) 13441-13453.

Susanto, R.D., Gordon, A.L., Zheng, Q., 2001. Upwelling along the coasts of Java and Sumatra and its relation to ENSO. *Geophys. Res. Lett.*, **28** (8) 1599-1602.

Sverdrup, H.U., Johnson, M.W., Fleming, R.H., 1942. "The Oceans: Their Physics, Chemistry and General Biology". Prentice-Hall, New York, 1087 pp.

Swallow, J.C., 1955. A neutral-buoyancy float for measuring deep currents. *Deep-Sea Res.*, **3** (1), 93-104.

Swallow, J. C., Caston, G.F., 1973. The preconditioning phase of MEDOC 1969. *Deep-Sea Res.*, **20**, 429-448.

Swallow, J.C., Fieux, M., 1982. Historical evidence for two gyres in the Somali Current. *J. Mar. Res.*, **40** suppl., 747-755.

Swallow, J.C., Fieux, M., Schott, F., 1988. The boundary currents east and north of Madagascar. 1: Geostrophic currents and transports. *J. Geophys. Res.*, **93** (C5) 4951-4962.

Swallow, J.C., Schott, F., Fieux, M., 1991. Structure and transport of the East African Coastal Current. *J. Geophys. Res.*, **96** (12) 22245-22257.

Swift, J., 1984. The circulation of the Denmark Strait and the Iceland-Scotland Overflow Waters in the North Atlantic. *Deep-Sea Res.*, **31** (11) 1339-1355.

Swift, J., Koltermann, P., 1988. The origin of Norwegian Sea Deep Water. *J. Geophys. Res.*, **93**, 3563-3569.

Swift, J., Aagaard, K., Malmberg, S., 1980. The contribution of the Denmark Strait overflow to the deep North Atlantic. *Deep-Sea Res.*, **27A**, 29-42.

Swingedouw, D., 2006. Origine et impact climatique d'un changement de circulation thermohaline au cours des prochains siècles dans le modèle IPSL-CM4, *Thesis*, University Pierre et Marie Curie, Paris, 186 pp.Talley, L.D., McCartney, M.S., 1982. Distribution and circulation of Labrador Sea water. *J. Phys. Oceanogr.*, **12**, 1189-1205.

Talley, L.D., Raymer, M.E., 1982. Eighteen Degree Water variability. *J. Mar. Res.*, **40** suppl., 725-775.

Talley, L.D., 1993. Distribution and formation of North Pacific Intermediate Water. *J. Phys. Oceanogr.*, **23**, 517-537.

Talley, L.D., 2003. Zonal asymmetries in Subantarctic Mode Waters, wind stress forcing, and subduction of Antarctic Intermediate Water. *Geophys. Res. Abs.*, **5**, 13297.

Talley, L.D., Sprintall, J., 2005. Deep expression of the Indonesian Throughflow: Indonesian Intermediate Water in the South Equatorial Current. *J. Geophys. Res.*, **110**, C10009, doi:10.1029/2004JC002826.

Talley, L. D., 2007. Hydrographic Atlas of the World Ocean Circulation Experiment (WOCE). Volume 2, Pacific Ocean. M. Sparrow, P. Chapman, J. Gould, eds., WCRP, International WOCE Project Office, Southampton, U.K., ISBN 0-904175-54-5, 327 pp.

Talley, L.D., 2011. Hydrographic Atlas of the World Ocean Circulation Experiment (WOCE), vol 3: Indian Ocean. Sparrow, M., Chapman, P., Gould, J., (Eds.). International WOCE Project Office, Southampton, U.K. http://www-pord.ucsd.edu/whp_atlas/indian_index.htm Online version.

Tchernia, P., 1980. "Descriptive Regional Oceanography". Pergamon, Oxford, 253 pp. + 19 plates.

Tchernia, P., Fieux, M., 1971. Résultats des observations hydrologiques exécutées à bord du N/O "Jean Charcot" pendant la campagne MEDOC 1969 (30 janvier-28 février) (18-31 mars), *Cah. Océanogr.*, **23**, supp.1, 677-686.

Tchernia, P., Jeannin, P.-F, 1983. "Quelques aspects de la circulation océanique Antarctique révélés par l'observation de la dérive d'icebergs (1972-1983)". Centre National d'Etudes Spatiales, Expéditions Polaires Françaises (CNRS), Muséum National d'Histoire Naturelle, Paris, 109 pp.

Terray, P., 2006. "Rôle de l'océan Indien dans le système couplé mousson-El Niño Oscillation Australe". Habilitation à Diriger des Recherches, Université Pierre et Marie Curie, Paris, Physique et chimie de l'océan et de l'atmosphère; climatologie, 176 pp.

Timmermans, M.L., and Garrett, C., 2006. Evolution of the deep water in the Canadian Basin in the Arctic Ocean. *J. Phys. Oceanogr.*, **36**, 866-874.

Tolmazin, D., 1985 Changing coastal oceanography of the Black Sea, Mediterranean effluent. *Progr. Oceanogr.*, **15**, 277-316.

Tomczak, M., Large, D.G.B., 1989 Optimum multiparameter analysis of mixing in the thermocline of the eastern Indian Ocean. *J. Geophys. Res.*,**94**, 16141-16149.

Tomczak, M., Godfrey, J.S., 1994. "Regional Oceanography: an Introduction". Pergamon, Oxford, 422 pp.

Toole, J.M., Warren, B.A., 1993. A hydrographic section across the subtropical south Indian Ocean. *Deep-Sea Res.*, **40**, 1973-2019.

Trenberth, K.E., Olson, J.G., Large, W.G. 1989. "A Global Ocean Wind Stress Climatology based on ECMWF Analyses". National Center for Atmospheric Reseach, Boulder, *NCAR Technical Note* NCAR/TN-338 + STR, 93 pp.

Trenberth, K.E., Olson, J.G., Large, W.G., 1990. The mean annual cycle in Global Ocean wind stress. *J. Phys. Oceanogr.*, **20**,1742-1760.

Trenberth, K.E., 1992. "Global Analyses from ECMWF and Atlas of 1000 to 10 mb Circulation Statistics". National Center for Atmospheric Reseach, Boulder, *NCAR Technical Note* NCAR/TN 373 + STR, 191 pp.

Trenberth, K.E., Caron, J., 2001. Estimates of Meridional Atmosphere and Ocean Heat Transports. *J. Climate*, **14**, (16) 433 - 3443.

Trenberth K.E., Jones P.D., Ambenje P. *et al.*, 2007. "Observations: Surface and Atmospheric Climate Change", Contribution of Working Group I to The Fourth Assessment Report of IPCC (Intergovernmental Panel on Climate Change). In: "*Climate Change 2007: The Physical Science Basis*", Cambridge University Press, Cambridge, UK, 336 pp.

Trenberth, K.E., Fasullo, J.T. Branstator, G., Phillips, A.S., 2014. Seasonal aspects of the recent pause in surface warming, *Nature Climate Change*, **4**, 911-916.

Tsuchiya, M., 1975. Subsurface countercurrents in the eastern equatorial Pacific Ocean, *J. Mar. Res.*, **33**, suppl., 145-175.

UNESCO, 1992. "Background papers and supporting data on the International Equation of State of Seawater 1980". UNESCO, Paris, 192 pp. (*UNESCO Technical Papers in Marine Science*, n°. 38).

UNESCO, 1983. "Algorithms for computation of fundamental properties of seawater". UNESCO, Paris, 1983, 54 pp. (*UNESCO Technical Papers in Marine Science*, n°44).

Van Aken, H.M., Brodjonegoro, I.S., Jaya, I., 2009. The deep water motion through the Lifamatola Passage and its contribution to the Indonesian Throughflow, *Deep-Sea Res. I*, **56** (8), 1203-1216.

Vincent, E.M., Lengaigne M., Menkes, C.E., Jourdain, N.C., Marchesiello, P., Madec, G., 2011. Interannual variability of the South Pacific Convergence Zone and implications for tropical cyclone genesis. Climate Dyn. 36 (9-10), 1881-1896.

Vivier, F., 1999. Le Courant des Malouines: que suggèrent les mesures? *Thesis,* University Pierre et Marie Curie-Paris 6, 235 pp. Vivier F., Provost C., 2012. Direct velocity measurements in the Malvinas Current. *J. Geophys. Res.*, **104** (C9) 21083-21103.

Voituriez, B., 1981. Les sous-courants équatoriaux nord et sud et la formation des dômes thermiques tropicaux. *Oceanologica Acta*, **4**, 497-506.

Voituriez, B., 2006. "Le Gulf Stream". UNESCO, Paris (*IOC Oceans Forum*), 209 pp. Vukovich, F.M., 1988. Loop current boundary variations. *J. Geophys. Res.*, **93**, 15585-15591.

Wacongne, S., Piton, B., 1992. The near-surface circulation in the northeastern corner of the South Atlantic Ocean. *Deep-Sea Res.*, **39**, 1273-1298.

Warren, B.A., 1981. Deep circulation of the world ocean. In: "Evolution of Physical Oceanography", B. Warren and C. Wunsch, eds., MIT Press, Cambridge, Massachusetts, 6-42.

Warren, B., Speer, K.G., 1991. Deep circulation in the eastern South Atlantic Ocean. *Deep-Sea Res.*, **38**, suppl. 1, S281-S322.

Warren, B., Wunsch, C., eds, 1981. "*Evolution of physical oceanography, Scientific Surveys in Honor of Henry Stommel*". MIT Press, Cambridge Massachusetts, 623 pp.

Watanabe, T., Wakatsuchi, M., 1998. Formation of 26.8-26.9 water in the Kuril Basin of the Sea of Okhotsk as a possible origin of North Intermediate Water, *J. Geophys. Res.*, **103** (C2) 2849-2865.

Webster, F., Fieux, M., 1984. TOGA overview. In: "Large scale Oceanographic Experiments and Satellites". C. Gautier and M. Fieux, eds., Reidel Publishing Company, 17-24.

Whitworth, T., Nowlin, W.D., 1987. Water masses and currents of the Southern Ocean at the Greenwich Meridian. *J. Geophys. Res.*, **92** (C6) 6462-6476.

Wijffels, S., Firing, E., Toole, J., 1995. The mean structure and variability of the Mindanao Current at 8°N. *J. Geophys. Res.*, **100** (C9) 18421-18435.

Wijffels, S., Toole, J.M., Davis, R., 2001. Revisiting the south Pacific subtropical circulation: A synthesis of World Ocean Circulation Experiment observations along 32°S. *J. Geophys. Res.*, **106**, (C9), 19481-19513.

Wijffels, S., Sprintall, J., Fieux, M., Bray, N., 2002. The JADE and WOCE I10/IR6 Throughflow sections in the southeast Indian Ocean. Part 1: Water mass distribution and variability. *Deep-Sea Res. II*, **49**, 1341-1362.

Wijffels, S., Meyers, G., 2004. An intersection of oceanic wave guides: Variability in the Indonesian Throughflow region. *J. Phys. Oceanogr.*, **34**, 1232-1253.

Wijffels, S.E., Meyers, G., Godfrey, J.S., 2008. A twenty year average of the Indonesian Throughflow: regional currents and the interbasin exchange. *J. Phys. Oceanogr.*, **38**, 1965-1978. Doi: 10.1175/JPO2008/3987.1

Wilson, W.D., Johns, W.E., 1997. Velocity structure and transport in the Windward Islands passages. *Deep-Sea Res. I*, **44** (3) 487-520.

WMO, 1998. "The Global Climate System Review (Dec. 93 – May 96)". J.M. Nichols, ed., World Climate Data and Monitoring Programme. World Meteorological Organization, Geneva, n° 856, 95 pp.

WOCE, 2001. "US WOCE Implementation Report, n°13". US WOCE Office, College Station, Texas, 64 pp.

Worthington, L.V., 1969. An attempt to measure the volume transport of Norwegian Sea overflow water through the Denmark Strait. *Deep-Sea Res.*, **16**, suppl., 421-432.

Worthington, L.V., 1981. The water masses of the world ocean: some results of a fine-scale census, in: "*Evolution of Physical Oceanography, Scientific Surveys in Honor of Henry Stommel*", B. Warren and C. Wunsch, eds., Cambridge, Mass. USA: MIT Press, Cambridge Massachusetts, 42-69.

Wust, G., 1936. Schichtung und Zirculation des Atlantischen Ozeans: Das Bodenwasser und die Gliederung der Atlantischen Tiefsee. *Wissenschaftliche Ergebnisse der Deutschen Atlantischen Expedition "Meteor", 1925-1927*, vol VI, 1, 3-107.

Wust, G., 1963. On the stratification and the circulation in the cold water sphere of the Antillean-Caribbean basins. *Deep-Sea Res.*, **10**, 165-187.

Wyrtki, K., 1961. "Physical oceanography of the Southeast Asian waters, Scientific Results of Maritime Investigations of the South China Sea and Gulf of Thailand 1959-1961". Scripps Institution of Oceanography, La Jolla, California, *NAGA report* n° 2, 195 pp.

Wyrtki, K., 1971. "Oceanographic Atlas of the International Indian Ocean Expedition". US National Science Foundation, Washington, DC, 531pp.

Wyrtki, K., 1973. An equatorial jet in the Indian Ocean. *Science*, 1973, **181**, 262-264.

Wyrtki, K., 1987. Indonesian throughflow and the associated pressure gradient, *J. Geophys. Res.*, **92**, 12,941-12,946.

Yashayaev, I., Bersch, M., Van Aken, H.M., 2007. Spreading of the Labrador Sea Water to the Irminger and Iceland basins. *Geophys. Res. Lett.*, **34**, L10602, doi:10.1029/2006GL028999.

Yashayaev, I., 2007. Hydrographic changes in the Labrador Sea, 1960-2005. *Progr. Oceanogr.*, **73**, 242-276.

Yashayaev, I., Loder, J.W., 2016. Recurrent resplenishment of Labrador Sea Water and associated decadal-scale variability. *J. Geophys. Res. Oceans*, **121**, 11, 8095-8114, doi:10.1002/2016JC012046.

You, Y., Tomczak, M., 1993. Thermocline circulation and ventilation in the Indian Ocean derived from water masses analysis. *Deep-Sea Res.*, **40**, 13-56.

Yu, L., Jiu, X., Weller, R., 2008. "Multidecadal global flux datasets from the objectively analyzed air-sea fluxes (OAFlux) projects: Latent and sensible heat fluxes, ocean evaporation, and related surface meteorological variables". Woods Hole Oceanographic Institution, Woods Hole, Massachusetts, *OAFlux Project Technical Report* (OA-2008-01).

Zweng, M.M, Reagan, J.R., Antonov, J.I., Locarnini, R.A., Mishonov, A.V., Boyer, T.P., Garcia, H.E., Baranova, O.K., Johnson, D.R., Seidov, D., Biddle, M.M., 2013. World Ocean Atlas 2013, Volume 2: Salinity. S. Levitus, Ed., A. Mishonov Technical Ed.; *NOAA Atlas NESDIS* 74, 39 pp.

Acronyms and abbreviations

ADCP Acoustic Doppler Current Profiler
ARGO Array of Real-time Geostrophic Oceanography
ARGOS Advanced Research and Global Observation Satellite (CNES/NOAA)
ATLAS Autonomous Temperature Line Acquisition System (NOAA)
AVHRR Advanced Very High Resolution Radiometer
AVISO Archivage, Validation et Interprétation des données des Satellites Océanographiques (*Archival, validation and interpretation of oceanographic satellite data*) (CNES, France)
BORHA Bouée de Recherche Habitée (*Manned research buoy*) (French, installed in the Gulf of Lyons)
CAMS/OPI Climate Anomaly Monitoring System/Outgoing longwave radiation Precipitation Index (NOAA NCEP)
CERSAT Centre d'Exploitation et de Recherche Satellitaire (*Center for Satellite Exploitation and Research*) (IFREMER, France)
CIRES Cooperative Institute for Research in Environmental Sciences (NCEP/NCAR)
CITHER Circulation Thermohaline (French Atlantic program)
CFC chlorofluorocarbon
CLIVAR Climate Variability and Predictability
CLS Collecte Localisation Satellites (ARGOS) (CNES/IFREMER, France)
CMDT Combined Mean Dynamic Topography
CNES Centre National d'Etudes Spatiales (*National Center for Space Studies*) (French)
CNRS Centre National de la Recherche Scientifique (France)
CTD Conductivity Temperature Depth
ECMWF European Center for Medium-Range Weather Forecasts
EQUALANT Equatorial Atlantic Experiment
ERA-40 ECMWF 40-year Reanalysis
ERBE Earth Radiation Budget Experiment
ERS European Remote Sensing Satellite
ESA European Space Agency
ESIP Earth Science Information Partner Federation (NASA)
ESRL Earth System Research Laboratory (NOAA)
FGGE First GARP Global Experiment
FUNCEME Fundacao Cearense de Meteorologia e recursos hidricos (*Foundation for Meteorological and Hydrological Resources in the State of Ceará*) (Brazil)
GARP Global Atmospheric Research Program
GATE GARP Atlantic Tropical Experiment
GEOSECS Geochemical Ocean Sections Study (1973-1976)
GHRSST Group for High-Resolution Sea-Surface Temperature (www.ghrsst.org)

GMS Geostationary Meteorological Satellites
GODAE Global Ocean Data Assimilation Experiment
GOOS Global Ocean Observing System (IOC and WMO)
GPCP Global Precipitation Climatology Project (WCRP), monthly precipitation datasets from 1979 to present combine observations and satellite precipitation data into 2.5°x2.5° global grids.
GPS Global Positioning System
IFM-GEOMAR Leibnitz Institute of Marine Sciences (Institut Für Meereskunde & Research Center for Marine Geosciences, Kiel, Germany)
IFREMER Institut Français de Recherche pour l'Exploitation de la Mer (*French Research Institute for Ocean Exploitation*)
INSTANT International Nusantara Stratification and Transport program (International Throughflow study)
IOC Intergovernmental Oceanographic Commission (UNESCO)
IOD Indian Ocean Dipole
IPCC Intergovernmental Panel on Climate Change (WMO, UNEP)
IPEV Institute Paul Émile Victor (France)
IRD Institut de Recherche pour le Développement (*Research Institute for Development*) (France)
IRI International Research Institute for Climate and Society (LDEO)
ISCCP International Satellite Cloud Climatology Project (NASA)
ITCZ Intertropical Convergence Zone
JADE Java Australia Dynamic Experiment (French Throughflow study)
JCOMMOPS WMO-IOC Joint Technical Commission for Oceanography and Marine Meteorology in-situ Observing Programmes Support Centre
KESS Kuroshio Extension Study System (URI, WHOI, U Hawaii)
LADCP Lowered Acoustic Doppler Current Profiler
LAS Live Access Server (PO.DAAC) (podaac-tools.jpl.nasa.gov/las)
LDEO Lamont-Doherty Earth Observatory
LEGI Laboratoire des Ecoulements Géophysiques et Industriels (*Laboratory of Geophysical and Industrial Flows*) (France)
LOCEAN Laboratoire d'Océanographie et du Climat : Expérimentation et Approches Numériques (*Laboratory of Oceanography and Climate: Experiments and Numerical Approaches*) (Paris, France)
LODYC Laboratoire d'Océanographie Dynamique et du Climat (*Laboratory of Dynamic Oceanography and Climate*) (later called LOCEAN, France)
MEDOC Méditerranée Occidentale (Study of deep convection in the western Mediterranean)
MERCATOR-OCEAN Center for operational oceanographic forecasts (CNRS, IFREMER, METEO FRANCE, SHOM, CNES, IRD, France)
NASA National Aeronautics and Space Administration
NCAR National Center for Atmospheric Research
NCEP National Centers for Environmental Prediction (NOAA)

NEMO Nucleus for European Modelling of the Ocean (CNRS, MERCATOR-OCEAN, UKMO, NERC)
NERC Natural Environment Research Council (U.K.)
NOAA National Oceanic and Atmospheric Administration (USA)
NODC National Oceanographic Data Center (NOAA)
OAFlux Objectively Analyzed air-sea Fluxes for the global Oceans (WHOI/NOAA/OCO)
OCO Office of Climate Observations (NOAA)
ODV Ocean Data View
ORCA Numerical circulation model
PIRATA Prediction and Research moored Array in the Tropical Atlantic
PMEL Pacific Marine Environmental Laboratory (NOAA)
PO.DAAC Physical Oceanography Distributed Active Archive Center (NASA)
POET PO.DAAC Ocean ESIP Tool
QUICKSCAT Quick Scatterometer (NASA)
RAFOS SOFAR, reversed (acoustic-tracked float system)
SADCP Shipboard Acoustic Doppler Current Profiler
SATMOS Satellite meteorology center of METEO-France
SAVE South Atlantic Ventilation Experiment
SHOM Service Hydrographique et Océanographique de la Marine (*Navy Hydrographic and Oceanographic Service*) (France)
SINODE Surface Indian Ocean Dynamic Experiment (France)
SMOS Soil Moisture Ocean Salinity mission (ESA)
SOFAR Sound Fixing And Ranging
SSM/I Special Sensor Microwave/Imager (NASA)
TIP Transport Indien Pacifique (*Pacific-Indian Transport*) (French program)
TOGA Tropical Ocean and Global Atmosphere (International program, 1985-1995)
TTO Transient Tracers in the Ocean
UKMO United Kingdom Meteorological Office
UMR Unité Mixte de Recherche (*Mixed Research Unit*) (France)
URI University of Rhode Island
WCRP World Climate Research Program
WHOI Woods Hole Oceanographic Institution
WHP WOCE Hydrographic Program
WMO World Meteorological Organization
WOCE World Ocean Circulation Experiment (International experiment, 1990-2000)

Water Mass Abbreviations

AABW	Antarctic Bottom Water
AAIW	Antarctic Intermediate Water
ABW	Arctic Bottom Water
ADW	Adriatic Sea Deep Water (Mediterranean Sea)
ADW	Arctic Deep Water (Arctic Ocean)
ASW	Arctic Surface Water (Arctic Ocean)
ASW	Arabian Sea Water (Indian Ocean)
AtW	Atlantic Water
BBW	Bay of Bengal Water
CBDW	Canadian Basin Deep Water
CDW	Cretan Deep Water
CNADW	Central North Atlantic Deep Water
CPDW	Circumpolar Deep Water
DAW	Deep Arctic Water
DSOW	Denmark Strait Overflow Water
DSW	Denmark Strait Overflow Water
EBDW	Eurasian Basin Deep Water
ENPCW	East North Pacific Central Water
ESPCW	Eastern South Pacific Central Water
EW	Equatorial Water
GSDW	Greenland Sea Deep Water
IBDW	Ionian Basin Deep Water
IDW	Indian Deep Water
IIW	Indonesian Intermediate Water
ISW	Iceland Scotland Overflow Water (Atlantic Ocean)
ISW	Indonesian Surface Water (Indian Ocean)
JSCW	Japan Sea Central Water
JSIW	Japan Sea Intermediate Water
JSPW	Japan Sea Proper Water
HSIW	High Salinity Intermediate Water (Japan Sea)
LCPDW	Lower Circumpolar Deep Water
LDW	Levantine Basin Deep Water
LIW	Levantine Intermediate Water
LNADW	Lower North Atlantic Deep Water
LSW	Labrador Sea Water
MW	Mediterranean Water
NACW	North Atlantic Central Water
NADW	North Atlantic Deep Water
NASTW	North Atlantic Subtropical Water
NEABW	Northeast Atlantic Bottom Water
NEPCW	Northeast Pacific Central Water

NICW	North Indian Central Water
NPCW	North Pacific Central Water
NPDW	North Pacific Deep Water
NPEW	North Pacific Equatorial Water
NPIW	North Pacific Intermediate Water
NPSAW	North Pacific Subarctic Water
NPSW	North Pacific Surface Water
NPTW	North Pacific Tropical Water
NSDW	Norwegian Sea Deep Water
NWPCW	Northwest Pacific Central Water
NWW	Northwest Indian Water
PDW	Pacific Deep Water
PGW	Persian Gulf Water
RSW	Red Sea Deep Water (Red Sea)
RSW	Red Sea Water (Indian Ocean)
SACW	South Atlantic Central Water
SAMW	Subantarctic Mode Water
SASTW	South Atlantic Subtropical Water
SAW	Surface Arctic Water
SEPCW	Southeast Pacific Central Water
SICW	South Indian Central Water
SISTW	South Indian Subtropical Water
SPEW	South Pacific Equatorial Water
SPHW	Summer Pacific Halocline Water
SPMW	Sub Polar Mode Water
SPTW	South Pacific Tropical Water
STMW	Subtropical Mode Water
SWPCW	Southwest Pacific Central Water
TWW	Tsushima Current Warm Water
UCPDW	Upper Circumpolar Deep Water
UNADW	Upper North Atlantic Deep Water
WSBW	Weddell Sea Bottom Water
WSDW	Weddell Sea Deep Water

Ocean Current Abbreviations

AC	Agulhas Current
ACC	Antarctic Circumpolar Current
AntC	Antilles Current
ARC	Agulhas Return Current
CarC	Caribbean Current
CIBC	Central Iceland Basin Current
DWBC	Deep Western Boundary Current
EACC	East African Coastal Current (or Zanzibar Current)
EAuckC	East Auckland Current
EGC	East Greenland Current
EIC	Equatorial Intermediate Current
EICC	East Indian Coastal Current
EKC	East Korean Warm Current
ERC	East Reykjanes Current
EUC	Equatorial Undercurrent
FC	Florida Current
GBUC	Great Barrier Reef Undercurrent
GCUC	Gabon Congo Undercurrent
GS	Gulf Stream
Irm	Irminger Current
LC	Loop Current
LEIC	Lower Equatorial Intermediate Current
NBCC	North Brazil Coastal Current
NBUC	North Brazil Undercurrent
NC	Norwegian Current
NEC	North Equatorial Current
NECC	North Equatorial Counter Current
NEUC	North Equatorial Undercurrent
NEMC	Northeast Madagascar Current
NEMC	Northeast Monsoon Current (Fig. 1-56, p 84)
NGCC	New Guinea Coastal Current
NGUC	New Guinea Coastal Undercurrent
NKC	North Korean Cold Current
NQC	North Queensland Current
PCC	Peru Coastal Current
PCUC	Peru Coastal Undercurrent
RHC	Ras Al Hadd Current
RTC	Rockall Trench Current
SAFC	Subarctic Front Current
SC	Somali Current
SEC	South Equatorial Current

SECC	South Equatorial Countercurrent
SEMC	Southeast Madagascar Current
SICC	South Indian Subtropical Countercurrent
SEUC	South Equatorial Undercurrent
SWMC	Southwest Monsoon Current
WGC	West Greenland Current
WICC	West Indian Coastal Current
ZC	Zanzibar Current (East African Coastal Current)
ZUC	Zanzibar Undercurrent

Other abbreviations

AAZ	Antarctic Zone
AD	Antarctic Divergence (in Antarctic Ocean)
AD	Angola Dome (in Atlantic Ocean)
AMOC	Atlantic Meridional Overturning Circulation
ASF	Antarctic Slope Front
AZ	Antarctic Zone
ENSO	El Nino-Southern Oscillation
GD	Guinea Dome
IOD	Indian Ocean Dipole
ITF	Indonesian Throughflow
ITCZ	Intertropical Convergence Zone
LH	Laccadive High
LL	Laccadive Low
MOC	Meridional Overturning Circulation
NAO	North Atlantic Oscillation
PF	Polar Front
PFZ	Polar Front Zone
SACCF	Southern Antarctic Circumpolar Current Front
SAF	Subantarctic Front
SAZ	Subantarctic Zone
SB	Southern Boundary
SGW	Somali Great Whirl
SLD	Sri Lanka Dome
SocE	Socotra Eddy
SOI	Southern Oscillation Index
SPCZ	South Pacific Convergence Zone
SSS	Sea Surface Salinity
SST	Sea Surface Temperature
STF	Subtropical Front
SZ	Southern Zone
THC	Thermohaline Circulation

Web sites

Annual to Decadal Variability in the Labrador Sea (Igor Yashayaev, Allyn Clarke, John Lazier, Robert Gershey, Peter Jones): www2.mar.dfo-mpo.gc.ca/science/ocean/woce/labsea/labsea_poster.html

ARGO floats. 3000 ARGO float positions: www.argo.ucsd.edu/statusbig.gif

AVISO Live Access Server. Mean dynamic topography: www.aviso.oceanobs.com/en/applications/ocean/large-scale/circulation/mean-dynamic-topography.html
or: www.aviso.altimetry.fr/en/data/products/auxiliary-products/mdt/mdt-direct-method.html

AVISO: The seasons of the Gulf Stream. www.aviso.altimetry.fr/en/data/ data-access/las-live-access-server/lively-data/2005/aug-17-2005-the-seasons-of-the-gulf-stream.html

EGO (European Gliding Observatories): www.ego-network.org/dokuwiki/doku.php?id=public:howagliderwork

eWOCE: Electronic Atlas of WOCE Data. (Reiner Schlitzer, AWI Alfred Wegener Institute, Bremerhaven, Germany): www.ewoce.org

Global Ocean Heat and Salt Content (Syd Levitus, John Antonov, Tom Boyer): www.nodc.noaa.gov/OC5/3M_HEAT_CONTENT/

GPCP: www.esrl.noaa.gov/psd/data/gridded/data.gpcp.html
or: http://www.gewex.org/data-sets-global-precipitation-climatology-project-gpcp/

INSTANT (International Nusantara Stratification and Transport program): ocp.ldeo.columbia.edu/res/div/ocp/projects/instant.shtml
or: www.marine.csiro.au/~cow074/

IRI/LDEO Climate Data Library. International Research Institute for Research and Society, Lamont-Doherty Earth Observatory, Columbia University: iridl.ldeo.columbia.edu.
or: iridl.ldeo.columbia.edu/SOURCES/.NOAA/.NODC/.WOA05/

ISCCP Global Surface and Atmospheric Radiative fluxes: isccp.giss.nasa.gov/projects/flux.html

JOHNS HOPKINS UNIVERSITY Applied Physics Laboratory: fermi.jhuapl.edu/

KESS (Kuroshio Extension System Study), Dynamics of ocean currents and Fronts: www.po.gso.uri.edu/dynamics/KESS

LAS Live Access Server (see podaac-tools.jpl.nasa.gov/las)

LAS/AVISO: Live Access Server program: las.aviso.oceanobs.com/las/getUI.do

MEDATLAS, Mediterranean and Black Sea data and products: www.ifremer.fr/medar/

MEER (Marine and Environmental Education and Research), University of California course in Seas and Oceans: www.meer.org/ebook/TOC-oceanography.htm

Mediterranean surface circulation (C. Millot, I. Taupier-Letage): www.ifremer.fr/lobtln/OTHER/schema_circulation-surface_Mediterranee.png

NOAA (National Oceanic and Atmospheric Administration) web sites:
www.noaa.gov/
or: www.esrl.noaa.gov/psd/enso/
or: www.nodc.noaa.gov/OC5/WOA05/pr_woa05.html

NOAA PMEL TAO Project Office, 2009. El Nino theme page: www.pmel.noaa.gov/tao/elnino/nino-home.html

OAFlux (Objectively Analyzed Air-Sea Fluxes for the Global Oceans): oaflux.whoi.edu

ODV (Ocean Data View), (Reiner Schlitzer, Alfred Wegener Institute, Bremerhaven, Germany): odv.awi.de

PIRATA (Prediction and Research Moored Array in the Atlantic): www.pmel.noaa.gov/pirata

PO DAAC (Physical Oceanography Distributed Active Archive Center) Live Access Server at the NASA Jet Propulsion Laboratory, Pasadena, CA): podaac-tools.jpl.nasa.gov/las

POET The data was originally retrieved using the POET tool developed at the Physical Oceanography Distributed Active Archive Center (PO. DAAC). The tool has been replaced by the Live Access Server (LAS).
Specific GHRSST data sets may be retrieved through:
podaactools.jpl.nasa.gov/las/UI.vm
For information on the ghrsst data sets:
www.ghrsst.org and podaac.jpl.nasa.gov

SCOW Scatterometer Climatology of Ocean Winds Atlas: cioss.coas.oregonstate.edu/scow

WOA05 WORLD OCEAN ATLAS 2005, Data Sets and Products:
www.nodc.noaa.gov/OC5/WOA05/pr_woa05.html
or: WOA05/woadata.html

WOCE Atlantic Ocean hydrographic Atlas (Koltermann P., *et al.*, 2011):
https://doi.org/10.21976/C6RP4Z
or: whp-atlas.ucsd.edu/atlantic_index.html

WOCE Indian Ocean hydrographic Atlas (Lynne D. Talley, 2011): https://doi.org/10.21976/C61595

WOCE Pacific Ocean hydrographic Atlas (Lynne D. Talley, 2007): whp-atlas.ucsd.edu/pacific_index.html

WOCE Southern Ocean hydrographic Atlas (Alejandro Orsi, Tom Withworth, III):
woceatlas.tamu.edu.
or: whp-atlas.ucsd.edu/indian_index.htm

WOD (World Ocean Database), NOAA/NODC: www.nodc.noaa.gov/OC5/WOD/pr_wod.html

Index

B

D

E

F

G

R

S

T

www.ingramcontent.com/pod-product-compliance
Ingram Content Group UK Ltd.
Pitfield, Milton Keynes, MK11 3LW, UK
UKHW062303290726
14090UKWH00017B/865

9 782759 820702